Multivariate Data
Analysis
with Readings

SECOND EDITION

Multivariate Data Analysis
with Readings

Joseph F. Hair, Jr.
LOUISIANA STATE UNIVERSITY

Rolph E. Anderson
DREXEL UNIVERSITY

Ronald L. Tatham
BURKE MARKETING RESEARCH

Macmillan Publishing Company
NEW YORK

Collier Macmillan Publishers
LONDON

Macmillan Publishing Company
866 Third Avenue, New York, New York 10022

Collier Macmillan Canada, Inc.

Library of Congress Cataloging in Publication Data
Hair, Joseph F.
  Multivariate data analysis with readings.

  Rev. ed. of: Multivariate data analysis with readings.
  Bibliography: p.
  Includes index.
  1. Multivariate analysis.   I. Anderson, Rolph E.
II. Tatham, Ronald L.   III. Title.
QA278.H34   1987   519.5'35   86–16361
ISBN 0–02–348980–4

Printing: 5 6 7 8   Year: 0 1 2 3 4 5 6

ISBN 0-02-348980-4

PREFACE

America and much of the world are experiencing revolutionary developments in computer technology. Prepackaged software programs for the computer are now abundantly available and easy to use. Several major universities are already requiring their matriculating students to purchase or have access to their own personal microcomputers for computer-assisted learning across their curriculum. At the same time, faculty members are being encouraged to use microcomputers in teaching their classes—whether in engineering, business administration, science, humanities, or the arts. In the late 1980's and early 1990's, we will see microcomputers become commonplace in our schools, work places, and homes. Not only research scholars, professional managers, and teachers, but average householders will utilize computers to help make decisions on home mortgages, stock investments, savings alternatives, etc. In short, a substantial number of us will become data analysts, and we increasingly will rely on multivariate data analysis to help solve analytical problems. While computers have taken away most of the number-manipulating drudgery in analyzing data, they have also ushered in a new family of tools called multivariate statistics which allow the study of people and things within the complexity of their natural "real world" setting. Prior to the development of high speed computers, multivariate statistical calculations were too complicated for practical application even though the theoretical and mathematical constructs had been known for decades.

Today, just as you need not be a mechanical engineer or automobile mechanic to operate a sophisticated car, you do not have to be a statistician in order to successfully apply powerful multivariate statistical techniques to analyze data with a computer. In *Multivariate Data Analysis with Readings*, we "talk" our way through the fundamentals on which the family of multivariate statistical techniques is based. Written for the nonspecialist in multivariate statistics, this book provides a simple introduction to multivariate analysis with emphasis on the practical application of these valuable tools. We specifically wrote this book for those who want a conceptual understanding of multivariate techniques—what they can do, when they should be used, and how they are interpreted—without becoming bogged down by symbols, formulas, or mathematical derivations. We try to avoid using jargon, technical "buzz words," and confusing terminology while maintaining accuracy and precision in our explanations. When new terms are introduced to clarify our discussions, we carefully define the unfamiliar terms at the beginning of the chapter. This approach enables readers to grasp the essentials of multivariate analysis and prepares them to make the transition from this introductory book to more complex presentations. We believe that *Multivariate Data Analysis with Readings* is the most practical and academically sound guide available to understanding and applying multivariate statistical techniques.

Although intended as a basic text for an introductory course in multivariate data analysis, some schools may want to use the book as a supplementary text for a research course, a quantitative models course, or a second course in statistics. Feedback from our first edition indicates that even readers with advanced statistical backgrounds find the book useful for review and convenient reference.

A new feature of the second edition is the broader scope of the readings. Our experience with the first edition demonstrated a strong interest for the text among disciplines outside of marketing/business. As a result, in the second edition we have included examples of applications of multivariate techniques not only in the field of marketing/ business, but also in psychology, education, and the behavioral sciences in general. We are confident these applications will be beneficial to the marketing/business adopter by providing a different perspective, and also to the behavioral science adopter, who may wish to review sample applications of the multivariate techniques in a more relevant context.

Multivariate Data Analysis with Readings is organized into three parts. The first is presented in chapter one and consists of an introduction to the field of multivariate data analysis. Each of the multivariate techniques covered in the text is succinctly described. Then guidelines are provided for selecting the appropriate multivariate tool for different types of data and for specific applications. Finally, a single data bank utilized throughout most of the book for expository purposes is described. Unique data bases were used for multidimensional scaling and conjoint analysis because the standardized data bank was not appropriate. Several additional data bases with multiple applications of the techniques are contained in the instructor's manual.

In the second and third major parts of the text, we classify multivariate statistics into dependence and interdependence techniques. Chapters two, three, four, and five focus on dependence techniques in which the variables can be divided into independent and dependent classifications. These techniques include: multiple regression analysis, multiple discriminant analysis, multivariate analysis of variance, and canonical correlation analysis. The last four chapters deal with the analysis of interdependence in which the variables or subsets of variables are not categorized as either independent or dependent. Instead, the entire set of variables is dealt with as a group, and an effort is made to give meaning to the set of variables, individuals, or objects. Techniques included here are factor analysis, cluster analysis, multidimensional scaling, and conjoint analysis.

The format for each chapter of the book consists of a text portion followed by articles carefully selected from the behavioral literature. In the text portion, we introduce the fundamentals of the technique and explain the various issues involved in applying it. Following the text material are articles that set forth the theory and illustrate its practical application. We believe that it is both instructive and reassuring for the reader to see how others have applied the multivariate tools.

We wrote this book for people who want an overall understanding of what multivariate techniques can do for them, when and how they can be applied, and how results are interpreted so that they can use the tools successfully and read the technical literature with confidence. We believe that we have achieved these objectives.

A number of people assisted us in completing the second edition of this text, including Jill Cavell, Randy Russ, Bill Simmons, and Ron Roullier, all of Louisiana State University. Special thanks are due Mike Baumgardner, who provided substantial input to the revisions of the MANOVA chapter. We are also indebted to the following individuals for their invaluable assistance:

Alan J. Bush, Texas A & M University
Alvin C. Burns, University of Central Florida
David Andrus, Kansas State University
Chaim Ehrman, University of Illinois at Chicago
Joel Evans, Hofstra University
John Lastovicka, University of Kansas
Thomas R. Gillpatrick, Portland State University
Jerry L. Wall, Northeast Louisiana University

J. F. H.
R. E. A.
R. L. T.

CONTENTS

CHAPTER ONE

Introduction   **1**

CHAPTER PREVIEW 1   KEY TERMS 1
What is Multivariate Analysis? 2
Multivariate Analysis Defined 3
Types of Multivariate Analysis 4
A Classification of Multivariate Techniques 6
Data Bank 11
SUMMARY 14   QUESTIONS 14   REFERENCES 15

CHAPTER TWO

Multiple Regression Analysis   **17**

CHAPTER PREVIEW 17   KEY TERMS 17
What is Multiple Regression Analysis? 20
How is Regression Analysis Used? 20
How Does Regression Analysis Help Us to Predict? 21
Using Multiple Regression Analysis to Predict 31
General Approaches to Regression Analysis 40
Multicollinearity 42
Illustration of a Regression Analysis 43
SUMMARY 52   QUESTIONS 53   REFERENCES 53

Readings:

Measuring the Quantity and Mix of a Product Demand 54
Social Interactions as Predictors of Children's Likability and Friendship Patterns: A Multiple Regression Analysis 63

CHAPTER THREE

Multiple Discriminant Analysis   **73**

CHAPTER PREVIEW 73   KEY TERMS 73
What Is Discriminant Analysis? 75
Assumptions of Discriminant Analysis 76
Hypothetical Example of Discriminant Analysis 77
Objectives of Discriminant Analysis 79
Geometric Representation 79
Application of Discriminant Analysis 81
A Two-Group Illustrative Example 92
A Three-Group Illustrative Example 101
SUMMARY 113   QUESTIONS 114   REFERENCES 114

Readings: **Alternative Approaches for Interpretation of Multiple Discriminant Analysis in Marketing Research   115**
The Personality Inventory for Children: Differential Diagnosis in School Settings   127
Validation of Discriminant Analysis in Marketing Research   133

CHAPTER   **Multivariate Analysis of Variance**   **145**
FOUR   CHAPTER PREVIEW   145   KEY TERMS   145
What is Multivariate Analysis of Variance?   147
Case 1: Difference Between Two Independent Groups   147
Case 2: Difference between k Independent Groups   153
Factorial Designs   161
MANOVA Counterparts of Other ANOVA Designs   162
ANCOVA and MANCOVA   162
Assumptions of ANOVA and MANOVA   163
SUMMARY   163   QUESTIONS   164   REFERENCES   164

Readings: **The Effects of Social Skills Training and Peer Involvement on the Social Adjustment of Preadolescents   165**
Assessment of Stress-Related Psychophysiological Reactions in Chronic Back Pain Patients   176

CHAPTER   **Canonical Correlation Analysis**   **187**
FIVE   CHAPTER PREVIEW   187   KEY TERMS   187
What is Canonical Correlation Analysis?   188
Objectives of Canonical Analysis   190
Application of Canonical Correlation   190
An Illustrative Example   192
How to Interpret Canonical Functions   200
SUMMARY   201   QUESTIONS   202   REFERENCES   202

Readings   **Using Canonical Correlation to Construct Product Spaces for Objects with Known Feature Structures   203**
Content and Response-Style in the Construct Validation of Self-Report Inventories: A Canonical Analysis   217
Factor Structure and Correlates of Ratings of Inattention, Hyperactivity, and Antisocial Behavior in a Large Sample of 9-Year-Old Children From the General Population   221

CHAPTER   **Factor Analysis**   **233**
SIX   CHAPTER PREVIEW   233   KEY TERMS   233
What Is Factor Analysis?   235
Purposes of Factor Analysis   235
An Illustrative Example   252
SUMMARY   260   QUESTIONS   261   REFERENCES   261

Readings: **The Organizational Content of Market Research Use   262**
A Factorial Analysis of the Authoritarian Personality   273
The Application and Misapplication of Factor Analysis in Marketing Research   278

CHAPTER   **Cluster Analysis**   **293**
SEVEN   CHAPTER PREVIEW 293   KEY TERMS 293
               What Is Cluster Analysis? 295
               How Does Cluster Analysis Work? 296
               Applications of Cluster Analysis 297
               An Illustrative Example 308
               SUMMARY 315   QUESTIONS 316   REFERENCES 316

Readings: **Cluster Analysis in Marketing Research: Review and Suggestions for
               Application 317**
               **Classification and Validation of Behavioral Subtypes of Learning-
               Disabled Children 338**

CHAPTER   **Multidimensional Scaling**   **349**
EIGHT   CHAPTER PREVIEW 349   KEY TERMS 349
               What Is Multidimensional Scaling? 351
               Basic Concepts and Assumptions of Multidimensional Scaling 354
               Input Multidimensional Scaling 355
               Methods of Multidimensional Scaling 360
               A Generalized Multidimensional Scaling Approach 365
               The Candy Bar Example Revisited 368
               SUMMARY 369   QUESTIONS 369   REFERENCES 369

Readings: **Product Positioning: An Application of Multidimensional
               Scaling 371**
               **Personal Space as a Function of Infant Illness: An Application of
               Multidimensional Scaling 382**
               **Alternative Perceptual Mapping Techniques: Relative Accuracy and
               Usefulness 392**

CHAPTER   **Conjoint Analysis**   **407**
NINE   CHAPTER PREVIEW 407   KEY TERMS 408
               What Is Conjoint Analysis? 408
               The Purposes of Conjoint Analysis 409
               An Experiment Using Conjoint Analysis 411
               Procedures Using Conjoint Analysis 413
               SUMMARY 419   QUESTIONS 419

Readings: **A Hybrid Utility Estimation Model for Conjoint Analysis 420**
               **Reliability and Validity of Conjoint Analysis and Self-Explicated
               Weights: A Comparison 428**
               **Levels of Aggregation in Conjoint Analysis: An Empirical
               Comparison 436**

               Index   **445**

CHAPTER ONE

Introduction

CHAPTER PREVIEW

Chapter One presents a simplified overview of multivariate data analysis. It stresses that multivariate analysis methods will increasingly influence not only the analytical aspects of research but also the design and approaches to data collection for decision-making/problem-solving situations. A classification of the several types of multivariate techniques is presented and general guidelines for their application are provided. The chapter concludes with a discussion of the data bank that is utilized throughout most of the text to illustrate the application of the techniques. Before starting the chapter, you should familiarize yourself with the definitions of key terms to facilitate your reading.

After studying the concepts presented in this introductory chapter, you should be able to:

- Explain what multivariate analysis is and when its application is appropriate
- Define and discuss the specific techniques included in multivariate analysis
- Determine which multivariate technique is appropriate for a specific research problem.
- Discuss the nature of measurement scales and their relationship to multivariate techniques.

KEY TERMS

Bivariate partial correlation   Simple (two-variable) correlations between the two sets of residuals (unexplained variances) that remain after the association of other independent variables is removed.

Dependence technique   A classification of statistical techniques distinguished by having a variable or set of variables identified as the dependent variable and the remaining variables as independent. An example is regression analysis.

Dependent variable   The presumed effect of or response to a change in an independent variable.

Dummy variable   A nonmetrically measured variable that has been transformed into a metric dummy variable by assigning a 1 or a 0 to a subject, depending upon whether it possesses or does not possess a particular characteristic.

Independent variable   The presumed cause of any change in a response or dependent variable.

Interdependence technique   A classification of statistical techniques for which the variables are not divided into dependent and independent groups. Rather, all variables are analyzed as a single set. An example is factor analysis.

Metric data   Also called *quantitative data,* these are measurements used to identify or describe subjects (or objects) not only on the basis of type or kind but also by the amount or degree to which the subject may be characterized by a particular attribute. For example, a person's age and weight are considered metric data. These are also known as *interval* and *ratio data.*

Nonmetric data   Also called *qualitative data,* these are attributes, characteristics, or categorical properties that can be used to identify or describe a subject or object. Examples of nonmetric data are sex (male, female) or occupation (physician, attorney, professor). These are also referred to as *nominal* and *ordinal* data.

Treatment   The independent variable that the researcher manipulates to see the effect (if any) on the dependent variables.

Univariate analysis of variance (ANOVA)   A statistical technique used to determine, on the basis of one dependent measure, whether samples are from populations with equal means.

What Is Multivariate Analysis?

With the growth of computer technology in recent years, remarkable advances have been made in the analysis of psychological, sociological, and other types of behavioral data. Computers have made it possible to analyze large quantities of complex data with relative ease. At the same time, the ability to conceptualize data analysis has also advanced. Much of the increased understanding and mastery of data analysis has come about through the study of statistics and statistical inference. Equally important has been the expanded understanding and application of a group of analytical statistical techniques known as *multivariate analysis.*

Multivariate analytical techniques are being applied on a widespread basis in industry, government, and university-related research centers. Several books and articles have been published on the theoretical and mathematical aspects of these tools. Few books, however, have been written for the researcher who is not a specialist in math or statistics. Fewer such books discuss applications of multivariate statistical methods of interest to behavioral scientists or business and government managers who want to expand their knowledge of multivariate analysis to understand better increasingly complex phenomena in their work environment. For example, businesspeople are interested in learning how to develop strategies to appeal to customers with varied demographic and psychographic characteristics in a marketplace with multiple constraints (legal, economic, competitive, technological, etc.). Multivariate techniques are required to study these multiple relationships adequately and obtain a more complete, realistic understanding for decision making.

Any researcher who examines only two variable relationships and avoids multivariate analysis is ignoring powerful tools that can provide potentially very useful information. As one researcher states: "For the purposes of . . . any . . . applied field, most of our tools are, or should be, multivariate. One is pushed to a conclusion that unless a . . . problem is treated as a multivariate problem, it is treated superficially [1]. According to statisticians Hardyck and Petrinovich [4]:

. . . multivariate analysis methods will predominate in the future and will result in drastic changes in the manner in which research workers think about problems and how they design their research. These methods make it possible to ask specific and precise questions of considerable complexity in natural settings. This makes it possible to conduct theoretically significant research and to evaluate the effects of naturally occurring parametric variations in the context in which they normally occur. In this way, the natural correlations among the manifold influences on behavior can be preserved and separate effects of these influences can be studied statistically without causing a typical isolation of either individuals or variables.

IMPACT OF THE COMPUTER REVOLUTION Widespread application of computers (first mainframe computers and, more recently, microcomputers) to process large, complex data banks has spurred the use of multivariate statistical methods. Today a number of prepackaged computer programs are available for multivariate data analysis and others are being developed [9,7,10,12,13]. In fact, many researchers have appeared who realistically call themselves *data analysts* instead of statisticians or (in the vernacular) "quantitative types." These data analysts have contributed substantially to the increase in the number of journal articles using multivariate statistical techniques. Even for people with strong quantitative training, the availability of prepackaged programs for multivariate analysis has facilitated the complex manipulation of data matrices that have long hampered the growth of multivariate techniques.

With several major universities already requiring entering students to purchase their own microcomputers before matriculating, students and professors will soon be analyzing multivariate data routinely for decisions of various kinds in diverse fields. Some of the prepackaged programs designed for mainframe computers (e.g., the SPSS package) are now available in a form suitable for microcomputers [11], and more will soon be available.

Multivariate Analysis Defined Multivariate analysis is not easy to define. Broadly speaking, it refers to all statistical methods that simultaneously analyze multiple measurements on each individual or object under investigation. Any simultaneous analysis of more than two variables can be loosely considered multivariate analysis. As such, multivariate techniques are extensions of univariate analysis (analysis of single variable distributions) and bivariate analysis (cross-classification, correlation, and simple regression used to analyze two variables).

One reason for the difficulty of defining multivariate analysis is that the term *multivariate* is not used consistently in the literature. To some researchers, *multivariate* simply means examining relationships between or among more than two variables. Others use the term only for problems where there are multiple variables, all of which are assumed to have a multivariate normal distribution. However, to be considered truly multivariate, all of the variables must be random variables that are interrelated in such ways that their different effects cannot meaningfully be interpreted separately. Some authors state that the purpose of multivariate analysis is to measure, explain and/or predict the degree of relationship among *variates* (weighted combinations of variables). Thus, the multivariate character lies in the multiple variates (multiple combinations of variables), not only in the number of variables or observations. For the purposes of this book, we will not insist on a rigid definition of multivariate analysis. Instead, both multivariable techniques and truly multivariate techniques will be discussed because the authors believe that knowledge of multivariable techniques is an essential first step in understanding multivariate analysis.

Types of Multivariate Techniques

Multivariate analysis is a relatively new but rapidly expanding approach to data analysis. Specific techniques included in multivariate analysis are (1) multiple regression and multiple correlation; (2) multiple discriminant analysis; (3) principal components analysis and common factor analysis; (4) multivariate analysis of variance and covariance; (5) canonical correlation analysis; (6) cluster analysis; (7) multidimensional scaling; and (8) conjoint analysis. The first four categories of techniques are frequently used by practicing statisticians, and all but the fourth are fairly well established in academic research literature. The four remaining techniques are less well known and have been applied only tentatively and experimentally by researchers. A separate chapter is devoted to each of these techniques. At this point, we will introduce each of the multivariate techniques briefly defining the technique and the objective for its application.

Multiple regression is the method of analysis that is appropriate when the research problem involves a single metric dependent variable presumed to be related to one or more metric independent variables. The objective of multiple regression analysis is to predict the changes in the dependent variable in response to changes in the several independent variables. This objective is most often achieved through the statistical rule of least squares.

Whenever the researcher is interested in predicting the level of the dependent variable, multiple regression is useful. For example, monthly expenditures on dining out (dependent variable) might be predicted from information regarding a family's income, size, and the age of the head of the household (independent variables). Similarly, the researcher might attempt to predict a company's sales from information on its expenditures for advertising, the number of salespeople, and the number of stores selling its products.

Multiple discriminant analysis (MDA) If the single dependent variable is dichotomous (e.g., male-female) or multichotomous (e.g., high-medium-low) and therefore nonmetric, the multivariate technique of *multiple discriminant analysis* is appropriate. Discriminant analysis is useful in situations where the total sample can be divided into groups based on a dependent variable that has several known classes. The primary objectives of multiple discriminant analysis are to understand group differences and predict the likelihood that an entity (individual or object) will belong to a particular class or group based on several metric independent variables. For example, discriminant analysis might be used to distinguish innovators from noninnovators according to their demographic and psychographic profiles. Other applications include distinguishing heavy product users from light users, males from females, national brand buyers from private label buyers, and good credit risks from poor credit risks. Even the Internal Revenue Service uses discriminant analysis to compare selected federal tax returns with a composite hypothetical normal taxpayer's return (at different income levels) to identify the most promising returns and areas for audit.

Multivariate analysis of variance (MANOVA) is a statistical technique that can be used to explore simultaneously the relationship between several categorical independent variables (usually referred to as *treatments*) and two or more metric dependent variables. As such, it represents an extension of univariate analysis of variance (ANOVA). Multivariate analysis of covariance (MANCOVA) can be used in conjunction with MANOVA to remove (after the experiment) the effect of any uncontrolled independent variables on the dependent variables. The procedure is similar to that involved in bivariate partial correlation. MANOVA is useful when the researcher designs an experimental situation (manipulation of several nonmetric treatment variables) to test hypotheses concerning the variance in group responses on two or more metric dependent variables.

Canonical correlation analysis can be viewed as a logical extension of multiple regression analysis. Recall that multiple regression analysis involves a single metric dependent variable and several metric independent variables. With canonical analysis the objective is to correlate simultaneously several metric dependent variables and several metric independent variables. Whereas multiple regression involves a single dependent variable, canonical correlation involves multiple dependent variables. The underlying principle is to develop a linear combination of each set of variables (both independent and dependent) in a manner that maximizes the correlation between the two sets. Stated in a different manner, the procedure involves obtaining a set of weights for the dependent and independent variables that provide the maximum simple correlation between the set of dependent variables and the set of independent variables.

The techniques discussed thus far have focused on multivariate methods applied to data that contain both dependent and independent variables. However, if the researcher is investigating the interrelations, and therefore the interdependence among all of the variables, without regard to whether they are dependent or independent variables, several other multivariate methods are appropriate. These methods include factor analysis, cluster analysis, multidimensional scaling, and conjoint analysis.

Factor analysis, including variations such as component analysis and common factor analysis, is a statistical approach that can be used to analyze interrelationships among a large number of variables and to explain these variables in terms of their common underlying dimensions (factors). The statistical approach involves finding a way of condensing the information contained in a number of original variables into a smaller set of dimensions (factors) with a minimum loss of information.

Cluster analysis is an analytical technique that can be used to develop meaningful subgroups of individuals or objects. Specifically, the objective is to classify a sample of entities (individuals or objects) into a small number of mutually exclusive groups based on the similarities among the entities. Unlike discriminant analysis, the groups are not predefined. Instead, the technique is used to identify the groups.

Cluster analysis usually involves at least two steps. The first is the measurement of some form of similarity or association between the entities in order to determine how many groups really exist in the sample. The second step is to profile the persons or variables in order to determine their composition. This step may be accomplished by applying discriminant analysis to the groups identified by the cluster technique.

In *multidimensional scaling,* the objective is to transform consumer judgments of similarity or preference (e.g., preference for stores or brands) into distances represented in a multidimensional space. If objects A and B are judged by the respondents as being most similar compared to all other possible pairs of objects, multidimensional scaling techniques will position objects A and B in such a way that the distance between them in multidimensional space is smaller than that between any other two objects. A related scaling technique, *conjoint analysis,* is concerned with the joint effect of two or more independent variables on the ordering of a single dependent variable. It permits development of stronger measurement scales by transforming rank order responses into metric effects. As will be seen, metric and nonmetric multidimensional scaling techniques produce similar-appearing results. The fundamental difference is that metric multidimensional scaling involves a preliminary transformation of the nonmetric data into metric data, followed by positioning of the objects using the transformed data.

A Classification of Multivariate Techniques

To assist you in becoming familiar with specific multivariate techniques, a classification of most of the multivariate methods is presented in Figure 1.1. This classification is based on three judgments the analyst must make about the nature and utilization of the data: (1) Can the variables be divided into independent and dependent classifications based on some theory? (2) If they can, how many variables are treated as dependent in a single analysis? (3) How are the variables measured? The selection of the appropriate multivariate technique to be utilized will depend upon the answers to these three questions.

There is a close relationship between the various multivariate methods, and they can be viewed as a family of techniques. When considering the application of multivariate statistical techniques, the first question to be asked is: Can the data variables be divided into independent and dependent classifications? The answer to this question indicates whether a dependence or interdependence technique should be utilized. Note that in the classification shown in Figure 1.1, the dependence techniques are on the left side and the interdependence techniques are on the right. A *dependence technique* may be defined as one in which a variable or set of variables is identified as the dependent variable to be predicted or explained by other independent variables. An example of a dependence technique is multiple regression analysis. In contrast, an *interdependence technique* is one in which no single variable or group of variables is defined as being independent or dependent. Rather, the procedure involves the analysis of all variables in the set simultaneously. Factor analysis is an example of an interdependence technique. Let's focus on dependence techniques first and use the classification in Figure 1.1 to select the appropriate multivariate method.

The different methods that constitute the analysis of dependence can be categorized by two things: (1) the number of dependent variables and (2) the type of measurement scale employed by the variables. Regarding the number of dependent variables, dependence techniques can be classified as those having either a single dependent variable or several dependent variables. Dependence techniques can be further classified as those with either metric (quantitative/numerical) or nonmetric (qualitative/categorical) dependent variables [the concept of metric or nonmetric variables will be discussed in the following section on measurement scales]. If the analysis involves a single dependent variable that is metric, the appropriate technique is either multiple correlation or multiple regression analysis. On the other hand, if the single dependent variable is nonmetric (categorical), then the appropriate technique is multiple discriminant analysis. In contrast, when the research problem involves several dependent variables, two other techniques of analysis are appropriate. If the several dependent variables are metric, the techniques of multivariate analysis of variance or canonical correlation analysis are appropriate. If the several dependent variables are nonmetric, then they can be transformed through dummy variable coding (0–1) and canonical analysis can be used.[1] Conjoint analysis is a special case, as denoted by the dashed lines in Figure

[1] Dummy variable coding will be discussed in greater detail later. Briefly, dummy variable coding is a means of transforming nonmetric data into metric data. It involves the creation of so-called dummy variables, in which 1's and 0's are assigned to subjects depending on whether they do or do not possess a characteristic in question. For example, if a subject is a male, assign him a 1; if the subject is a female, assign her a 0—or the reverse.

FIGURE 1.1 A classification of multivariate methods.

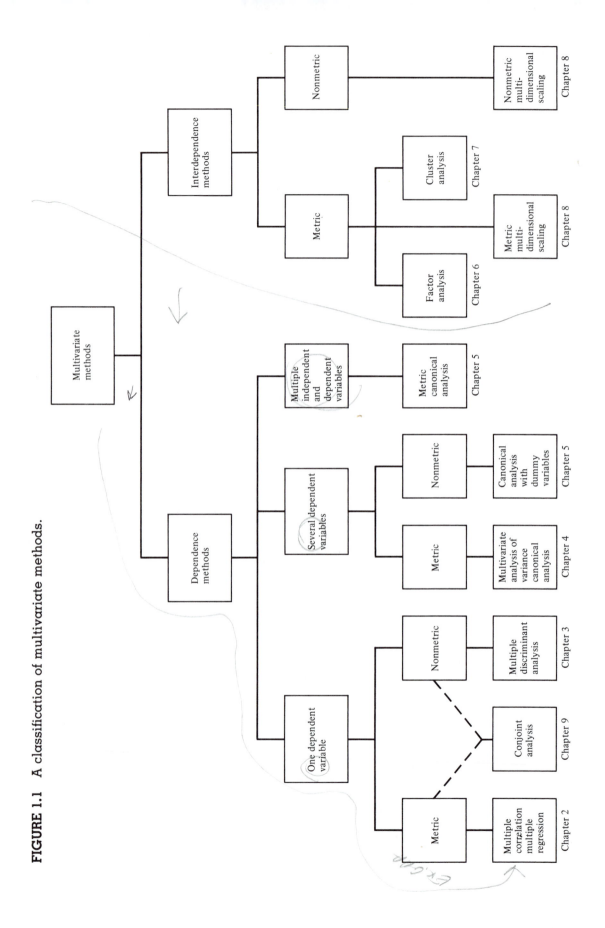

1.1. It is a dependent procedure that may treat the dependent variable either as non-metric or metric, depending on the circumstances. This will be covered in Chapter 9.

Interdependence techniques are shown on the right side of Figure 1.1. Readers will recall that with interdependence techniques the variables cannot be classified as either dependent or independent. Instead, all of the variables are analyzed simultaneously in an effort to give meaning to the entire set of variables or subjects. As with dependence techniques, the interdependence techniques can also be classified as either metric or nonmetric. Generally, factor analysis and cluster analysis are considered to be metric interdependence techniques. However, nonmetric data may be transformed through dummy variable coding for use with factor analysis and cluster analysis. Both metric and nonmetric approaches to multidimensional scaling have been developed.

Measurement Scales

Most data analysis involves the partitioning, identification, and measurement of variation in a dependent variable due to one or more independent variables. The key word here is *measurement* because the researcher cannot partition or identify variation unless it can be measured. Measurement is important not only in data analysis but also in the selection of the appropriate multivariate method of analysis. In the next few paragraphs we shall discuss the concept of measurement as it relates to data analysis and particularly to the various multivariate techniques.

There are two basic kinds of data: nonmetric (qualitative) and metric (quantitative). Nonmetric data are attributes, characteristics, or categorical properties that can be used to identify or describe a subject. While nonmetric data differ in type or kind, metric data measurements are made so that subjects may be identified as differing in amount or degree. Metrically measured variables reflect relative quantity or distance, whereas nonmetrically measured variables do not. As seen in Figure 1.2, nonmetric data are measured with nominal or ordinal scales and metric variables with interval and ratio scales.

Measurement with a nominal scale involves assigning numbers that are used to label or identify subjects or objects. Nominal scales provide the least precise measurement, since the data consists merely of the number of occurrences in each class or category of the variable being studied. Therefore, the numbers or symbols assigned to the objects have no quantitative meaning beyond indicating the presence or absence of the attribute or characteristic under investigation. Examples of nominally scaled data include an individual's sex, religion, or political party. In working with these data, the analyst might assign numbers to each category, for example, 1's for females and 0's for males. The numbers here only represent categories or classes and do not imply amounts of an attribute or characteristic.

Ordinal scales are the next higher level of measurement precision. Variables can be ordered or ranked with ordinal scales in relation to

FIGURE 1.2 Types of data and measurement scales.

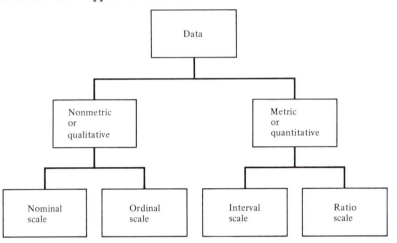

the amount of the attribute possessed. Every subclass can be compared with another in terms of a "greater than" or "less than" relationship. For example, different levels of an individual consumer's satisfaction with several new products can be illustrated on an ordinal scale. The following scale shows a respondent's view of three products. The respondent is more satisfied with A than B and more satisfied with B than C.

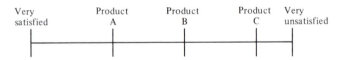

Numbers utilized in ordinal scales such as these are nonquantitative, since they indicate only relative positions in an ordered series. There is no measure of how much satisfaction the consumer receives in absolute terms, nor does the researcher know the exact difference between points on the scale of satisfaction. Most scales in the behavioral sciences fall into this ordinal category.

Interval scales and ratio scales (both metric) provide the highest level of measurement precision. Thus, they permit nearly all mathematical operations to be performed. These two scales have constant units of measurement, so differences between any two adjacent points on any part of the scales are equal. The only real difference between interval and ratio scales is that interval scales have an arbitrary zero point, while ratio scales have an absolute zero point. The most familiar interval scales are the Fahrenheit and Celsius temperature scales. Both have a different arbitrary zero point, and neither indicates a zero amount or lack of temperature, since we can register temperatures below the zero point of each scale. Therefore, it is not possible to say that any value on an interval scale is some multiple of some other point on the scale. For example, an 80°F day cannot correctly be said to be twice as hot as a 40°F day because we know that 80°F, using a different scale, such as Celsius, is 26.7°C. Similarly, 40°F, using Celsius, is 4.4°C. Although 80°F *is* indeed twice 40°F, one cannot state that the heat of 80°F is twice the heat of 40°F because, using different scales, the heat is not twice as great, that is, 4.4°C times 2 ≠ 26.7°C.

Ratio scales represent the highest form of measurement precision, since they possess the advantages of all lower scales plus an absolute zero point. All mathematical operations are allowable with ratio scale measurements. The bathroom scale or other common weighing machines are examples of these scales, for they have an absolute zero point and can speak in terms of multiples when relating one point on the scale to another; for example, 100 pounds is twice as heavy as 50 pounds.

Understanding the different types of measurement scales is important in determining which multivariate technique is most applicable to the data. Recall that two of the considerations in developing the classification of multivariate methods presented in Figure 1.1 were how the variables are measured and whether they can be divided into independent and dependent classifications. These two considerations are summarized in matrix form in Figure 1.3. The matrix can be used to select the appropriate multivariate technique by variable and data types. For example, if the independent variables are metrically scaled and the dependent variable(s) is nonmetric, then the appropriate techniques are multiple discriminant analysis and/or canonical correlation with dummy variables. Thus, by identifying the variable type (independent or dependent) and the data type (metric or nonmetric), one can select the appropriate multivariate technique to utilize in analyzing the data.

Data Bank   To explain and illustrate each of the multivariate techniques more fully, a single data bank is utilized throughout the book. The data

FIGURE 1.3   Selection of multivariate techniques by data and variable types.

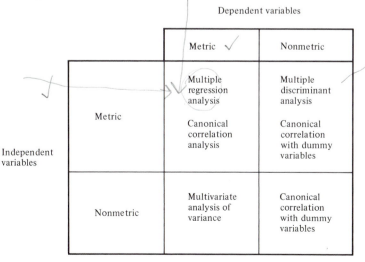

bank is a hypothetical set of data generated by the authors to meet the assumptions required in the application of multivariate techniques. The data bank consists of 50 observations on 12 separate variables. We will assume that the data were obtained from the Hair, Anderson, Tatham Company (HATCO). HATCO is interested in identifying (and, it is hoped, predicting) which individuals will be knowledgeable and highly motivated salespeople. The data provided in the data bank will enable HATCO to develop a profile of the type of individual most likely to be knowledgeable and successful, and will also make it possible to evaluate two different approaches to training. A brief description of the data base variables is provided in Table 1.1, in which the variables are also classified as to whether they are independent or dependent and metric or nonmetric. A listing of the data bank is provided in the accompanying instructor's manual for those who wish to reproduce the solutions reported in this book.

During the first 3 months of employment, HATCO obtains measures of social-psychological attitudes from all of its employees. These measures represent the first six variables of the data set. The six variables are measures of self-esteem, locus of control, alienation, social responsibility, Machiavellianism, and political opinion.

During the past several years, HATCO has been testing two different methods of sales training to supplement regular reading assignments and lectures in the training of 50 salespeople. The two different methods of sales training being tested are (1) role playing and (2) case study analysis. Prior to the training programs, each of the trainees is randomly assigned to one of the two sales training methods. Thus, in our example,

TABLE 1.1
Description of Data Base Variables

	Variable Description	Variable Type
X_1	Self-esteem	Independent/metric
X_2	Locus of control	Independent/metric
X_3	Alienation	Independent/metric
X_4	Social responsibility	Independent/metric
X_5	Machiavellianism	Independent/metric
X_6	Political opinion	Independent/metric
X_7	Knowledge	Dependent/metric
X_8	Motivation	Dependent/metric
X_9	Type of training Role playing (coded 1) Case study (coded 0)	Independent/nonmetric
X_{10}	Top Salesperson Award Recipient (coded 1) Nonrecipient (coded 0)	Dependent/nonmetric
X_{11}	Time of day Morning (coded 1) Afternoon (coded 0)	Independent/nonmetric
X_{12}	Customer satisfaction High (coded 3) Medium (coded 2) Low (coded 1)	Dependent/nonmetric

we will assume that 25 of the trainees were assigned to the role-playing situation and 25 to the case study situation. These two methods of training are represented by variable 9, type of training. Also, some receive training in the morning, others in the afternoon (variable 11).

At the end of the training program, each of the sales trainees is given two tests. One test measures overall knowledge about the company's products and the other test measures motivation toward achievement in the trainee's sales career. Variables 7 and 8 represent the results for these tests. HATCO has also been monitoring the trainee's performance following the training period. Company records indicate that 30 of the trainees were awarded the Top Salesperson of the Month Award after completion of the training program. Variable 10 records whether or not a sales trainee received this award. Variable 12 is a measure of customer satisfaction with this salesperson's performance.

Variables 7, 8, 9, and 10 have already been described in sufficient detail. However, additional information needs to be provided on variables 1 to 6 to clarify their interpretation and facilitate understanding of their use in multivariate techniques. The following definitions, adapted from Robinson and Shaver [8], may be utilized:

X_1 **Self-esteem** This variable is designed to measure attitudes toward the self along a favorable to unfavorable dimension. For example, high self-esteem means that the individual respects himself and considers himself worthy, but does not necessarily consider himself better than others and definitely does not consider himself worse. A person with high self-esteem does not feel he is the ultimate in perfection; on the contrary, he recognizes his limitations and expects to grow and improve. In contrast a person with low self-esteem has the opposite of the characteristics describing a person with high self-esteem. This variable is scored so that high values indicate a high self-esteem score.

X_2 **Locus of control** This variable concerns the degree to which individuals perceive themselves as being in control of their lives and the events that influence it. Some people tend to attribute the influences on their lives to forces within their own control, such as their own work efforts and skills. Others explain their experiences in life by reference to factors beyond their control, such as luck, chance, or strong forces they cannot overcome. The first group has an internal locus of control, the second an external locus of control. This variable is scored so that a high value indicates an external locus of control and a low value an internal locus of control.

X_3 **Alienation** Alienation is defined as the subjective feeling of estrangement from society and its culture. The concept of alienation is defined and measured by a number of separate components, including powerlessness, normlessness, meaninglessness, cultural estrangement, social estrangement, and work estrangement. Persons who exhibit alienation believe they can be characterized by these six components. This variable is scored so that a high value indicates an individual who is not alienated and a low value an individual who tends to be alienated.

X_4 **Social responsibility**   This variable assesses a person's traditional social responsibility and orientation toward helping others even when there is nothing to be gained by doing so. The results of this variable are likely to be the opposite of alienation. A high score on this variable indicates a person who exhibits high social responsibility, and vice versa.

X_5 **Machiavellianism**   This variable measures a person's general strategy for dealing with people, especially the degree to which they feel that other people are manipulable in interpersonal situations. Persons who score high on the Machiavellian scale tend to have a cool detachment, which makes them less emotionally involved with other people, with sensitive issues, or with saving face in embarrassing situations. A highly Machiavellian individual also tends to be more manipulative and impersonal in dealing with others. A high score indicates a person characterized by high Machiavellianism, and vice versa.

X_6 **Political orientation**   This variable measures an individual's political orientation, focusing particularly on a continuum of opinions from radicalism to conservatism. It consists of a number of measures relating to topics of religion, welfare, unionism, government, morals, racial tolerance, contraception, and the legal system. A high score on this variable indicates that an individual tends to be more conservative, and a low score that he or she is more liberal.

As noted earlier, these data bank variables will be utilized throughout the book when examples of actual results of computer programs are discussed. Two exceptions should be noted. In some instances, other hypothetical problems for HATCO are used as examples, and in Chapters 8 and 9 a unique set of examples is provided.

Summary

This chapter has introduced the exciting, challenging topic of multivariate data analysis. The following chapters discuss each of the techniques in sufficient detail to enable the novice data analyst to understand what a particular technique can achieve, when and how it should be applied, and how the results of its application are to be interpreted. End-of-chapter readings from academic literature further demonstrate the application and interpretation of the techniques.

QUESTIONS

1. In your own words, define multivariate analysis.

2. Name several factors that have contributed to the increased application of techniques for multivariate data analysis in recent years.

3. List and describe the eight multivariate data analysis techniques described in this chapter.

4. Explain why and how the various multivariate methods can be viewed as a family of techniques.

5. Why is knowledge of measurement scales important to an understanding of multivariate data analysis?

REFERENCES

1. Gatty, R. "Multivariate Analysis for Marketing Research: An Evaluation," *Applied Statistics,* Vol. 15, November 1966, pp. 157–72.
2. Green, P. E., *Analyzing Multivariate Data.* Hinsdale, Ill.: Holt, Rinehart, & Winston, 1978.
3. Green, P. E., and J. Douglas Carroll, *Mathematical Tools for Applied Multivariate Analysis.* New York: Academic Press, 1978.
4. Hardyck, C. D., and L. F. Petrinovich, *Introduction to Statistics for the Behavioral Sciences,* 2nd ed. Philadelphia: Saunders Co., 1976.
5. Kachigan, S. K., *Multivariate Statistical Analysis.* New York: Radius Press, 1982.
6. McGraw-Hill, Inc., *SPSS-X Users's Guide,* 2nd ed. Chicago, 1986.
7. McGraw-Hill, Inc., *SPSS-X Advanced Statistics Guide,* Chicago, 1986.
8. Robinson, J. P., and P. R. Shaver, *Measures of Social Psychological Attitudes,* rev. ed. Survey Research Center, Institute of Social Research, The University of Michigan, Ann Arbor, 1979.
9. SAS Institute, Inc., *SAS User's Guide: Basics,* Version 5, ed. Cary, N. C., 1985.
10. SAS Institute, Inc., *SAS User's Guide: Statistics,* Version 5, Cary, N. C., 1985.
11. *SPSS PC.* Chicago: McGraw Hill, 1985.
12. *STATPRO PC,* Wadsworth Professional Software, Boston, 1985.
13. *SYSTAT.* SPSS, Inc., Chicago, 1985.

CHAPTER TWO

Multiple Regression Analysis

CHAPTER PREVIEW

In this chapter, we describe multiple regression analysis as it is used to solve important research problems, particularly in business. Guidelines are presented for judging the appropriateness of multiple regression for various types of problems. Suggestions are provided for interpreting the results of its application, from a managerial as well as a statistical viewpoint. Many readers who are already knowledgeable about multiple regression procedures can simply skim this chapter. But for those who are less familiar with the subject, this chapter will provide a valuable background for the study of multivariate data analysis. It is helpful to review and familiarize yourself with the key terms before beginning to read the chapter.

Multiple regression analysis is a general statistical technique used to analyze the relationship between a single dependent variable and several independent variables. After studying the overview of regression analysis presented in this chapter, you should be able to do the following:

- Determine when regression analysis is the appropriate statistical tool to use in analyzing a problem.
- Understand how regression analysis helps us make predictions.
- Understand the least squares concept of prediction.
- Be aware of the important assumptions underlying regression analysis.
- Interpret the results of regression from both a statistical and a managerial viewpoint.
- Explain the difference between stepwise and simultaneous regression.
- Use dummy variables with an understanding of their interpretation.

KEY TERMS

Coefficient of determination (r^2) Measures the proportion of the variation of the dependent variable about its mean that is explained by the independent or predictor variable(s). The coefficient can vary between 0 and 1. If the regression model is properly applied and estimated, the higher the value of r^2, the greater the explanatory power of the regression equation, and therefore the better the prediction of the criterion variable.

Collinearity   A concept that expresses the relationship between two (collinearity) or more variables (multicollinearity). Two variables are said to exhibit complete collinearity if their correlation coefficient is 1 and complete lack of collinearity if their correlation coefficient is 0.

Correlation coefficient (r)   Indicates the strength of the association between the dependent and independent variables. The magnitude of the coefficient is not easy to interpret (see the definition of coefficient of determination), but the sign ($+$ or $-$) indicates the direction of the relationship. The correlation coefficient varies from -1 to $+1$, with $+1$, for example, indicating a direct perfect relationship, 0 indicating no relationship, and -1 indicating a reversed relationship (as one grows larger, the other grows smaller).

Criterion variable   Dependent variable.

Degrees of freedom   Calculated from the total number of observations minus the number of parameters estimated from those data. These parameter estimates are restrictions on the data, since, once made, they define the population from which the data are assumed to have been drawn. For example, in estimating the random error in a two-variable regression model, we must have estimated two parameters, β_0 with b_0 and β_1 with b_1, since the measure of error is $\Sigma(Y -$ prediction$)^2$, which is $\Sigma(Y - b_1X)^3$. Therefore, with n values of 6 we have $(n - 2)$ degrees of freedom for the estimation of random error.

Dummy variables   An independent variable used to account for the effect that different levels of a variable produce upon a dependent variable. Variables that are sometimes treated as dummy variables in regression analysis are sex and race. To account for L levels of such a variety, $L - 1$ dummy variables are needed. For example, we could represent the variable sex as two variables, X_1 and X_2. When the respondent is male, $X_1 = 1$ and $X_2 = 0$. When the respondent is female, $X_1 = 0$ and $X_2 = 1$. However, when $X_1 = 1$, X_2 must be 0, so we need only one dummy variable, X_1, to represent the variable sex. For students familiar with matrix algebra, b_0 of the regression equation cannot be estimated because there would be redundant information. If X_1 is 0, we know that the value of X_2 has to be 1, so we do not need both values. We always have one dummy variable less than the number of levels of the variables we are using.

Error or residual   Seldom will our predictions be perfect. We assume that random error will occur, but we assume that this error is an estimate of the true random error in the population, not just the error in prediction for our sample. We assume that the error in the population that we are estimating is distributed with an average value of 0 and a constant variance. The error in predicting our sample data is called the *residual*.

Homoscedasticity   When the variance of the error terms (*e*) appears constant over a range of x values, the data are said to be homoscedastic and therefore satisfy the assumption of homoscedasticity. The assumption of equal variance of the population error (ϵ), where ϵ is estimated from e, is critical to the proper application of linear regression. When the error terms have increasing or modulating variance, the data are heteroscedastic. The discussion examining the residuals (ϵ) in this chapter will further illustrate this point.

Intercept   The value on the Y axis (criterion variable axis) where the line defined by the regression equation $Y = b_0 + b_1X_1$ crosses the axis. It is described by the constant term b_0 in the equation. The intercept may have no managerial interpretation and may serve only to aid in making predictions. For example, if the X_1 variable has the value 0, the Y value would be predicted to be $Y = b_0$. If it is not possible for X to have a value of 0, this result may have no practical interpretation.

Linearity   Used to express the concept that the model possesses the properties of additivity and homogeneity. In the population model $Y = \beta_0 + \beta_1X_1 + \epsilon$, the effect of a change of 1 in X_1 is to add β_1 (a constant) units of Y. The model $Y = \beta_0 + \beta_2$ is not additive because a unit change in X_1 does not increase Y by β_1 units; rather, it increases Y by $(X_1 + 1)\beta_2$ units (an amount that varies for different levels of X).

Parameter   A quantity (measure) characteristic of the population. For example, μ and σ^2 are the symbols used for the population parameters, mean (μ) and variance (σ^2). These are typically estimated from sample data where the arithmetic average of the sample is the estimate of the population average and the variance of the sample is used to estimate the variance of the population.

Partial correlation coefficient   Measures the strength of the relationship between the criterion or dependent variable and a single predictor variable when the effects of the other predictor variables in the model are held constant. For example r_v, $X_2 \cdot X_1$ measures the variation in X associated with X_2 when the effect of X_1 on both Y and X_2 is held constant.

Partial F values   When a variable (say, Xa) is added to a regression equation after many other variables have already been entered into the equation, its contribution may be very small because (X_a) is highly correlated with the variables already in the equation. The partial F test is simply an F test for the additional contribution of a variable above the contributions of those variables already in the equation. A partial F may be calculated for all of the variables by simply pretending that each, in turn, is the last to enter the equation. This gives the additional contribution of each variable above all of the others in the equation.

Predictor variable   Independent variable.

Regression coefficient   The numerical value of any parameter estimate that is directly associated with the independent variables; for example, in the model $Y = b_0 + b_1X_1$ the value b_1 is the regression coefficient for the variable X_1. In the multiple predictor model (e.g., $Y = b_0 +$

$b_1X_1 + b_2X_2$) the regression coefficients are *partial* because each takes into account not only the relationship between Y and X_1 and Y and X_2, but also the relationship between X_1 and X_2.

Standardization   The process whereby raw data are transformed into new measurement variables with a mean of 0 and standard deviation of 1; the appropriate formula is $(X_i - \bar{X})/S_x$. When data are transformed in this manner, the b_0 term in the regression equation assumes a value of 0. After the data have been standardized, the term *beta coefficient* is often used to denote the regression coefficient. When two or more independent variables are measured on different units (e.g., expenditures in dollars and education in years), standardized coefficients allow the researcher to compare the relative effect on the dependent variable of each independent variable.

Zero slope   The presence of zero slope in a regression equation indicates that the regression line is horizontal, and therefore Y does not vary with X. The availability of such a regression equation to predict values of Y from values of X does not increase the predictive accuracy of the researcher. That is, if the equation describes a line with zero slope, all values of X would be associated with the same value of Y.

What Is Multiple Regression Analysis?

Multiple regression analysis is a statistical technique that can be used to analyze the relationship between a single dependent (criterion) variable and several independent (predictor) variables. The objective of multiple regression analysis is to use the several independent variables whose values are known to predict the single dependent value the researcher wishes to know. As noted in Chapter One, multiple regression analysis is a dependence technique. Thus, to use it, you must be able to divide the variables into a single dependent variable and several independent variables. Regression analysis is also the statistical tool that should be used when both the dependent and the independent variables are metric. However, under certain circumstances, it is possible to use dummy-coded independent variables in the analysis. To summarize, to use multiple regression analysis, (1) the data must be metric and (2) before deriving the regression equation, the researcher must decide which variable is to be dependent and which are to be independent.

How Is Regression Analysis Used?

As noted earlier, multiple regression analysis is a general statistical technique that can be used to examine the relationship between a single dependent variable and a set of independent variables. The following are four different purposes for which it can be used:

1. Determine the appropriateness of using the regression procedure with the problem. For example, is the regression approach appropriate for attempting to predict company sales from the expenditures for advertising? The results obtained from the analysis may be inter-

preted in such a way as to suggest whether the application was appropriate.

2. Examine the statistical significance of the attempted prediction. If we used a sample of patrons in a restaurant and attempted to predict their monthly expenditures for dining out from information on their family income, family size, and the age of the head of the household, is our prediction any better than what we might expect to achieve by chance?

3. Examine the strength of the association between the single dependent variable and the one or more independent variables. When collinearity among the independent variables is minimal (or has been removed by factor analysis), we can identify the extent to which each of the independent variables is related to the dependent variable. For example, we can determine which variable is more important in predicting the number of ounces of hand lotion used in a household—the number of children in the household, the age of the female head of the household, or the family's income.

4. Predict the values of one variable from the values of others. Can we predict the number of cases of detergent we will sell from our knowledge of the number of competitive brands in each store carrying our detergent and the median family income in each store's trade area?

In all of the previous examples, one variable was given a rather special status: it became the variable we wished to predict (the dependent variable). The variables we select in our attempt to predict the criterion variable are called *independent*, or *predictor*, variables. In using regression analysis, a decision has to be made regarding the number of predictor variables to include in the equation. In making this decision, we assume that each additional predictor variable gives more information and therefore a better prediction about the criterion variable. Otherwise it would not be included in the analysis.

How Does Regression Analysis Help Us to Predict?

The objective of regression analysis is to help predict a single dependent variable from the knowledge of one or more independent variables. When the problem involves a single dependent variable that is predicted by a single independent variable, the statistical technique is referred to as *simple regression*. When the problem involves a single dependent variable predicted by two or more independent variables, it is referred to as *multiple regression analysis*. The following section is divided into three parts to help you understand how regression helps us to predict. The three topics covered are (1) prediction using a single measure—the average, (2) prediction using two measures—simple regression, and (3) prediction using several measures—multiple regression.

Prediction Using a Single Measure—the Average

Let's start with a simple measure. Assume that we surveyed eight families and asked how many credit cards were held by all family members.

The data are shown in columns 1 and 2 of Table 2.1. If we were asked to predict how many credit cards a family holds using only these data, we could simply use the arithmetic average of seven credit cards as an acceptable predictor. Our prediction typically would be stated as follows: "The average number of credit cards held by a family is seven."

One question left unanswered is: How accurate is this prediction? The customary way to assess the adequacy of using the average as a predictor is to examine the errors that are made when it is used for this purpose. For example, if we predict that family 1 has seven credit cards, we overestimate by three. Thus, the error is +3. If this procedure were followed for each family, in some instances our estimate would be too high, in others too low, and in still others it would give the correct number of cards held. By simply adding the errors, we might expect to obtain a measure of the prediction accuracy. However, we would not—the errors would always sum to zero. Therefore, we would not have a measure of the adequacy of our prediction. To overcome this problem, we can square each error and then add the results together to obtain the sum of squared errors. The result, referred to as the *sum of squared errors*, provides a good measure of the prediction accuracy of the arithmetic average. We wish to obtain the smallest possible sum of squared errors, since this would mean that our predictions would be more accurate. For a single set of observations, no other measure of central tendency (including other more sophisticated statistical techniques) will produce a smaller sum of squared errors than will the arithmetic average. Therefore, for a single set of observations, the average is the best predictor of the number of credit cards held by families. (*Note:* the sum of squared errors for our example problem is 22.)

Prediction Using Two Measures—Simple Regression

As researchers and businesspeople, we are always interested in improving our predictions. In the preceding section, we learned that with a

TABLE 2.1 HATCO Survey Results for Average Number of Credit Cards	Family Number	Actual Number of Credit Cards	Average Number of Credit Cards*	Error†	Errors Squared
	1	4	7	+3	9
	2	6	7	+1	1
	3	6	7	+1	1
	4	7	7	0	0
	5	8	7	−1	1
	6	7	7	0	0
	7	8	7	−1	1
	8	10	7	−3	9
		56		0	22

* Average number of credit cards = 56 ÷ 8 = 7.
† Error refers to the difference between the actual number of credit cards held by a family and the estimated number of cards held (seven) using the arithmetic average as a predictor.

single set of measures the average is the best predictor. But in our example survey we also collected information on other measures. Let's determine if knowledge of another measure—the number of people in each family—will help our predictions. This procedure involves two measures and is referred to as *simple regression*.

Simple regression is another procedure for describing data (just as the average describes data) and it uses the same rule—minimizing the sum of the squared errors of prediction. We know that (without using family size) we can describe the number of credit cards held as seven. Another way to write the prediction is as follows:

$$\text{Predicted number of} = \text{Average number of}$$
$$\text{credit cards held} \qquad \text{credit cards held}$$

or

$$\hat{Y} = \bar{y}$$

Using our additional information on family size, we could try to improve our predictions. If we assume that family size is related to the number of credit cards held by the family, the difference in the number of credit cards held is associated with the difference in family size (see Table 2.2), and we can improve our prediction using regression. We could describe the relationship as follows:

$$\text{Number of credit cards held} = \text{Change in number of credit cards} \times \text{(Family size)}$$
$$\text{held associated with}$$
$$\text{unit change in family size}$$

or

$$\hat{Y} = b_1 X_1.$$

For example, if we find that for each additional member in a family the number of credit cards increases (on the average) by two, we would predict that families of four would have 8 credit cards and families

TABLE 2.2 HATCO Survey Results Relating Number of Credit Cards to Family Size	Family Number	Number of Credit Cards* (Y_i)	Family† Size (X_i)	Prediction	Errors Squared
	1	4	2	4.81	.66
	2	6	2	4.81	1.42
	3	6	4	6.76	.58
	4	7	4	6.76	.06
	5	8	5	7.73	.07
	6	7	5	7.73	.53
	7	8	6	8.7	.40
	8	10	6	8.7	1.69
					5.50

* Average number of credit cards = $(\Sigma\ Y)/n = 56/8 = 7$.
† Average family size = $(\Sigma\ Y)/n = 34/8 = 4.25$.

of five would have 10 credit cards. Thus, the number of credit cards = 2X (family size). However, we often find that the prediction is improved by adding a constant value because the following relationship may be found:

Family Size	Number of Credit Cards
1	4
2	6
3	8
4	10
5	12

It can be observed that "number of credit cards = 2 × (family size)" is wrong by two credit cards in every case:

Family Size	Number of Credit Cards	Predicted Number of Credit Cards	Error
1	4	2	−2
2	6	4	−2
3	8	6	−2
4	10	8	−2
5	12	10	−2

Therefore, changing our description to

$$\text{Number of credit cards} = 2 + 2 \times (\text{family size})$$

gives us perfect predictions in all cases.

We will take this approach with our sample of eight families and see how well the description fits our data. The procedure followed is as follows:

Predicted number
 of credit cards = Constant + Change in number × (Family size)
                              of credit cards
                              with differing
                              family size

or

$$\hat{Y} = b_0 + b_1 X_1.$$

If the constant term (b_0) does not help us predict, the process of minimizing the sum of squared errors will give an estimate of the constant term to be zero. The terms b_0 and b_1 are called *regression coefficients*.

Using a mathematical procedure,[1] we find the values of b_0 and b_1 such that the sum of the squared errors of prediction is minimized. For this example, the appropriate values are:

$$Y = 2.87 + .97 \text{ (family size)}$$

Since we have used the same criterion (least squares), we can determine whether our knowledge of family size has helped us predict credit card holdings. The sum of squared errors using the average was 22, and using our new procedure it is 5.50 (see Table 2.2, column 5). Using the least squares criterion, we see that our new approach, simple regression, is better than using the average. This description indicates that for each additional family member, the credit card holdings are higher on the average by .97. The constant 2.87 is an artificial term that is useful for helping us predict, but it has no interpretation in credit card holdings because we do not assume that if family size was zero, credit card holdings would be 2.87 (obviously a ridiculous proposition).

Major Assumptions

We have shown how improvements in prediction are possible, but in doing so we had to make several assumptions about the relationship between the variable to be predicted and the variables we wanted to use for predicting. In the following sections we discuss these assumptions, which include the assumption of a statistical relationship, the assumption of equal variance of the criterion variables, and the assumption of uncorrelated errors.

STATISTICAL RELATIONSHIP. Since we are dealing with sample data representing human behavior, we are assuming that our description of credit card holdings is statistical, not functional. For example:

$$\text{Total cost} = \text{variable cost} + \text{fixed cost.}$$

If the variable cost is $2 per unit, the fixed cost is $500, and we produce 100 units, we assume that the total cost will be exactly $700 and that any deviation from $700 is caused by our inability to measure cost because the relationship between the costs is fixed. This is called a *functional relationship*.

In our credit card example, we found two families with two members, two with four members, and so on, who had different numbers of credit cards. More than one value of the criterion variable will usually be observed for any value of a predictor variable. The criterion variable is assumed to be a random variable (see Key Terms), and for a given

---

[1] The mechanics of deriving the regression coefficients such that the sum of the squared errors is minimized is left to other more technically oriented texts dealing with regression. See [1,2,3, and 6].

predictor we can only hope to estimate the average value of the criterion variable associated with it. In our example, the two families with four members held an average of 6.5 credit cards, and our prediction was 6.76. Our prediction is not as accurate as we would like, but it is better than just using the average of 7 credit cards. The error is assumed to be the result of random behavior among credit card holders.

In summary, a functional relationship calculates an exact value, while a statistical relationship estimates an average value. Throughout this book, we will be concerned with statistical relationships. Both of these relationships are displayed in Figure 2.1.

EQUAL VARIANCE OF THE CRITERION VARIABLES.   In most situations, we have many different values for the criterion variable at each value for the predictor variable, just as we had two different numbers of credit cards held for our two families with four members. Since our prediction is based upon actual values from our predictions, we assume that at each level of the predictor variable the values of the criterion variable all have the same variance (homoscedasticity). This situation is illustrated in Figure 2.1b by showing the distributions of the criterion variable to be identical at the two levels of the predictor variable. The

FIGURE 2.1   Comparison of functional and statistical relationships.

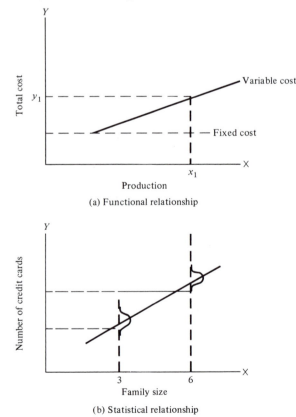

(a) Functional relationship

(b) Statistical relationship

illustration indicates that families of different sizes have many different number of credit cards, but these numbers tend to vary from the average held by families of size x_1, by the same amount that the number held by families of x_2, and so on, vary from their average number held. That is, the variances of Y at x_1, x_2, x_3, and x_4 are all equal (homoscedastic). If this is not true, our predictions will be better at some levels of the criterion variable than at others. However, since our rule for determining how good our predictions are is based on the equated deviations over all levels of the predictor variable, we could be misled into thinking that our predictions are equally descriptive at all levels.

LACK OF CORRELATION OF ERRORS.   Our predictions were not perfect in our credit card example, and we will rarely find a situation where they are. However, we do want to find that any errors we make in prediction are uncorrelated with each other. For example, if we found a pattern that suggests that every other error is positive while the alternative error terms are negative, we would know that some unexplained systematic relationship exists in the criterion variable. If such a situation exists, we cannot be confident that our prediction errors are independent of the levels at which we are trying to predict. In our credit card example, we would like to have just as strong a belief that any error is randomly distributed when predicting for a family of two as for a family of six. That is, the error will not always be positive at one level of prediction and negative at another.

Fixed Versus Random Predictors

When examining a regression model such as the one we have discussed up to this point, we have assumed a situation in which the levels of the predictor variables are fixed. For example, we wish to know the impact on the preference of three levels of sweetener in a cola drink. We make up three batches of cola and have a number of persons sample each. We then predict the preference rating for each cola, using level of sweetener as the predictor. We have fixed the level of sweetener and are interested in its effect at these levels. We do not assume the three levels to be a random sample for a large number of possible levels of sweetener. A *random predictor variable* is one in which the levels of the predictor are selected at random. Our interest is not just in the levels examined but rather in the larger population of possible predictor levels from which we selected a sample.

Most regression models based on survey data are random effects models. As an illustration, a survey was conducted to help assess the relationship between age of the respondent and frequency of visits to physicians. The predictor variable "age of respondent" was randomly selected from the population and the inference regarding the population is of concern, not just knowledge of the individuals in the sample.

The estimation procedures for models using both types of predictor variables are the same except for the error terms. In the random effects model, a portion of the random error comes from the sampling of the predictors. However, the statistical procedures based on the fixed model are quite robust, so using the statistical analysis as if you were dealing with a fixed model (as most analysis packages assume) may still be appropriate as a reasonable approximation.

Prediction Using Several Measures: Multiple Regression Analysis

We previously demonstrated how simple regression helped improve our prediction of credit card holdings. By using data on family size, we predicted the number of credit cards a family would have more accurately than we could by simply using the arithmetic average. This result raises the question of whether we could improve our prediction even further by using additional data obtained from the families. Would our prediction of the number of credit cards be improved if we used data not only on family size but also on family income?

To improve further our prediction of credit card holdings, let's use additional data obtained from our eight families. The variable we shall add is family income (see Table 2.3). We simply expand our simple regression model as follows:

Number of credit cards = Constant number of credit cards independent of family size and income + (Change in credit card holdings associated with unit change in family size) × (Family size) + (Change in credit card holdings associated with unit change in family size) × (Family income)

or

$$Y = .482 + .63X_1 + .216X_2.$$

We can again find our error by predicting Y and subtracting our prediction from the actual value as in columns 5 and 6 of Table 2.3. The total sum of squared errors is 3.05 for our prediction, using both

TABLE 2.3 HATCO Survey Results Relating Number of Credit Cards to Family Size and Family Income	Family Number	Number of Credit Cards (Y)	Family Size (X₁)	Family Income (X₂)	Prediction	Error Squared
	1	4	2	14	4.76	.59
	2	6	2	16	5.20	.64
	3	6	4	14	6.03	.00
	4	7	4	17	6.68	.10
	5	8	5	18	7.53	.22
	6	7	5	21	8.18	1.38
	7	8	6	17	7.95	.00
	8	10	6	25	9.67	.11
						3.05

family size and family income, compared to 5.50 (Table 2.2) using family size and 22 (Table 2.1) using the arithmetic average. We assume at this point that some improvement in prediction has been found.

WHAT NEW ASSUMPTIONS HAVE WE MADE? We have added one more predictive variable to predict better the number of credit cards held by families. When doing this, we must keep in mind all of the previously cited assumptions, and must concern ourselves with any possible interaction and/or correlation among our predictor variables because we have assumed that they do not interact and are uncorrelated. The rationale for considering independence, interaction, and/or correlation among the predictor variables is discussed in the following paragraphs.

INDEPENDENCE. The ideal situation would be to find data such as those shown in Table 2.4a. In attempting to predict credit card holdings, we would find that the following equation describes the relationship:

$$Y = .5X_1 + .2X_2.$$

where X_1 = family size and X_2 = family income. We can conclude that by holding family income constant, credit card holdings change on the average by .5 for each additional family member. Conversely, by holding family size constant, credit cards holdings increase on the average by .2 for a $1 increase in family income. Figure 2.2a illustrates why we can confidently make these statements. There is no correlation or interaction between the two predictor variables, family size and family income. For each level of family size, a change in family income of $1 produces a constant change in average credit card holdings. For each level of family income, a change in family size of two members produces a constant change in average credit card holdings. We can now examine the effects of our interpretation when there is interaction or correlation between our two predictor variables.

INTERACTION. The data shown in Table 2.4b reveal interaction between family size and family income, as illustrated by Figure 2.2b. It becomes evident that at larger family sizes, level of income has no effect on the number of credit cards held. For a family of two, a change in income from $10 to $20 is associated with a four-unit change in credit cards. This is not true for families of four, where there is no difference

TABLE 2.4	(a) Independent			(b) Interacting			(c) Correlated		
Numerical Illustration of Independent, Interacting, and Correlated Predictors	Credit Cards	Family Size	Family Income	Credit Cards	Family Size	Family Income	Credit Cards	Family Size	Family Income
	3	2	10	2	2	10	2	2	10
	4	4	10	8	4	10	4	4	20
	5	2	20	6	2	20	6	2	10
	6	4	20	8	4	20	8	4	20

FIGURE 2.2 Examples of independent, interacting, and correlated predictors.

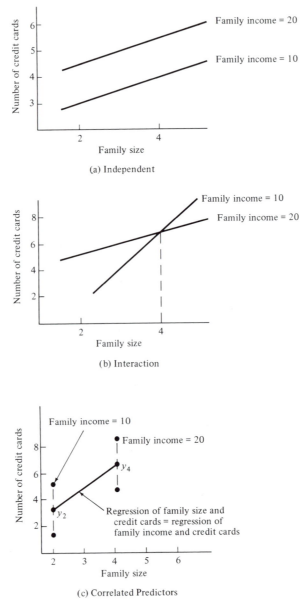

(a) Independent

(b) Interaction

(c) Correlated Predictors

in the number of credit cards held. If we used multiple regression to predict credit card holdings, we would find the following predictive equation:

$$Y = -3 + 2X_1 + .2X_2.$$

This is the best least squares equation available. It cannot be interpreted in the same way we interpreted the equation when we had no interaction. In this interactive situation, the coefficient for the effect of family size (2) reflects an average effect over both levels of family income and does not represent a constant effect. Similarly, the coefficient for family income (.2) represents an average effect of family income over the two levels of family size and, again, is not a constant effect.

CORRELATION.   The data shown in Table 2.4c reflect perfect correlation between the two predictor variables, family size and family income. The simple correlation between these two variables is 1.0, as indicated in the following correlation matrix:

	Correlation Matrix		
	Number of Credit Cards	**Family Size**	**Family Income**
Number of Credit Cards	1.00	.60	.45
Family Size	.60	1.00	1.00
Family Income	.45	1.00	1.00

Since this correlation is perfect, we suspect that no additional information will be gained by using both variables. Figure 2.2c shows that the change in credit card holdings associated with a change in family size from two to four is exactly the same as the average change associated with a change in income from 10 to 20. In a situation such as the one previously illustrated, only the predictor variable with the highest simple correlation (family size) would be used to estimate the number of credit cards held by a family. Other variables that are highly correlated with the single best predictor would be discarded. The rule of thumb often used to determine the cutoff point for intercorrelation among predictor variables is that no predictor should be included that is more closely related to the best predictor than it is to the dependent variable.

Using Multiple Regression to Predict

In each illustration, we have been concerned with predicting the number of credit cards held by families and have seen that the arithmetic average, family size simple regression, or family size and family income multiple regression all provide predictions. Further, we have assessed the accuracy of our predictions by examining the sum of the squared errors of prediction. This is an oversimplification, but it will serve to acquaint you with the essential meaning of regression analysis. Now we must examine our predictive ability from a statistical viewpoint (as an estimator of population characteristics).

In our example of multiple regression, we used a sample of eight families and attempted to predict the number of credit cards they held using two other measures (family size and family income). Rarely would we want to know only about these eight families. Instead, in most cases, we would use a sample to develop a predictive model for all families having credit cards. From this viewpoint, our predictions about the eight families are only a way of obtaining the predictive model

($Y = b_0 + b_1X_1 + b_2X_2$). We are now concerned with how well we think this model predicts for all families not included in the sample we used to develop the predictive equation. We believe that a relationship between family size, family income, and number of credit cards exists in the population of all families having credit cards, and our sample of eight families allows us to infer what this relationship is in the population.

Let's review the four purposes of multiple regression analysis:

1. Determining the appropriateness of our predictive model.
2. Examining the statistical significance of our model.
3. Predicting with the model.
4. Examining the strength of association between the variables.

All of these topics will be approached from the viewpoint of understanding the characteristics of the population based on the sample data rather than by simply examining the sample data. For many illustrations, we will use the simple model (predicting number of credit cards held from family size) because it is easier to visualize than the more complicated model with two predictor variables (family size and family income).

Determining the Appropriateness of Our Predictive Model

After calculating the regression coefficients and looking at the predictions made with our equation, we are faced with dilemma. Have we met the assumptions of the model? Are there errors in prediction that suggest that we should look for new predictive variables? When the model is completed, we must examine its appropriateness in order to answer these questions. This can be accomplished by examining the errors in prediction—called the *residuals*.

We should note that residuals (the differences between the observed values and the values our model predicts) are an artifact of the particular predictive model we are using; they are not equivalent to the random error in the population. These residuals should reflect the properties of the population random error if the model is appropriate. Analysis of residuals may therefore be used to examine the appropriateness of the predictive model in terms of:

1. The linearity of the phenomenon measured.
2. The constant variance of the error terms.
3. The independence of the error terms.
4. The normality of the error term distribution.
5. The addition of other variables.

LINEARITY OF THE PHENOMENON.   We can examine the linearity of the phenomenon in two ways: by plotting the residuals and by partitioning the error. Plotting the residuals against the predictor variables (e.g., family size) often reveals the shortcomings of inappropriately applying a linear model. The plot should show the residuals falling randomly, with relatively equal dispersion about zero and no strong tendency to be either greater or less than zero.

	Family	Average Number of Credit Cards per Family Size	Prediction	Family Size	(Prediction Error Squared)
TABLE 2.5	1	5	4.8	2	.036
HATCO Survey	2	5	4.8	2	.036
Results for	3	6.5	6.7	4	.063
Average Number of	4	6.5	6.7	4	.063
Credit Cards and	5	7.5	7.7	5	.048
Family Size	6	7.5	7.7	5	.048
	7	9	8.7	6	.096
	8	9	8.7	6	.096
					.486

Partitioning the error consists of examining the residuals to estimate how much of the variation is due to lack of fit of the model and how much is due to random error. That is, how well do we predict the average value of Y for each value of X? To do this, we must have repeated observations for at least two of the X values (family size) and, it is hoped, for all of them. For every family size (x) for which we have multiple observations, we use the average number of credit cards rather than the actual data. In our study we had two families of each size, so we would use the data shown in Table 2.5. We would use our predictive model ($Y = b_0 + b_1X_1$) and calculate the sum of squared errors for the averages. Since we originally stated that the model will predict the average number of credit cards held, this analysis, using only the means, directly examines this prediction. From our data (Table 2.5), the squared errors of prediction are .486. Since our sum of squared errors using the original data was 5.50, we can see that we are predicting the average number of credit cards for each family size rather well. The error in predicting the average values for each family size is called a *lack of fit* error, and the variation of the individual observations about these averages is called *pure* error. Figure 2.3 illustrates lack of fit and pure error. We cannot possibly hope to predict each observation (pure error), but we do hope to predict all of the averages (lack of fit).

Intuitively, we see that if the pure error is significantly less than the lack of fit error, the appropriateness of the predictive model is questionable. We can test the appropriateness of our model with the F test, which simply compares the lack of fit to the pure error.[2] For

---

[2] Since we have based the analysis on the four average values for number of credit cards held, we have estimated four parameters and therefore have only the eight observations minus the four averages ($8 - 4 = 4$ degrees of freedom) for the numerator of the F test. The denominator has two degrees of freedom because we have used the four averages to estimate the equation parameters of β_0 and β_1 (i.e., the number of average values minus the number of parameters is $4 - 2 = $ two degrees of freedom). In practice, one would not attempt an analysis with so small a data set. An often used rule of thumb is to have 10 times more observations than variables in the predictive model. But in practice, the technique is frequently applied with five times more observations than variables in the model.

FIGURE 2.3   Illustration of pure error and lack of fit error.

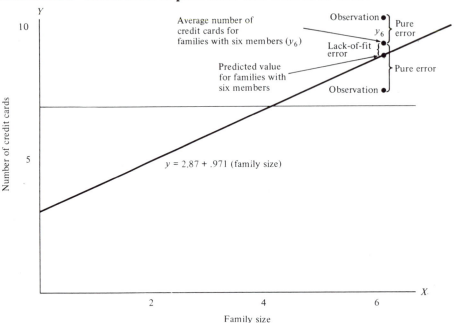

our data, the F value (.486 ÷ 2 d.f.)/(5.014 ÷ 4 d.f.) = .193. An F value approaching 1 indicates that a linear model is appropriate, while large values of F suggest that the phenomenon is not linear. If the model fits the data perfectly, lack of fit error equals 0 and the only error is the pure error—a rare occurrence.

CONSTANT VARIANCE OF THE ERROR TERM.   A plot of residuals against the predicted criterion variable indicates that the variance is not constant if it displays anything other than a random pattern. Figure 2.4a displays a hypothetical random distribution of residuals. Thus, the variance of the error term is constant and a linear regression model is appropriate.

INDEPENDENCE OF THE ERROR TERMS.   Plotting the residuals against time, even if time is not a value under consideration for the model, will reveal whether the sequencing of the measurements has affected the outcome of the experiment. Again, the pattern should appear random. Figure 2.4b displays a residual plot that exhibits an association between the residuals and the sequencing of the measurements. That is, early measurements are negative, but later ones increase and become positive.

NORMALITY OF THE ERROR TERM DISTRIBUTION.   Three procedures can be used to test the normality of the error term distribution. The simplest method is to construct histograms of the residuals in order to check visually whether the residuals appear to have a normal distribution. Alternatively, one might determine the percentages of the residuals

FIGURE 2.4   Residual plots.

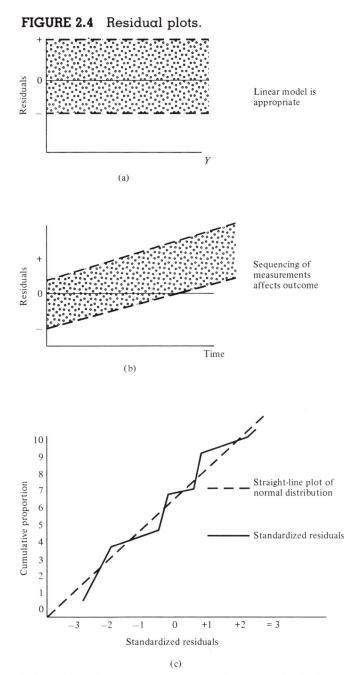

(a)

(b)

(c)

(Adapted from John Neter and William Wasserman, *Applied Linear Statistical Models*, Richard D. Irwin, Inc., 1974.)

falling within $\pm$ 1 SE or $\pm$ 2 SE. A third procedure would be to plot the cumulative standardized residuals on normal probability paper, as in Figure 2.4c. Any departure from normality may be seen by comparing the cumulative standardized residuals with the straight line representing the perfect normal distribution. To do this, the analyst would examine the plots to determine if they are reasonably comparable (without extensive deviations).

ADDITION OF OTHER VARIABLES.   After examining our predictive model, we might consider another variable such as family income, as illustrated earlier. One way to decide whether or not to add the new variable is to take only errors (residuals) from our predictions using family size and try to predict these errors using our new variable(s). If the new variable(s) explain(s) a significant portion of the residual variation, it (they) should be examined carefully for inclusion.

In summary, residuals give the analyst a quick method for determining how well the model approximates the assumptions upon which it is based and what types of errors are being encountered. It also provides a vehicle for examining other variables as candidates for inclusion.

Examining the Statistical Significance of Our Model

If we were to take repeated samples of eight families and ask them how many family members and credit cards they have, we would seldom get exactly the same values for $Y = b_0 + b_1 X_1$ from all samples. We would expect chance variation to cause differences among many samples. Usually we take only one sample and base our predictive model on it. We can test certain hypotheses concerning our predictive model in order to ensure that it represents the population of all families having credit cards rather than just our one sample of eight people. These tests may take one of two basic forms: a test of coefficients and a test of the variation explained (coefficient of determination).

TESTS OF COEFFICIENTS   In the sample model, we said that the number of credit cards equaled 2.87 + .971 (family size). We would test two hypotheses:

Hypothesis 1. The intercept (constant term) value of 2.87 arose by sampling error, and the real constant term appropriate to the population is zero. With this hypothesis, we would simply be testing whether the constant term should be considered appropriate for our predictive model. If it is found not to differ significantly from zero, we would assume that the constant term should not be used for predictive purposes. The appropriate test is the *t* test, which is commonly available on computerized regression analysis programs. From a practical point of view, this test is seldom necessary. If the data used to develop the model did not include some observations with all of the predictors measured at zero, the constant term is "outside" the data and only acts to position the model. It is then not necessary to test the constant term.

Hypothesis 2. The coefficient .971 indicates that an increase of one unit in family size is associated with an increase in the average of credit cards held by .971. We can test whether this coefficient differs significantly from 0. If it could have occurred because of sampling error, we would conclude that family size has no impact on the number of credit cards held. Note that this is not a test of any exact value of the coefficient, but rather of whether it should be used at all. Again, the appropriate test is the t test.

COEFFICIENT OF DETERMINATION   A quick way to obtain an approximation of how well the line we have fitted describes family holdings of credit cards is to examine the amount of variation our predictive model explains. For example, we could take the average number of credit cards held by our sampled families and use it as our best estimate of the number held by any family. We know that this is not an extremely accurate estimate, and we examined its accuracy by calculating the squared sum of errors in prediction (sum of squares = 22). This is a measure of how well the average explains the purchases observed. Now that we have fitted a regression model using family size, does it explain the variation better than the average? We know that it is somewhat better because the sum of squared errors is now 5.50. We can look at how well our model does by examining this improvement.

Sum of squared errors in prediction using the average = 22.0
− Sum of squared errors in prediction using family size =   5.5
Sum of squared errors explained by regression = 16.5

Therefore, we explained 16.5 squared errors by changing from the average to a regression model using family size. This is an improvement of 75 percent (16.5 ÷ 22 = .75) over use of the average.
The ratio of

$$\frac{\text{Sum of squared errors explained by regression}}{\text{Sum of squared errors about the average}}$$

is called the *coefficient of determination*. It is represented by the symbol r^2. If the regression model using family size perfectly predicted all families' holdings of credit cards, this ratio would equal 1. If using family size gave no better predictions than using the average, the ratio (r^2) would be very close to 0.

We often use the coefficient of correlation (r) to assess the relationship between Y and X. However, other than having a sign ($+r$, $−r$) to denote the slope of the regression line, it does little to clarify the nature of the association. For example, if $r^2 = .75$, we know that 75 percent of the variation in Y is explained by introducing the variable X. The corresponding value of $r = +.86$ offers the sign ($+$) as additional information but may mislead some analysts to believe a stronger relationship exists.

To test the hypothesis that the amount of variation explained by the regression model above that explained by using the average for prediction did not occur by chance (i.e., that r^2 is greater than 0), the F ratio is used. The test statistic F is the ratio

$$\frac{\text{Sum of squared error explained by regression} \div \text{degrees of freedom}}{\text{Sum of squared error about the average} \div \text{degrees of freedom}} = F$$

Two important features of this ratio should be noted:

1. Each sum of squares divided by its appropriate degrees of freedom is simply the variance of the prediction errors.
2. Intuitively, one knows that if the ratio of the explained variance to the variance about the mean is high, the use of family size must be of significant value in explaining the number of credit cards held by families.

For our example, the F ratio is $(16.5 \div 1/5.50 \div 6) = 18.1$. When compared to the tabled F statistic of 1 and six degrees of freedom of 5.99 (which would occur with a probability of .05), it leads us to reject the hypothesis that the reduction in error we obtained by using family size to predict credit card holdings was a chance occurrence. This means that finding a sample showing that we can explain 18 times as much variation when using family size as when using the average is not very likely to happen by chance (less than 5 percent of the time) when family size is not related to the number of credit cards in the population.

Predicting with the Model

After satisfying ourselves that using family size to predict the number of credit cards held offers predictions significantly better than those afforded by the average, we examine the actual predictions. This is easily done by plugging in the appropriate values, as in the case of a family size of six for our example:

$$Y_1 = b_0 + b_1 X_1$$
$$Y_1 = 2.87 + .971(6) = 8.7.$$

Since 8.7 is the estimated mean value for the number of credit cards held when the family size is six, we must look at the potential error in this estimate to understand its value. If we were to take repeated samples at $X = 6$, we know that we would observe differing values of Y (our original data support this statement). When predicting as was just described, we are predicting an average value of Y that might occur for a given value of X. In our problem, we could predict that for families of six, the average number of credit cards held would be 8.7. This does not mean that we would expect every family of six to have 8.7 credit cards. It simply means that our best estimate of the average number of credit cards held by six-member families is 8.7. We can place confidence limits about the estimated value of Y, just as we can place them about other estimates from samples. These confidence limits would give us a range within which we expect the average for our sample values to fall relative to the true population characteristics.

Since we estimated our model from sample values, the estimates b_0 and b_1 are based on values of both Y and X. If the values of X (in our problem, family size) differ greatly from the average family size ($X = 4.25$) from sample to sample, the estimates of b_0 and b_1 can vary greatly. As we make estimates of Y for families of sizes further from X, the confidence range on the predictions becomes wider. As a rule of thumb, if you wish to predict from a regression model, select values of X such that your area of prediction is near the average (X) value of X. This means that if you are especially interested in predicting for medium-sized families of three, four, or five members, try to include families both larger and smaller in the sample. Additionally, one assumes that the predictions are valid only in the following situations:

1. The conditions and relationships measured at the time the sample of families is taken do not change materially. For example, if most companies suddenly start charging a monthly fee for credit cards, your predictions could be very poor because families may change their credit card holdings.
2. The model is not to be used for estimation beyond the range of the X values (family size) found in the sample. For instance, if the largest family in your sample had 6 members, it might be unwise to predict the number of credit cards held by families with 10 members.

In the former case, one is assuming that the model continues to fit the modeled relations over time. In the latter case, one assumes (dangerously) that values of X beyond those measured will have the same relation to Y as the X values included in the regression model. In both situations, there are few ways to validate these assumptions legitimately other than to reestimate the models with new data.

Examining the Strength of Association Among the Variables

USING THE REGRESSION COEFFICIENTS   We would like to know which variable, family size or family income, is most helpful in predicting the number of credit cards held by a family. Unfortunately, the regression coefficients (b_0, b_1, and b_2) do not really give us this information. To illustrate why, we can use a rather obvious case. Suppose we wanted to predict teenagers' monthly expenditures on records (Y), using two predictor variables; X_1 is parents' income in thousands of dollars and X_2 is the teenager's monthly allowance measured in dollars. We found the following model by a least squares procedure:

$$Y = -.01 + X_1 + .001X_2$$

You might assume that X_1 is more important because its coefficient is 1,000 times larger than the coefficient for X_2. This is not true, however.

A \$10 increase in the parents' income produces a $1 \times \$10 \div \$1,000$ change in average record purchases (since the X_1 value is measured in thousands of dollars). This change is .01 in the average number of records. A change of \$10 in the teenager's monthly allowance produces a (.001) (\$10) change in average record expenditures or a .01 change in the average number of records (since the teenager's allowance was measured in dollars).

A \$10 change in the parents' income produced the same effect as a \$10 change in the teenager's allowance. Both variables are equally important, but the regression coefficients do not directly reveal this fact. We can resolve this problem by using a modified regression coefficient called the *beta coefficient.*

BETA COEFFICIENT   If each of our predictor variables had been standardized (see Key Terms) before we estimated the regression equation, we would have found different regression coefficients. The coefficients resulting from standardized data are called *beta coefficients.* Their value is that they eliminate the problem of dealing with different units of measurement (as illustrated previously), and they reflect the relative impact on the criterion variable of a change in 1 SD in either variable. Now we have a common unit of measurement, and the coefficients tell us which variable is most influential.

Three cautions must be given when using beta coefficients. First, they should be used as a guide to the relative importance of individual independent variables only when collinearity is minimal. Second, the beta values can be interpreted only in the context of the other variables in the equation. For example, a beta value for family size reflects its importance only in relation to family income, not in any absolute sense. If another predictor variable was added to the equation, the beta coefficient for family size would probably change, because there would likely be some relationship between family size and the new predictor variable.

The third caution is that the levels (e.g. families of size five, six, and seven) affect the beta value. Had we found families of size 8, 9, and 10, the value of beta would likely change. In summary, use beta only as a guide to the relative importance of the predictor variables included in your equation, and only over the range for which you actually have sample data.

General Approaches to Regression Analysis

There are many approaches you may use when attempting to determine the best predictive model using regression analysis. The three most common are backward elimination, stepwise forward estimation, and all-possible-subsets regression. These procedures and some general guidelines for using them will be discussed next.

Backward Elimination

Backward elimination is largely a trial-and-error procedure for finding the best regression estimates. It involves computing a regression equation with all of the variables, and then going back and deleting those independent variables that do not contribute significantly to it. The steps are as follows:

1. Compute a single regression equation using all of the predictor variables that interest you.
2. Calculate a partial F test (see Key Terms) for each variable as if it were to be used after the variance accounted for by all other predictor variables is removed.
3. Eliminate those predictor variables with a partial F value that indicates that they are not making a significant contribution.
4. After eliminating variables, reestimate the regression model using only the remaining predictor variables.
5. Return to step 2 and continue the process until you identify all variables of interest and determine their contributions. This is time-consuming, but with adequate computer facilities it is a satisfactory process for many researchers.

Stepwise Forward Estimation

Stepwise forward estimation also allows you to examine the contribution of each predictor variable to the regression model. However, rather than deleting variables, as in the backward elimination procedure, each variable is considered for inclusion prior to developing the equation. The specific are as follows:

1. Start with the simple regression model in which only the one predictor most highly correlated with the criterion variable is used. The equation would be $Y = b_0 + b_1X_1$.
2. Examine the partial correlation coefficients (see Key Terms) to find an additional predictor variable that explains both a significant portion and the largest portion of the error remaining from the first regression equation.
3. Recompute the regression equation using the two predictor variables, and examine the partial F value for the original variable in the model to see if it still makes a significant contribution, given the presence of the new predictor variable. This ability to eliminate variables already in the model distinguishes the stepwise model from simple forward addition models. If the original variable still makes a significant contribution, the equation would be $Y = b_0 + b_1X_1 + b_2X_2$.
4. Continue this procedure by examining all predictors not in the model to determine if one should be included in the model. If a new predictor is included, examine all predictors previously in the model to judge if they should be kept. A potential bias in the stepwise procedure results from the consideration for selection of only one variable at a time. Suppose variables X_3 and X_4 *together* would explain a significant portion of the variance (each given the presence of the other), but neither is significant by itself. In this situation, neither would be considered for the final model.

All-Possible-Subsets Regression

The all-possible-subsets procedure is exactly what the name suggests. All possible single-variable, two-variable, three-variable, and other models are examined. With 10 variables to be considered, you must look at $10 + 45 + 120 + 210 + 252 + 210 + 120 + 45 + 10 + 1 = 686$ models. The computer cost is sometimes prohibitive. Obviously, the best model (by whatever criterion) must be found, since the researcher can examine all of them.

Multicollinearity A key assumption not examined up to this point is the assumption of independence of the variables used as predictors. If the predictor variables are not independent, the regression coefficients may be incorrectly estimated and have the wrong signs.
The following example illustrates this point:

Respondent	Dependent D	Predictor A	Predictor B
1	5	6	13
2	3	8	13
3	9	8	11
4	9	10	11
5	13	10	9
6	11	12	9
7	17	12	7
8	15	14	7

The two regressions using A and B separately are

$$D = -5 + 1.5 \, (A)$$
$$D = 30 - 2 \quad (B)$$

It is clear that the relationship between A and D is positive, while the relationship between B and D is negative. The multiple regression is

$$D = 50 - 1 \, (A) - 3 \, (B)$$

It would now appear to the casual observer that the relationship between A and D is *negative* when in fact we know it is not. The sign of A is wrong in an intuitive sense but reflects the extremely strong negative correlation between A and B.
The model with a wrong sign may be intuitively unsatisfactory, but it is the best predictive equation using these data. In this case, what options does the researcher have?

• Omit one or more highly correlated predictor variables and seek others to help in the prediction.

- Use the model with the highly correlated predictors for prediction only (i.e., make no attempt to interpret the partial regression coefficients).
- Use the simple correlations between each predictor and the dependent variable to understand the predictor–dependent variable relationship.
- Use a more sophisticated method of analysis such as Bayesian regression (or a special case—ridge regression) or regression on principal components to obtain a model that more clearly reflects the simple effects of the predictors. These procedures are mentioned here for reference only since they are beyond the scope of this text.

Illustration of a Regression Analysis

In Chapter One you were introduced to a problem in which HATCO had obtained the following measures:

Variable Description
X_1 Self-esteem
X_2 Locus of control
X_3 Alienation
X_4 Social responsibility
X_5 Machiavellianism
X_6 Political opinion
$Y = X_7$ Knowledge

To demonstrate the use of multiple regression, we will show the procedures used by HATCO to attempt to predict the level of knowledge achieved by sales personnel using these six social-psychological attitude measures obtained from each employee.

The simplified procedure we will follow to demonstrate the use of multiple regression is displayed in Figure 2.5. It represents a step-by-step procedure to be followed in the application and interpretation of stepwise regression analysis. Most canned computer programs for stepwise multiple regression will automatically go through this sequence.

Explanation of Results (Single-Variable Model)

Table 2.6 displays all of the correlations among the six independent variables and their correlations with the dependent variable (Y). Examination of the correlation matrix indicates that predictor 5 is most closely correlated with the dependent variable (.68). Our first step is to build the regression equation using this best predictor. Note that the correlation of predictor 1 with the dependent variable is .67. However, predictor 1 is correlated (.61) with predictor 5. This is your first clue that the use of both predictor variables 5 and 1 might not be appropriate, since they are almost as highly correlated with each other as they are with the dependent variable. The results of this first step typically

FIGURE 2.5 Simplified stepwise regression procedure used by HATCO.

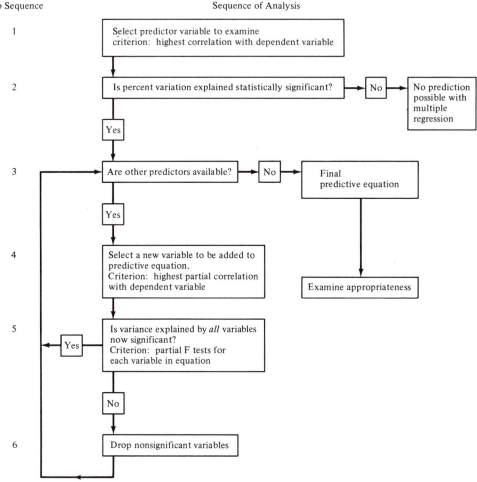

Step Sequence   Sequence of Analysis

1 — Select predictor variable to examine criterion: highest correlation with dependent variable

2 — Is percent variation explained statistically significant? → No → No prediction possible with multiple regression

Yes

3 — Are other predictors available? → No → Final predictive equation

Yes

4 — Select a new variable to be added to predictive equation. Criterion: highest partial correlation with dependent variable

Examine appropriateness

5 — Is variance explained by *all* variables now significant? Criterion: partial F tests for each variable in equation → Yes

No

6 — Drop nonsignificant variables

TABLE 2.6	Variables		Predictors						Dependent
Correlation Matrix:		$X(1)$	$X(2)$	$X(3)$	$X(4)$	$X(5)$	$X(6)$	Y
HATCO Data								
Predictors	$X(1)$	1.00						
	$X(2)$	−.38	1.00					
	$X(3)$	.−51	.52	1.00				
	$X(4)$	.020	.25	−.13	1.00			
	$X(5)$	.61	.48	.06	.24	1.00		
	$X(6)$	.04	.15	−.05	.77	.16	1.00	
	Y	.67	.04	.57	.19	.68	.21	1.00

appear as shown in Table 2.7. The concepts from Table 2.7 with which you should be familiar follow.

MULTIPLE R.   Multiple R is the correlation coefficient (at this step) for the simple regression of predictor 5 and the dependent variable. It has no + or − sign because in multiple regression the signs of the individual variables may differ, so this coefficient reflects only the degree of association.

R SQUARE.   R square is the correlation coefficient squared; it is also referred to as the *coefficient of determination*. This value indicates the percentage of total variation of Y explained by X_5. The total sum of squares (73.1 + 83.7 = 156.8) is the squared error that would occur if we used only the mean of the dependent variable as our predictor. Using predictor X(5) reduces this error by 46.6 percent (73.1 ÷ 156.8 = 46.6%).

STANDARD ERROR OF THE ESTIMATE.   The standard error of the estimate is another measure of the accuracy of our predictions. It is the square root of the sum of the squared errors divided by the degree of freedom.

TABLE 2.7
Example Output:
Step 1 of HATCO
Multiple
Regression
Example

Step 1:	
Variable entered	X(5)
Multiple R	.682
Multiple R-square	.466
Std. error of est.	1.320

Analysis of Variance

	Sum of Squares	DF	Mean Square	F Ratio
Regression	73.1	1	73.1	41.9
Residual	83.7	48	1.74	

Variables in Equations

Variable	Coefficient	Std. Error of Coeff.	Std. Reg. Coeff.	F to Remove	Variables not in the Equation		
Y-intercept	4.19						
X(5)	.85	.13	.68	41.9		Partial corr.	F Ratio
					X(1)	.438	11.176
					X(2)	−.445	11.635
					X(3)	.728	53.211
					X(4)	.039	.072
					X(6)	.150	1.096

It represents an estimate of the standard deviation of the points (y values) around the regression line, that is, it is a measure of variation around the regression line. Standard error = sum of squared errors ÷ degrees of freedom. The standard error of the predictions is used in estimating the size of the confidence interval for the predictions. See Neter and Wasserman [4] for details regarding this procedure.

VARIABLES IN THE EQUATION (STEP 1).   The value .85 is the b_1 regression coefficient calculated from the new data. The standardized regression coefficient or beta value of .68 is the b_1 value calculated from standardized data. Note that with only one independent variable the beta coefficient equals the correlation coefficient. The beta value allows you to compare the effect of X_5 on Y to the effect on Y of other variables at each stage, since this reduces b_1 to a comparable unit, the number of standard deviations. (At this step we have no other variable available for comparison.)

STANDARD ERROR OF THE COEFFICIENT.   The standard error of the coefficient is the standard error of the estimate of b_1. The value of b_1 divided by the standard error (.85/.13 = 6.5) is the calculated t value for a t test of the hypothesis $\beta_1 = 0$. A smaller standard error implies more reliable prediction. Thus, we would like to have small standard errors and therefore smaller confidence intervals. This is also referred to as the *standard error of the regression coefficient*; it is an estimate of how much the regression coefficient (b) will vary from sample to sample of the same size taken from the same population. That is, if one were to take multiple samples of the same size from the same population and use them to calculate the regression equation, this would be an estimate of how much the regression coefficient would vary from sample to sample.

F TO REMOVE. F to remove is the partial F value (and in step 1 is identical to the overall F value). This F statistic exceeds the F value in Table 2.6 for 1 and 48 degrees of freedom at the 99 percent confidence level. This is acceptable to HATCO's management as statistically significant.

For step 3 in the sequence of analysis, the section of the table headed "Variables Not in the Equation" will illustrate the meaning. For the illustration, we will use variable X_3, which is not yet in the regression model.

PARTIAL CORRELATION.   The value .728 represents the partial correlation of X_3 with Y, given that X_5 is already in the regression model. It is an indication of the variation in Y not accounted for by X_5 that can be accounted for by X_3. Remember, the partial correlation coefficient can be misinterpreted. This does not mean that we explain 72.8 percent of the previously unexplained variance. It means that 53 percent ($72.8^2 = 53\%$, the partial coefficient of determination) of the unexplained (not the total) variance and can now be accounted for by X_3.

Since 47 percent was explained by X_5, $(1 - .47) \times .53 = 28.3$ percent of the total variance could be explained by adding variable X_3. A Venn diagram illustrates this concept:

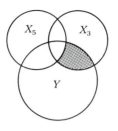

The cross-hatched area of X_3 as a proportion of the shaded area of Y represents the partial correlation of X_3 with Y given X_5. This represents the variance in X_3 (after removal of the effects of X_5 on X_3) in common with the remaining variance in Y (after removing the effects of X_5 on Y).

F. The column of F numbers represents the partial F values for all of the variables not yet in the equation. These partial F values are calculated as a ratio of the additional sum of squared errors explained by including a particular variable and the sum of squared errors left after adding that same variable. If this F level is not significant at the appropriate risk level, a variable usually would not be allowed to enter the regression model. The tabled F value at $\alpha = .01$ and 2 and 47 degrees of freedom is $F = 2.09$. Looking at the column of partial F values, note that these values for variables X_1, X_2, and X_3 are all larger than the tabled value. Therefore, all three variables could be considered for inclusion in the model.

Recall that the simple correlation of variable X_1 with the dependent variable (Y) was .67 and that for variable X_3 was .57. Therefore, you may have thought that variable X_1 would be included in the model next. But in deciding which additional variables to include in the equation, we would first select the independent variable that exhibits the highest partial correlation with the dependent variable (not the highest simple correlation). The partial correlation coefficient for X_3 is the largest (.728), and therefore X_3 will be considered for addition to the model before X_1.

We now know that a significant portion of the variance in the dependent variable is explained by predictor variable 5. We can also see that predictor variable 3 has the highest partial correlation with the dependent variable and that the F ratio for this partial correlation is significant at the .01 level. We can now look at the new model using both variables 5 and 3. The output is shown in Table 2.8.

Explanation of a Two-Variable Model—Variable X_3 Added

The multiple R value and R squared values have both increased in size. The R square has increased by the 28 percent we predicted when

TABLE 2.8
Example Output:
Step 2 of HATCO
Multiple
Regression
Example

Variable entered	$X(3)$
Multiple R	.865
Multiple R-square	.749
Std. error of est.	.914

Analysis of variance

	Sum of Squares	DF	Mean Square	F Ratio
Regression	117.5	2	58.78	70.35
Residual	39.2	47	.83	

Variables in Equation

Variable		Coeffi-cient	Std. Error of Coeff.	Std. Reg. Coeff.	F to Remove	Variables Not in the Equation		
Y-intercept		−.855				Par-tial Corr.		F Ratio
$X(3)$	3	.338	.046	.533	53.2			
$X(5)$	5	.819	.092	.652	79.6			
						$X(1)$	.002	.00
						$X(2)$	−.006	.00
						$X(4)$	.211	2.16
						$X(6)$	.279	3.89

we examined the partial correlation coefficient for X_3 of .728. The increase in R square of 28 percent is derived by multiplying the 53.4 percent of the variation that was not explained after step 1 by the partial F squared: $53.4 \times (.728)^2 = .28$. That is, of the 53.4 percent unexplained after step 1, $(.728)^2$ of this variance was explained by adding variable X_3, yielding a total variance explained of $.466 + (.534 \times (.728)^2) = .749$.

The value of b_1 has changed very little. This is a further clue that variables X_5 and X_3 are relatively independent. If the effect on Y of X_3 was independent of the effect of X_5, the b_1 coefficient would not change at all.

F TO REMOVE. The two values of F presented here are again partial F values. The F value for X_5 is now 79.6, where it was 41.9 in step 1. The F value for X_5 at this step indicates the ratio of the sum of squares due to regression added by including X_5 in the regression model as if X_3 had been in the equation first. We can therefore examine the contribution of all variables entered at earlier steps as if they were to be entered after the variable entered at this step. Note that the F value for X_3 (53.2) is the same value shown for X_3 in step 1 under the heading "Variables Not in the Equation" (see Table 2.7).

Since predictors 3 and 5 both make significant contributions to the explanation of the variation in the dependent variable, we can ask, Are other predictors available? Looking at the partial correlations for the variables not in the equation in Table 2.8, we see that predictor 6 has the highest partial correlation (.279). This variable would explain 7.7 percent of the heretofore unexplained variance $(.279)^2 = .077)$ or 2 percent of the total variance $(1 - .749) \times .077 = .022)$. This is a very modest contribution of the explanatory power of our predictive equation in spite of the significance of the F value for predictor 6 at the 0.5 α level. (*Note:* The tabled F value for 3 and 46 degrees of freedom at $p = .05$ is 2.81, while the F value of predictor 6 is 3.89.)

THREE-VARIABLE MODEL. We decide to enter predictor 6 into the regression equation; the results are shown in Table 2.9. As we predicted, the value of R-square increases by 2 percent. In addition, examination of the partial correlations for variables 1, 2, and 4 indicates that no additional value will be gained by adding them to the predictive equation. These partial correlations are all very small and have partial F values associated with them that would not be statistically significant.

We can now examine our predictive equation that includes variables 3, 5, and 6. The section of Table 2.9 headed "Variables in the Equation" yields the prediction equation in the column labeled "Coefficient." From this column, we read the constant term in the equation or Y-intercept of −1.684 and the coefficients of variables 3, 5, and 6, respectively, to be .343, .790, and .171. This equation would be written

$$Y = -1.684 + .343X_3 + .790X_5 + .171X_6.$$

TABLE 2.9
Example Output: Step 3 of HATCO Multiple Regression Example

Variable entered	X(6)
Multiple R	.877
Multiple R-square	.769
Std. error of est.	.887

Analysis of variance

	Sum of Squares	DF	Mean Square	F Ratio
Regression	120.6	3	40.2	51.08
Residual	36.2	46	0.78	

Variables in Equation

Variable		Coefficient	Std. Error of Coeff.	Std. Reg. Coeff.	F to Remove	Variables not in the Equation	
Y-intercept		−1.684					
X(3)	3	.343	.045	.541	57.9	Partial Corr.	R Ratio
X(5)	5	.790	.090	.629	76.4		
X(6)	6	.171	.086	.142	3.8		
						X(1) .017	.013
						X(2) −.019	.017
						X(4) .004	.001

To examine the appropriateness of this equation, we have already considered statistical significance. A look at the residual plots reinforces our judgment on the appropriateness of this predictive equation. Figures 2.6 and 2.7 show a plot of the residuals against the predicted values and a plot of the normalized residuals against an expected normal distribution. The errors appear to be reasonably normally distributed, which reassures us that the previously presented equation is appropriate. The vertical dashed lines in Figure 2.6 show + and − 1 SE, as illustrated earlier in Figure 2.4.

A Final Digression: Dummy Variables

To this point, all of our illustrations have implicitly assumed metric measurement for both predictor and criterion variables. The criterion variable may legitimately be measured as a dichotomous (0,1) variable. In such cases, regression analysis is comparable to discriminant analysis. Using dichotomous measures as predictor variables, however, results in interpretational differences. The following example will help clarify this concept.

Assume that we want to predict the number of credit cards held by families and that our measurement consists of

$$Age\ of\ Household$$
$$X_1 = 1\ \text{if age} <40,\ \text{else}\ X_1 = 0$$
$$X_2 = 1\ \text{if age} \geq 40,\ \text{else}\ X_2 = 0$$

FIGURE 2.6   Plot of predicted values against residuals.

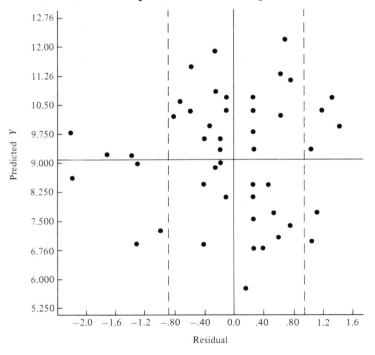

FIGURE 2.7 Normal probability plot of residuals.

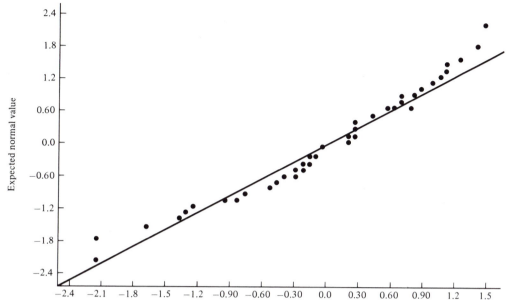

Both variables (X_1 and X_2 are not necessary because when $X_1 = 0$ age must be ≥ 40 by definition. Correspondingly, if we had also measured

Household Income
$X_3 = 1$ if income $<15{,}000$, else $X_3 = 0$
$X_4 = 1$ if income $\geq 15{,}000$ and $<25{,}000$, else $X_4 = 0$
$X_5 = 1$ if income $\geq 25{,}000$, else $X_5 = 0$

Only two of the variables X_3, X_4, X_5 are necessary because the zero level of two variables defines the presence of the third variable. Assume the following data:

Credit Cards Owned	Age (X_1)	Income (X_2)	Income (X_3)	Number of Children (NC)
3	1	1	0	1
3	1	1	0	3
4	1	0	1	2
5	1	0	1	4
6	1	0	0	2
5	1	0	0	1
1	0	1	0	3
2	0	1	0	3
3	0	0	1	2
2	0	0	1	1
4	0	0	0	2
5	0	0	0	2

The regression model one might estimate from this data using only age is as follows:

$$\text{Credit cards} = 2.833 + 1.5 \text{ (age)}$$

This model simply shows the following:

When age = 0, credit cards − 2.833
When age = 1, credit cards − 4.333

It quickly becomes obvious that a two-level dummy variable model shows the average of one group and the difference between that average and the average of the other group.

	Averages	
Group 0	2.8333	Difference = 4.333 − 2.8333 = 1.5
Group 1	4.333	

The full model, which includes a continuous variable, has a similiar interpretation:

$$\text{Credit cards} = 3.632 + 1.5X_1 - 3.915X_2 - 1.676X_3 + .353 \text{ (N.C.)}$$

We could write this model as two separate models. Two of the six possible models are as follows:

When age $\geq$40 and income $\geq$\$25,000 (since $X_1 = 0$ $X_2 = 0$ $X_3 = 0$)
Credit cards = 3.632 + .353 (N.C.)

When age $\geq$40, income $\geq$\$15,000 and <\$25,000
Since $X_1 = 0$, $X_2 = 0$, $X_3 = 1$)
Credit cards = 1.956 + .353 (N.C.)

All six combinations of the levels of the dummy variables can be used to show a different model. However, all of these models have the same coefficient for N.C. (number of children) and a different intercept term that essentially shows the difference between the average number of credit cards held by people at different combinations of the dummy variables.

Summary

This chapter presents a simplified introduction to the rationale and fundamental concepts underlying multiple regression analysis. It emphasizes that multiple regression analysis can describe and predict the relationship between two or more intervally scaled variables. Also, multiple regression analysis, which can be used to examine the incremental and/or total explanatory power of many variables, is a great improvement over the sequential analysis approach necessary with univariate techniques. Both stepwise and simultaneous approaches can be used to estimate a regression equation, and under certain circumstances nonmetric dummy coded variables can be included in a regression equation. This chapter will give you a fundamental understanding of how regression works and what can be achieved through its use. Also, familiarity with the concepts presented in this chapter will help you better understand the more complex and detailed technical presentations in other text books.

QUESTIONS

1. How would you explain the relative importance of the predictor variables used in a regression equation?

2. Why is it important to examine the assumption of linearity when using regression?

3. Could you find a regression equation that would be acceptable as statistically significant and yet offer no acceptable interpretational value to management? How could this happen?

4. What is the difference in interpretation between the regression coefficients associated with interval-scaled predictor variables as opposed to dummy (0,1) predictor variables?

5. What are the differences between interactive and correlated predictor variables? Do any of these differences affect your interpretation of the regression equation?

REFERENCES

1. Draper, Norman, and Harry Smith, *Applied Regression Analysis*. New York: Wiley, Inc., New York, 1966.
2. Huang, David S., *Regression and Econometric Methods*. John Wiley & Sons, Inc., 1970.
3. Johnston, J., *Econometric Methods*, 3rd ed. New York: McGraw-Hill, 1984.
4. Neter, John, and William Wasserman, *Applied Linear Statistical Models*, Homewood, Ill.: Irwin, Inc., 1984.
5. SAS Institute, *SAS User's Guide: Basics*, Version 5 ed. Cary, N.C., 1985.
6. SAS Institute, *SAS User's Guide: Statistics*, Version 5 ed. Cary, N.C., 1985.
7. McGraw-Hill, *SPSS-X Users's Guide*, 2nd ed. Chicago, 1986.
8. McGraw-Hill, *SPSS-X Advanced Statistics Guide*, Chicago, 1986.

SELECTED READINGS

Measuring the Quantity and Mix of Product Demand

GORDON A. WYNER
LOIS H. BENEDETTI
BART M. TRAPP

Introduction

Market planners often need to forecast product demand at various alternative price levels to develop pricing strategies and set sales goals for their firms. Frequently there are no adequate historical data that can be used to project future market behavior. Sometimes the market's history is too short, or the historical range of prices is not indicative of what future prices are likely to be, or the terms on which the product is to be offered are substantially different from what has previously been the case. Traditional econometric techniques for building models to estimate demand and price elasticity are not appropriate.

This paper presents an approach for building a model to estimate demand in a situation where sales history is not relevant. A general research design is presented and followed by an application of the approach in a nationwide study. The development of the approach and the empirical study

"Measuring the Quantity and Mix of Product Demand," Gordon A. Wyner, Lois H. Benedetti, and Bart M. Trapp, Vol. 48 (Winter 1984), pp. 101–109. Reprinted from the *Journal of Marketing*, published by the American Marketing Association.
Gordon A. Wyner is Vice President and Director of Marketing Sciences, M/A/R/C, Inc., Dallas.
Lois H. Benedetti was Manager, Future Customer Needs Analysis, GTE Telephone Opearating Group, Stamford CT, and is now a marketing consultant.
Bart M. Trapp is Group Manager, Marketing Information Systems, GTE Service Corporation, Irving TX.

relating to the residential telephone market were supported by GTE Corporation.

The research design deals with an important class of purchase decision making situations, those in which the individual purchaser selects from an array of available products and has the option of choosing as many or as few of each type (including zero) as he or she desires. The decision to purchase can reflect a mix of products, varying in quantity, terms (lease versus purchase), brand and style. This is an important distinction because a number of research methods have been developed to measure product demand experimentally in the case where the purchaser selects a single, preferred product from those available. These methods, which can be categorized generally as traditional conjoint analysis models, are not appropriate in this context. The current study extends beyond these first choice models to capture multiple purchases in three senses of the term. First, it takes into account the quantity of a particular product that will be demanded. Second, it allows for the possibility that more than one type (or brand) of product can be chosen by an individual. Third, it can reflect a mix of purchases on different terms. Thus, it provides a method for estimating the impact of product features, including price, on the number and type of units demanded.

Our approach to model building is comprised of several components, of which no single one is unique. By combining and implementing the components, a new, systematic research strategy is developed. Some developments in conjoint analysis in these directions are in the process of being made. McAlister (1978), for example, has modeled the consumer's choice of a subset of a bundle of items, and Mahajan, Green and Goldberg (1982) have pursued the issue of cross-elasticity among multiple products. The methodology reported here falls within the general area of extensions of conjoint analysis, designed specifically to answer questions relating to quantity demanded and cross impact.

The approach presented here may also be viewed in relation to a class of marketing planning models generally known as new product models (Blackburn and Clancy 1982). These models forecast demand for new package goods products using a simulated or actual test market to estimate initial trial and repeat purchases. Typically, they do not treat competitive products or prices as variables to be measured. Rather, a single price and competitive structure is employed in a concept test format. Thus, although these models can estimate aggregate demand, their capabilities for analyzing changes in prices or in competitive product entries are limited. Costly replicate experiments would be required to capture alternative prices and competitive sets.

Research Approach

Analytical Model Specification

The point of departure for any modeling building venture is to specify the relevant variables and structural relationships among them. In its most general form a demand model relates the quantity demanded of a particular product to the price of that product and the prices of other substitutable products. A product is defined as the goods itself, the terms under which it is acquired (that is, whether it is to be leased or purchased), and its associated brand. Thus, in this model the same physical product could appear twice, once for an outright purchase, another time for a rental if both possibilities are relevant. Different brands or product types would be represented as different products.

Given this specification, suppose there are 2K products on the market, half of which represent purchased products, half leased.[1] In general, the objective is to estimate the parameters of the following kind of model:

$$Y_{1\rho} = f(X_{1\rho}, X_{1L}, X_{2\rho}, X_{2L}, \ldots X_{\kappa\rho}, X_{\kappa L})$$
$$Y_{1L} = f(X_{1\rho}, X_{1L}, X_{2\rho}, X_{2L}, \ldots X_{\kappa\rho}, X_{\kappa L})$$

[1] It is conceivable that there are more than two possible sets of terms for products to be modeled. For instance—there could be purchases of new and used products, each of which has its own price range. This would require adding another equation to the model.

$$Y_{2\rho} = f(X_{1\rho}, X_{1L}, X_{2\rho}, X_{2L}, \ldots X_{\kappa\rho}, X_{\kappa L})$$
$$Y_{1L} = f(X_{1\rho}, X_{1L}, X_{2\rho}, X_{2L2L}, \ldots X_{\kappa\rho}, X_{\kappa L})$$
.
.
.
$$Y_{\kappa\rho} = f(X_{1\rho}, X_{1L}, X_{2\rho}, X_{2L}, \ldots X_{\kappa\rho}, X_{\kappa L})$$
$$Y_{\kappa L} = f(X_{1\rho}, X_{1L}, X_{2\rho}, X_{2L}, \ldots X_{\kappa\rho}, X_{\kappa L})$$

where

$Y_{i\rho}$ represents the quantity demanded of product i on a purchase basis, Y_{iL} represents the quantity demanded of product i on a lease basis, $X_{i\rho}$ represents the price of product i, and X_{iL} represents the lease rate of product i.

The variables include all products about which demand estimates are to be made. The demand for any one of them is, at least potentially, a function of the price and lease rate of that product and the prices and lease rates of all other available products. Empirical analysis will determine whether these relationships are strong, weak or perhaps nonexistent. This specification allows one to test for both self-elastic effects (e.g., the impact of the price of product 1 ($x_{i\rho}$) on the demand for product ($y_{i\rho}$) and cross-elastic effects (e.g., the effect of the price of product 2 ($x_{2\rho}$) on the demand for product ($y_{i\rho}$).

For ease of presentation, in this model a one-to-one correspondence is made between independent and dependent variables. One and only one independent variable, the price or lease rate, is included for each product. Other product attributes, such as style, color or advanced features, could be included. These would complicate the practical implementation of the approach only and could be incorporated in cases where they are theoretically justified.[2]

Data Collection Design

Given the lack of relevant historical data, a data collection approach must be developed to simulate the purchase decision making process. The researcher first determines the appropriate prod-

[2] It should also be noted that these price–demand relationships reflect equilibrium levels, assuming that the consumer has full information about the available alternatives and makes an immediate buying decision based on this information. Thus, this paper does not deal explicitly with the speed with which these levels would be reached, although there are other models that can be linked with these to address those issues (e.g., Bass 1969).

ucts and price ranges to include in the model. An experimentally designed set of product arrays are then prepared for evaluation by research subjects. Each product in a set must be described realistically to the potential consumer in terms of its features and price. The consumer/respondent is then presented one product set at a time and given an opportunity to record the number of products of each type that he or she would acquire. The response indicates the selection of product type, quantity, and whether they will be leased or purchased. Thus, each of these essential components of the decision making process is represented in the outcome of the research task.

To ensure realism in the simulated decision making process, the interviewer reviews with the respondent his/her current products and their prices and the implied cost of the products selected in the experiment. Respondents are instructed to respond in the way they think they would if given the choice of the available products in a real market environment. Thus, they employ whatever objectives or constraints *they* would normally use. This approach discourages unrealistic responses, while allowing for the possibility of an overall increase or decrease in total expenditures. This is preferable to using a fixed budget constraint in either the data collection or the data analysis stage, which assumes that overall expenditures remain constant. To assume a fixed budget is especially unwise when the model incorporates new price ranges, products or product terms that may change buying behavior substantially from what has occurred in the past, whether in the direction of an increase or decrease in expenditures and demand.

The following simple example illustrates the process. Suppose an office products manufacturer had two kinds of typewriters, electronic and electromechanical. Each was available on a lease or purchase basis. The entire set of price/lease combinations needed for analysis would be generated in the following manner. The relevant ranges of prices and lease rates for this market would be specified at discrete levels (e.g., Table 1).[3] If each of the four products were evaluated at each of three levels, a total of 81 scenarios would be needed to represent all possible combinations.

[3] A decision would be made by the researcher as to what repair service would be included under each option. For this illustration, assume that service is included under a rental option but must be acquired, at prevailing market rates, in the case of purchase. The respondent would be instructed to take this into account in evaluating the products.

Note that each scenario requires four responses so that a total of 324 judgments would be required. Clearly the respondent's burden can grow large quickly as the number of products and price levels increases. Fortunately, fractional factorial experimental designs are available that reduce the number of combinations while maintaining the capability of isolating the effects of each independent variable. For example, a subset of 18 combinations can be used to represent the full 81 and still yield independent estimates of the effects, provided that the subset represents an "orthogonal array" (Addleman 1962). In most cases there is no need to estimate the parameters of the model for each individual consumer. Thus each consumer does not need to react to the entire set; subsets of the required set of experimental combinations can be allocated to an independent sample of respondents. The analysis can then be conducted at an aggregate level.

Homogeneity of response across consumer segments need not be assumed. If there is good reason to believe that demand functions will vary for different categories of consumers (e.g., lower income and upper income groups), this can be built into the design and analysis. The required set of experimental combinations (whether all possible ones or a fractional design) must be administered to a sample drawn from each segment of interest. A separate analysis for each segment can then be conducted.

One of the experimentally designed scenarios presented to purchase decision makers might look like Table. 2. The interviewer asks how many of each type would be acquired and records the answers. He or she then derives the total cost implied by the answers. Next, this figure is compared to the decision maker's current equipment cost, and the respondent is given the opportunity to change answers to make them more realistic, if necessary. The revised answers, in this case 3, 1, 1 and 0, become the input for data analysis.

Statistical Estimation

Several options are available for estimation of the demand models, depending upon theoretical and practical considerations. In theory the parameters of the models can be related to several different levels of aggregation. At the most microscopic level they can represent the relationships between price and demand for each individual purchaser

TABLE 1   Product/Range of Prices and Lease Rates

	Electric/ Purchase Price	Electric/ Lease Rate	Electronic/ Purchase Price	Electronic/ Lease Rate
Levels:	$600 $700 $800	$200/yr. $250/yr. $300/yr.	$1,000 $1,200 $1,400	$300/yr. $400/yr. $500/yr.

TABLE 2   Sample Price/Lease Configuration

	Price Scenario			
Interviewer reads these instructions:	Electric/ Purchase $700	Electric/ Lease $250/yr.	Electronic/ Purchase $1,200	Electronic/ Lease $300/yr.
How many would you acquire of each?	3	2	1	0
The one-time cost of these is $3,300 and the recurring annual cost is $500.	$2,100	$500	$1,200	$0
Compare this to your current cost of equipment.				
Do you wish to change any of your answers?	3	1	1	0
The one-time cost of these is $3,300 and the recurring annual cost is $250.	$2,100	$250	$1,200	$0

in relation to each product. Often, however, there is no need to measure demand curves for individual decision makers. It is sufficient that the data are collected in a way that reflects the purchase decisions of individuals. In this case, the estimation can be done at an aggregate level such as the overall market level or at targeted segment levels. The parameters then represent aggregate market demand behavior.

Practical constraints may limit flexibility if the number of variables and price levels is large. If split sample techniques are used, then clearly the analysis will have to be conducted at some level of aggregation.

Once the level of analysis has been determined, the specific statistical model would be chosen. For most applications, multiple regression analysis would be the appropriate technique.

Empirical Application

With the advent of deregulation of the residential telecommunications market, the market planner must take into account several new elements that render previous market data and forecasting models virtually useless. For several reasons it is conceivable that telephone companies may want to sell their installed base of telephones, which have

historically been offered only on a rental basis, and actively promote the sale, rather than rental, of new telephones. Since these scenarios depart dramatically from the market situation that has prevailed for many years, historical rental data and models could not be employed, and thus the model presented in this paper was implemented.

Analytical Model

In this application two market scenarios were specified, one called *Inservice*, which reflects the current installed base of telephones, and another called *Inward*, which reflects the acquisition of telephones among the population moving into a telephone company territory.

INSERVICE. In this scenario four types of telephones were specified as comprising the full array of available products, each of which could be acquired on one of three different financial terms: a lease for any of the models in one's home, purchase of any of these models and purchase of a new model.[4] Thus, a total of 12 products, according to our general model specification, was identi-

[4] Because of the proprietary nature of this research, the types of telephones are identified only by number, from the lowest to the highest priced.

fied. Since the only feature to be explicitly evaluated (other than telephone type itself) was price, 12 independent variables were identified (a lease rate, used purchase and new purchase price for each of four telephone types).

INWARD. In this scenario five different types of telephones were identified as being exhaustive of those likely to be made available to incoming customers. Each of these could be acquired on either a lease or (new) purchase basis and thus 10 products in all were specified in the model.

Data Collection Design

In the Inservice scenario the relevant range of prices and lease rates were specified with 9 of the 12 at three levels and 3 at two levels. Since the number of possible experimental combinations implied by these numbers of levels is enormous, a fractional factorial design was employed. A set of 32 combinations allowing for orthogonal main effects estimates was selected (Addleman 1962). Similarly, in the Inward scenario, the range of prices and lease rates was specified. In this case, nine were set at three levels, one at two. Another 32 combination fractional factorial design was used to assure unconfounded main effects estimation.

Sets of price/lease combinations were developed as described earlier. Realism in the data collection was achieved by reviewing with respondents/customers their current telephone instruments and informing them of the specifics of the potential new ways in which equipment

and service would be offered. Actual examples of telephones as well as pictures were used to assure proper identification of each model. Use of a formal budget constraint was considered and rejected because of the very real possibility of an overall decrease in consumer expenditures under these new scenarios.

An extensive pretest (of about 100 respondents) was used to determine the maximum number of price/lease combinations to give to each person to evaluate. For each of the two scenarios reported here only four combinations were given to any one individual. This reduced the respondent burden to a minimum and reduced the chances for respondent error in the data. For each Inservice combination, 12 responses were obtained. For each Inward combination, 10 responses were made. Thus, each person would provide a total of 88 responses ($12 \times 4 + 10 \times 4$). To assure that there would be no bias in the allocation of individuals to price/lease combinations, a representative national sample (of about 150) was selected for each of the sets of four combinations.

In administering the price/lease combinations, customers were asked to record their answers as in the example in Table 2. The guidelines described above for assuring realistic responses were implemented. Customers were advised of the cost implications of their decisions and allowed to change their initial response if they wished to.

Although every effort was made to simulate the situation that consumers will actually experience in the market, it is possible that respondents received and evaluated more information about these products than they would in the real world. A definitive test of this issue is beyond the scope

TABLE 3 Variables in the Inservice Data Matrix

Dependent Variables: The Number of Telephones Demanded												Independent Variables: The Price or Lease Rate of the Telephone											
Telephone Type												Telephone Type											
1			2			3			4			1			2			3			4			
Leased	Purchased used	Purchased new	Leased	Purchased used	Purchased new	Leased	Purchased used	Purchased new	Leased	Purchased used	Purchased new	Lease rate	Price—used	Price—new	Lease rate	Price—used	Price—new	Lease rate	Price—used	Price—new	Lease rate	Price—used	Price—new

of this paper. One approach that should be considered in future studies, is to vary the information introduced in the experiment systematically. In this way, the impact of different knowledge levels on product demand can be measured.

Statistical Estimation

PREPARATION OF THE DATA MATRICES. The responses for each national sample were cumulated across individuals. Thus, for each of the 32 price/lease combinations (in both the Inservice and Inward scenarios) the response data represented the total numbers of telephones to be acquired of each type, by each method of acquisition. In the Inservice case, the entire cumulated data matrix for all telephone types was 32×24. The 32 rows represented the different experimental combinations, i.e., the number of observations. The first 12 columns represent the responses to the combinations, i.e., the dependent variables for each telephone type. The second 12 columns represent the prices and lease rates of the 12 telephone types, i.e., the independent variables (see Table 3). Each dependent variable observation represents the national market response based on the answers of about 150 customers. In similar fashion, the Inward scenario data matrix, 32×20, was constructed (see Table 4). Thus, the analysis would provide parameter estimates of overall market demand as a function of prices and lease rates, taking into account the individual level acquisition decisions of about 1,200 customers in all.

Prior to estimation, each set of dependent variable observations was converted from a metric of total number of telephones demanded to the mean number per household. This was done to standardize across the national samples, which differed slightly in number (150 ± 5) and to provide a convenient metric for later population projections.

SPECIFICATION OF REGRESSION EQUATIONS. In some cases it is useful to aggregate across dependent variables prior to estimation. If, for example, the business objectives relate to varying the mix of leased versus purchased telephones, then the identity of specific telephone types need not be preserved in the analysis stage. In the analysis results presented here, the demand for the specific telephone types were summed prior to estimation. Thus, five estimated equations are shown. For the Inservice scenario they reflect total telephones leased, total purchased used and total purchased new. For the Inward scenario they represent total purchased new and total leased.

The data matrices for these equations were constructed by extracting and summing the appropriate dependent variable columns of the original matrices (Tables 3 and 4). The total leased in the Inservice case, for example, was derived from the sum of the number of telephone types 1, 2, 3 and 4 that would be leased. The total purchased new and the total purchased used were derived, in similar fashion, from the remaining columns of the original matrix (Table 3). All of the independent price/lease variables are entered into each equation. This allows us to isolate the effects of prices of particular instrument types in generating total demand.

TABLE 4 Variables in the Inward Data Matrix

Dependent Variables: The Number of Telephones Demanded										Independent Variables: The Price or Lease Rate of the Telephone									
Telephone Type										Telephone Type									
1		2		3		4		5		1		2		3		4		5	
Leased	Purchased new	Leased	Purchased new	Leased	Purchased new	Leased	Purchased new	Leased	Purchased new	Lease rate	Price—new	Lease rate	Price—new	Lease rate	Price—new	Lease rate	Price—new	Lease rate	Price—new

Results

INSERVICE SCENARIO. As would be anticipated, the demand for leased telephones under the Inservice scenario is highly sensitive to the lease rates for rented instruments. There is a statistically significant negative relationship between the rates of each of the four telephone types (1, 2, 3 and 4) and aggregate lease demand (Table 5). The greatest impact on demand results from lease rate changes in the lowest level instrument type, 1, which reflects the large installed base of these low-end telephones and the appeal of these products to lessors. Two significant cross impacts are apparent: increases in new purchase prices for types 2 and 4 will increase demand for leasing. Leasing does not appear to be sensitive to purchase prices for used, Inservice instruments. None of the regression coefficients representing the prices of used telephones are statistically significant at the .05 level.

These results show clearly and quantitatively the impact of offering new products on a purchase basis at the same time that the Inservice products are made available on lease or purchase terms. Consumers will forego the possibility of having a new telephone and continue to lease their current ones if the new purchase price is too high. The overall model yields an $R^2 = .80$ and is statistically significant at $p < .01$. The signs of the coefficients are easily interpretable—negative self-elastic effects, positive cross-elastic effects.[5]

For forecasting purposes, the magnitude of the coefficients reflects the impact of price/lease rate changes in the number of telephones demanded. For instance, the absolutely largest coefficient, $-.0125$, indicates that for each dollar increase in telephone type 1 lease rates, the average number of leased telephones per household will fall by .0125. In a market approaching 10 million households, this translates into about 125,000 telephones.

The demand for new telephones is also heavily

driven by the pricing of low-end products. While both new and used purchase prices have a significant impact for type 1 instruments, the cross-impact effect (due to used purchase prices) is about twice the magnitude of the self-elastic impact (Table 5). As with the lease model, the statistical fit is good, $R^2 = .71$, $p < .01$.

Compared to the demand for leased and newly purchased instruments, the demand for Inservice telephones is much more price sensitive. Of the 12 potential impacts on demand, 7 are statistically significant at the .05 level, including all 4 used purchase price variables (Table 5). If, for example, the marketing objective is to reduce leasing and increase purchase of Inservice telephones, these results can be used to determine by how much rates and prices should be changed. Trial and error (starting with current rates and modifying them, for example) can be used to reach a solution, or alternatively, an optimization model can be used, given a specific objective function.

INWARD SCENARIO. When consumers move into a new home, the aggregate demand for telephones is most strongly influenced by the cost of only three of the possible ten instrument types: the purchase prices of the two lowest priced models (1 and 2) and the lease rate of the second lowest model (2). The new purchase prices of types 1 and 2 strongly influence the new purchase demand but also the demand for leased instruments. Higher purchase prices will shift more consumers toward leasing (Table 6).

The lease rate for model 2 also has both a self- and cross-elastic effect. Higher lease rates will reduce leasing demand but increase new purchase demand. By setting lease rates and prices appropriately, the planner can vary the mix of lease and purchased instruments in the direction that is most financially advantageous. The overall fit of both Inward scenario models is fairly good, $R^2 = .56$ and $.62$, and $p = .03$ and $.01$, for lease and purchase, respectively.

A comparison of the magnitude of the coefficients in the two buying scenarios reveal that the market is more price sensitive in the Inservice than in the Inward case. The largest significant coefficient in the Inward scenario is .0100, and most others are closer to .0080. In the Inservice case, several significant coefficients are above .0100, with one exceeding twice that amount, .0228.

---

[5] In this and all subsequently reported results, the linear effects of the independent variables were estimated. For two variables in the Inservice scenario and one in the Inward, the effects were assumed linear in the design, i.e., only two price levels were evaluated. For those variables set at three levels, visual inspection of the data clearly revealed that where there were effects, they were linear. Dummy variable coding was considered and rejected due to the good statistical fit of the linear models and the ease of use of the linear coefficients for forecasting demand.

TABLE 5 Regression Results for Inservice Mode

Independent Variables		Dependent Variables					
		Total Leased		Total Purchased New		Total Purchased Used	
Telephone Type		B	P	B	P	B	P
1	Lease rate	−.0125	<.01	−.0006	.85	.0076	.02
1	Price—used	.0022	.47	.0189	<.01	−.0228	<.01
1	Price—new	.0010	.55	−.0095	<.01	.0084	<.01
2	Lease rate	−.0034	.01	.0022	.22	.0016	.30
2	Price—used	.0018	.28	.0031	.22	−.0048	.04
2	Price—new	.0025	.04	−.0026	.16	.0036	.03
3	Lease rate	−.0024	.03	.0023	.15	−.0001	.97
3	Price—used	.0034	.18	.0042	.27	−.0068	.05
3	Price—new	.0006	.63	−.0012	.51	−.0023	.14
4	Lease rate	−.0028	<.01	.0014	.30	.0015	.22
4	Price—used	.0023	.17	.0023	.36	−.0050	.03
4	Price—new	.0024	.01	−.0028	.04	−.0014	.24
	Intercept	.3129		.2453		1.2837	
	Mean	.3968		.4346		.8255	
	R^2	.80		.71		.81	
	F	6.23		3.94		6.83	
	P	<.01		<.01		<.01	
	df	12, 19		12, 19		12, 19	
	n	32		32		32	

TABLE 6 Regression Results for Inward Mode

Independent Variables		Dependent Variables			
		Total Leased		Total Purchased New	
Telephone		B	P	B	P
1	Lease rate	−.0018	.72	.0006	.90
1	Price—new	.0076	.05	−.0100	<.01
2	Lease rate	−.0082	<.01	.0076	.01
2	Price—new	.0088	.03	−.0088	.02
3	Lease rate	−.0030	.21	.0027	.21
3	Price—new	.0015	.70	−.0009	.79
4	Lease rate	−.0031	.13	.0025	.17
4	Price—new	−.0035	.22	−.0036	.16
5	Lease rate	−.0018	.74	.0023	.25
5	Price—new	−.0005	.42	.0004	.74
	Intercept	.3759		1.4406	
	Mean	.6020		1.1177	
	R^2	.56		.62	
	F	2.72		3.46	
	P	.03		.01	
	df	10, 21		10, 21	
	n	32		32	

Conclusions

Based on the statistical results—the regression diagnostics, the significant and interpretable coefficients—we conclude that this modeling approach is effective in measuring price–demand relationships. The use of experimentally designed scenarios that simulate actual buying decisions was successful in supporting a market planning model in a case where no satisfactory alternative was available.

The overall benefit from a marketing perspective is to be found in applying the results to alternative pricing configurations, determining the financial impact and selecting the best approach from a business standpoint. Rather than rely on ad hoc estimates of prospective demand, the method provides quantitative estimates of the relevant parameters needed to forecast demand systematically. In a case where historical data are inadequate for the task, this experimental approach can fill a significant gap.

This approach is likely to be useful in a number of substantive areas. The specifics of the design will vary depending upon the type of output needed, the complexity of the market and the data collection requirements. It is possible, though, to suggest some typical research situations in which this technique would be useful.

Consumer Mix Problems

Many consumer products are purchased in quantity reflecting various brands and product types. The purchase decision maker may not consciously try to achieve an optimal mix, but in effect, consumer behavior may reflect some sort of mix criterion. Examples include beer purchasers who buy for their households that include some regular beer drinkers, super premium drinkers and some lite beer drinkers; gasoline purchasers who sometimes buy at low price, self-service stations and other times at high price, full-service stations; investors who place some of their money in guaranteed, fixed return certificates and some in variable rate, riskier instruments.

Industrial Mix Problems

Often in industrial purchase decision making a conscious effort is made to cultivate several sources of supply so as to avoid being locked into any one supplier. Examples include commodity chemical purchases and crude oil purchases. Again, no presumption is made that decision makers are entirely rational or follow some optimizing strategy. The hypothesis is simply that there are empirically measurable relationships between price, brand and other characteristics, and quantity demanded.

Lease/Purchase Problems

A frequently occurring industrial marketing problem relates to strategies for offering equipment on a lease or purchase basis. Obvious examples include computers, software, office equipment such as typewriters and word processors, and corporate fleets of automobiles and aircraft. Rigorous demand analysis can be used to fine tune pricing strategies that call for precise estimates of the financial impact of a shift in the ratio of leased to purchased equipment.

Clearly this demand-analysis technique has its limitations. If a market is populated with dozens of products or competitors, it will be nearly impossible to develop a tractable research design. If awareness of available products is highly variable across the consuming population, this complexity will have to be factored in. Yet, there are numerous areas where it is likely to be more realistic and useful than other available approaches.

References

Addleman, Sidney (1962), "Orthogonal Main Effects Plans for Asymmetrical Factorial Experiments," *Technometrics*, 4 (February), 21–46.

Bass, Frank (1969), "A New Product Growth Model for Consumer Durables," *Management Science*, 15 (January), 215–227.

Blackburn, Joseph and Kevin Clancy (1982), "Litmus: A New Product Planning Model," in *Marketing Planning Models*, A. A. Zoltners, ed., North Holland Publishing Co., TIMS/Studies in the Management Sciences, 18, 43–61.

Mahajan, V., P. Green and S. Goldberg (1982), "A Conjoint Model for Measuring Self- and Cross-Price Demand Relationships," *Journal of Marketing Research*, 19 (August), 334–342.

McAlister, Leigh (1979), "Choosing Multiple Items from a Product Class," *Journal of Consumer Research*, 6 (December), 213–224.

Social Interactions as Predictors of Children's Likability and Friendship Patterns: A Multiple Regression Analysis[1]

FRANK M. GRESHAM

Introduction

Social skills training with children has received an abundance of attention from behavioral researchers during the past 5 years. This is evidenced by the large number of review articles (Cartledge and Milburn, 1978; Combs and Slaby, 1977; Gresham, 1981a; Van Hasselt et al., 1979) and books (Bellack and Hersen, 1979; Cartledge and Milburn, 1980; Rathjen and Foreyt, 1980) on this topic. These reviews generally conclude that social skills training techniques have been effective in teaching social skills to children.

Much less attention has been directed toward the investigation of specific social skillls assessment measures. As such, there is a paucity of data concerning the reliability and validity of various social skills assessment measures. At least three reviews have been written on the assessment of children's social skills (Foster and Ritchey, 1979; Green and Forehand, 1980; Gresham, 1981b). These reviews agree that the most frequently used methods of assessing children's social skills are sociometric measures, behavioral observations, and teacher ratings. Most social skills training studies have used only one type of social skill assessment method as the primary selection and outcome measure (Gresham, 1981a). In the behavioral literature, behavioral observations are used in approximately two-thirds of the studies and sociometric measures in less than 10% of the studies. Multiple-outcome measures have been used in only 25% of the social skills training

studies (Gresham, 1981d). Teacher nominations (Bolstad and Johnson, 1972; Evers and Schwarz, 1973; O'Connor, 1969, 1972) and teacher ratings (Milburn, 1974; Reardon et al., 1978; Stodden, 1981) have been used less frequently.

Green and Forehand (1980) maintain that the construct of social skills is in need of basic development. These authors suggest that much more research should be conducted in establishing the construct, concurrent, and predictive validity of various social skills assessment techniques.

The primary purpose of the present study is to investigate the criterion-related (both concurrent and predictive) validity of children's social interactions as measures of social skill. The temporal reliability of various social skills measures was also investigated to determine the stability of behavioral and sociometric measures. Past research has investigated the concurrent validity of social interaction and sociometric measures using correlational methods (Gottman, 1977; Gottman et al., 1975; Hartup et al., 1967). Since these measures tend to be intercorrelated, a stepwise multiple regression analysis which would take into account these intercorrelations to produce the best model for predicting social acceptance seemed appropriate. The following section briefly reviews the sociometric and behavioral measures used in the current investigation.

Sociometric and Behavioral Measures

Behavioral researchers are increasingly employing sociometric methods as selection, outcome, and social validation measures in social skills training investigations (Gottman et al., 1976; Gresham and Nagle, 1980; La Greca and Santogrossi, 1980; Oden and Asher, 1977). Sociometrics have not been considered traditional behavioral assessment devices because they do not specify antecedent behaviors which may lead to high or low levels of peer acceptance. The major advantage of sociometrics is their use as selection and social validation measures. Using peers as judges to validate the effectiveness of social skills training is a means of establishing social validity (Van Houten, 1979; Wolf, 1978).

Sociometrics are usually presented in *peer nomination* or *peer rating* scale formats. In the peer nomination format, children are asked to choose a specified number of peers according to

"Social Interactions as Predictors of Children's Likability and Friendship Patterns: A Multiple Regression Analysis," Frank M. Gresham, Vol. 4 (1) 1982, pp. 39–53. Reprinted from the *Journal of Behavioral Assessment*, published by the Plenum Publishing Corporation.
Frank M. Gresham is Professor in the Department of Psychology, Louisiana State University.
[1] The current data were gathered as part of a larger research project concerning social skills assessment and training in children.

a specified criterion. For example, children may be asked to nominate their three best friends and/or their three favorite work partners in the classroom. With peer ratings, children are asked to rate their classmates on three-point or five-point Likert scales according to specified criteria (e.g., play partners, work partners, etc.). A child's score is the average of the ratings received from same-sex peers since there appears to be considerable sex bias in children's sociometric ratings (Singleton and Asher, 1977).

There is some evidence to suggest that peer nominations and peer ratings assess different aspects of sociometric status. Hymel and Asher (1977) speculated that peer nominations measure who a child's best friends are, whereas peer ratings measure a child's overall acceptability or likability in a peer group. To support this assumption, Hymel and Asher found that of the 23 children in their sample who received no peer nominations, 11 received fairly high peer ratings (above 3.00 on a five-point rating scale). Gresham (1981c) found support for this distinction between likability and friendship in a factor analytic investigation of nomination and sociometric measures. The distinction between likability (acceptance) and friendship is an interesting one that requires further investigation.

Naturalistic observation is perhaps the most face valid method of assessing children's social skills (Asher and Hymel, 1980). Observations in naturalistic environments have the advantage over sociometric measures in that they can identify the antecedents and consequences of certain social behaviors. Observations possess some evidence of concurrent validity as they correlate moderately with some sociometric measures (Hartup et al., 1967) and teacher nominations (Bolstad and Johnson, 1972). Direct observations have also shown relatively strong correlations among themselves, such as the correlation between rates of giving and rates of receiving positive and negative social reinforcement in preschool populations (Charlesworth and Hartup, 1967; Hartup et al., 1967; Keller and Carlson, 1974).

The foregoing review suggests some relationship between two diverse methods of assessing children's social skills: sociometrics and behavioral observations. The following study investigates the validity of several categories of social behavior in predicting a sociometric status using stepwise multiple regression procedures.

Method

Subjects

Forty children (18 boys and 22 girls) selected from 14 third- and fourth-grade classrooms in a southeastern metropolitan city served as subjects. Subjects were chosen on the basis of low scores on two sociometric rating scales which measured the degree to which children like to "play with" and "work with" each other. Initially, the 4 lowest-rated children (2 boys and 2 girls) in each of the 14 classrooms were selected, yielding 56 children. From this original pool of 56 children, 40 were randomly selected for participation.

Low-status children were selected because this study was part of a larger investigation concerning social skills training with children. Selected children were rated by 462 peers in 14 classrooms on two Likert-type sociometric scales (play with and work with) which ranged in value from 1 (not all) to 5 (very much). Composite scores on the two scales (play with and work with) were used to select the lowest rated children. The classrooms had an average enrollment of 31 children, with approximately equal sex representation. Only same-sex ratings were used to select subjects for participation.

Social Skill Measures

Both sociometric and observational measures were used in the current project. Descriptions of each of these measures are given below.

SOCIOMETRIC MEASURES.   Three categories of peer nominations and two categories of peer ratings constituted the sociometric measures. Peer nominations consisted of best friends, play with, and work with nominations. Nominations were collected from the 462 children in the 14 classrooms previously described. Children in each of the classrooms were asked to nominate their three best friends, three people they would most like to play with, and three people they would most like to work with, respectively. Only same-sex nominations were used for analysis.

On the peer rating scales (previously described), children in each of the 14 classrooms were asked to rate every member of the class on a 1 to 5 scale according to how much they liked to play with and work with them. Only same-sex ratings were used for analysis.

OBSERVATIONAL MEASURES.   Four categories of social interaction in the classroom comprised the observational data. The four categories were (1) rates of initiating positive peer interaction, (2) rates of initiating negative peer interaction, (3) rates of receiving positive peer interaction, and (4) rates of receiving negative peer interaction. Each of the four categories had four subcategories of physical, verbal, cooperative/uncooperative, and nonverbal. These categories were selected on the basis of past research which suggests that rates of initiating and receiving positive and negative social interaction are correlated and predictive of social acceptance (Charlesworth and Hartup, 1967; Greenwood et al., 1977; Hartup et al., 1967; Keller and Carlson, 1974).

Only the major categories of initiating and receiving positive and negative interaction were used for analysis. A fine-grained coding of the subcategories would have been desirable, however, subcategories were collapsed within each category to facilitate recording and observer agreement estimates.

Procedure

All measures were collected three times over a period of 6 weeks. Assessments took place in late March (Time 1), mid-April (Time 2), and early May (Time 3). Thus, each of the 40 subjects had a total of 27 measures collected on him or her (9 measures for each of the three times), yielding 1080 observations for the entire sample.

Peer rating sociometrics were collected by giving sheets of paper to each child in the 14 classrooms with each class member's name typed on separate lines. A series of five faces ranging from frowns to smiles with corresponding numbers (1–5) ran across the page horizontally. Each face represented one point on the five-point scale. Every child was rated on the "play with" and "work with" scales, respectively. Subsequent to the completion of the peer rating scales, the three peer nomination measures were administered. Children were asked to write on a piece of paper their three best friends, three people they would most like to play with, and three people with whom they would most like to work. Ratings and nomi-

nations were collected 3 weeks later (Time 2) and 3 weeks after that, at Time 3. Teachers in each of the 14 classrooms administered the sociometrics for all three time periods.

Trained observers collected behavioral observations for Times 1, 2, and 3. Five observers were trained over a period of 12 days (5 hr of training) in the observational code to an average interobserver agreement of 80%. Each target child was observed for a total of 10 min over a period of 2 days (5 min each day) for each of the three time periods. A sequential time-sampling procedure (Thomson et al., 1974) was used and observations were collected in the morning and sampled from small group activities, group discussion, lecture, and individual work at desk.

Five-minute observations for each day were divided into 30 10-sec intervals and target children's behavior was recorded sequentially over the 30 intervals. Behavior was observed at the end of each 10-sec interval and recorded as occurring or not occurring relative to the four categories. Observers recorded the behavior during the next 5 sec and then moved on to the next target child for observation. For example, if a class had four target children, Child 1's behavior would be observed for 10 sec, and if it occurred at the 10th second of that interval, it was recorded. The observer then located Child 2 and recorded his/her behavior if it occurred at the 10th second, and so on, until all four children in that class had been observed for 30 intervals. The same procedure was repeated for the second 5-min observation day for all target children. Rates of behavior were based on 60 intervals of observation (30 intervals each day). The average rates of behavior over the 2-day observation were used for data analysis for all three time periods.

To protect against observer bias and drift, observers were kept blind to the purpose of the investigation and observers were rotated during reliability checks. Reliability checks were conducted randomly four times during each observation period. Percentage agreement was calculated on an interval-by-interval basis using the following formula:

$$\text{Agreement} = \frac{\text{Number of agreements}}{\substack{\text{Number of agreements} \\ + \text{ Number of disagreements}}} \times 100.$$

Agreement estimates ranged from 0.61 to 0.83 for Time 1, 0.72 to 0.85 for Time 2, and 0.63 to 0.93

Table I Agreement Estimates for Behavior Classes for Three Time Periods

Behavior Class	Time 1	Time 2	Time 3	Median
Initiating positive peer interaction	0.82	0.84	0.93	0.84
Initiating negative peer interaction	0.83	0.82	0.93	0.82
Receiving positive peer interaction	0.61	0.72	0.63	0.63
Receiving negative peer interaction	0.83	0.85	0.92	0.85

for Time 3. Table I presents overall and median reliability scores for each class of behaviors for each of the three time periods.[2]

Results

All data were analyzed using the correlation and stepwise procedures of the Statistical Analysis System (Statistical Analysis System Institute, 1979). Table II presents the stability coefficients for behavioral and sociometric measures over the three time periods.

From Table II it is apparent that the highest stability coefficients were from Time 2 to Time

[2] Percentage agreement was calculated for occurrences plus nonoccurrences of target behavior classes.

3 ($\bar{X} = 0.58$) and the least stable coefficients were from Time 1 to Time 3 ($\bar{X} = 0.33$). Surprisingly, the average stability coefficients for the behavioral measures ($\bar{X} = 0.413$), the sociometric ratings ($\bar{X} = 0.406$), and the sociometric nominations ($\bar{X} = 0.413$) were of receiving positive and negative interaction ($r = -0.38$, $P < 0.01$). These findings support previous work with preschoolers that has demonstrated significant relationships between rates of giving and rates of receiving positive and negative social reinforcement (Hartup et al., 1967; Keller and Carlson, 1974).

In summary, the correlational data in Table III offer evidence for the validity of the behavioral measures of initiating and receiving positive and negative peer interaction in predicting sociometric status as measured by sociometric ratings and nominations. Moreover, the different pattern of correlations between the behavioral measures and the sociometric measures suggests that peer ratings and peer nominations may be assessing separate aspects of sociometric status (Hymel and Asher, 1977).

Although the correlations in Table III shed some light on the relationship between behavioral and sociometric measures, the correlation procedure does not take into account the degree of intercorrelation between the behavioral measures in predicting sociometric status. To compensate for this intercorrelation, five separate stepwise multiple regression analyses using the four social interaction measures as predictors and the five sociometric measures as criteria were performed.

Table IV presents the multiple regression analyses for the play with (PWR) and work with

TABLE II Stability Coefficients for Behavioral and Sociometric Measures[a]

Measures	Time 1–Time 2	Time 2–Time 3	Time 1–Time 3
Initiating positive peer interaction	0.39	0.80	0.37
Initiating negative peer interaction	0.23	0.56	0.27
Receiving positive peer interaction	0.45	0.67	0.30
Receiving negative peer interaction	0.20	0.40	0.32
Play with rating	0.30	0.59	0.39
Work with rating	0.33	0.63	0.20
Play with nomination	0.62	0.52	0.32
Work with nomination	0.34	0.50	0.28
Best friends nomination	0.19	0.53	0.22

[a] Coefficients of 0.31 or higher significant at the 0.05 level.

TABLE III Composite Correlation Matrix and P Values[a]

	PWR	WWR	PW	WW	BF	IP	IN	RP	RN
PWR		0.88	0.44	0.52	0.55	0.29	−0.30	0.32	−0.36
		0.0001	0.004	0.0007	0.0002	0.06	0.05	0.04	0.02
WWR			0.38	0.44	0.48	0.24	−0.26	0.41	−0.31
			0.01	0.004	0.001	0.13	0.10	0.009	0.04
PW				0.84	0.74	0.44	−0.47	0.24	−0.32
				0.0001	0.0001	0.004	0.002	0.13	0.04
WW					0.76	0.36	−0.41	0.27	−0.31
					0.0001	0.05	0.008	0.09	0.04
BF						0.35	−0.39	0.23	−0.29
						0.02	0.01	0.14	0.07
IP							−0.34	0.67	−0.36
							0.03	0.0001	−0.02
IN								−0.41	0.59
								0.008	0.0001
RP									−0.38
									0.01
RN									

[a] PWR, play with rating scale; WWR, work with rating scale; PW, play with peer nomination; WW, work with peer nomination; BF, best friends peer nomination; IP, initiating positive interaction; IN, initiating negative interaction; RP, receiving positive interaction; RN, receiving negative interaction. Correlation coefficients computed on linear composite scores of measures from all three time periods (Time 1 + Time 2 + Time 3). $N = 40$ (composite scores).

(WWR) rating scales. As can be seen from Table IV, receiving negative peer interaction was the best predictor of sociometric status as measured by the PWR [$F(1,38) = 5.81, P < 0.02$], accounting for over 13% of the variance. The addition of receiving positive peer interaction accounted for an additional 4% of the variance, yielding a significant two-variable model [$F(2,37) = 3.82, P < 0.03$].

The addition of initiating positive and initiating negative peer interaction in three- and four-variable models, respectively, did not add significantly to the prediction.

Using the work with rating scale (WWR) as the criterion, Table IV shows that receiving positive interaction was the best predictor of sociometric status [$F(1,38) = 7.48, P < 0.009$]. The addition

TABLE IV Stepwise Regression for Sociometric Ratings Composite (Time 1 + Time 2 + Time 3)

Step No.	Behavior Variable Entered[a]	Multiple R^2 Value	df	F Value	Significance Level
(a) Play with rating scale					
1	Receiving negative peer interaction	0.132	1,38	5.81	$P < 0.02$
2	Receiving positive peer interaction	0.171	2,37	3.82	$P < 0.03$
3	Initiating positive peer interaction	0.177	3,36	2.58	$P < 0.06$
4	Initiating negative peer interaction	0.180	4,35	1.93	$P < 0.12$
(b) Work with rating scale					
1	Receiving positive peer interaction	0.165	1,38	7.48	$P < 0.009$
2	Receiving negative peer interaction	0.194	2,37	4.46	$P < 0.01$
3	Initiating positive peer interaction	0.199	3,36	2.98	$P < 0.04$
4	Initiating negative peer interaction	0.200	4,35	2.18	$P < 0.09$

[a] Behavior variables are a composite of all three time periods (Time 1 + Time 2 + Time 3).

of receiving negative peer interaction in a two-variable model accounted for an additional 3% of the variance [$F_{(2,37)} = 4.46$, $P < 0.01$]. The three-variable model of receiving positive, receiving negative, and initiating negative peer interaction accounted for almost 20% of the variance in WWR [$F_{(3,36)} = 2.98$, $P < 0.04$]. Initiating negative peer interaction in a four-variable model did not add significantly to the prediction ($P > 0.05$).

Table V presents the stepwise multiple regression analyses for the play with (PW), work with (WW), and best friends (BF) peer nomination measures. As can be seen from Table V, the social interaction measures best predicted the play with nomination measure. Initiating negative peer interaction was the best predictor of the PW nomination, accounting for over 22% of the variance [$F_{(1,38)} = 10.92$, $P < 0.002$]. The four-variable model accounted for over 34% of the variance [$F_{(4,35)} = 4.59$, $P < 0.004$].

Using the WW nomination as the criterion, initiating negative peer interaction was the best predictor, accounting for almost 17% of the variance [$F_{(1,38)} = 7.70$, $P < 0.008$]. Only the addition of initiating positive and receiving negative interaction in a three-variable model added significantly to the prediction [$F_{(3,36)} = 3.03$, $P < 0.04$].

Finally, as in the PW and WW measures, initiating negative peer interaction was the best predictor of the BF nomination measure [$F_{(1,38)} =$ 6.64, $P < 0.01$]. The addition of initiating and receiving positive interaction accounted for nearly 21% of the variance [$F_{(3,36)} = 3.12$, $P < 0.03$].

In summary, the social interaction measures accounted for similar amounts of variance using the rating scale and nomination sociometrics. The one exception to this is the play with nomination, from which the interaction measures predicted over 34% of the variance. Also, it is interesting to note that for all of the nomination measures, initiating negative peer interaction was the best predictor, which was followed by initiating positive interaction in every case. Overall, the social behaviors which involved *receiving* interactions (positive and negative) best predicted peer acceptance (peer rating scales) and the behaviors that involved *initiating* (positive and negative) best predicted friendship patterns (peer nomination measures).

Discussion

The behaviors of initiating and receiving positive and negative social interaction significantly predicted children's acceptance and friendship patterns. These data support previous correlational research which has shown relationships between positive and negative social interactions and sociometric status (Charlesworth and Hartup, 1967;

TABLE V   Stepwise Regression for Sociometric Nominations Composite (Time 1 + Time 2 + Time 3)

Step No.	Behavior Variable Entered[a]	Multiple R^2 Value	df	F Value	Significance Level
	(a) Play with rating scale				
1	Initiating negative peer interaction	0.223	1,38	10.92	$P < 0.002$
2	Initiating positive peer interaction	0.314	2,37	8.47	$P < 0.0009$
3	Receiving positive peer interaction	0.344	3,36	6.30	$P < 0.001$
4	Receiving negative peer interaction	0.344	4,35	4.59	$P < 0.004$
	(b) Work with nomination				
1	Initiating negative peer interaction	0.168	1,38	7.70	$P < 0.008$
2	Initiating positive peer interaction	0.198	2,37	4.59	$P < 0.01$
3	Initiating negative peer interaction	0.201	3,36	3.03	$P < 0.04$
4	Receiving positive peer interaction	0.201	4,35	2.21	$P < 0.08$
	(c) Best friends nomination				
1	Initiating negative peer interaction	0.148	1,38	6.64	$P < 0.01$
2	Initiating positive peer interaction	0.200	2,37	4.65	$P < 0.01$
3	Receiving positive peer interaction	0.206	3,36	3.12	$P < 0.03$
4	Receiving negative peer interaction	0.207	4,35	2.29	$P < 0.07$

[a] Behavior variables are a composite of all three time periods (Time 1 + Time 2 + Time 3).

Gottman et al., 1975; Hartup et al., 1967). Additionally, the current research supports previous investigations that have demonstrated the reciprocal nature of giving and receiving positive and negative social reinforcement (Hartup et al., 1967; Keller and Carlson, 1974).

The correlations between the interaction measures and the sociometric measures offer evidence for the concurrent validity of the behavioral classes used (i.e., initiating and receiving positive and negative social interaction) in this investigation. The positive interactions included the subclasses of physical (embracing, shaking hands, patting, etc.), verbal (positive verbalizations), cooperation (sharing, working with others, etc.), and nonverbal (smiling, laughing, nods, etc.). Negative interactions included these same four subclasses, except the interactions were coded as negative (grabbing, hitting, verbal aggression, not sharing, frowns, etc.). The correlations in Table III suggest that the general behavioral classes of social interaction were slightly better predictions of children's friendship patterns (as measured by peer nominations) than they were of children's acceptance patterns (as measured by peer ratings). Caution should be exercised in intrepreting these results because of the nature of the sample (poorly accepted children) and the rather limited number of behavior classes used. It could be that different correlation patterns would be found with different samples and different combinations of behavioral measures.

It would have been desirable to use a more molecular categorization of social behaviors as predictors rather than the general behavioral classes used in the present investigation. Other studies have used such microbehaviors in regression studies of social skill with delinquent (Spence, 1981) and adult (Romano and Bellack, 1980) populations. The present study did not attempt such a microanalysis because of the relatively low frequency of some behavior subcategories (e.g., positive physical interactions and negative nonverbal interactions) and the problems with obtaining adequate levels of interobserver agreement. Future investigations should assess the relative contributions or weightings of specific behaviors in predicting sociometric status.

The stepwise multiple regression analyses present a more accurate picture of the relationship between behavioral and sociometric measures since the predictors (social interaction classes) were moderately intercorrelated. The best predictors of children's acceptance (PWR and WWR) involved the process of *receiving* social interactions (positive and negative). Thus, it appears from these data that overall acceptance (or unacceptance) in the peer group is best predicted by the reception of social interaction from peers. For the PWR and WWR, reception of social interaction accounted for 17.1 and 19.4% of the variance, respectively. Initiation of social interaction accounted for only 0.9% and 0.6% in peer ratings, respectively.

Conversely, the best predictors of children's friendship patterns (PW, WW, and BF) involved the process of *initiating* social interactions (positive and negative). For each of the three nomination measures, initiation of social interaction accounted for 31.4, 19.8, and 20% of the variance, respectively. Reception of social interaction accounted for only 0.3, 0.3, and 0.7% of the variance, respectively. Overall, initiating negative interaction was a much better predictor of friendship patterns than initiating positive interaction.

Some tentative implications for social skills assessment and training can be identified from these data. First, it appears that poorly accepted children may be identified more by the rates at which they receive social interactions from others than by the rates at which they initiate interactions with others. From simply a correlational standpoint, the data suggest that peer acceptance is more of a passive process in which interactions are received from others noncontingently. Surely this is not the case, since children probably receive positive or negative social interactions contingently. Future observational research should investigate antecedent behaviors or response chains that occasion the reception of positive and negative social interactions.

The present data also suggest that the best predictors of children's friendship patterns involve the active process of initiating social interactions. This supports past research that indicates that friendship-making skills involve active behaviors of greeting, asking for and giving information, extending offers of inclusion, and effective leave taking (Gottman et al., 1975, 1976; Oden and Asher, 1977). It is unknown from the present study which specific behaviors under the behavioral class of initiating social interaction best predicted friendship.

Past efforts to identify the relationship between social interactions and sociometric status have used correlational analyses which do not take into account the degree of intercorrelation between predictors. Moreover, past research has tended to use nomination measures (Hartup et al., 1967) instead of peer rating measures as criteria. The present investigation corrects some of these past deficiencies by using stepwise regression analyses and investigating peer rating as well as peer nomination measures. Hopefully, future research will heed Green and Forehand's (1980) advice and investigate the topographic components and the contextual factors relating to children's social skills. This study, along with the work of Green et al. (1980), reinforces the notion that social competence is multidimensional and should be assessed behaviorally using multifactored assessment techniques.

References

Asher, S. R., and Hymel, S. Children's social competence in peer relations: Sociometric and behavioral assessment. In J. D. Wine and M. D. Smye (Eds.), *Social Competence.* New York: Guilford Press, 1981.

Bellack, A. S., and Hersen, M. (Eds.). *Research and Practice in Social Skills Training.* New York: Plenum Press, 1979.

Bolstad, O. D., and Johnson, S. M. Self-regulation in the modification of disruptive classroom behavior. *Journal of Applied Behavior Analysis,* 1972, 5, 443–454.

Cartledge, G., and Milburn, J. F. The case for teaching social skills in the classroom. *Review of Educational Research,* 1978, 48, 133–156.

Cartledge, G., and Milburn, J. F. (Eds.). *Teaching Social Skills to Children: Innovative Approaches.* New York: Pergamon Press, 1980.

Charlesworth, R., and Hartup, W. W. Positive social reinforcement in the nursery school peer group. *Child Development,* 1967, 38, 993–1002.

Combs, M. L., and Slaby, D. A. Social skills training with children. In B. B. Lahey and A. E. Kazdin (Eds.). *Advances in Clinical Child Psychology,* Vol. 1. New York: Plenum Press, 1977.

Evers, W. L., and Schwarz, J. C. Modifying social withdrawal in preschoolers: The effects of filmed modeling and teacher praise. *Journal of Abnormal Child Psychology,* 1973, 1, 248–256.

Foster, S. L., and Ritchey, W. L. Issues in the assessment of social competence in children. *Journal of Applied Behavior Analysis,* 1979, 12, 625–638.

Gottman, J. M. Toward a definition of social isolation in children. *Child Development,* 1977, 48, 513–517.

Gottman, J. M., Gonso, J., and Rasmussen, B. Social interaction, social competence, and friendship in children. *Child Development,* 1975, 46, 709–718.

Gottman, J. M., Gonso, J., and Schuler, P. Teaching social skills to isolated children. *Journal of Abnormal Child Psychology,* 1976, 4, 179–197.

Green, K. D., and Forehand, R. Assessment of children's

social skills: A review of methods. *Journal of Behavioral Assessment,* 1980, 2, 143–157.

Green, K. D., Forehand, R., Beck, S. J. and Vosk, B. *Child Development,* 1980, 51, 1149–1156.

Greenwood, C. R., Walker, H. M., and Hops, H. Issues in social interaction/withdrawal assessment. *Exceptional Children,* 1977, 43, 490–499.

Gresham, F. M. Social skills training with handicapped children: A review. *Review of Educational Research,* 1981a, 51, 139–176.

Gresham, F. M. Assessment of children's social skills. *Journal of School Psychology,* 1981b, 19, 120–133.

Gresham, F. M. Validity of social skills measures for assessing social competence in low-status children: A multivariate investigation. *Developmental Psychology,* 1981c, 17, 390–398.

Gresham, F. M. *Social skills training with children and adolescents,* Unpublished manuscript. Baton Rouge: Louisiana State University, 1981d.

Gresham, F. M., and Nagle, R. J. Social skills training with children: Responsiveness to modeling and coaching as a function of peer orientation. *Journal of Consulting and Clinical Psychology,* 1980, 48, 718–729.

Hartup, W. W., Glazer, J. A., and Charlesworth, R. Peer reinforcement and sociometric status. *Child Development,* 1967, 38, 1017–1024.

Hymel, S., and Asher, S. R. *Assessment and training of isolated children's skills.* Paper presented at the meeting of the Society for Research in Child Development, New Orleans, 1977 (ERIC Document Reproduction Service No. ED 136–930).

Keller, M. F., and Carlson, P. M. The use of symbolic modeling to promote social skills in preschool children with low levels of social responsiveness. *Child Development,* 1974, 45, 912–919.

La Greca, A. M., and Santogrossi, D. A. Social skills training with elementary school students: A behavioral group approach. *Journal of Consulting and Clinical Psychology,* 1980, 48, 220–227.

Milburn, J. F. *Special education and regular class teacher attitudes regarding social behaviors of children: Steps toward the development of a social skills curriculum,* Unpublished doctoral dissertation. Columbus: The Ohio State University, 1974.

O'Connor, R. D. Modification of social withdrawal through symbolic modeling. *Journal of Applied Behavior Analysis,* 1969, 2, 15–22.

O'Connor, R. D. Relative effects of modeling, shaping, and the combined procedures for modification of social withdrawal. *Journal of Abnormal Psychology,* 1972, 79, 327–334.

Oden, S., and Asher, S. R. Coaching children in social skills for friendship making. *Child Development,* 1977, 48, 495–506.

Rathjen, D. P., and Foreyt, J. P. *Social competence: Interventions for children and adults.* New York: Pergamon Press, 1980.

Reardon, R. C., Hersen, M. Bellack, A. S., and Foley, J. M. *Measuring social skill in grade school boys,* Unpublished manuscript. Pittsburgh: University of Pittsburgh, 1978.

Romano, J. M., and Bellack, A. S. Social validation of a component model of assertive behavior. *Journal of Consulting and Clinical Psychology,* 1980, *48,* 478–490.

Singleton, L. C. and Asher, S. R. Peer preferences and social interaction among third-grade children in an integrated school district. *Journal of Educational Psychology,* 1977, *69,* 330–336.

Spence, S. H. Validation of social skills of adolescent males in an interview conversation with a previously unknown adult. *Journal of Applied Behavior Analysis,* 1981, *14,* 159–168.

Statistical Analysis System Institute. *Statistical Analysis System user's guide.* Raleigh, N.C.: SAS Institute, 1979.

Stodden, V. A. *The factor structure and discriminant validity of Social Behavior Assessment for emotionally disabled and normal students,* Unpublished specialist thesis. Ames: Iowa State University, 1981.

Thomson, C., Holmberg, M., and Baer, D. M. A brief report on a comparison of time sampling procedure. *Journal of Applied Behavior Analysis,* 1974, *7,* 623–626.

Van Hasselt, V. B., Hersen, M., Whitehill, M. B., and Bellack, A. S. Social skill assessment and training for children: An evaluation review. *Behavior Research and Therapy,* 1979, *17,* 413–437.

Van Houten, R. Social validation: The evolution of standards of competency for target behaviors. *Journal of Applied Behavior Analysis,* 1979, *12,* 581–591.

Wolf, M. M. Social validity: The case for subjective measurement or how behavior analysis is finding its heart. *Journal of Applied Behavior Analysis,* 1978, *11,* 203–214.

CHAPTER THREE

Multiple Discriminant Analysis

CHAPTER PREVIEW

Much has been written on the multivariate statistical technique of multiple discriminant analysis. This chapter discusses a complex and sophisticated technique—multiple discriminant analysis—without resorting to statistical jargon and mathematical formulas or glossing over important concepts. The chapter has two major objectives: (1) to introduce the underlying nature, philosophy, and conditions of discriminant analysis and (2) to demonstrate its application and interpretation with an illustrative example. Before proceeding further, it will be helpful to define the key terms.

Students who understand the most important concepts in the area of discriminant analysis should be able to:

- State the circumstances under which a linear discriminant function rather than multiple regression should be used.
- Identify and compare the three major stages in the application of discriminant analysis.
- Describe the two computational approaches for discriminant analysis and state when each should be used.
- Tell how to interpret the nature of a linear discriminant function, that is, identify independent variables with significant discriminatory power.
- Explain the usefulness of the classification matrix methodology.
- Describe how to develop a classification matrix.
- Explain a chance model.
- Differentiate between the hit-ratio and multiple regression's R^2.
- Justify the use of a split sample approach to validating the discriminant function.

KEY TERMS

Analysis sample When constructing classification matrices, the original sample should be divided randomly into two groups, one for developing the discriminant function and the other for validating it. The group used to compute the discriminant function is referred to as the *analysis sample*.

Categorical variable Referred to by some as a *nonmetric, nominal, binary, qualitative,* or *taxonomic variable*. When a number or value is assigned to a categorical variable, it serves merely as a label or means of identification. The number on a football jersey is an example.

Centroid   The mean value for the discriminant Z-scores for a particular category or group. A two-group discriminant analysis has two centroids, one for each of the groups.

Classification matrix   Also referred to as a *confusion, assignment,* or *prediction matrix.* It is a matrix containing numbers that reveal the predictive ability of the discriminant function. The numbers on the diagonal of the matrix represent correct classifications and the off-diagonal numbers incorrect classifications.

Cutting score   The criterion (score) against which each individual's discriminant score is judged to determine into which group the individual should be classified. When the analysis involves two groups, the hit ratio is determined by computing a single "cutting" score. Those entities whose Z-scores are below this score are assigned to one group, while those whose scores are above it are classified in the other group.

Discriminant function   A linear equation in the following form:

$$Z = W_1X_1 + W_2X_2 + \ldots + W_nX_n$$

where

Z = discriminant score
W = discriminant weight
X = independent variable

Discriminant loadings   Referred to by some as *structure correlations,* they measure the simple linear correlation between the independent variables and the discriminant function.

Discriminant score   Referred to as a *Z-score;* defined by the previous equation.

Discriminant weight   Referred to by some as a *discriminant coefficient,* its size is determined by the variance structure of the original variables. Independent variables with large discriminatory power usually have large weights and those with little discriminatory power usually have small weights; collinearity among the independent variables will cause an exception to this rule.

Hit ratio   The percentage of statistical units (individuals, respondents, objects, etc.) correctly classified by the discriminant function.

Hold-out sample   Also referred to as the *validation sample,* it is the group of subjects held out of the total sample when the function is computed.

Linear combination   Also referred to as *linear composites, linear compounds,* and *discriminant variates,* they represent the weighted sum of two or more variables.

Metric variable   A variable with a constant unit of measurement. If a variable is scaled from 1 to 9, the difference between 1 and 2 is the same as that between 8 and 9.

Tolerance   The proportion of the variation in the independent variables that is not explained by the variables already in the model
(function). It can be used to protect against multicollinearity. A tolerance of 0 means that a predictor (independent variable) under consideration is a perfect linear combination of variables already in the
model (equation). A tolerance of 1 means that a predictor is totally
independent of other predictors already in the model. The default
option in most computer packages sets the minimum acceptable
tolerance at .01. This allows quite a bit of redundancy or multicollinearity in the predictors. In short, it means that if at least 1 percent
of the variation in the response variable remains unexplained by
the predictors already included in the function, the predictor variable
under consideration will be allowed to enter the function.

What Is Discriminant Analysis?

In attempting to choose an appropriate analytical technique, we sometimes encounter a problem that involves a categorical dependent variable and several metric independent variables. For example, we may
wish to distinguish individuals who are good credit risks from those
who are bad risks, based on family income and size.

Discriminant analysis is the appropriate statistical technique when
the dependent variable is categorical (nominal or nonmetric) and the
independent variables are metric. In many cases, the dependent variable
consists of two groups or classifications, for example, male versus female
or high versus low. In other instances, more than two groups are involved, such as a three-group classification involving low, medium,
and high classifications. Discriminant analysis is capable of handling
either two groups or multiple groups (three or more). When two classifications are involved, the technique is referred to as two-group discriminant analysis. When three or more classifications are identified, the
technique is referred to as *multiple discriminant analysis (MDA)*.

Discriminant analysis involves deriving the linear combination of
the two (or more) independent variables that will discriminate best
between the a priori defined groups. This is achieved by the statistical
decision rule of maximizing the between-group variance relative to
the within-group variance; this relationship is expressed as the ratio
of between-group to within-group variance. The linear combinations
for a discriminant analysis are derived from an equation that takes
the following form:

$$Z = W_1X_1 + W_2X_2 + W_3X_3 + \ldots + W_nX_n$$

where

Z = the discriminant score
W = the discriminant weights
X = the independent variables

Discriminant analysis is the appropriate statistical technique for testing the hypothesis that the group means of the two or more groups

are equal. To do so, discriminant analysis multiplies each independent variable by its corresponding weight and adds these products together (see the preceding equation). The result is a single composite discriminant score for each individual in the analysis. By averaging the discriminant scores for all of the individuals within a particular group, we arrive at the group mean. This group mean is referred to as a *centroid*. When the analysis involves two groups, there are two centroids; with three groups there are three centroids, and so forth. The centroids indicate the most typical location of an individual from a particular group, and a comparison of the group centroids shows how far apart the groups are along the dimension being tested.

The test for the statistical significance of the discriminant function is a generalized measure of the distance between the group centroids. It is computed by comparing the distribution of the discriminant scores for the two or more groups. If the overlap in the distribution is small, the discriminant function separates the groups well. If the overlap is large, the function is a poor discriminator between the groups. The distributions of discriminant scores shown in Figure 3.1 further illustrate this concept. For example, the top diagram represents a function that separates the groups well, whereas the lower diagram shows a function that is a relatively poor discriminator between groups A and B (note that the shaded areas represent probabilities of misclassifying statistical units from A and B).

Analogy with Regression and ANOVA

The application and interpretation of discriminant analysis is much the same as in regression analysis. That is, a linear combination of metric measurements for two or more independent variables is used to describe or predict the behavior of a single dependent variable. The key difference is that discriminant analysis is appropriate for research problems in which the dependent variable is categorical (nominal or nonmetric), whereas in regression the dependent variable is metric.

Discriminant analysis is also comparable to analysis of variance (ANOVA). In discriminant analysis the single dependent variable is categorical and the independent variables are metric. The opposite is true of ANOVA. ANOVA involves metric dependent variables and a single categorical independent variable.

Assumptions of Discriminant Analysis

It is desirable that certain conditions be met for proper application of discriminant analysis. The assumptions for deriving the discriminant function are multivariate normality of the distributions and unknown (but equal) dispersion and covariance structures for the groups. When classification accuracies are determined, we must also assume equal costs of misclassification, equal a priori group probabilities, and known dispersion and covariation structures. There is evidence, however, that discriminant analysis is not very sensitive to violations of these assump-

FIGURE 3.1 Univariate representation of discriminant Z-scores.

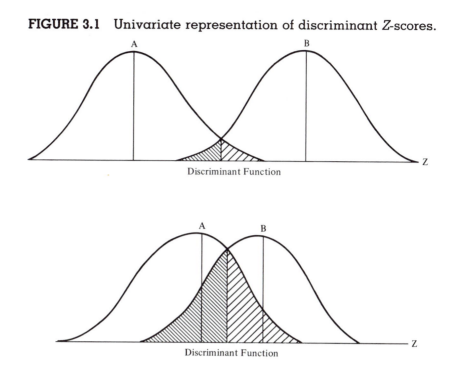

tions unless the violations are extreme [8]. This is particularly true with large sample sizes.

Hypothetical Example of Discriminant Analysis

Let's refer to a HATCO research problem to demonstrate the underlying logic of MDA. Suppose HATCO wants to find out whether one of its new products—a new and improved food mixer—will be commercially acceptable. In carrying out the investigation, HATCO is primarily interested in indentifying (if possible) those consumers who would purchase the new product and those who would not purchase it. In statistical terminology, then, HATCO would like to minimize the number of errors they would make in deciding which consumers would buy the new food mixer and which would not. To assist in identifying potential purchasers, HATCO has devised three rating scales to be used by consumers to evaluate the new product. Consumers would evaluate its durability, performance, and style with the three scales. Rather than relying on each scale as a separate measure, HATCO believes that a weighted combination of all three would better predict whether a consumer is likely to purchase the new product.

MDA can be utilized to obtain a weighted combination of the three scales. Then this weighted combination can be used to predict the likelihood of a consumer purchasing the product.

In addition to determining whether persons who are likely to purchase the new product can be distinguished from those who would not, HATCO would like to know which characteristics of its new product are useful in differentiating purchasers from nonpurchasers. That

		Ratings on Characteristics (0 = Very Poor; 10 = Excellent)		
Purchase Intention	**Subject Number**	**X_1 Durability**	**X_2 Performance**	**X_3 Style**
"Would	1	8	9	6
purchase"	2	6	7	5
	3	10	6	3
	4	9	4	4
	5	7	8	2
Mean rating		8.0	6.8	4.0
"Would not	6	5	4	7
purchase"	7	3	7	2
	8	4	5	5
	9	2	4	3
	10	2	2	2
Mean rating		3.2	4.4	3.8
Difference between mean ratings		4.8	2.4	0.2

TABLE 3.1 HATCO Survey Results for Evaluation of New Consumer Product

is, which of the three characteristics of the new product—durability, performance, or style—best separates purchasers from nonpurchasers? For example, if the response "would purchase" is always associated with a high durability rating and the response "would not purchase" is always associated with a low durability rating, HATCO could conclude that durability is a characteristic that works well in separating purchasers from nonpurchasers. In contrast, if HATCO found that about as many persons with a high rating on style said they would purchase as said they would not, then style is a characteristic that discriminates poorly between purchasers and nonpurchasers.

In Table 3.1, we assume that the ratings represent the judgments of a panel of 10 housewives who are potential purchasers of the new mixer, and that a particular price was specified for it (in rating the product, each housewife would be implicitly comparing it with products already on the market). Also, after the product was evaluated by the housewives they were asked to complete a buying intention scale. From the results of this scale, five respondents were classified in the "would purchase" group and five in the "would not purchase" group. As noted in Table 3.1 the difference between mean ratings for "would purchase" and "would not purchase" on the characteristic of durability is high (8.0 − 3.2 = 4.8). Thus, durability appears to discriminate well between the "would purchase" and "would not purchase" groups and is likely to be an important characteristic to potential purchasers.[1]

[1] This conclusion is based only on differences in the means and could possibly change after consideration of the standard deviations of the two sets of data. That is, with large standard deviations the difference between the means may not be statistically significant.

On the other hand, the characteristic of style has a difference between mean ratings for the "would purchase" and "would not purchase" groups of only 0.2 (4.0 − 3.8 = 0.2). Therefore, we would expect this characteristic to be less discriminating in terms of a decision to purchase or not to purchase.

The MDA technique follows a procedure very similar to that shown in the hypothetical example. It identifies the areas (characteristics) where the greatest difference exists between the groups; derives a discriminant weighting coefficient for each variable to reflect these differences; and then assigns each individual to a group using the weights and each individual's ratings on the characteristics.

Objectives of Discriminant Analysis

A review of the objectives for applying discriminant analysis should further clarify its nature. These include:

1. Determining if statistically significant differences exist between the average score profiles of the two (or more) a priori defined groups.
2. Establishing procedures for classifying statistical units (individuals or objects) into groups on the basis of their scores on several variables.
3. Determining which of the independent variables account most for the differences in the average score profiles of the two or more groups [5].

As can be noted from these objectives, discriminant analysis is useful when the analyst is interested either in understanding group differences or in correctly classifying statistical units into groups or classes. Discriminant analysis, therefore, can be considered either a type of profile analysis or an analytical predictive technique. In either case, the technique is most appropriate where there is a single categorical dependent variable and several metrically scaled independent variables.

Geometric Representation

A graphic illustration of a two-group analysis will help to explain further the nature of discriminant analysis [5]. Figure 3.2 represents a scatter diagram and projection that shows what happens when a two-group discriminant function is computed. Let's assume that we have two groups, A and B, and two measurements, X_1 and X_2 on each member of the two groups. We can plot in a scatter diagram the association of variable X_1 with variable X_2 for each member of the two groups. Group membership is identified by the use of dots and circles. In Figure 3.2 the dots represent the variable measurements for the members of group A and the circles represent the variable measurements for group B. The ellipses drawn around the dots and circles would enclose some prespecified proportion of the points, usually 95 percent or more in each group. If we draw a straight line through the two points where the ellipses intersect and then project the line to a new Z axis, we can say that the overlap between the univariate distributions A' and B' (represented by the shaded area) is smaller than would be obtained

FIGURE 3.2 Graphic illustration of two-group discriminant analysis.

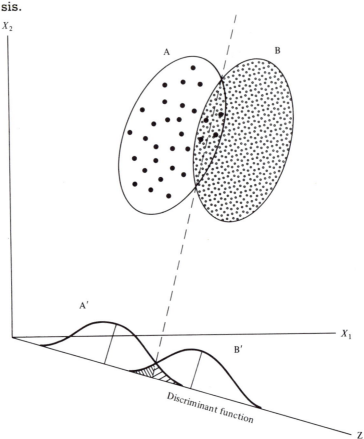

by any other line drawn through the ellipses formed by the scatter plots [5].

The important thing to note about Figure 3.2 is that the Z-axis expresses the two-variable profiles of groups A and B as single numbers (discriminant scores). By finding a linear combination of the original variables X_1 and X_2, we can project the result as a discriminant function. For example, if the dots and circles are projected onto the new Z-axis as discriminant Z-scores, the result condenses the information about group differences (shown in the $X_1 X_2$ plot) into a set of points (Z-scores) on a single axis.

To summarize, for a given discriminant analysis problem, the computer derives a linear combination of the independent variables. The result is a series of discriminant scores for each individual in each group. The discriminant scores are computed based on the statistical rule of maximizing the variance between the groups and minimizing the variance within them. If the variance between the groups is large relative to the variance within them, we say that the discriminant function separates the groups well.

Application of Discriminant Analysis

The application of discriminant analysis can be divided into three major stages: (1) derivation, (2) validation, and (3) interpretation. The *derivation stage* involves determining whether or not a statistically significant function can be derived to separate the two (or more) groups. The *validation stage* involves developing a classification matrix to evaluate further the predictive accuracy of the discriminant function. The *interpretation stage* involves determining which of the independent variables contribute the most to discriminating between the groups. Each of these stages will be discussed in the following sections.

Stage One: Derivation

The derivation stage consists of several separate steps. The steps are variable selection, sample division, the computational method, and statistical significance.

VARIABLE SELECTION. To apply discriminant analysis, the analyst first must specify which variables are to be independent and which one is to be dependent. Recall that the dependent variable is categorical and the independent variables are metric.

The analyst should focus on the dependent variable first. The number of dependent variable groups (categories) can be two or more, but these groups must be mutually exclusive and exhaustive. In some cases, the dependent variable may be two groups (dichotomous), such as good versus bad. In other cases, the dependent variable may involve several groups (multichotomous), such as predicting the occupation— for example, physician, attorney, or professor.

The preceding examples of categorical variables were true dichotomies (or multichotomies). There are some situations, however, where discriminant analysis is appropriate even if the dependent variable is not a true categorical variable. We may have a dependent variable that is of ordinal or interval measurement that we wish to use as a categorical dependent variable. In such cases, we would have to create an artificial dichotomy. For example, if we had a variable that measured the average number of cola drinks consumed per day, and the individuals responded on a scale from zero to eight or more per day, we could create an artificial trichotomy (three groups) by simply designating those individuals who consumed 0, one, or two cola drinks per day as light users, those who consumed three, four, or five per day as medium users, and those who consumed six, seven, or eight, or more as heavy users. Such a procedure would create a three group categorical variable in which the objective would be to discriminate between light, medium, and heavy users of colas.

Any number of artificial categorical groups can be developed. Most frequently the approach would involve creating two, three, or four categories. But a larger number of categories could be established if the need arose. When three or more categories are created, the possibility arises of examining only the extreme groups in a two-group discriminant analysis. This procedure is called the *polar-extremes approach*.

The polar extremes approach involves comparing only the extreme two groups and excluding the middle group (or groups) from the discriminant analysis. The analyst could examine the light and heavy users of cola drinks and exclude the medium users. This approach can be used any time the analyst wishes to examine only the extreme groups. However, the analyst may want to try this approach when the results of a regression analysis are not as good as anticipated. Such a procedure may be helpful because it is possible that group differences may appear even though regression results were poor. That is, the polar extremes approach with discriminant analysis can reveal differences that are not as prominent in a regression analysis of the full data set [5]. Such manipulation of the data naturally would necessitate caution in interpreting one's findings.

After a decision has been made on the dependent variable, the analyst must decide which independent variables to include in the analysis. Independent variables usually are selected in two ways. The first approach involves identifying variables either from previous research or from the theoretical model that is the underlying basis of the research question. The second approach is intuitive. It involves trying to extend the researcher's knowledge and intuitively selecting variables for which no previous research or theory exists, but which logically might be related to predicting the groups for the dependent variable.

SAMPLE DIVISION.   When applying discriminant analysis, the analyst wants to test the validity of the discriminant function that has been derived. A number of procedures have been suggested for doing this, but the most popular one involves developing the discriminant function on one group and then testing it on a second group. The usual procedure is to divide the total sample of respondents randomly into two groups. One of these groups, referred to as the *analysis sample*, is used to develop the discriminant function. The second group, referred to as the *holdout sample*, is used to test the discriminant function. This method of validating the function is referred to as the *split-sample* or *cross-validation approach* [4].

The justification for dividing the total sample into two groups is that an upward bias will occur in the prediction accuracy of the discriminant function if the individuals used in developing the classification matrix are the same as those used in computing the function. That is, the classification accuracy will be higher than is valid for the discriminant function if it was used to classify a separate sample. The implications of this upward bias are particularly important when the analyst is concerned with the external validity of the findings.

No definite guidelines have been established for dividing the sample into analysis and holdout groups. The most popular procedure is to divide the total group so that one-half of the respondents are placed in the analysis sample and the other half are placed in the holdout sample. However, no hard-and-fast rule has been established here, and some researchers prefer a 60–40 or 75–25 split between the analysis and holdout groups.

When selecting the individuals for the analysis and holdout groups, a proportionately stratified sampling procedure is usually followed. If the categorical groups for the discriminant analysis are equally represented in the total sample, an equal number of individuals is selected. If the categorical groups are unequal, the sizes of the groups selected for the holdout sample should be proportionate to the total sample distribution. If a sample consists of 50 males and 50 females, the holdout sample would have 25 males and 25 females. If the sample contained 70 females and 30 males, then the holdout samples would consist of 35 females and 15 males.

Several additional comments need to be made regarding the division of the total sample into analysis and holdout groups. One is that if the analyst is going to divide the sample into analysis and holdout groups, the sample must be sufficiently large to do so. Again, no hard-and-fast rules have been established, but it seems logical that the analyst would want at least 100 in the total sample to justify dividing it into the two groups. One compromise procedure the analyst can select if the sample size is too small to justify a division into analysis and holdout groups is to develop the function on the entire sample and then use the function to classify the same group used to develop the function. This procedure results in an upward bias in the predictive accuracy of the function but is certainly better than not testing the function at all.

Recall that the most frequent procedure utilized in validating the discriminant function is to divide the groups randomly into analysis and holdout samples once. This procedure involves developing a discriminant function with the analysis sample and then applying it to the holdout sample. Other researchers have suggested, however, that greater confidence could be placed in the validity of the function by following this procedure several times [4]. Instead of randomly dividing the total sample into analysis and holdout groups once, the analyst would randomly divide the total sample into analysis and holdout samples several times, each time testing the validity of the function through the development of a classification matrix and a hit ratio. Then the several hit ratios would be averaged to obtain a single measure.

Other more sophisticated methods have been suggested for validating discriminant functions. The selection of a validation method depends upon the amount of time and facilities available and the degree to which the researcher wishes to follow a rigorous analysis procedure. A summary of more rigorous methods may be found in Crask and Perreault [1].

COMPUTATIONAL METHOD. Two computational methods can be utilized in deriving a discriminant function: the simultaneous (direct) method and the stepwise method.

The *simultaneous method* involves computing the discriminant function so that all of the independent variables are considered concurrently. Thus, the discriminant function(s) is computed based upon the entire set of independent variables, regardless of the discriminating power of each independent variable. The simultaneous method is appropriate when, for theoretical reasons, the analyst wants to include all of the independent variables in the analysis and is not interested in seeing intermediate results based only on the most discriminating variables [12].

The *stepwise method* is an alternative to the simultaneous approach. It involves entering the independent variables into the discriminant function one at a time on the basis of their discriminating power. The stepwise approach begins by choosing the single best discriminating variable. The initial variable is then paired with each of the other independent variables one at a time, and a second variable is chosen. The second variable is the one that is best able to improve the discriminating power of the function in combination with the first variable. The third and any subsequent variables are selected in a similar manner. As additional variables are included, some previously selected variables may be removed if the information they contain about group differences is available in some combination of the other included variables. Eventually, either all independent variables will have been included in the function or the excluded variables will have been judged as not contributing significantly to further discrimination [12].

The stepwise method is useful when the analyst wants to consider a relatively large number of independent variables for inclusion in the function. By sequentially selecting the next best discriminating variable at each step, variables that are not useful in discriminating between the groups are eliminated and a reduced set of variables is identified. The reduced set typically is almost as good as, and sometimes better than, the complete set of variables.

STATISTICAL SIGNIFICANCE. After the discriminant function has been computed, the analyst must assess its level of significance. The conventional criterion of .05 or beyond is often used. Many researchers believe that if the function is not significant at or beyond the .05 level, there is little justification for going further. Some social scientists and business analysts, however, disagree. Their decision rule for continuing is the cost versus the value of the information, and higher levels of risk (e.g., significance levels >.05) may be acceptable for their purposes. For example, they may decide to examine discriminant functions that are significant at the .2 or even the .3 level if the circumstances justify it.

Stage Two: Validation

The validation stage involves several major considerations: the reason for developing classification matrices, cutting score determination, constructing classification matrices, chance models, and classification accuracy relative to chance.

WHY CLASSIFICATION MATRICES ARE DEVELOPED. As has been noted, one of the standard outputs of a discriminant analysis is a measure of the statistical significance of the function. For the SPSS [11, 12] package, the statistic is a Chi-Square. For the BMD package [3] it is the Mahalanobis D^2 statistic. These statistical tests assess the significance of the discriminant functions, but they do not tell you how well the function predicts. For example, suppose the two groups are significantly different beyond the .01 level. With sufficiently large sample sizes, the group means (centroids) could be virtually identical and we still would have statistical significance. In short, these statistics suffer the same drawbacks of classical tests of hypotheses. Thus, the level of significance of these statistics is a very poor indication of the function's ability to discriminate between the two groups. To determine the predictive ability of a discriminant function, the analyst must construct classification matrices.

To clarify further the usefulness of the classification matrix procedure, we shall relate it to the concept of an R^2 in regression analysis. Most of us have probably read academic articles in which the author has found statistically significant relationships, and yet has explained only 10 percent (or less) of the variance; i.e., $R^2 = 0.10$. Usually this R^2 is significantly different from zero simply because the sample size is large. With multiple discriminant analysis, the hit ratio (percentage correctly classified) is analogous to regression's R^2. The hit ratio reveals how well the discriminant function classified the statistical units; the R^2 indicates how much variance the regression equation explained. The F-test for statistical significance of the R^2 is, therefore, analogous to the chi-square (or D^2) test of significance in discriminant analysis. Clearly, with a sufficiently large sample size in discriminant analysis, we could have a statistically significant difference between the two (or more) groups and yet correctly classify only 53 percent (when chance is 50 percent, with equal group sizes) [10].

CUTTING SCORE DETERMINATION.   If the statistical test indicates that the function discriminates significantly, it is customary to develop classification matrices to provide a more accurate assessment of the discriminating power of the function. Before a classification matrix can be constructed, however, the analyst must determine the cutting score. The *cutting score* is the criterion (score) against which each individual's discriminant score is judged to determine into which group the individual should be classified.

In constructing classification matrices, the analyst will want to determine the *optimum cutting score* (also referred to as a *critical Z value*). The optimal cutting score will differ depending on whether the sizes of the groups are equal or unequal. If the groups are of equal size, the optimal cutting score will be halfway between the two group centroids. The cutting score is therefore defined as:

$$Z_{CE} = \frac{\overline{Z}_A + \overline{Z}_B}{2}$$

where

Z_{CE} = critical cutting score value for equal group sizes
$\overline{Z}_A$ = centroid for group A
$\overline{Z}_B$ = centroid for group B

If the groups are not of equal size, a weighted average of the group centroids will provide an optimal cutting score, calculated as follows:

$$Z_{CU} = \frac{N_A \overline{Z}_A + N_B \overline{Z}_B}{N_A + N_B}$$

where

Z_{CU} = critical cutting score value for unequal group sizes
N_A = number in group A
N_B = number in group B
$\overline{Z}_A$ = centroid for group A
$\overline{Z}_B$ = centroid for group B

Both of the preceding formulas assume that the distributions are normally distributed and the group dispersion structures are known.

The concept of an optimal cutting score is illustrated in Figures 3.3 and 3.4. The optimal cutting score for equal groups is shown in Figure 3.3. The effect of one group being larger than the other is illustrated in Figure 3.4. Both the weighted and unweighted cutting scores are shown. It is apparent that if group A is much smaller than group B, the optimal cutting score will be closer to the centroid of group A than to that of group B. Also, note that if the unweighted cutting score were used, none of the individuals in group A would be misclassified, but a substantial portion of those in group B would.

The optimal cutting score also must consider the cost of misclassifying an individual into the wrong group. If the costs of misclassifying an individual are approximately equal, the optimal cutting score will be the one that will misclassify the fewest number of individuals in

FIGURE 3.3 Optimal cutting score with equal sample sizes.

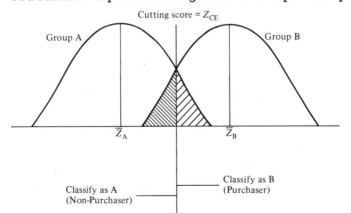

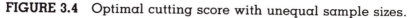

FIGURE 3.4 Optimal cutting score with unequal sample sizes.

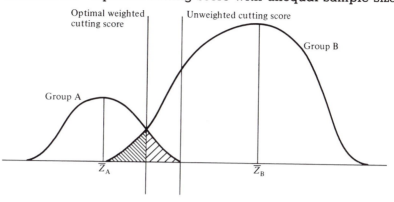

all groups. If the misclassification costs are unequal, the optimum cutting score will be the one that minimizes the costs of misclassification.

More sophisticated approaches to determining cutting scores are discussed by Massey [9]. The approaches are based upon a Bayesian statistical model and are appropriate when the costs of misclassification into certain groups are very high, when the groups are of grossly different sizes, or when one wants to take advantage of a priori knowledge of group membership probabilities.

In practice, when calculating the cutting score, it is usually not necessary to insert the raw variable measurements for every individual into the discriminant function and obtain the discriminant score for each person to use in computing the $\overline{Z}_A$ and $\overline{Z}_B$ (group A and B centroids). In many instances the computer program will provide the discriminant scores as well as the $\overline{Z}_A$ and $\overline{Z}_B$ as regular output. When the analyst has the group centroids and sample sizes, he or she must merely substitute the values in the appropriate formula to obtain the optimal cutting score.

CONSTRUCTING CLASSIFICATION MATRICES. To validate the discriminant function through the use of classification matrices, the sample should be randomly divided into two groups. One of the groups (the analysis sample) is used to compute the discriminant function. The other group (the holdout or validation sample) is retained for use in developing the classification matrix. The procedure involves multiplying the weights generated by the analysis sample by the raw variable measurements of the holdout sample to obtain discriminant scores for the holdout sample. Then the individual discriminant scores for the holdout sample are compared with the critical cutting score value and classified as follows:

1. Classify an individual into group A if $Z_n < Z_{ct}$.
2. Classify an individual into group B if $Z_n > Z_{ct}$.

where

Z_n = the discriminant Z-score for the nth individual
Z_{ct} = the critical cutting score value

The results of the classification procedure are presented in matrix form, as shown in Table 3.2. The entries on the diagonal of the matrix represent the number of individuals correctly assigned to their group. The numbers off the diagonal represent the incorrect classifications. The entries under the column labeled "Actual Total" represent the number of individuals actually in each of the two groups. The entries at the bottom of the columns represent the number of individuals assigned to the two groups by the discriminant function. The percentage correctly classified for each group is shown at the right side of the matrix and the overall percentage correctly classified (hit ratio) is shown at the bottom. For example, in our HATCO problem that attempted to predict which housewives would purchase a new consumer product, the number of individuals actually in and correctly assigned to actual group 1, "Would Purchase," was 22. The number incorrectly assigned to actual group, 2, "Would Not Purchase," was 3. Similarly, the number of correct classifications to actual group 2 was 20 and the number of incorrect assignments to actual group 1 was 5. Thus, the percentage classification accuracy of the discriminant function for actual groups 1 and 2 would be 88 and 80 percent, respectively. The overall classification accuracy (hit ratio) would be 84 percent.

One final classification procedures topic should be discussed. A t-test is available to determine the level of significance for the classification accuracy. The formula for a two group analysis (equal sample size) is:

$$t = \frac{p - .5}{\sqrt{\frac{.5(1 - .5)}{N}}}$$

where

p = proportion correctly classified
N = sample size

This formula can be adapted for use with more groups and unequal sample sizes.

TABLE 3.2 Classification Matrix for HATCO's New Consumer Product

| Actual Group | Predicted Group | | Actual Total | Group Classification Percentage |
	Would Purchase (1)	Would Not Purchase (2)		
(1)	22	3	25	88
(2)	5	20	25	80
Predicted Total	27	23	50	

Percent correctly classified (hit ratio) = (100) (22 + 20/50) = 84%.

CHANCE MODELS.   This section describes the procedure involved in determining the percentage of individuals that would be correctly classified by chance. As noted earlier, the predictive accuracy of the discriminant function is measured by the hit ratio, which is obtained from the classification matrix. The analyst may ask, what is considered an acceptable level of predictive accuracy for a discriminant function, and what is not considered acceptable? For example, is 60 percent an acceptable level, or should one expect to obtain 80 to 90 percent predictive accuracy? To answer this question, the analyst must first determine the percentage that could be classified correctly by chance (without the aid of the discriminant function).

When the sample sizes of the groups are equal, the determination of the chance classification is rather simple, obtained by dividing 1 by the number of groups. The forumla is $C = 1 \div$ number of groups. For instance, in a two-group function the chance probability would be .50; for a three-group function the chance probability would be .33, and so forth.

The determination of the chance classification for situations where the group sizes are unequal is somewhat more involved. Let's assume that we have a sample in which 75 subjects belong to one group and 25 to the other. We could arbitrarily assign all the subjects to the larger group and achieve a 75 percent classification accuracy without the aid of a discriminant function. It could be concluded that unless the discriminant function achieves a classification accuracy higher than 75 percent, it should be disregarded because it has not helped us improve our prediction accuracy.

Determining the chance classification based on the sample size of the largest group is referred to as the *maximum chance criterion*. it is determined by computing the percentage of the total sample represented by the largest of the two (or more) groups. For example, if the group sizes are 65 and 35, the maximum chance criterion is 65 percent correct classifications. Therefore, if the hit ratio for the discriminant function does not exceed 65 percent, it has not helped us predict, based on this criterion.

The maximum chance criterion should be used when the sole objective of the discriminant analysis is to maximize the percentage correctly classified [10]. Situations in which we are concerned only about maximizing the percentage correctly classified are rare. Usually the analyst uses discriminant analysis to identify correctly members of both groups. In cases where the sample sizes are unequal and the analyst wants to classify members of both groups, the discriminant function defies the odds by classifying a subject in the smaller group. The chance criterion should take this into account [10]. Therefore, another chance model— the *proportional chance criterion*—should be used in most situations.

The proportional chance criterion should be used when group sizes are unequal and the analyst wishes to identify correctly members of the two (or more) groups. The formula for this criteria is

$$C \text{ proportional} = p^2 + (1 - p)^2$$

where

p = the proportion of individuals in group 1
$1 - p$ = the proportion of individuals in group 2

Using the group sizes from our earlier example (75 and 25), the proportional chance criterion would be 62.5 percent compared to 75 percent. Therefore, a prediction accuracy of 75 percent might be acceptable because it is above the 62.5 percent proportional chance criterion.

It should be noted that these chance model criteria are useful only when computed with holdout samples (split sample approach). If the individuals used in calculating the discriminant function are the ones being classified, the result will be an upward bias in the prediction accuracy [4]. In such cases, both of these criteria would have to be adjusted upward to account for this bias.

CLASSIFICATION ACCURACY RELATIVE TO CHANCE.   The question of classification accuracy is crucial. If the percentage of correct classifications is significantly larger than would be expected by chance, an attempt can be made to interpret the discriminant functions in the hope of developing group profiles. However, if the classification accuracy is no greater than can be expected by chance, whatever structural differences appear to exist merit little or no interpretation [4]. That is, differences in score profiles would provide no meaningful information for identifying group membership.

The question, then, is, how high should the classification accuracy be relative to chance? For example, if chance is 50 percent (two-group, equal sample sizes), does a classification (predictive) accuracy of 60 percent justify moving to the interpretation stage? No general guidelines have been developed to answer this question. Ultimately the decision depends on the cost in relation to the value of the information. If the costs associated with a 60 percent predictive accuracy (relative to 50 percent by chance) are greater than the value to be derived from the findings, there is no justification for interpretation. If the value is high relative to the costs, 60 percent accuracy would justify moving on to interpretation.

The cost versus value argument offers little assistance to the neophyte data analyst. Therefore, the authors suggest the following criterion: the classification accuracy should be at least 25 percent greater than that achieved by chance. For example, if chance accuracy is 50 percent, the classification accuracy should be 62.5 percent. If chance accuracy is 30 percent, the classification accuracy should be 37.5 percent. This criterion provides only a rough estimate of the acceptable level of predictive accuracy. The criterion is easy to apply with groups of equal size. With groups of unequal size, an upper limit is reached when the maximum chance model is used to determine chance accuracy. This does not present too great a problem, however, since under most circumstances the maximum chance model would not be used with unequal group sizes (see the section on chance models).

Stage Three: Interpretation

If the discriminant function is statistically significant and the classification accuracy is acceptable, the analyst should continue to Stage Three, which focuses on making substantive interpretations of the findings. This process involves two distinct phases. The first phase involves examining the discriminant functions to determine the relative importance of each independent variable in discriminating between the groups. Three methods of determining the relative importance have been proposed: (1) standardized discriminant weights, (2) discriminant structure correlations, and (3) partial F-values. The second phase involves examining the group means for each important variable to profile the differences in the groups.

DISCRIMINANT WEIGHTS. The traditional approach used in interpreting discriminant functions involves examining the sign and magnitude of the standardized discriminant weight (sometimes referred to as a *discriminant coefficient*) assigned to each variable in computing the discriminant functions. Independent variables with relatively larger weights contribute more to the discriminating power of the function than do variables with smaller weights. Therefore, when the sign is ignored, each weight represents the relative contribution of its associated variable to that function. The sign merely denotes that the variable makes either a positive or a negative contribution [12].

The interpretation of discriminant weights is analogous to the interpretation of beta weights in regression analysis and is therefore subject to the same criticisms. For example, a small weight may indicate either that its corresponding variable is irrelevant in determining a relationship or that it has been partialed out of the relationship because of a high degree of multicollinearity. Another problem with the use of discriminant weights is that they are subject to considerable instability. These problems suggest caution in using weights to interpret the results of discriminant analysis.

DISCRIMINANT LOADINGS. In recent years, loadings have increasingly been used as a basis for interpretation because of the deficiencies in utilizing weights. Discriminant loadings, referred to sometimes as *structure correlations*, measure the simple linear correlation between each independent variable and the discriminant function. The discriminant loadings reflect the variance that the independent variables share with the discriminant function, and can be interpreted like factor loadings in assessing the relative contribution of each independent variable to the discriminant function (Chapter 6 further discusses factor loadings interpretation).

Discriminant loadings (like weights) may be subject to instability. Loadings are considered relatively more valid than weights as a means of interpreting the discriminating power of independent variables. The analyst still must be cautious when using loadings for interpreting discriminant functions.

PARTIAL *F*-VALUES.   As discussed earlier, two computational approaches can be utilized in deriving discriminant functions—simultaneous and stepwise. When the stepwise method is selected, an additional means of interpreting the relative discriminating power of the independent variables is available through the use of the partial *F*-values. This is accomplished by examining the absolute sizes of the significant *F*-values and ranking them. Larger *F*-values indicate greater discriminating power.

In practice, rankings using the *F*-values approach is the same as those using the weights, but the *F*-values indicate the associated level of significance for each variable.

WHICH INTERPRETATION METHOD TO USE.   Several methods for interpreting the nature of discriminant functions have been discussed. Which method should the analyst use? Since most discriminant analysis problems necessitate the use of a computer, the analyst frequently must use whichever method is available on the system packages. The loadings approach is somewhat more valid than the use of weights and should be utilized whenever possible. The analyst should not hesitate to use the weights if they are the only method available, but they should be interpreted in light of their limitations. Additional material on interpretation approaches is contained in the readings at the end of this chapter.

PROFILING GROUP DIFFERENCES.   When the analyst has identified the independent variables that make the greatest contribution in discriminating between the groups, the next step is to profile the characteristics of the groups based on group means. For example, referring back to the HATCO survey data presented in Table 3.1, we see that the mean rating on "Durability" for the "Would Purchase" group is 8.0, while the comparable mean rating for the "Would Not Purchase" group is 3.2. Thus, a profile of the characteristics of these two groups shows that the "Would Purchase" group rates the perceived durability of the new product substantially higher than the "Would Not Purchase" group.

A Two-Group Illustrative Example

To illustrate the application of a two-group discriminant analysis, we shall use variables drawn from the HATCO data bank introduced in Chapter 1. You should recall that variable 10 is a categorical variable that recorded whether or not a salesperson received recognition for outstanding performance. Such recognition comes in the form of a top salesperson award, which HATCO awards on a monthly basis. It is the company's belief that the award motivates the sales force to become more productive. HATCO would like to determine which social-psychological characteristics have the largest effect on distinguishing the difference between performance levels. After determining which relationships between the social-psychological tests, and performance levels are significant, HATCO will be able to assess which tests are most efficient in evaluating new sales recruits. In order to test for the relationship, the first six variables from the data bank will be used to discriminate between those who have received the award and those who have not. By developing a discriminant function using existing salesperson records, HATCO hopes to be able to identify new sales recruits with high potential.

To perform this analysis, we will once again choose from the statistical packages. Because of the widespread availability of SAS and SPSS, we will focus our discussion on these two packages. Some analysts prefer SPSS to SAS because of its stepwise procedures, the amount of statistics it provides, and the wider range of discriminant procedures it allows. SAS, however, has been superior in data manipulation, in its ability to analyze holdout samples, and in its easier-to-read printout. A current update of SPSS now allows for the application of holdout procedures, and recent revisions of SAS include stepwise and several other discriminant procedures. While these changes make the two packages equally appealing, the SPSS package is still favored by many researchers because of their familiarity with its procedures. Exhibit 3.1 shows the SPSSX statements necessary to activate the discriminant analysis procedure.

Previous discussion has emphasized the need for validating a split sample or holdout approach. The statements given in Exhibit 3.1 allow for the use of a holdout sample. It should be noted, however, that with only 50 observations in the data set, a holdout sample would not normally be applied and is used only for illustrative purposes in this example. To assure randomness in the selection of the holdout sample, a uniform random variable was created for each observation. It was then necessary to determine what percentage of the data set would be used in the holdout sample. Since the data set was so small, it was decided to use a 44 percent holdout. A dummy variable was then created so that all observations with uniform random variables greater than a specified value were assigned a value of 1, while all others were assigned a value of 0. Any holdout sample size can be specified by simply varying the cutoff point for the dummy variable. (*Note:* In using SPSSX, the cutoff value may be greater than the actual percentage needed. In the example, a cutoff value of .63 was used,

---

EXHIBIT 3.1
SPSSX Program

```
DATA LIST FILE = HATCO
        /1 ID 1–3 X1 12–17 X2 19–24 X3 26–31 X4 33–38 X5 40–45
        X6 47–52
        X7 54–59 X8 61–66 X9 73 X10 70 X11 75 X12 77
COMPUTE SET = UNIFORM(1) GT .63
DISCRIMINANT GROUPS = X10(0,1)/
            VARIABLES = X1 to X6/
            SELECT = SET(0)/
            METHOD = MAHAL/
            PRIORS = SIZE?
OPTION 7
STATISTICS 1 2 5 6 7 8 10 13 14
```

which resulted in a 44 percent holdout sample. Of the 50 observations, 22 were assigned to the holdout sample and the remaining 28 were used to develop the discriminant function.)

There are several discriminant procedures available in the SPSS package, but in this example the Mahalanobis procedure was used. The Mahalanobis procedure is based on generalized squared Euclidean distance that adjusts for unequal variances. The major advantage of this procedure is that it is computed in the original space of the predictor variables rather than as a collapsed version, which is used in Fisher's method. The use of the Mahalanobis procedure is particularly critical as the number of predictor variables increases, since Fisher's method results in a considerable reduction in dimensionality. The loss in dimensionality causes a loss of information, since it decreases predictor variability. In general, Mahalanobis is the preferred procedure when one is interested in the maximal use of available information. The Mahalanobis procedure in the SPSS package performs a stepwise discriminant analysis that is similar to a stepwise regression analysis. This stepwise procedure is designed to develop the best one-variable model, followed by the best two-variable model, followed by the best three-variable model, and so forth, until no other variables meet the desired selection rule. The selection rule in this procedure is to maximize Mahalanobis distance D^2 between groups.

The priors statement is used to set prior probabilities for classification purposes. If no priors statement is given, the SPSS package automatically defaults to equal probabilities; in other words, each group has an equal chance of occurring. In this case, we know that the dependent variable consists of two groups, one with a sample size of 30 and the other with a sample size of 20. If we were assured that the top salesperson award would always occur in these proportions, we could set the prior probabilities to be 60 and 40 percent. In most cases, however, we cannot be sure what the population proportions are. If the assumption is made that the sample was selected in an unbiased manner, the best estimate of population proportions is the sample proportions. By specifing PRIORS = SIZE, the SPSS package substitutes the sample proportions for the prior probabilities.

Let us begin our analysis of the two-group HATCO top salesperson discriminant analysis problem by examining Table 3.3. This table shows the unweighted group means for each of the independent variables,

TABLE 3.3 Group Means and Standard Deviations for the HATCO Discriminant Analysis	Group Means	X_{10}	X_1	X_2	X_3	X_4	X_5	X_6	N
	Nonrecipients	0	5.42	6.87	13.80	16.22	6.10	4.76	11
	Recipients	1	8.60	3.23	17.92	10.64	5.91	5.56	17
	Total		7.35	4.66	16.31	16.48	5.48	5.25	28
	Group Std. Dev.								
	Nonrecipients	0	1.87	2.31	2.03	1.61	1.62	1.05	11
	Recipients	1	2.16	1.89	1.78	2.52	1.02	1.64	17
	Total		2.56	2.71	2.75	2.19	1.27	1.50	28

based on the 28 observations used in the development of the discriminant function. (Recall that we are using a holdout group of $N = 22$; this is the analysis group, with $N = 28$.) Table 3.4 shows the univariate analysis of variance used in testing the means of the individual variables between groups.

Remember, the objective of this analysis was to determine which variables are most efficient in discriminating between the two groups. For this reason, a stepwise procedure was used. If the objective of the analysis had simply been to determine the discriminating capabilities of the entire battery of test scores, without regard to the capabilities of each test component, all variables would have been entered into the model immediately.

Recall that the stepwise procedure begins with all of the variables excluded from the model and selects the variable that maximizes the Mahalanobis distance between the groups. In this example, a minimum F value of 1.0000 was also required for entry. This limitation allowed only X_1, X_2, X_3, X_5, and X_6 to be considered for entry into the initial model. The maximum Mahalanobis D^2 is associated with X_3. After X_3 has entered the model, the remaining variables are evaluated on the basis of the distance between their means after the variance associated with X_3 is removed. Variables X_1 and X_5 were the only variables with F values greater than 1.0 after X_3 had been included in the model (as shown in Table 3.5). Variable X_1 will be the next variable to enter the model because it has the highest F to Enter (7.8985). It is very likely that X_5 will also enter the model because of its large F to Enter (6.4873). Variables X_2, X_4, and X_6 did not enter the model because their F to Enter values were below 1 (X_6 was excluded after X_1 entered the model). Note that in cases where more than two variables are entered into the model, the variables already in the model are also evaluated for possible removal. A variable may be removed if multicollinearity exists between the independent variables.

Table 3.6 provides the overall discriminant analysis results for the two-group model based on recipients versus nonrecipients of the outstanding sales person award. The summary table at the top indicates that two variables entered the model—X_1 and X_3—and both are significant discriminators based on the Wilks' lambda and the minimum D-squared. The multivariate aspects of the model are reported under the heading "Canonical Discriminant Functions." Note that the discriminant function is highly significant (.0000) and displays a canonical correlation of .8223487. One interprets this correlation by squaring it

TABLE 3.4.
Test for Equality of Group Means

WILKS' LAMBDA (U-STATISTIC) AND UNIVARIATE F-RATIO
WITH 1 AND 26 DEGREES OF FREEDOM

VARIABLE	WILKS' LAMBDA	F	SIGNIFICANCE
X1	0.54053	22.10	0.0001
X2	0.86683	3.994	0.0562
X3	0.42479	35.21	0.0000
X4	0.98412	.4195	0.5229
X5	0.86110	4.194	0.0508
X6	0.94833	1.417	0.2447

TABLE 3.5   Results from Step 1 of Stepwise Discriminant Analysis Model

```
AT STEP    1, X3        WAS INCLUDED IN THE ANALYSIS.
                                  DEGREES OF FREEDOM   SIGNIF.    BETWEEN GROUPS
WILKS' LAMBDA         0.42479       1    1     26.0
EQUIVALENT F         35.2066             1     26.0   0.0000

MINIMUM D SQUARED     5.05530                                         0          1
EQUIVALENT F         35.2066             1     26.0   0.0000

---------------- VARIABLES IN THE ANALYSIS AFTER STEP    1 ----------------

VARIABLE  TOLERANCE   F TO REMOVE    D SQUARED        BETWEEN GROUPS
X3        1.0000000     35.207

---------------- VARIABLES NOT IN THE ANALYSIS AFTER STEP    1 ----------------
                         MINIMUM
VARIABLE  TOLERANCE   TOLERANCE  F TO ENTER    D SQUARED      BETWEEN GROUPS
X1        0.9967657   0.9967657    7.8031       7.798461        0          1
X2        0.7694328   0.7694328     .38401
X4        0.9959233   0.9959233     .43219
X5        0.9121816   0.9121816    6.4873       7.335891        0          1
X6        0.9977694   0.9977694     .33898

F STATISTICS AND SIGNIFICANCES BETWEEN PAIRS OF GROUPS AFTER STEP    1
EACH F STATISTIC HAS    1 AND       26.0 DEGREES OF FREEDOM.
                   GROUP       0
```

$(.8223487)^2 = .6763$ and concluding that 67.63 percent of the variance in the dependent variable (X_{10}) can be accounted for (explained) by this model, which includes only three independent variables. The standardized canonical discriminant function coefficients are the weights that will be used in validating the function (stage two). The loadings are reported under the heading "Structure Matrix" and are ordered from top to bottom by the size of the loading. Group centroids are reported at the bottom of the table; they represent the mean of the individual Z-scores for each group. The loadings will be discussed later under the interpretation phase.

Group centroids can be used to interpret discriminant function results from a global or overall perspective. Reference to Table 3.6 reveals that the group centroid for the nonrecipients (group 0) is −1.49602, while that for recipients (group 1) is 1.29655; recall that these are reported in the form of Z-scores. These aggregate Z-scores (group centroids) can be interpreted as the number of standard deviations each group is away from the average of both (all) groups (with standardization, the average of all groups is zero). Looking at Figure 3.5, we can see a plot of the centroids showing each group's deviation from the overall mean of the two groups. To show that the overall mean is zero, multiply the number in each group by its centroid and add the result (e.g., −1.49602 × 13 + 1.29655 × 15 = 0).

The *second stage* in a discriminant analysis is *validation* of the function. To accomplish this, we must develop a classification matrix to assess the predictive accuracy of the function. The computer program develops the classification matrix (if requested) for both the analysis and holdout samples (the results are shown in Table 3.7). Before devel-

TABLE 3.6 HATCO Two-Group Stepwise Discriminant Analysis Results

SUMMARY TABLE

STEP	ACTION ENTERED REMOVED	VARS IN	WILKS' LAMBDA	SIG.	MINIMUM D SQUARED	SIG.	BETWEEN GROUPS	LABEL
1	X3	1	.42479	.0000	5.05530	.0000	0 1	
2	X1	2	.32374	.0000	7.79846	.0000	0 1	

CANONICAL DISCRIMINANT FUNCTIONS

FUNCTION	EIGENVALUE	PERCENT OF VARIANCE	CUMULATIVE PERCENT	CANONICAL CORRELATION	::	AFTER FUNCTION	WILKS' LAMBDA	CHI-SQUARED	D.F.	SIGNIFICANCE
					::	0	0.3237426	28.195	2	0.0000
1*	2.08887	100.00	100.00	0.8223487	::					

* MARKS THE 1 CANONICAL DISCRIMINANT FUNCTIONS REMAINING IN THE ANALYSIS.

STANDARDIZED CANONICAL DISCRIMINANT FUNCTION COEFFICIENTS

```
           FUNC  1

X1         0.59405
X3         0.77135
```

STRUCTURE MATRIX:
POOLED WITHIN-GROUPS CORRELATIONS BETWEEN DISCRIMINATING VARIABLES
AND CANONICAL DISCRIMINANT FUNCTIONS
(VARIABLES ORDERED BY SIZE OF CORRELATION WITHIN FUNCTION)

```
           FUNC  1

X3         0.80514
X1         0.63792
X2        -0.37942
X5         0.18990
X4        -0.05513
X6         0.02404
```

CANONICAL DISCRIMINANT FUNCTIONS EVALUATED AT GROUP MEANS (GROUP CENTROIDS)

```
GROUP      FUNC  1

  0       -1.49602
  1        1.29655
```

FIGURE 3.5 Plot of group centroids ($\bar{Z}$)

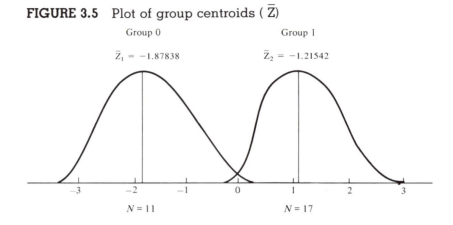

oping the classification matrix, the cutting score must be determined. Recall that the cutting score is the criterion against which each individual's discriminant Z-score is judged to determine into which group the individual should be classified. Several options are possible. In this case, we selected the priors = size option because the group sizes are not equal.

To explain further the importance of cutting score determination, let us focus on how the prior probabilities are used in the calculation of the cutting score. If the two groups were of equal size, the cutting score would simply be the average of the two centroids. Since the two groups are of unequal size, a weighted average must be used to

TABLE 3.7 Classification Matrices for Two-Group Discriminant Function for Both Analysis and Holdout Samples

CLASSIFICATION RESULTS FOR CASES SELECTED FOR USE IN THE ANALYSIS –

ACTUAL GROUP	NO. OF CASES	PREDICTED GROUP MEMBERSHIP 0	1
GROUP 0	13	13 100.0%	0 0.0%
GROUP 1	15	2 13.3%	13 86.7%

PERCENT OF "GROUPED" CASES CORRECTLY CLASSIFIED: 92.86%

(13 + 15 / 28 = 92.86%)

CLASSIFICATION RESULTS FOR CASES NOT SELECTED FOR USE IN THE ANALYSIS –

ACTUAL GROUP	NO. OF CASES	PREDICTED GROUP MEMBERSHIP 0	1
GROUP 0	7	6 85.7%	1 14.3%
GROUP 1	15	3 20.0%	12 80.0%

PERCENT OF "GROUPED" CASES CORRECTLY CLASSIFIED: 81.82%

(6 + 12 / 22 = 81.82%)

account for the differences in variance. The weighted average is calculated as follows:

$$Z_{cu} = \frac{N_A Z_A + N_B Z_B}{N_A + N_B}$$

where

Z_{cu} = critical cutting score value for unequal group sizes
N_A = number in group A
N_B = number in group B
Z_A = centroid for group A
Z_B = centroid for group B

By substituting the appropriate numbers in the formula, we can obtain the critical cutting score (assuming equal costs of misclassification).

$$Z_{cu} = \frac{(13)(-1.49602) + (15)(1.29655)}{28} = 0$$

Group sizes used in the preceding calculation are based on the data set used in the analysis sample and do not include the holdout sample. It should be noted that by using this approach, the cutting score will always be zero (0) when the data are standardized. Since the critical cutting score is 0, the procedures for classifying individuals are as follows:

1. Classify an individual as a nonrecipient if his or her discriminant score is negative.
2. Classify an individual as a recipient if his or her discriminant score is positive.

Using these criteria, the computer program developed classification matrices for the observations included in the development of the discriminant function (analysis sample), as well as those in the holdout sample. The results are shown in Table 3.7. The 92.86 percent accuracy for the analysis sample is higher than the 81.82 percent accuracy for the holdout sample, as anticipated. This demonstrates the upward bias in not using a holdout sample to validate the discriminant function.

The 81.82 percent classification accuracy is quite high. For illustrative purposes, however, let us compare it with the priori chance of classifying individuals correctly without the discriminant function. The proportional chance criterion is the appropriate chance model to use for our HATCO example. We have unequal group sizes and we want to identify members of both groups correctly. The formula is:

$$C_{pro} = p^2 + (1 - p)^2$$

where

C_{pro} = the proportional chance criterion
  p = proportion of individuals in group 1
$1 - p$ = proportion of individuals in group 2

Substituting the appropriate numbers in the formula, we obtain:

$$C_{pro} = (.46)^2 + (.54)^2$$
$$= .21 + .29$$
$$= 50\%$$

The maximum chance criterion can also be calculated. The maximum chance criterion is simply the percentage correctly classified if all observations were placed in the group with the greatest probability of occurrence. Since one group occurs 54 percent of the time, we could be correct 54 percent of the time if we assigned all observations to this group. Since the maximum chance criterion is larger than the proportional chance criterion, our model should outperform the 54 percent level.

The classification accuracy of 81.82 percent is substantially higher than the proportional chance criterion of 50 percent and the maximum chance criterion of 54 percent. Note that the proportional chance criterion is compared to the percentage correctly classified in the holdout sample, which reduces the upward bias caused by classifying the same individuals as used in computing the function. Remember that caution should always be used in the application of a holdout sample to such a small data set. The holdout sample was used in this case not as an example of good research technique for small data sets, but solely for illustrative purposes. If such results were obtained from a much larger sample size, this discriminant function would be considered a valid predictor of recipients versus nonrecipients.

After validating the function, the next phase is *interpretation*. This phase involves examining the function to determine the relative importance of each independent variable in discriminating between the groups. Recall that Table 3.6 contains the discriminant weights and loadings for the function. The independent variables were screened by the stepwise procedure, and only two were significant enough to be included in the function—X_1 and X_3. For interpretation purposes, we will rank the independent variables in terms of their relative discriminatory power. The rankings are based on the absolute sizes of either the loadings or the weights. Signs do not affect the rankings; they indicate a positive or negative relationship with the dependent variable. Since the loadings are considered somewhat more valid than the weights, we shall use them in our example. The weights can be used for comparison purposes.

When using the discriminant loadings interpretation approach, we need to know which variables are significant discriminators. With stepwise procedures, this is easy because criteria can be established ahead of time to prevent nonsignificant variables from entering the function. Then all variables included in the function are evaluated. In simultaneous discriminant analysis, however, all variables are entered in the function, and generally any variables exhibiting loadings ± .30 or higher are considered significant. Other considerations are possible, however (see Chapter Six).

The analyst usually is interested in substantive interpretations of the individual variables. This is accomplished by identifying the variables that are significant in the discriminant function and understanding how they are scored (i.e., what high scores and low scores mean). The three significant variables and their scoring procedures are X_1 = self-esteem = high score = high self-esteem; X_2 = locus of control = high score = external locus of control; and X_3 = alienation = high score = not alienated. For more detail on these variables, see Chapter 1. From Table 3.6 we can use the structure matrix information (loadings) and determine the ranking of these variables in terms of their discriminatory value. It reveals that X_3 discriminates the most and X_1 the least. But remember that they are both significant discriminators between recipients and nonrecipients of the outstanding salesperson award. Referring back to Table 3.3, we note that the means for recipients on variables X_1 and X_3 are higher than those for nonrecipients, and that for X_2 the mean is relatively lower. Using this information, along with the knowledge that they are significant discriminators, we can interpret these findings substantively to profile the characteristics of recipients and nonrecipients. Specifically, our findings suggest that recipients exhibit high self-esteem and internal locus of control, and are not alienated. With information such as this, HATCO could select new salespersons that fit the recipients' profile and expect to be more successful in hiring a good sales force.

A Three-Group Illustrative Example

To illustrate the application of a three-group discriminant analysis, we once again use the HATCO data set. The purpose of this example is to determine the relationship between variables X_1 to X_8 and the level of customer satisfaction (variable X_{12}). To test for this relationship, a discriminant analysis is performed using customer satisfaction (X_{12}) as the dependent variable and the social-psychological test scores (variables X_1–X_6) plus the measures of knowledge and motivation (variables X_7 and X_8) as the independent variables. The resulting discriminant model, like the two-group top salesperson model just discussed, is predictive and could help HATCO develop evaluative criteria for potential sales trainees.

The procedures used in this three-group model are similar to those used in the preceding example shown in Exhibit 3.1. SPSS is used, a holdout sample is retained for illustrative purposes, and a stepwise Mahalanobis D^2 procedure with a varimax rotation is applied. The primary difference is in specifying three groups instead of two by replacing "GROUPS=VAR10(0,1)" with "GROUPS=VAR12(1,3)" in the SPSS program statements. (Notice that only the highest and lowest group values are specified.)

In the previous example, we were concerned with discriminating between only two groups, so we were able to develop a single canonical function and cutting score to divide the groups. In this example, it is necessary to develop two separate canonical functions to distinguish between the three groups. The first function separates one group from the other two, and the second separates the remaining two groups. By using the discriminant scores from both functions, we can develop a two dimensional territorial map on which we plot the group centroids and each observation's discriminant scores. The map is used to develop group boundaries that will enable us to classify the holdout observations in much the same way as the cutting score was applied in the two-group case. Such a map is shown in Figure 3.8.

As in the previous example, we begin our analysis by reviewing the group means and standard deviations to see if the groups are significantly different on any single variable. Table 3.8 gives the group means and standard deviations, while Table 3.9 displays Wilks' lambda and univariate F-ratio's (simple ANOVAs) for each of the independent variables. Review of the significance levels of the individual variables shown in Table 3.9 reveals that on a univariate basis, all of the variables except X_4 and X_6 display significant differences between the group means. We do not know whether the differences are between groups 1 and 2, 2 and 3, or 1 and 3. But we do know that significant differences exist.

The stepwise procedure is performed in the same manner as in the two-group example. The data in Table 3.10 show that the first variable to enter the model is X_7. Review of the F To Enter data reveals that of the variables not in the model after step 1, all but X_5 have F values greater than 1 and can thus be expected to be considered for inclusion in the model in later steps (recall that we set our cutoff for F to Enter at 1.0 or larger).

The information provided in Table 3.11 summarizes the five steps of the three-group discriminant analysis. Note that variables X_7, X_3, X_6, X_8, and X_1 all were entered into the discriminant model. By comparing these results with the univariate results shown in Tables 3.8 and 3.9, one can see that it is not always possible to predict from univariate results which variables will be included in a discriminant function.

Since this is a three-group discriminant analysis model, it is necessary to calculate two canonical discriminant functions in order to discriminate between the three groups. The variables are now entered into the canonical discriminant procedure and linear composites are formulated. Note that the discriminant functions are based only on the variables included in the discriminant model (X_1, X_3, X_6, X_7, and X_8). However, after the linear composites are calculated, the procedure correlates all eight independent variables with the canonical discriminant functions to develop a structure (loadings) matrix. This procedure enables us to see where the discrimination would occur if all eight variables were included in the model (that is, if none were excluded by multicollinearity or lack of statistical significance). Much of this discussion is based upon concepts presented in the chapters on canonical correlation (Chapter Five) and factor analysis (Chapter Six), so

TABLE 3.8 Group Means and Standard Deviations for HATCO Three-Group Discriminant Function

GROUP MEANS

X12	X1	X2	X3	X4	X5	X6	X7	X8
1	4.28358	4.73825	14.14100	10.24158	4.54350	5.52742	7.37658	7.74067
2	6.12167	6.18333	14.66489	11.01111	6.13733	5.09922	9.19289	9.70467
3	9.39080	3.76800	18.37640	10.76590	6.56140	5.47070	10.84670	10.95720
TOTAL	6.46471	4.84481	15.65935	10.63413	5.65716	5.38481	9.02329	9.34845

GROUP STANDARD DEVIATIONS

X12	X1	X2	X3	X4	X5	X6	X7	X8
1	1.90385	1.51746	1.84766	1.65066	1.11235	1.19927	0.96513	1.06885
2	1.46312	3.04327	2.39572	1.76195	1.59156	1.12305	0.73566	0.81905
3	2.04672	2.12081	1.07472	3.01705	0.74727	2.26990	0.84440	1.66260
TOTAL	2.81937	2.36473	2.60929	2.16049	1.46103	1.56349	1.70333	1.83352

TABLE 3.9
Test for Equality of
Group Means
Wilks' Lambda
(U-Statistic) and
Univariate F-Ratio
with 2 and 31
Degrees of
Freedom

```
WILKS' LAMBDA (U-STATISTIC) AND UNIVARIATE F-RATIO
WITH   2 AND        28 DEGREES OF FREEDOM
```

VARIABLE	WILKS' LAMBDA	F	SIGNIFICANCE
X1	0.39712	21.25	0.0000
X2	0.83395	2.788	0.0787
X3	0.45955	16.46	0.0000
X4	0.97642	.3381	0.7160
X5	0.60751	9.045	0.0009
X6	0.98566	.2037	0.8169
X7	0.24119	44.05	0.0000
X8	0.42449	18.98	0.0000

you may wish either to skim these chapters now or to come back to this topic after you have read them.

The linear composites are similar to a regression line (i.e., they are a linear combination of variables). Just as a regression line tries to explain the maximum amount of variation in its dependent variable, these linear composites attempt to explain the variations or differences in discriminant dependent categorical variables. The first linear composite is developed to explain (account for) the largest amount of variation (difference) in the discriminant groups. The second linear composite, which is orthogonally independent of the first, explains the largest percentage of the remaining (residual) variance after the variance for the first composite function is removed.

After the linear composites are developed, they can be "rotated" to redistribute the variance (this concept will be more fully explained in Chapter Six). Basically, rotation preserves the original structure and reliability of the discriminant models while at the same time making them easier to interpret substantively. In the present application, we choose the most widely used procedure of varimax rotation.

Table 3.12 contains the results for the canonical discriminant functions. Note that the functions are statistically significant, as measured

TABLE 3.10 First Step in Three-Group Discriminant Function

```
AT STEP    1, X7        WAS INCLUDED IN THE ANALYSIS.

                                 DEGREES OF FREEDOM    SIGNIF.   BETWEEN GROUPS
                                  1    2
WILKS' LAMBDA        0.24119                    28.0
EQUIVALENT F         44.0460           2        28.0    0.0000

MINIMUM D SQUARED    3.64801                                      2          3
EQUIVALENT F         17.2800      1            28.0    0.0003


---------------- VARIABLES IN THE ANALYSIS AFTER STEP    1 ----------------


VARIABLE   TOLERANCE   F TO REMOVE    D SQUARED        BETWEEN GROUPS

X7         1.0000000     44.046


---------------- VARIABLES NOT IN THE ANALYSIS AFTER STEP    1 ----------------
                        MINIMUM
VARIABLE   TOLERANCE   TOLERANCE   F TO ENTER    D SQUARED      BETWEEN GROUPS
X1         0.9488504   0.9488504     2.9273      4.690654         1        2
X2         0.9081858   0.9081858     4.3235      4.400225         1        2
X3         0.9809406   0.9809406     7.7392      4.737995         1        2
X4         0.8874617   0.8874617     1.0258      4.284716         2        3
X5         0.7724625   0.7724625      .84678
X6         0.8595584   0.8595584     2.0182      3.921183         2        3
X8         0.9355201   0.9355201     1.9177      3.945953         2        3
```

TABLE 3.11 Summary Table After Five Steps in HATCO Three-Group Discriminant
Function

SUMMARY TABLE

STEP	ACTION ENTERED REMOVED	VARS IN	WILKS' LAMBDA	SIG.	MINIMUM D SQUARED	SIG.	BETWEEN GROUPS		LABEL
1	X7	1	.24119	.0000	3.64801	.0003	2	3	
2	X3	2	.15330	.0000	4.73800	.0002	1	2	
3	X6	3	.13136	.0000	6.22273	.0002	1	2	
4	X8	4	.10460	.0000	8.94129	.0000	1	2	
5	X1	5	.08772	.0000	9.06520	.0001	1	2	

by the chi-square statistic, and that the first function accounts for 93.88 percent of the variance. Below the summary information are the discriminant function coefficients (weights) for the predictive models and the unrotated structure matrix. Since we have chosen to rotate the linear composites, there is no need to interpret the structure loadings at this point. After the first function is extracted, the chi-square is recalculated. The results show that significant differences are present in the remaining variance. If more groups are used in the model, (e.g., a four-group discriminant analysis), additional canonical discriminant functions would be possible, and the chi-square statistic would be continually recalculated on the residual variance to test for significant differences until the maximum number of canonical discriminant functions were extracted (maximum number = number of groups minus one).

Validation of Canonical Discriminant Functions

Before discussing the interpretation of the functions, we must determine that the functions are valid predictors. This is accomplished in the same fashion as with the two-group discriminant model, that is, by examination of the classification matrices. Reference to Table 3.14 shows that the discriminant functions are valid. The hit ratio for the analysis sample is 100 percent, whereas that for the holdout sample is 73.68 percent. These results again point out the upward bias that is likely without a holdout sample. Both of these hit ratios are high. But to evaluate the effectiveness of the model completely, we can again compare these hit ratios to the maximum chance and proportional chance criteria.

The maximum chance criterion is simply the hit ratio obtained if we assign all of the observations to the group with the highest probability of occurrence. In the present sample of 50 observations, 17 were in group 1, 16 in group 2, and 17 in group 3. Based on this information, we can see that the highest probability would be 34 percent (17/50 = 34%). Thus, the maximum chance criterion would be 34 percent because, if we simply classified all observations into group 1 or 3, we would be correct 34 percent of the time. Based on the maximum chance criterion, therefore, our model is very good.

The proportional chance criterion is calculated by squaring the proportions of each group, as shown in Table 3.15.

TABLE 3.12 Multivariate Results for HATCO Three-Group Discriminant Function

CANONICAL DISCRIMINANT FUNCTIONS

FUNCTION	EIGENVALUE	PERCENT OF VARIANCE	CUMULATIVE PERCENT	CANONICAL CORRELATION	: AFTER : FUNCTION	WILKS' LAMBDA	CHI-SQUARED	D.F.	SIGNIFICANCE
					0				
1*	6.87249	93.88	93.88	0.9343316	1	0.0877205	63.274	10	0.0000
2*	0.44806	6.12	100.00	0.5562563		0.6905789	9.6259	4	0.0472

* MARKS THE 2 CANONICAL DISCRIMINANT FUNCTIONS REMAINING IN THE ANALYSIS.

STANDARDIZED CANONICAL DISCRIMINANT FUNCTION COEFFICIENTS

	FUNC 1	FUNC 2
X1	0.37288	0.44953
X3	0.49049	0.88933
X6	-0.50483	-0.61601
X7	-0.78966	-0.17005
X8	0.22560	-0.99480

STRUCTURE MATRIX:
POOLED WITHIN-GROUPS CORRELATIONS BETWEEN DISCRIMINATING VARIABLES
AND CANONICAL DISCRIMINANT FUNCTIONS
(VARIABLES ORDERED BY SIZE OF CORRELATION WITHIN FUNCTION)

	FUNC 1	FUNC 2
X7	0.67437*	-0.21495
X1	0.46787*	-0.17504
X8	0.43744*	-0.30133
X2	-0.06332	-0.61071*
X3	0.39041	0.53567*
X5	0.31038	-0.43405*
X6	-0.00563	-0.17886*
X4	-0.00895	-0.10352*

TABLE 3.13 Results for Varimax Rotated Three-Group Discriminant Function

```
VARIMAX ROTATION TRANSFORMATION MATRIX

                  FUNC   1      FUNC   2

% VARIANCE        86.95         13.05

FUNC   1        0.95971      -0.28098
FUNC   2       -0.28098      -0.95971


ROTATED CORRELATIONS BETWEEN DISCRIMINATING VARIABLES
                  AND CANONICAL DISCRIMINANT FUNCTIONS
(VARIABLES ORDERED BY SIZE OF CORRELATION WITHIN FUNCTION)

                  FUNC   1      FUNC   2

X7              0.70760*      0.01681
X8              0.50449*      0.16628
X5              0.41984*      0.32935
X1              0.39984*     -0.29945

X3              0.22417      -0.62379*
X2              0.11083       0.60390*
X6             -0.05566      -0.17007*
X4              0.02050       0.10186*


ROTATED STANDARDIZED DISCRIMINANT FUNCTION COEFFICIENTS
                  BASED ON ROTATION OF STRUCTURE MATRIX

                  FUNC   1      FUNC   2

X1              0.23407      -0.52755
X3              0.22085      -0.99132
X6             -0.65758      -0.44935
X7              0.80563      -0.05868
X8              0.49602       0.89134


CANONICAL DISCRIMINANT FUNCTIONS EVALUATED AT GROUP MEANS (GROUP CENTROIDS)

     GROUP      FUNC   1      FUNC   2

       1       -2.69620       0.36051
       2        0.25395       0.96200
       3        3.00683      -1.29841
```

Since C_{max} is greater than C_{pro}, the maximum chance criterion is the criterion to outperform. The hit ratio of 73.68 percent exceeds the C_{max} criterion substantially, so we again conclude that the discriminant model is valid.

Table 3.16 contains additional classification data for the three-group discriminant analysis. The observation number is shown on the left side of the table. The "Sel" column indicates whether an observation was selected to be included in the analysis or holdout group. The "yes" indicates that the observation is in the analysis sample and a "no" indicates that it is in the holdout sample. In the "Actual Group" column, a 1 indicates group 1 (low satisfaction), a 2 group 2 (medium satisfaction), and a 3 group 3 (high satisfaction). The asterisk beside the numbers indicates that a particular observation was misclassified by the discriminant function. The "Highest Probability" column shows the group assignment of an observation by the model that is most likely using the discriminant function whereas the "2nd Highest" column shows the second most likely assignment using the discriminant function. The discriminant scores for each observation on each function are shown on the right side of the table. (*Note:* When there is one

TABLE 3.14 Classification Matrices for Analysis Sample and Holdout Sample for HATCO Three-Group Discriminant Function

CLASSIFICATION RESULTS FOR CASES SELECTED FOR USE IN THE ANALYSIS -

ACTUAL GROUP	NO. OF CASES	PREDICTED GROUP MEMBERSHIP 1	2	3
GROUP 1	12	12 100.0%	0 0.0%	0 0.0%
GROUP 2	9	0 0.0%	9 100.0%	0 0.0%
GROUP 3	10	0 0.0%	0 0.0%	10 100.0%

PERCENT OF "GROUPED" CASES CORRECTLY CLASSIFIED: 100.00%

(12 + 9 +10 / 31 = 100 %)

CLASSIFICATION RESULTS FOR CASES NOT SELECTED FOR USE IN THE ANALYSIS -

ACTUAL GROUP	NO. OF CASES	PREDICTED GROUP MEMBERSHIP 1	2	3
GROUP 1	5	5 100.0%	0 0.0%	0 0.0%
GROUP 2	7	2 28.6%	3 42.9%	2 28.6%
GROUP 3	7	0 0.0%	1 14.3%	6 85.7%

PERCENT OF "GROUPED" CASES CORRECTLY CLASSIFIED: 73.68%

(5 + 3 + 6 / 19 = 73.68%)

discriminant function (two groups), classification of cases is based on the values for the single function. When there are several groups, a case's values on all functions are considered simultaneously.)

Interpretation of Multiple Discriminant Analysis

The final stage of any discriminant analysis is always interpretation. Interpretation of solutions usually involves two approaches. One is to evaluate and plot the group centroids. The other is to evaluate and plot the loadings of the linear composites (sometimes referred to as *attribute vectors*).

TABLE 3.15 Calculation of Chance Criteria	Maximum chance criteria
	Group 1: 17/50 = 34 percent
	Group 2: 16/50 = 32 percent
	Group 3: 17/50 = 34 percent
	$C_{max} = 34$ percent
	Proportional chance criteria
	$C_{pro} = (p_1)^2 + (p_2)^2 + (p_3)^2$
	$C_{pro} = (.34)^2 + (.32)^2 + (.34)^2$
	$C_{pro} = .3336$ or 33.36 percent

ultiple Discriminant Analysis

TABLE 3.16 Classification Data for Three-Group Discriminant Function

CASE SEQNUM	MIS VAL	SEL	ACTUAL GROUP	HIGHEST PROBABILITY GROUP P(D/G) P(G/D)			2ND HIGHEST GROUP P(G/D)		DISCRIMINANT SCORES...	
1	YES		1	1	0.6329	0.9513	2	0.0437	-1.9706	-0.1255
2	NO		2	2	0.3202	0.9780	3	0.0213	1.6363	1.5677
3	YES		2	2	0.9539	0.9665	1	0.0320	0.0073	0.7786
4	YES		1	1	0.9526	0.9951	2	0.0049	-2.9112	0.5863
5	YES		2	2	0.1498	0.9979	1	0.0021	0.5178	2.8927
6	YES		1	1	0.8541	0.9741	2	0.0259	-2.2282	0.0500
7	YES		1	1	0.8399	0.9609	2	0.0391	-2.2188	0.7082
8	NO		2	2	0.1167	0.8062	1	0.1938	-1.0330	2.5865
9	YES		1	1	0.0444	0.9999	2	0.0001	-4.5199	2.0641
10	YES		1	1	0.5475	0.8270	2	0.1730	-1.6741	1.7604
11	YES		1	1	0.8189	0.9980	2	0.0020	-3.0630	-0.1544
12	NO		1	1	0.3698	0.6435	2	0.3565	-1.3449	0.7650
13	NO		1	1	0.9009	0.9979	2	0.0021	-3.1528	0.3476
14	NO		1	1	0.1431	0.9925	2	0.0075	-3.1121	2.2879
15	YES		2	2	0.5586	0.9642	3	0.0352	1.3329	0.9802
16	YES		2	2	0.3211	0.8930	3	0.1069	1.7612	0.9767
17	YES		1	1	0.2074	0.9999	2	0.0001	-4.4547	0.5919
18	YES		2	2	0.5501	0.9792	1	0.0207	-0.0890	2.0001
19	YES		2	2	0.3117	0.9702	1	0.0298	-0.2949	2.3869
20	YES		1	1	0.5579	0.9934	2	0.0066	-2.9749	1.4043
21	NO	**	2	1	0.2488	0.9072	2	0.0928	-2.1571	1.9391
22	YES		1	1	0.6364	0.8777	2	0.1223	-1.7866	0.6369
23	YES		2	2	0.1209	0.5109	1	0.4862	-0.8095	-0.7973
24	YES		3	3	0.2646	0.7286	2	0.2711	1.4694	-0.7552
25	NO	**	2	1	0.0189	0.6339	2	0.3661	-1.7925	3.0289
26	NO		3	3	0.8143	0.9977	2	0.0023	2.5277	-1.7242
27	NO	**	3	2	0.6490	0.9797	3	0.0195	1.1785	1.0612
28	NO	**	2	3	0.3947	0.9704	2	0.0295	1.6789	-1.6075
29	YES		3	3	0.7470	0.9999	2	0.0001	3.6271	-1.7441
30	YES		3	3	0.1875	0.9993	2	0.0007	4.3282	-0.0326
31	YES		1	1	0.1924	0.9279	2	0.0720	-1.6291	-1.1083
32	NO		3	3	0.0572	1.0000	2	0.0000	5.3526	-1.7668
33	NO	**	2	3	0.2146	0.8436	2	0.1559	1.2528	-1.3283
34	YES		3	3	0.8993	0.9987	2	0.0013	3.3330	-0.9729
35	YES		3	3	0.5571	0.9993	2	0.0007	2.5593	-2.2553
36	NO		2	2	0.7407	0.9938	3	0.0045	0.9059	1.3807
37	NO		1	1	0.9914	0.9945	2	0.0055	-2.8108	0.2965
38	YES		2	2	0.5992	0.9330	1	0.0603	-0.0601	-0.0001
39	YES		3	3	0.5745	0.9999	2	0.0001	3.4096	-2.2712
40	YES		2	2	0.2970	0.8927	1	0.0857	-0.0802	-0.5600
41	NO		3	3	0.3378	0.9973	2	0.0027	1.9583	-2.3332
42	YES		3	3	0.1188	1.0000	2	0.0000	4.9505	-0.6096
43	YES		3	3	0.5479	0.9987	2	0.0013	2.3625	-2.1861
44	NO		1	1	0.0103	0.8568	2	0.1336	-1.1563	-2.2439
45	NO		3	3	0.4255	1.0000	2	0.0000	3.9774	-2.1741
46	NO		3	3	0.7925	0.9998	2	0.0002	3.6889	-1.2875
47	YES		3	3	0.2633	0.8650	2	0.1347	1.3737	-1.2571
48	YES		1	1	0.3414	0.9983	2	0.0017	-2.9234	-1.0878
49	YES		3	3	0.8738	0.9903	2	0.0097	2.6735	-0.9000
50	NO		3	3	0.5944	0.9999	2	0.0001	3.3643	-2.2537

PLOTTING GROUP CENTROIDS.   Group centroids can be plotted to demonstrate the results of a multiple discriminant analysis from a global perspective. Plots are usually prepared for the first two or three discriminant functions (assuming they are statistically significant and valid predictive functions). Using the results for the three-group canonical discriminant function shown at the bottom of Table 3.13, we can illustrate how group centroids are plotted in reduced discriminant space (so called because all of the functions and thus all of the variances are not plotted).

The values are plotted in Figure 3.6, showing the position of each group. This graphic presentation shows that there appear to be differences in the groups on the eight predictor variables. But it does not satisfactorily explain what these differences are. Circles enclosing the distribution of the observations around their respective centroids can be drawn to clarify group differences further, but this procedure is beyond the scope of this text (see [2], pp. 411–413).

PLOTTING THE DISCRIMINANT LOADINGS.   To depict the differences in the groups on the eight predictor variables, the analyst can compare and

FIGURE 3.6 Plot of group centroids in reduced discriminant space.

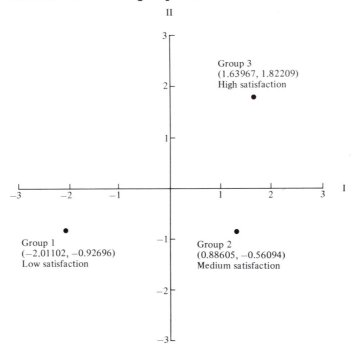

contrast the individual group means, as shown earlier with the two-group example. Another approach that is often used with multiple discriminant analysis solutions is to plot the discriminant loadings. The simplest approach is to plot actual loadings on a graph. One can plot either the unrotated loadings from the structure matrix information given in Table 3.12 or the rotated correlations (loadings) provided in Table 3.13. The preferred approach would be to plot the rotated loadings. An even more accurate approach, however, involves what is called *stretching* the vectors.

Before explaining the process of stretching, we must define what a vector is in this context. A *vector* is merely a straight line drawn from the origin (center) of the graph to the coordinates of a particular variable vector. The length of each vector is therefore an indicator of the relative importance of each variable in discriminating among the groups. To stretch a vector, one multiplies each discriminant loading (after rotation) by its respective univariate F-value. For example, the discriminant loadings for variable X_5 are .41984 for function 1 and .32935 for function 2. These values are multiplied by 9.045 (the univariate F-value for X_5 from Table 3.9) to obtain the appropriate coordinates for plotting the vector for variable X_5,—Machiavellianism.

The plotting process always involves all of the variables included in the model by the stepwise procedure (in our example, variables X_1, X_3, X_6, X_7, and X_8). But the analyst frequently plots the variables not included in the discriminant function if their respective univariate F-ratios are statistically significant. This procedure shows the importance of collinear variables that were not included in the final stepwise model. The significance of including these variables in the plotting process is demonstrated in our example. Variable X_6 is included in our final model because it is significant with a small portion of the residual variance. Variable X_5, for example, is much more important but was eliminated due to collinearity.

The plots of the stretched attribute vectors for the rotated discriminant loadings are shown in Figure 3.7. By plotting the vectors using this procedure, one causes them to point to the groups having the highest mean level on a respective predictor and away from the groups having the lowest mean score. Thus, interpretation of the plots in Figure 3.7 indicates that the first discriminant function is the primary source of differences between the highly satisfied (group 3) and less satisfied customers (group 2) versus the least satisfied customers (group 1). The directions of the vectors indicate that this dimension (function 1) corresponds most closely to variables X_8 (Motivation) and X_5 (Machiavellianism), and fairly closely to X_7 (Knowledge). Thus, the distinguishing characteristics of salespersons with satisfied customers are that they are relatively more highly motivated, more knowledgeable, and more Machiavellian in their outlook.

FIGURE 3.7 Plot of stretched attribute vectors (variables) in reduced discriminant space.

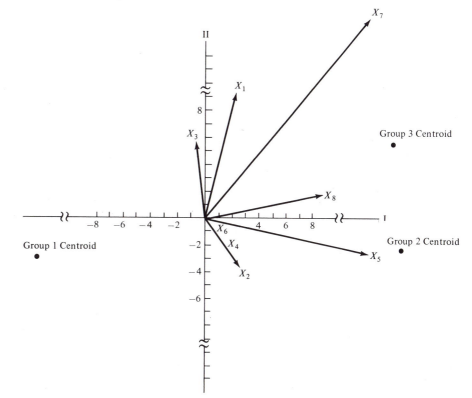

Visual inspection of the plots for function 2 indicates that variables X_1, X_2, and X_3 are closely associated with it, and that it is the primary source of differences between salespersons with highly satisfied customers and all other salespersons (groups 2 and 3). Moreover, the length and direction of the vectors indicate that the distinguishing characteristics of the salespersons with the most highly satisfied customers are that they have high self-esteem and an internal locus of control, and are not alienated.

It should be noted that the group centroids shown in Figure 3.7 were also stretched by multiplying them by the approximate F-value associated with each of the two discriminant functions. If the loadings are stretched one must also stretch the centroids in order to plot them accurately on the same graph. The approximate F-values for each discriminant function can be obtained by multiplying each discriminant function's eigenvalue (see Table 3.12) by $(n - k) / (k - 1)$, as shown:

$$F_1 = \frac{50 - 3}{3 - 1}\,(3.89058) = 91.42863$$

$$F_2 = \frac{50 - 3}{3 - 1}\,(.49417) = 11.612995$$

where

n = sample size
k = number of groups

For more details on this procedure see [2], pp. 415–416.

For those who do not wish to stretch the attribute vectors and centroids, there is an alternative in the "territorial maps" provided by the computer program that is part of the SPSS package. It does not include the vectors, but it does plot the centroids, as shown in Figure 3.8. Each group centroid is designated by the asterisk, and group boundaries are established by a group's corresponding number. For example, the group 3 centroid is bounded by 3's, the group 2 centroid by 2's, and so forth.

At this point, you may feel that all of this plotting is too complicated. What other procedure is simpler, yet effective? The easiest approach is to use the rotated correlations (loadings) provided in Table 3.13 and the group centroids. The asterisks beside the loadings show which are significant for each function (e.g., X_7, X_8, X_5, and X_1) for function 1. Note that X_6 is not significant but has an asterisk on function 2 because it is higher here than on function 1. Thus, just as with the plotting, we have identified variables X_5 and X_8 as being associated with the first function. To determine which groups each function discriminates between, simply look at the group centroids and see where the differences lie. This is merely a distance assessment, not a statistical measure, but it usually is sufficient. For example, looking at function 1, we see that the centroid for group 1 is -2.69620, for group 2 it is

FIGURE 3.8 Territorial map for three-group discriminant function.

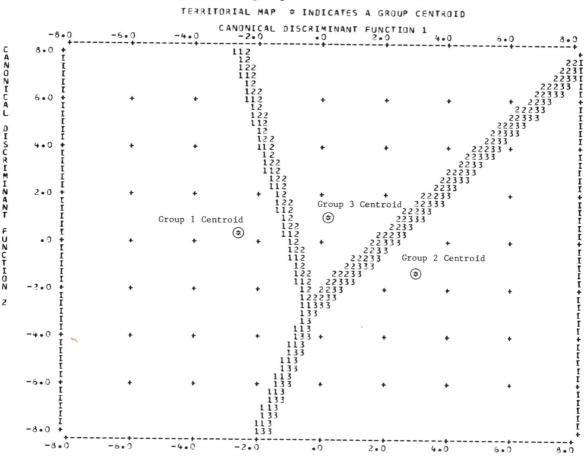
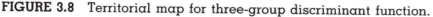

.25395, and for group 3 it is 3.00689. From this we conclude that the primary source of the differences for this function is between groups 2 and 3 versus group 1. A similar approach can be used with function 2. However, since function 1 represents substantially more variance than function 2, one should also note the contribution of variable X_6 to it, since this variable loaded almost as high on function 1 as on function 2.

Summary

The underlying nature, concepts, and approach to multiple discriminant analysis have been presented. Basic guidelines for its application and interpretation were included to clarify further the methodological concepts. Illustrative examples for both two-group and three-group solutions were presented based on the HATCO data base. These applications demonstrated the major points you need to be familiar with in applying discriminant analysis.

Multiple discriminant analysis helps you to understand and explain research problems that involve a single categorical dependent variable and several metric independent variables. A mixed data set (both metric and nonmetric) is also possible for the independent variables if the nonmetric variables are dummy coded (0–1). The result of a discriminant analysis can assist you in profiling the intergroup characteristics of the subjects and in assigning them to their appropriate groups. Potential applications of discriminant analysis to both business and nonbusiness problems are numerous.

Some of the concepts presented in this chapter are based upon material discussed in Chapters 5 and 6. Thus, it is recommended that this chapter be studied in conjunction with these two.

QUESTIONS

1. How would you differentiate between multiple discriminant analysis, regression analysis, and analysis of variance?

2. What criteria could you use in deciding whether or not to stop a discriminant analysis after the derivation stage? After the validation stage?

3. What procedure would you follow in dividing your sample into analysis and holdout groups? How would you change this procedure if your sample consisted of fewer than 100 individuals or objects?

4. How would you determine the optimum cutting score?

5. How would you determine whether or not the classification accuracy of the discriminant function is sufficiently high relative to chance classification?

6. How does a two-group discriminant analysis differ from a three-group analysis?

7. Why would an analyst stretch the loadings and centroid data in plotting a discriminant analysis solution?

REFERENCES

1. Crask, M., and W. Perreault, "Validation of Discriminant Analysis in Marketing Research," *Journal of Marketing Research*, vol. XIV (February 1977), pp. 60–68.
2. Dillon, William R. and M. Goldstein, *Multivariate Analysis: Methods and Applications.* New York: Wiley, 1984.
3. Dixon, W. J., *Biomedical Computer Programs.* Los Angeles: University of California Press, 1979.
4. Frank, R. E., W. F. Massey, and D. G. Morrison, "Bias in Multiple Discriminant Analysis," *Journal of Marketing Research*, vol. 2, no. 3 (August 1965), pp. 250–58.
5. Green, Paul F., and Donald S. Tull, *Research for Marketing Decisions.* Englewood Cliffs, N.J.: Prentice-Hall, Inc., 1979.
6. Green, P. E., *Analyzing Multivariate Data.* Hinsdale, Ill.: Holt, Rinehart & Winston, Inc., 1978.
7. Green, P. E., and J. Douglas Carroll, *Mathematical Tools for Applied Multivariate Analysis.* New York: Academic Press, 1978.
8. Harris, R. J., *A Primer of Multivariate Statistics.* New York: Academic Press, 1975.
9. Massey, W. F., "Bayesian Multiple Discriminant Analysis," Working Paper No. 58, Graduate School of Business, Stanford University, July 1965.
10. Morrison, Donald G., "On the Interpretation of Discriminant Analysis," *Journal of Marketing Research*, vol. 6, no. 2 (May 1969), pp. 156–63.
11. McGraw-Hill, *SPSS-X Users's Guide*, 2nd ed. Chicago, 1986.
12. McGraw-Hill, *SPSS-X Advanced Statistics Guide.* Chicago, 1986.
13. SAS Institute, *SAS User's Guide: Basics*, Version 5 ed. Cary, N. C., 1985.
14. SAS Institute, Inc., *SAS User's Guide: Statistics*, Version 5 ed. Cary, N. C., 1985.

SELECTED READINGS

Alternative Approaches for Interpretation of Multiple Discriminant Analysis in Marketing Research

WILLIAM D. PERREAULT, JR.,
DOUGLAS N. BEHRMAN,
GARY M. ARMSTRONG

Introduction

The essence of marketing management, and more specifically marketing research, is to bring order to the many factors that influence the informed choice of a marketing mix. This issue is perhaps more pronounced in the management of market segmentation strategies, which rely on identifying important but sometimes subtle distinctions among markets. As a result, marketers have aggressively developed or adopted analytical models that help them to evaluate differences among market segments. While many useful approaches have been developed, multiple discriminant analysis is one of the most widely touted of the procedures [24,27,28,31,40].

Although marketing researchers now have two decades of experience with multiple discriminant analysis [4,22] and have beneficially applied it in a wide variety of situations [3–6,12,15,17, 24,26,28,29,37,40,41,45,47], its full potential as an input to marketing decision making has yet to be realized. In particular, most applications

"Alternative Approaches for Interpretation of Multiple Discriminant Analysis in Marketing Research," William D. Perreault, Jr., Douglas N. Behrman, and Gary M. Armstrong, Vol. 7 (1979), pp. 151–173. Reprinted from the *Journal of Business Research*, published by Elsevier North Holland, Inc. The authors are professors of Business Administration at the University of North Carolina, Chapel Hill.

using MDA to *profile* market segments and identify the salient ways in which they differ have relied heavily or totally on interpretive approaches which provide information that is incomplete or even misleading.

Thus, the purpose of this article is to point out the pitfalls of the most frequently used approaches to interpretation and to present a number of new and useful approaches. Specifically, the problems of (1) identifying which of the multiple predictor variables contributes significantly to the multivariate relationships, (2) interpreting the nature of the underlying discriminant functions, and (3) identifying exactly how the groups differ, in a pairwise fashion, on each of the predictors are considered. To develop and clarify more fully the limitations and advantages of different approaches to these interpretive problems, they are applied and illustrated in the context of an innovator segmentation study.

A Study of the Purchase of an Innovation

The diffusion and adoption of new products is critical to a firm's success. Marketers have devoted considerable attention to the diffusion process in general [43] and the innovator segment in particular [16,23,33]. Furthermore, it is an area where discriminant analysis has frequently been used to profile the innovator segment [3,12,17,39,- 42,44,45].

This study concerns the introduction of a major innovation by a large manufacturer of heavy farm equipment.[1] The company recently introduced a new type of equipment that substantially changed the process of harvesting hay. The major custom-

[1] The firm prefers to remain anonymous. The authors appreciate the willingness of management to allow the data to be used in this report.

115

ers for the firm's products are commercial farms. Decision makers concerned with marketing the products of the firm were especially interested in how customers who had purchased the innovation differed from those who bought the traditional offering. While the analysis reported here is concerned with customers' perceptions of select factors that influenced their purchase decisions, the total project was broader in scope and considered such characteristics as customers' information and influence sources, attitudes toward the product, and characteristics of their agribusiness operations. The information was collected to serve as a partial basis for developing a promotional campaign to other potential customers in the innovator segment. This study is chosen for the illustrations in this paper because it demonstrates the problems with traditional approaches to interpretation of discriminant analysis and because the alternative approaches advocated provide interesting insights concerning the contrast between the purchasers of the innovation and purchasers of the established product.

Data for the study were collected by a mail questionnaire which was sent to 499 of the company's customers who had purchased either the conventional equipment or the innovation in the 18 months the product had been in the market. Approximately half of the questionnaires were sent to customers in the company's Midwest sales region and half to customers in its Southeast sales region. Within each region, there was an equal representation of customers who had purchased

the conventional and those who had purchased the new product. A total of 209 questionnaires was returned, 163 (33%) of which were complete with respect to all the variables considered in this study.

The four criterion groups for the discriminant analysis were the four segments defined by the type of product purchased (innovation vs. conventional) and sales region (Southeast vs. Midwest). Region was explicitly considered in evaluating segment differences because previous experience had shown that customers in the different regions often reacted differently to marketing stimuli in the past (due to differences in weather, farm size, and other factors). The independent variables consisted of the customers' indication of how important each of seven characteristics of the machine (listed in Table 1) had been in influencing the purchase decision. The customers rated each factor on a six-point scale anchored by "not important" (1) and "very important" (6). The means and standard deviations of the responses to each of the independent variables by each of the four market segment groups are provided in Table 1. The next section reports the analysis and interpretation of these market segment profiles based on the discriminant function analysis.

Analysis

When the discriminant function analysis is computed for the seven purchase criteria and the four

TABLE 1 Means and Standard Deviations for Predictor Variables by Market Segment Groups

Predictors (Decision Factors)	Segments			
	Conventional Southeast (1)	Innovation Southeast (2)	Conventional Midwest (3)	Innovation Midwest (4)
1. Speed of Operation	4.27	5.08	4.50	5.43
	(1.60)	(1.18)	(1.52)	(0.90)
2. Operating Expense	4.60	5.31	4.44	4.84
	(1.41)	(0.92)	(1.39)	(1.34)
3. Initial Cost	4.60	4.22	5.05	4.50
	(1.73)	(1.47)	(1.07)	(1.47)
4. Physical Effort Required	4.06	5.57	4.05	5.60
	(1.67)	(1.09)	(1.65)	(0.87)
5. Ease of Hay Storage	4.00	5.68	4.58	5.31
	(1.95)	(0.76)	(1.47)	(1.36)
6. Reliability of Operation	5.33	5.40	5.38	5.17
	(1.47)	(0.91)	(1.25)	(1.16)
7. Dealer Service	5.21	5.06	5.67	4.60
	(1.51)	(1.35)	(0.58)	(1.52)
Group sizes	33	45	34	51

market segments, the multivariate hypothesis of no mean differences is rejected.[2] Even after the variance explained by the first discriminant function is partitioned out, the second discriminant function is significant (see Table 2). Thus, most of the discussion which follows deals with the interpretation of these two significant discriminant functions.

Variable Contribution

TRADITIONAL APPROACHES. Because there is significant multivariate discrimination among the segment groups, it is of interest to determine the "contribution" of each of the evaluation variables to the overall discrimination. Toward this objective, most marketing researchers in published reports have evaluated (1) the group means and the associated univariate (ANOVA) F-ratios, or (2) the magnitude of the standardized discriminant function coefficients. But these approaches can be and usually are misleading if the predictors are intercorrelated, as is generally the case in marketing research applications. The probability levels for the univariate F-ratios ignore the interdependence of the predictors. Moreover, *despite their widespread use*, the standardized coefficients can be quite misleading. When two or more predictors are correlated, the "weight" (coefficient) may be split between the two of them making the associated coefficients appear relatively small. Alternatively, the coefficient for one variable may be inflated (to consider both predictors), while the other variable is assigned a near zero coefficient. As is demonstrated in the next problem, it is important to evaluate such subtle but important artifacts in the univariate F-ratios or coefficients by also evaluating the discriminant function loadings (the product moment correlations between each of the predictor variables and the discriminant function composite score [5]) and covariance controlled partial F-ratios.

Discriminant Loadings

The problem of relying on the standardized coefficients for interpretation is readily observed when the loadings and coefficients for the purchase influence factors in this analysis (Table 3) are con-

[2] There are a number of excellent introductions to the rationale and computational logic of discriminant analysis [20,30,31], so that material is not repeated in detail here.

trasted. The discriminant function loadings reflect common variance among the predictors and thus tend to be a more useful interpretative aid than the coefficients. For example, the first function standardized coefficients associated with Operating Expense and Ease of Storage are equivalent in magnitude. This might suggest that they contribute equally to the discrimination. But a more accurate perspective emerges when the discriminant loadings are considered. The correlation between Operating Expense and the function composite score is only .25, while the correlation for Ease of Storage is almost twice as large (.49). Similarly, the coefficient for Dealer Service on the first function is −.46, suggesting that it is more salient than Ease of Storage, which has a coefficient of only .27. Yet the loadings show that the latter variable is much more substantially related to the function. As in many other applications, reliance on the standardized coefficients for interpretation in this study would have been substantially misleading.

Covariance Controlled Partial F-Ratios

While the loadings help to reveal which of the variables are, in a relative sense, most important to the overall function, they still do not allow the researcher to determine if any one variable is a significant contributor to the discriminant model, given the relationships that exist among all of the predictors. The univariate F-ratios do not meet this objective because they ignore interrelationships among the variables, and thus are confounded by variables that contribute to the discriminant relationship in a redundant way. However, a good approach to this problem is to compute a covariance-controlled partial F-ratio for each variable. The essence of this approach is to partition out the variance in the variable of interest which is already explained by the other variables. This allows the researcher to determine (1) if significant group differences remain after the impact of other variables is considered, or (2) if a variable which appeared not to be a discriminator is in fact significant when relationships with other variables are considered. The advantage of this approach, as contrasted with the traditional univariate procedure, is seen when one compares the univariate F-ratios and the partial F-ratios for each of the purchase influence considerations in Table 3.

TABLE 2 Tests of Significance and Related Statistics for Discriminant Functions

Tests of Discriminant Roots	Eigenvalue	Wilks' Lambda	Multivariate F-Ratio	Degrees of Freedom	Probability	Canonical R
Roots 1 through 3	.71	.51	5.58	21;439	.001	.64
Roots 2 through 3	.09	.87	1.89	12;307	.035	.29
Roots 3 through 3	.05	.95	1.69	5;154	.139	.22

TABLE 3 Discriminant Analysis Statistics for the Predictor Variables (Decision Factors)

Decision Factors	Standardized Coefficients		Discriminant Loadings		Univariate F		Partial F		Potency Index
	Function 1	Function 2	Function 1	Function 2	Ratio[a]	Prob.	Ratio[b]	Prob.	
1. Speed of Operation	.36	.53	.41	.31	6.98	.00	3.55	.02	.16
2. Operating Expense	.27	-.74	.25	-.49	3.55	.02	3.03	.03	.08
3. Initial Cost	-.36	.51	-.19	.27	2.16	.10	3.49	.02	.04
4. Physical Effort Required	.64	.24	.70	-.02	18.17	.00	7.34	.00	.43
5. Ease of Hay Storage	.27	-.44	.49	-.37	11.03	.00	3.45	.02	.22
6. Reliability in Operation	-.20	-.01	-.04	-.24	.34	.79	.65	.58	.008
7. Dealer Service	-.46	-.42	-.29	-.41	4.52	.01	4.76	.00	.09

[a] The degrees of freedom are 3 and 159.
[b] The degrees of freedom are 3 and 153.

When the nonindependent univariate F-ratios are considered, the null hypothesis of no mean differences is rejected for five of the purchase influences. On the other hand, the probability levels might lead one to conclude that Initial Cost and Reliability in Operation are not significant. But evaluation of the comparable partial Fs, which indicate when a variable contributes significantly to the model *when all of the interrelationships are considered*, reveals that this conclusion is not correct. Specifically, although the univariate F for Initial Cost is not significant, the variable is found to make a significant contribution to the model when suppressing or moderating effects of the other predictors are controlled. Thus, this additional procedure confirms that the Reliability in Operation variable is not important in the model (because of the insignificant partial F), but on the other hand demonstrates that Initial Cost should not be discounted as a discriminator among segments.

This general approach of evaluating covariance controlled F-ratios is also frequently useful in discriminant analysis when the researcher suspects that some extraneous measure(s), not of primary concern in a particular analysis, may have a substantial impact on the nature or significance of the discriminant results. The research may address this suspicion by covarying out the effect of such variables from the discriminant predictors of interest before the function is derived and its significance tested. For example, in this study it was of interest to management to determine if the net profitability of the purchaser's farm operation was an underlying variable which tended to "explain away" the strong differences that were observed between the market segments on the decision factors. However, when the analysis was redone on the seven decision factors, controlling for net profit, the significance of the results and the structure of the discriminant coefficient and loading matrices were virtually unchanged. Thus, the differential importance of the influences on the purchase of the innovation and the conventional product were not simply attributable to the profitability of the different types of customers.

Another aspect of the partial F approach should be mentioned. Unlike the loadings or the coefficients, the partial F is an aggregative measure in that it summarizes information across the different discriminant functions. A complementary approach to developing a summary index of contribution across a number of significant discriminant functions is discussed in the next section.

Potency Index

When there are several discriminant functions, as in the study considered here, it is sometimes useful to develop a summary or composite index of the relative discriminating potency of each of the variables. In applied research, a useful "potency index" may be computed as:

$$PI_j = \sum_{k=1}^{m} \left[L_{jk}^2 \left(\lambda_k / \sum_{k=1}^{m} \lambda_k \right) \right] \quad (1)$$

where:

PI_j = the potency index for the jth variable,
m = the number of significant discriminant functions,
L_{jk} = the loading of the jth variable of the kth discriminant function,
λ_k = the eigenvalue for the kth function.

This index is somewhat similar to the communality estimates which are used in factor analysis.

The computational logic of the index has both intuitive and interpretive appeal. The squared loading of a variable on a function represents the variance in the multivariable function, which is explained by that single variable. The ratio of the eigenvalue for a function to the sum of all the significant eigenvalues represents the portion of the total explainable variance accounted for by a particular function. Nevertheless, a limitation of this index is that it is only a relative measure and has no meaning in an absolute sense. More precisely, while all of the eigenvalues could be quite small (i.e., the model is not a strong discriminator), the ratio of an eigenvalue to the total could be the same as for a model with larger eigenvalues.

The potency index for each variable in this study is given in the final column of Table 3. As an illustration, the index for Physical Effort Required is high because of its large loading on the first (strongest) discriminant function. This suggests that it is roughly five times as important in contributing to discrimination as Operating Expense, which has an index of .08. Although this is frequently a useful heuristic index, it can be misleading if used without considering the limitation noted.

Taxonomy of Variables and Dimensional Interpretation

To this point, the analysis has focused on single purchase influence variables and their contribution relative to other variables included in the analysis. Yet frequently the researcher is also interested in characterizing the predictive composite function or functions to identify their substantive meaning. In this study, the question was to determine whether or not the underlying functions represented conceptually meaningful evaluation dimensions which differentiated among the groups.

The loadings discussed previously are also helpful in "labeling" the discriminant functions, just as loadings are helpful in interpreting factor analysis or canonical correlation [1]. At an intuitive level, many researchers feel more comfortable in evaluating a loading than a discriminant function coefficient because a loading is, quite simply, the familiar correlation coefficient. It is somewhat surprising, therefore, that this practice has not become more common in discriminant analysis. Perhaps the most likely reason is that the most frequently used computer programs for MDA (such as those described in [13, 32]) do not compute or display the loadings.

But the extent to which this type of interpretation will be successful depends both on the logical interrelationships of the variables as they combine to explain discrimination among the groups in the analysis, and on the number of discriminant functions. Also, interpretation of a potentially meaningful discriminant function is facilitated when the loadings associated with the function are quite low, and when few variables load substantially on more than one function. This allows the researcher to identify taxonomies of variables that simultaneously characterize the dimension of interest. Any potential interpretation will be greatest when it is clear which variables should be considered together and evaluated for common underlying meaning or theme.

Consider the loadings (Table 3) for the two functions derived in this analysis. The pattern of split loadings found in this analysis is not particularly suggestive concerning the nature of the underlying dimensions which discriminate between the groups. There is a strong loading on the first function for the decision factor Physical Effort Required, and its loading on the second function is near zero. But two other factors, Speed of Operation and Ease of Hay Storage, have moderate loadings on both the first and second function. Similarly, only two variables, Operation Expense and Dealer Service, correlate over .4 with the second function, and they also have correlations of about .25 with the first function. Yet in discriminant analysis, as in factor analysis, the structure of the matrix of loadings can sometimes be improved for interpretation by the use of rotation. If the rotation (transformation) of the discriminant function matrix is orthonormal, the *total discrimination of the initial MDA solution is preserved* [9, 21]. Kaiser's Normalized Varimax [25] is one particularly useful transformation procedure in this type of analysis [38].

In this study, the results of a Varimax rotation of the loadings matrix are provided in Table 4. With the exception of one variable, the problem of split loadings has been reduced. For example, Speed of Operation originally correlated .41 with the first function and .31 with the second function. After rotation, the loadings are .51 and −.06, respectively. Similarly, Ease of Hay Storage originally had loadings of .49 and −.37. The structure of the relationship of the variable to the functions is now much simpler. The correlations are .09 and −.61. One variable proved to be an exception to the general improvement. Physical Effort Required moved from a simple structure to a more complex structure.

But, more important, when the set of rotated loadings is considered simultaneously, an understandable taxonomy emerges on the first function. Speed of Operation, Physical Effort Required, and Dealer Service all load over .45 on the first function and none of the other variables load over .2. These three variables tend to constitute a "performance dimension." Higher scores will be realized on this performance dimension when an individual's purchase decision is more substantially influenced by the Speed of Operation of the equipment and the Physical Effort Required, but less influenced by concern about Dealer Service (which perhaps implies that they expect service will not be required). Thus, those with high scores on this dimension are more concerned with the performance characteristics of the equipment than are those with low scores.

On the second function, Operating Expense, Physical Effort Required, and Ease of Storage all have substantial negative correlations, and Initial Cost has a positive correlation. This appears to be a "Cost-Over-Time" dimension. A high score

TABLE 4 Varimax Rotated Discriminant Function Coefficients and Loadings

Purchase Decision Factors	Rotated Standardized Coefficients		Rotated Discriminant Loadings	
	Function 1	Function 2	Function 1	Function 2
1. Speed of Operation	.63	.13	.51	−.06
2. Operating Expense	−.33	−.72	−.17	−.52
3. Initial Cost	.11	.62	.06	.33
4. Physical Effort Required	.62	−.28	.48	−.50
5. Ease of Hay Storage	−.17	−.51	.09	−.61
6. Reliability in Operation	−.15	.13	−.20	−.14
7. Dealer Service	−.62	.03	−.49	−.09

results when delayed expenses (cost of operation, labor, storage) are considered less important, but initial cost is more important. Conversely, the low end reflects greater concern for longer term expenses, and less concern with Initial Cost. Thus, the transformation of the discriminant function loadings matrix facilitates substantive interpretation by identifying logical taxonomies of the variables which describe the underlying dimensions of discrimination among the groups. As is shown in the next section, this procedure helps us to more clearly understand the innovator segment and specify appropriate promotional emphasis.

Segment Differences and Characteristic Profiles

Market Segments in Discriminant Space

Now that the contributions of different variables have been determined and the nature of the underlying discriminant functions is more clearly understood, attention may be focused on the basic objective in using MDA in this study: developing a better understanding of the distinctive characteristics of the market segment groups being analyzed.

Toward this objective, it has been common in marketing research to compute the centroids for the different segments on the discriminant functions and then to plot the segments in the discriminant space (i.e., with the centroids as the coordinates of a segment). The positions of the groups, relative to the functions and to each other, may provide insights into segment similarities and differences. Typically, such plotting is based on the unrotated centroids (see for example [28]), but this procedure is equally applicable with the ro-

tated analysis if it is more clear. Instead of rotating the loadings, the matrix of standardized coefficients is rotated directly. Then the rotated centroids may be computed in the usual fashion. It is also possible to use analogies between geometry and correlational analysis to insert vectors in the discriminant space to depict the original predictor variables [24, 36]. Interpretation may be enhanced when the variable vectors and group centroids are simultaneously plotted in the discriminant space. The graphic representation helps to bring the various aspects of the analysis together in a concise and clear fashion.

Figure 1 provides a graphic representation of the results of the *rotated* discriminant solution and depicts the relationships among the market segment groups. The two innovator segments are closely positioned in the lower right quadrant of the plot. Thus, they tend to be relatively high on the Performance dimension, and relatively low on the Cost-Over-Time dimension. Conversely, the two conventional purchase segments are positioned together low on the Performance dimension and high on the Cost-Over-Time dimension. This suggests that the innovator segments were more concerned in their purchase decision with the performance characteristics of the equipment and were more concerned with longer term expenses than with initial costs. The proximity of the pair of innovator segments and of the pair of conventional purchase segments in the discriminant space suggests that, across regions, these decision factors were not very distinct.

Examination of the variable vectors also suggests more specific aspects of the relationships between the original predictors and the market segments. The length of the vectors (proportional to the potency indexes found in Table 3) depicts the relative importance of the variables in the solu-

FIGURE 1 Geometric Representation of Rotated Discriminant
Function Solution for Innovator Groups.

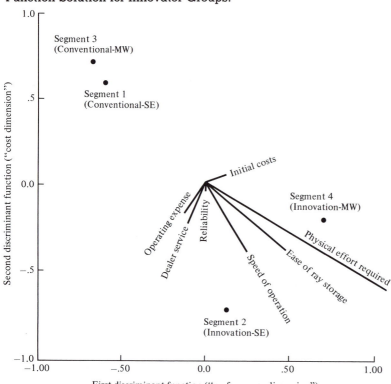

tion. In general, the variable vector points toward those segments in which the variable has a higher mean and away from those segments where the variable is less prominent. For example, Physical Effort Required is seen to be an important discriminator and is more prominent in the innovator groups than in the conventional segments.

Other visual representations of discriminant solutions have been less widely used but offer potential. For example, when there are more than two significant discriminant functions, hyperspace plotting [2, 11] may help to provide a basis for visual clustering of the market segments with respect to the discriminant dimensions. Alternatively, additional information about the solution may be displayed by delineating the discriminant space "territories" which separate the segments [32], or by projecting data elipses which are suggestive of violations of discriminant analysis computational assumptions [14].

In summary, plotting of discriminant solutions is an important way of communicating the results

of the analysis in a nontechnical way and of providing assistance in interpretation. Yet the among-segment relationships suggested by such graphics should be validated with statistical analyses so that the possibility of improper conclusions is reduced or eliminated. Also, as illustrated in the following section, graphic representations sometimes suggest relationships that can be evaluated with more rigorous procedures.

Statistical Analysis of Pairwise Group Differences

While graphic procedures help in considering omnibus relationships suggested by the discriminant analysis, the researcher is typically also concerned with determining which pairs of segments differ from each other, and on what specific variables. Besides the shortcomings of the univariate F-ratios discussed earlier in this paper, the test involved in this ratio is not very helpful with this problem because it is concerned only with whether or not *any* subsets of the segments differ significantly, and not with *which* segments are responsible for the significant departure from the null hypothesis. Thus, alternative approaches to this interpretive problem are needed.

When the market segment groups are based on a cross-classified market grid, as in this research (type of purchase by region), the multivariate general linear hypothesis procedure for partitioning variance [35] is helpful in evaluating further the nature of segment differences. For example, in this study it was desirable to test for evaluative differences due to the type product purchased, region, and their interaction. The multivariate tests for this analysis are shown in Table 5. The test statistics in Table 5 are consistent with the conclusion resulting from visual examination of the plot of the discriminant solution (Figure 1). The variables discriminate ($p < .001$) between the segments on the type of purchase, but there is not discrimination ($p < .111$) based on regional differences. Furthermore, because the centroids for the two innovator groups in the plot are further apart than are the centroids for the segments that purchased the conventional product, one might wonder if there is an interaction between type of product and region. The test (Table 5) of the multivariate interaction, however, is not significant, and thus it appears that the *visual* separation is due to sampling variation rather than systematic variation. Thus, market region need not be an important consideration in developing the promotional strategy.

Multivariate Confidence Intervals about Segment Differences

Where there are more than two segments among which variable differences are of interest, additional tests may be made to isolate variable specific between-segment differences based on "contrasts" among the segments. While some researchers might attempt to attack this problem with an exhaustive set of pairwise t-tests, that approach is even less appropriate in multivariate discriminant analysis than it would be if applied to data from a univariate experimental design. More significant differences than are real are likely to result due to the artifact that the probability statements do not take into consideration the multitude of variables or the exhaustive number of between-group tests. One must use a procedure that is based on multivariate confidence intervals about the differences in the mean values of the two (or more) segments to have confidence in the probability levels from such exhaustive contrasting of segments.

The estimation of confidence intervals for group differences in discriminant analysis may be thought of as a multivariate analogue to the Scheffe test, which is used in experimental research for post hoc comparisons of cell means [30]. The basic procedure for computing the confidence intervals is generalized to the discriminant analysis case from work done in developing simultaneous tests and confidence intervals in multiple regression models [30]. In regression related procedures, such tests are usually used to evaluate linear combinations of the regression coefficients, but when applied in the context of discriminant analysis (or multivariate analysis of variance [46]), they allow the researcher to consider group differences with respect to individual predictor variables. For example, in this study it was of interest to examine selection criteria differences between innovative and conventional segments within each of the territories. The simultaneous confidence intervals for these two pairs of comparisons are shown in Table 6.

Examination of Table 6 reveals that the confidence intervals in the Southeast territory for Physical Effort Required and Ease of Hay Storage do

TABLE 5 Partitioning of Variance for Multivariate Differences in Market Segment Groups

Source of Variation[a]	Multivariate F-Ratio	Degrees of Freedom	Probability
Type of purchase	15.42	7;153	.001
Sales region	1.71	7;153	.111
Interaction of purchase and sales region	1.58	7;153	.147

[a] The order of elimination of the effects for the significance tests [39] was as follows: the interaction was based on eliminating the Type of Purchase and Region; the test of Type of Purchase was based on eliminating Region, and the test of Region was based on eliminating Type of Purchase.

TABLE 6 Multivariate Simultaneous Confidence Intervals for Contrasts between Purchase Types within Sales Region[a]

	Multivariate Confidences Intervals				
Variable	Southeast Region: Conventional vs. Innovator Contrast			Midwest Region: Conventional vs. Innovator Contrast		
1. Speed of Operation	$-2.247 \leq -.816 \leq$	.615		$-2.314 \leq -.931 \leq$	.452
2. Operating Expense	$-2.124 \leq -.705 \leq$	.714		$-1.772 \leq -.402 \leq$	.968
3. Initial Cost	$-2.245 \leq .384 \leq$	2.469		$-1.025 \leq .549 \leq$	2.123
4. Physical Effort Required	$-2.975 \leq -1.517 \leq -$	.059		$-2.957 \leq -1.549 \leq -$	.141
5. *Ease of Hay Storage*	$-3.256 \leq -1.689 \leq -$	.122		$-2.238 \leq -.725 \leq$	.788
6. Reliability of Operation	$-1.397 \leq -.067 \leq$	1.267		$-1.079 \leq -.206 \leq$	1.491
7. Dealer Service	$-1.342 \leq -.145 \leq$	1.632		$-.368 \leq -1.069 \leq$	2.506

[a] These intervals are based on alpha = .05. Estimates of contrasts are noted in the middle of the confidence interval.

not include zero, and thus are substantially different between the two groups. In the Midwest, only Physical Effort Required is identified as significantly different between the two segments. All the other multivariate intervals include zero. These findings highlight the fact that multivariate confidence intervals tend to be quite conservative. When the intervals suggest a difference, the researcher can be quite confident that the difference is in fact substantive. Thus, in this problem, marketing managers know that Physical Effort Required, and to a lesser extent Ease of Hay Storage, are particularly important evaluative criteria for the innovator group.

Summary and Discussion

In this study of a commercial innovation, discriminant analysis has helped to reveal the nature of differences in the innovator and noninnovator market segments in terms of the importance of various factors which influence the purchase decision. Speed of Operation, Operating Expense, Initial Cost, Physical Effort Required to use the product, Ease of Hay Storage, and Dealer Service were all significant discriminators. Moreover, the variables tended to discriminate between innovators and noninnovators in terms of two meaningful underlying dimensions: Cost-Over-Time and Performance. Innovators were more concerned with the performance characteristics of the equipment and placed more emphasis on the long-term costs instead of the short-term costs—associated with the product. Furthermore, analysis revealed that these differences were independent of the region

of the country in which the purchaser's business operated. Additional research will be required to determine whether similar evaluative dimensions are equally salient to innovators of other commercial products in general. However, these results suggested to marketing managers that promotional efforts designed to reach the untapped portion of the innovator segment should emphasize the operating performance of the new product and the fact that the initial cost was offset by long-term savings. Specific features of the product, especially the low physical effort required, were highlighted, and accentuated these basic themes.

From a methodological perspective, however, these results are also interesting in that they were not revealed by the approaches traditionally used in interpreting discriminant analysis. The univariate F-ratios incorrectly suggested that one of the variables was not significant, and the standardized coefficients did not accurately reflect the relative importance of the variables in overall discrimination. Furthermore, the underlying conceptual organization of the dimensions that discriminate significantly between the segments was not apparent until the discriminant loadings matrix was rotated. In addition, alternative procedures for interpretation also revealed that the differences in the evaluation characteristics between the innovator and noninnovator segments generalized across different regions of the country and were not simply attributable to the net profitability of the customer's business. Finally, the analysis demonstrates the use of multivariate confidence intervals for variable by variable, segment by segment comparisons.

Thus, this paper proposes and illustrates a

TABLE 7 Summary of Traditional and Suggested Techniques for Interpretation of Multiple Discriminant Analysis

Aspect of Interpretation	Traditional Approaches	Major Problems with Traditional Approach	Suggested Alternative Approaches
I. Determination of the contribution of individual variables to overall discrimination	Compute univariate F-ratios and probabilities	Probability levels ignore interdependence of the predictors	Compute covariate controlled partial F-ratios
	Compare standardized discriminant coefficients	Weights may be split due to predictor intercorrelations; weights do not reflect common variance	Compute discriminant loadings
		Weights for only one function at a time	Evaluate potency index
II. Identifying meaning and "labeling" of discriminant functions	Evaluate relative magnitude of standardized coefficients (or discriminate loadings)	Weights not proportional to variance explained in total function; (split loadings may make it difficult to characterize dimensions)	Orthonormal rotation of discriminant function loadings matrix[a]
III. Evaluation of between-segment differences on specific variables	Univariate ANOVA; Exhaustive pairwise t-tests	Does not indicate between which segments the differences occur; tests not independent (lack of error control)	Test for multivariate main effects and interaction between segments[b]; multivariate confidence intervals for variable specific comparisons
	Plot segment centroids in discriminant space (unrotated)	Rotated solution may be a better basis for interpretation	Plot rotated variable vectors and centroids

[a] If the rotation is orthonormal, the total discrimination of the original (unrotated) solution will be preserved; this is generally not true with other types of rotation and they should be used only with caution.
[b] This approach is applicable when the market segments (or criterion groups) are based on crossed market grid dimensions.

number of procedures that supplement or complement traditional approaches to the interpretation of disciminant analysis. These approaches are helpful, not only in determining the contribution of individual variables to the overall discrimination, but also in characterizing the underlying dimensions of differentiation and in identifying specific pairwise differences in groups. To facilitate review, the major points illustrated and proposed in this paper are summarized in Table 7.

The use of discriminant analysis to *profile* market segments is the primary concern here. This does not imply that the use of discriminant analysis for purposes of classification is unimportant, but rather implicitly acknowledges that detailed attention has already been devoted to this topic in the marketing literature [10, 18, 19, 29, 31]. However, even when the emphasis of MDA research is on profiling segments, it is sensible also to consider the results of the associated classification analysis to be certain that the tests of profile differences do not just reflect trivial differentials isolated by the power of the analysis. For example, in this research, when all four groups are used as the classification criteria, 51% of the respondents are properly classified (based on a jackknife analysis [10] to reduce sample bias). However, most of the misclassifications were across regions; i.e., some of the innovators from the Southeast were classified into the innovator group from the Midwest, and vice versa. This is to be expected from the finding that region was not important for these evaluative criteria. When the jackknife classification is recomputed for the two-group criterion (innovator vs. noninnovator), the percentage correctly classified (81%) provides additional support that the findings are operationally useful.

The basic proposition advocated is that the various approaches are *complementary*, and combinations of them may provide management insights that might otherwise be lost. Similarly, the likelihood of an erroneous conclusion, based on incomplete information, is substantially reduced. Finally, a given research problem may not require use of all of the procedures.

References

1. Alpert, Mark I., and Peterson, Robert A., On the Interpretation of Canonical Analysis, *J. Marketing Res.* 9 (February, 1972): 187–192.
2. Andrews, D. F., Plots of Higher-Dimensional Data, *Biometrics* 28 (March, 1972): 125–136.
3. Armstrong, Gary M., and Feldman, Lawrence P., Self Concept Characteristics of Innovators and Opinion Leaders, *Proceedings of the Southern Marketing Association*, 1975.
4. Banks, Seymore, Why People Buy Particular Brands, in *Motivation and Market Research*, Robert Ferber and Hugh G. Wales, eds., Irwin, Homewood, Ill., 1958, pp. 277–93.
5. Bargmann, Rolf E., Interpretation and Use of a Generalized Discriminant Function, in *Essays in Probability and Statistics*, R. C. Bose, ed., University of North Carolina Press, 1970.
6. Brody, Robert P., and Cunningham, Scott M., Personality Variables and the Consumer Decision Process, *J. Marketing Res.* 5 (February, 1968): 50–57.
7. Churchill, Gilbert A., Jr., Ford, Neil M., and Ozanne, Urban B., An Analysis of Price Aggressiveness in Gasoline Marketing, *J. Marketing Res.* 7 (February, 1970): 36–42.
8. Claycamp, Henry J., Characteristics of Thrift Deposit Owners, *J. Marketing Res.*, 2 (May, 1965): 163–170.
9. Cliff, Norman, and Krus, David J., Interpretation of Canonical Analysis: Rotated vs. Unrotated Solutions, *Psychometrika* 41 (March, 1976): 35–42.
10. Crask, Melvin R., and Perreault, William D., Jr., Validation of Discriminant Analysis in Marketing Research, *J. Marketing Res.* 14 (February, 1977): 60–68.
11. Darden, William R., and Flascher, Alan B., Visual Presentation of Marketing Simuli in Hyperspace, *J. Marketing Res.* 11 (November, 1974): 456–461.
12. Darden, William R., and Reynolds, Fred D., Backward Profiling of Male Innovators, *J. Marketing Res.* 11 (February, 1974): 79–85.
13. Dixon, Wilfred, J., ed., *BMD: Biomedical Computer Programs*, University of California Press, Berkeley, 1973, 211–254.
14. Dixon, W. F., and Jennrich, R. I., Computer Graphical Analysis and Discrimination, in *Discriminant Analysis and Applications*, T. Cacoullos, ed., Academic Press, New York, 1973, 161–172.
15. Evans, Franklin B., Psychological and Objective Factors in the Prediction of Brand Choice: Ford versus Chevrolet, *J. Business* 32 (October, 1959): 340–369.
16. Feldman, Laurence P., and Armstrong, Gary M., Identifying Buyers of a Major Automotive Innovation, *J. Marketing* 39 (January, 1975): 47–53.
17. Frank, Ronald E., Massey, William F., and Morrison, Donald G., The Determinants of Innovative Behavior with Respect to a Branded, Frequently Purchased Food Product, in *Reflections on Progress in Marketing*, L. George Smith, ed., American Marketing Association, Chicago, 1964, 312–323.
18. Frank, Ronald E., Massey, William F., and Morrison, Donald G., Bias in Multiple Discriminant Analysis, *J. Marketing Res.* 2 (August, 1965): 250–258.
19. Green, Paul E., Bayesian Classification Procedures in Analyzing Customer Characteristics, *J. Marketing Res.* 1 (May, 1964): 44–50.
20. Green, Paul E., and Tull, Donald S., *Research for Marketing Decisions*, Prentice-Hall, Englewood Cliffs, 1975.
21. Hall, Charles E., Rotation of Canonical Variates in Multivariate Analysis of Variance, *J. Experimental Education* 38 (Winter, 1969): 31–38.

22. Harvey, John R., What Makes a Best Seller, in *Motivation and Market Research*, Robert Ferber and Hugh G. Wales, eds., Irwin, Homewood, Ill., 1958, 361–381.

23. Jacoby, Jacob, Personality and Innovation Proneness, *J. Marketing Res.* 8 (May, 1971): 244–247.

24. Johnson, Richard M., Market Segmentation: A Strategic Management Tool, *J. Marketing Res.* 8 (February, 1971): 13–18.

25. Kaiser, Henry F., The Varimax Criterion for Analytic Rotation in Factor Analysis, *Psychometrika* 23 (September, 1958): 187–200.

26. King, William R., Marketing Expansion—A Statistical Analysis, *Management Sci.* 9 (July, 1963): 563–573.

27. Levine, Phil, Locating Your Customers in a Segmented Market, *J. Marketing* 39 (October, 1975): 72–73.

28. Massey, William F., Discriminant Analysis of Audience Characteristics, *J. Advertising Res.* 5 (March, 1965): 39–48.

29. Montgomery, David B., New Product Distribution: An Analysis of Supermarket Buyer Decisions, *J. Marketing Res.* 12 (August, 1975): 255–264.

30. Morrison, Donald F., *Multivariate Statistical Methods*, McGraw-Hill, New York, 1967.

31. Morrison, Donald G., On the Interpretation of Discriminant Analysis, *J. Marketing Res.* 6 (May, 1969): 156–163.

32. Nie, Norman H., Hull, D. Hadlai, Jenkins, Jean G., Steinbrenner, Karin, and Brent, Dale H., eds., *SPSS: Statistical Package for the Social Sciences*, Englewood Cliffs, McGraw-Hill, 1975.

33. Ostlund, Lyman E., Identifying Early Buyers, *J. Advertising Res.* 12 (April, 1972): 25–30.

34. Overall, John E., and Klett, C. James, *Applied Multivariate Analysis*, McGraw-Hill, New York, 1972.

35. Perreault, William D., Jr., and Darden, William R., Unequal Cell Sizes in Marketing Experiments: Use of the General Linear Hypothesis, *J. Marketing Res.* 12 (August, 1975): 333–342.

36. Perreault, William D., Jr., and Darden, William R., GRAFIT: Computer Based Graphics for Interpretation of Multivariate Analysis, *J. Marketing Res.* 12 (August, 1975): 343–345.

37. Perreault, William D., Jr., French, Warren A., and Harris, Clyde E., Jr., Use of Multiple Discriminant Analysis to Improve the Salesman Selection Process, *J. Business* 50 (January, 1977): 50–62.

38. Perreault, William D., Jr., and Spiro, Rosann L., An Approach for Improved Interpretation of Multivariate Analysis, *Decision Sci.* 9 (July, 1978): 402–413.

39. Pessemier, Edgar A., Burger, Philip C., and Tigert, Douglas J., Can New Product Buyers Be Identified? *J. Marketing Res.* 4 (November, 1967): 349–354.

40. Rao, Tanniru, R., Is Brand Loyalty a Criterion for Marketing Segmentation: Discriminant Analysis, *Decision Sci.* 4 (July, 1973): 393–404.

41. Reynolds, Fred D., and Martin, Warren, S., A Multivariate Analysis of Intermarket Patronage: Some Empirical Findings, *J. Business Res.* 2 (April, 1974): 193–195.

42. Robertson, Thomas S., and Kennedy, John N., Prediction of Consumer Innovators Application of Multiple Discriminant Analysis, *J. Marketing Res.* 5 (February, 1968): 64–69.

43. Rogers, Everett M., *Diffusion of Innovations*, Free Press of Glencoe, New York, 1962.

44. Uhl, Kenneth, Andrus, Roman, and Poulsen, Lance, How Are Laggards Different? An Empirical Inquiry, *J. Marketing Res.* 7 (February, 1970): 51–54.

45. Utterback, James M., Successful Industrial Innovations: A Multivariate Analysis, *Decision Sci.* 6 (January, 1975): 65–77.

46. Wilkinson, Leland, Response Variable Hypotheses in the Multivariate Analysis of Variance, *Psyc. Bulletin* 82 (May, 1975): 408–412.

47. Wind, Yoram, Industrial Source Loyalty, *J. Marketing Res.* 7 (November, 1970): 450–457.

The Personality Inventory for Children: Differential Diagnosis in School Settings

STEVEN J. DEKREY
STEWART EHLY

The assessment of schoolchildren to determine unique educational needs has evolved into a comprehensive multidisciplinary process. The determination of a handicapping condition follows extensive assessment by educational specialists. Implicit in the multidisciplinary approach is the assumption that certain groups of children are so different from their average peers that categorical distinctions are possible and that the cognitive, perceptual, psycholinguistic, and social behaviors of such children vary significantly among groups (McCarthy & Paraskevopoulus, 1969).

One important consideration in the diagnostic process is the measurement of socioemotional or personality characteristics. The measurement and documentation of socioemotional status is necessary to meet inclusionary criteria for students with a primary handicapping condition involving emotional disabilities and to meet the exclusionary criteria of both mental disabilities and learning disabilities. Although a thorough assessment of behavioral and emotional factors is necessarily

"The Personality Inventory for Children: Differential Diagnosis in School Settings," Steven J. DeKrey and Stewart Ehly, Vol. 3 (1985), pp. 45–53. Reprinted from the *Journal of Psychoeducational Assessment*, published by Grune & Stratton, Inc. Steven J. DeKrey is on the faculty at Northwestern University. Stewart Ehly is on the faculty at the University of Iowa.

implied by the exclusionary and inclusionary criteria for differential diagnosis, Mowder (1980) has noted the inadequacies of current personality assessment devices in meeting this necessity. Mowder therefore recommends a multidimensional approach, which requires a comprehensive collection of *all* relevant data. Gerken (1979) supports such an approach, arguing for a purposeful process for collecting relevant data.

The development of well-validated personality assessment measures to collect relevant data has been slow and sporadic (Achenbach, 1978). Freeman (1971) contends that the delay in the development of personality assessment measures for children is due primarily to the lack of an acceptable theory to guide the development of assessment techniques. Most objective personality measures are based on ideas of adult psychopathology and depend on the competence of the respondent to read, understand, and complete a written instrument (Wirt & Broen, 1956). Children are considered incapable of supplying reliable self-reports. As a result, the assessment of personality has often relied on the responses from knowledgeable observers (parents and teachers).

Advances in empirical techniques for the assessment of school-aged children have fostered the development of an innovative measure, the Personality Inventory for Children (PIC). The scale construction methods were modeled after the MMPI (Hathaway & McKinley, 1951). The PIC includes 600 items that compose both empirical scales based on the statistical capabilities of the items and content scales selected on a rational basis. A total of 33 scales have been developed for the PIC to assist in diagnosing a variety of childhood disorders and behavioral syndromes.

Sixteen of the 33 scales are included on the profile sheet. Those 16 were found most helpful to clinicians. The 16 profile scales have been demonstrated to differentiate among the three most common special-education classifications of learning disabilities, mental disabilities, and emotional disabilities (DeKrey & Ehly, 1981). Three scales, Achievement (ACH), Intellectual Screening (IS), and Development (DVL), relate strongly to problems children have learning in school (Lachar & Gdowski, 1979). The PIC item sequence has recently been modified to allow several shortened versions to be scored. These scores include four new broadband factor scales and 14 short versions of original scales. One application of the items, including Parts I and II, uses only 280 of the original 600 items (Lachar, Gdowski, & Snyder, 1982). Although the original PIC demonstrated value in assisting differentiation among three educational disability groups in an earlier study (DeKrey & Ehly, 1981), the existence of the new, potentially more popular version requires proper validation before use in a school setting can be recommended.

The purpose of this study was to determine whether Parts I and II of the shorter version of the Personality Inventory for Children (PIC-S) comprise a relevant instrument for the assessment of educational handicaps in school-aged children. The combined discriminative ability of the 12 shortened clinical scales, the two original-length scales, the remaining two validity scales, and the four new factor scales and various combinations were assessed for four educational classifications.

Method

Subjects

To obtain an adequate number of subjects, two independent school systems were sampled. Each of the systems was in a midwestern city, with respective enrollments of approximately 2,000 and 3,000 students. The final sample included 95 male elementary school-aged (K–6) students attending school in one of the two districts. Sufficient samples, for statistical purposes, of females from each educational classification were not available. Participants from the educational classifications were identified with the assistance of school district administrators. Groups were represented as follows: 32 from regular-education classes (average age = 9.1, range = 6–12), 23 from learning-disabilities classes (average age = 10.3, range 7–12), 20 from educable mentally disabled classes (average age = 9.5, range = 7–13), and 20 from emotionally disabled classes (average age = 9.7, range = 6–13). Male elementary students were chosen, using a random selection procedure, from the total population of each educational classification.

Instrument

The revised-order PIC Administration Booklet (W-1521), Parts I, II, III, and IV, was published in 1981. The booklet revision includes all items on

Parts I, II, III, and IV that were selected for the original 600 items of the Personality Inventory for Children (Wirt, Lachar, Klinedinst, & Seat, 1977). Part I contains 131 original items and allows scoring of four new broad-band factor scales and the original Lie scales. The new broad-band factors are Undisciplined/Poor Self-Control, Social Incompetence, Internalization, Somatic Symptoms, and Cognitive Development. Part II contains an additional 149 items that allow scoring of shortened versions of 14 of the original profile scales, with the Development Scale appearing in its entirety. The profile scales are Lie, F, Defensiveness, Adjustment, Achievement, Intellectual Screening, Development, Somatic Concern, Depression, Family Relations, Delinquency, Withdrawal, Anxiety, Psychosis, Hyperactive, and Social Skills. Parts III and IV contain the remaining items allowing scoring of original scales. The current study used Parts I and II, a shortened version of the PIC. This combination allowed T scores with a mean of 50 and a standard deviation of 10 to be produced for each subject on 16 narrow-band scales and the four new broad-band factors. Norm tables for male students were applied to subject data.

Procedure

Each inventory was completed by the selected student's female caretaker; in all but one case, this was the biological mother. Every parent returned PIC-S materials by the end of the data-collection phase. The procedures assured a sample consistent with the established PIC norms.

Scores from the 20 scales were analyzed in an attempt to differentiate among the three groups of special-education students and a control group of regular-education students. The special-education groups represent the common disability classifications used by the Iowa State Department of Public Instruction (learning-disabled, mentally disabled, and emotionally disabled). The PIC-S scales were constructed as though they were independent measures and were treated independently in the analysis.

Results

Means and standard deviations of the four educational groups on all variables are noted in Table 1. Mean T scores for the research groups on each

of the 20 PIC-S variables were calculated. In Figure 1, the mean scores for the validity scales and the broad-band factor scales are presented. Figure 2 depicts the mean scores for the 12 clinical scales.

It is important to note that the PIC-S was designed to measure problematic areas, with a high score suggesting possible problems. Low scores have little interpretable meaning beyond predicting an absence of a problem. For example, in both figures the regular-education students are found to have low mean scores on all 20 variables; thus as a group, they evidence an absence of problems measured by the PIC-S. From the results given, distinct profiles characterize each group, with differences occurring on many of the scales.

A discriminant analysis was conducted, first using all 20 variables, then the 12 clinical scales plus the adjustment scale, and finally the four broad-band factors, to determine whether the PIC-S could be used to discriminate among the four educational groups in a school setting. Three significant functions were derived when using the discriminant analysis on all 20 PIC-S variables. The resulting functions were subjected to a chi-square test as part of a program to determine significance of the functions. A Wilks' lambda test and univariate F tests were conducted to determine significance of each of the 20 variables. The eigenvalues, relative variance, and canonical correlation for each of the three functions appear in Table 2. These results confirm the presence of three distinct functions.

The process of describing these functions included interpretation of structure coefficients, standardized canonical correlations, and known group characteristics. The functions were described as follows:

Function 1. General school maladaption (Factor IV, DVL, PSY, IS, ACH, Factor II, SSK, F, ADJ, D, Factor I)
Function 2. Conformity (Factor I, DLQ, ADJ, SSK, D, HPR, DEF)
Function 3. Social interaction (HPR, Factor II)

Function 1 is the most important, and alone it represents 52% of the discrimination power. Function 2 represents 33%, and Function 3 represents 15%.

Group centroids for each of the four groups in the form of z scores are reported in Table 3. These scores are interpreted to be the number of standard deviations for each group from the aver-

TABLE 1 Means (and Standard Deviations) for the 20 PIC-S Variables for Four Educational Groups

Group	Lie Scale	F Scale	Defensiveness Scale	Adjustment Scale	Achievement Scale
I	44.4 (9.3)	43.8 (7.0)	43.0 (8.8)	45.8 (8.7)	45.3 (9.3)
II	40.8 (8.7)	49.6 (12.4)	41.8 (12.5)	61.5 (13.5)	68.6 (9.2)
III	46.7 (9.0)	63.4 (14.9)	39.7 (13.6)	60.5 (12.9)	67.6 (9.6)
IV	38.5 (8.6)	61.9 (16.0)	31.4 (12.4)	80.0 (11.4)	60.4 (11.6)

Group	Intellectual Screening Scale	Development Scale	Somatic Concern Scale	Depression Scale	Family Relations Scale
I	44.8 (8.5)	43.6 (8.1)	46.7 (10.1)	43.7 (5.5)	43.8 (5.3)
II	67.6 (14.5)	62.6 (10.9)	48.3 (8.8)	49.9 (8.3)	51.2 (13.0)
III	91.8 (21.9)	66.1 (11.4)	52.4 (13.7)	52.3 (9.6)	54.1 (10.7)
IV	63.6 (20.8)	56.0 (9.5)	50.4 (11.3)	64.1 (15.6)	52.0 (10.1)

Group	Delinquency Scale	Withdrawal Scale	Anxiety Scale	Psychosis Scale	Hyperactivity Scale
I	46.4 (6.8)	42.3 (8.3)	45.4 (7.1)	44.0 (8.1)	49.3 (9.6)
II	51.5 (12.1)	45.7 (9.5)	52.1 (7.8)	50.8 (12.9)	49.3 (13.4)
III	59.9 (16.0)	52.1 (13.1)	51.2 (9.3)	67.8 (13.4)	49.8 (16.2)
IV	69.0 (19.0)	51.5 (9.2)	59.7 (13.3)	67.6 (16.2)	66.2 (16.7)

Group	Social Skills Scale	Factor I (Undisciplined)	Factor II (Social Incompetence)	Factor III (Internalization)	Factor IV (Cognitive Development)
I	41.3 (9.9)	48.8 (11.0)	43.0 (8.4)	45.9 (10.7)	43.8 (10.1)
II	51.6 (13.6)	53.8 (15.9)	52.3 (12.7)	50.2 (9.4)	63.8 (15.2)
III	56.2 (13.4)	59.1 (16.3)	58.0 (12.6)	53.1 (13.7)	82.9 (16.3)
IV	67.1 (12.2)	75.2 (13.0)	64.0 (11.6)	57.3 (14.0)	61.0 (16.1)

Note: Group I = Regular education (N = 32); Group II = Learning Disabilities (N = 23); Group III = Mental disabilities (N = 20); Group IV = Emotional disabilities (N = 20).

age of all groups on each function. The group centroids demonstrate that a significant degree of discrimination exists for the three.

Using the classification phase of the discrimination analysis program where actual group membership was compared with predicted group membership, 90% of the cases were correctly classified through use of the 20 PIC-S variables. Emotionally disabled children produced the most misclassifications (20% of Group IV), with 10% misclassified in both Groups II and III.

Univariate F tests conducted for all 20 variables to determine significance of mean group differences revealed that 18 of the 20 variables vary significantly among some groups (p = .01) (see Table 4). Follow-up testing using the Scheffé method resulted in 52 significant differences (p = .01) out of 60 tests of ordered means. Interpretable differences, using criteria developed by Lachar and Gdowski (1979), of group means are reported in Table 5.

Discussion

The purpose of this investigation was to examine the applicability of a shortened version of the Personality Inventory for Children to educational assessments. The results of the classification procedure using the three functions from the 20 PIC-S variables illustrate the measurement capabilities of the PIC-S in educational settings. On the PIC-S, children in regular education are characterized by lack of inflation of all variables. Learning-disabled children reveal high scores on adjustment, achievement, and development and absence of an extreme score on intellectual screening. Mentally disabled children are characterized by an extremely high score on intellectual screening and high scores on adjustment, achievement, development, and Factor IV (cognitive development). Emotionally disabled children are characterized by interpretable scores on hyperactivity and Fac-

FIGURE 1. Mean PIC-S group scores for the validity scales and the broad-band factor scales. Group I, regular education,----; Group II, learning disabilities, ——; Group III, mental disabilities, = ; Group IV, emotional disabilities, ••••.

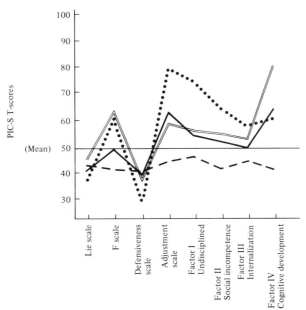

FIGURE 2 Mean PIC-S group scores for the 12 clinical scales. Group I, regular education, ----; Group II, learning disabilities, ——; Group III, mental disabilities, = ; Group IV, emotional disabilities, ••••.

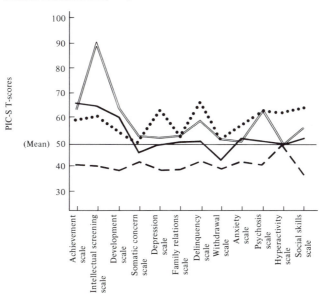

TABLE 2 Eigenvalues, Percent of Variance, and Canonical Correlations Using 20 PIC-S Variables

Function	Eigenvalue	Percent of Variance	Canonical Correlation
1	2.641	51.95	.8517
2	1.676	32.96	.7914
3	.767	15.09	.6589

tor I (undisciplined). Inflation was noted for adjustment and achievement.

The procedure resulted in a 90% accuracy rate by using the 20 coefficients for group assignment. This rate gains increased importance when compared to the random rate for no discrimination of only about 25%. Over 85% of the ordered mean comparisons of the 20 variables reached statistical significance at the .01 level using the conservative Scheffé method. Such results confirm the existence of significant group differences on 18 of the 20 PIC-S variables.

The findings of the present investigation illustrate the measurement potential of the PIC-S to provide objective confirmation of group assignment. Function 1, general school maladaption, appears to operate as a severity rating of a child's overall school adaptability. High scores on this function would predict difficulties in adjusting to the demands of a regular classroom. Interpretation of Function 2 is more difficult, with mixed predictive ability of the component variables. Group centroids and function characteristics suggest that the function is related to a behavioral component not measured by the school maladaption function. Function 3 scores appear to relate to an absence of poor socialization and denote the degree to which the students seek socialization. The descriptions given to all functions are tentative; more study of larger samples is clearly indicated.

The evidence should not be viewed as supporting the replacement of current assessment proce-

TABLE 4 Univariate F Tests for the 20 PIC-S Variables

Variable (Scale)	F	Significance
Adjustment	36.74	.0000
Intellectual Screening	34.37	.0000
Achievement	33.01	.0000
Factor IV	32.15	.0000
Development	27.20	.0000
Psychosis	22.95	.0000
Social Skills	19.41	.0000
Depression	17.91	.0000
Factor II	16.46	.0000
Factor I	15.63	.0000
F	14.61	.0000
Delinquency	13.10	.0000
Anxiety	9.77	.0001
Family Relations	5.66	.0013
Withdrawal	5.63	.0014
Defensiveness	4.55	.0051
Factor III	4.08	.0091
Lie	3.54	.0179
Somatic Concern	1.26	.2933

dures. The strength of the demonstrated discrimination should be interpreted as strong validity evidence for the application of the PIC-S for screening and confirmation of existing educational assessment methods. In addition to assisting with diagnostic decisions, the PIC-S provides considerable behavioral information that would be helpful for individually designed educational programming.

It is important to note that the results were obtained from a sample of male elementary students from a homogenous geographic setting. Results are specific to males only, and females could yield contrasting data. The impact of single state's (Iowa) special-education guidelines also poses generalization restrictions on the findings. Such limitations are common to many initial validation studies that use an accessible sampling population. As the use of the PIC and the PIC-S increases, additional validation efforts should result. The

TABLE 3 Group Centroids for the Four Groups Using 20 PIC-S Variables

Group	Function 1	Function 2	Function 3
I (Regular-education)	−2.04947	.57041	.27906
II (Learning-disabled)	.11319	−1.17950	−1.28855
III (Mentally disabled)	1.97757	1.86583	−.16126
IV (Emotionally disabled)	1.17142	−1.42206	1.19659

TABLE 5 Group Mean Scores Classified
According to Clinically Relevant T Scores

Variable	Criteria		
	Mean T Score		Group(s)
Lie	$\geq$ 59		None
F	$\geq$ 100		None
Defensiveness	$\geq$ 69		None
Adjustment	$\geq$ 59		II, III, IV
Achievement	$\geq$ 60		II, III, IV
Intellectual			
Screening	$\geq$ 70	($\geq$ 89[a])	III
Development	$\geq$ 60		II, III
Somatic Concern	$\geq$ 70		None
Depression	$\geq$ 70		None
Family Relations	$\geq$ 60		None
Delinquency	$\geq$ 80		None
Withdrawal	$\geq$ 70		None
Anxiety	$\geq$ 70		None
Psychosis	$\geq$ 80		None
Hyperactivity	$\geq$ 60		IV
Social Skills	$\geq$ 70		None
Factor I	$\geq$ 70		IV
Factor II	$\geq$ 70		None
Factor III	$\geq$ 70		None
Factor IV	$\geq$ 70		III

[a] Second order level of significance (severity).

generalizability of the present investigation would greatly improve if sample restrictions were removed by replication attempts. A cross-validation sample tested by existing function coefficients would increase the acceptability of the present validity evidence.

The appeal of a valid, cost-efficient measure such as the PIC or PIC-S is increased as educational resources become restricted. The application of the PIC or PIC-S to screen for handicapping conditions could be used to reduce the number of complete and costly educational evaluations. When used by a competent diagnostician, a more appropriate allocation of professional time should result.

References

Achenbach, T. M.(1978). Psychopathology of childhood: Research problems and issues. *Journal of Consulting and Clinical Psychology, 46,* 759–776.

DeKrey, S. J. & Ehly, S. W. (1981). Factor/cluster classification of profiles from Personality Inventory for Children. *Psychological Reports, 48,* 843–846.

Freeman, M. (1971). A reliability study of psychiatric diagnosis in childhood and adolescence. *Journal of Child Psychology and Psychiatry, 12,* 43–54.

Gerken, K. C. (1979). Translating data into decision making. In *The identification of emotionally disabled pupils: Data and decision making.* Des Moines, IA: Division of Special Education, Iowa Department of Public Instruction and Midwest Regional Resource Center.

Hathaway, S. R. & McKinley, J. C. (1951). *The Minnesota Multiphasic Personality Inventory Manual* (rev. ed.) New York: Psychological Corporation.

Lachar, D. & Gdowski, C. L. (1979). *Actuarial assessment of child and adolescent personality: An interpretive guide for the Personality Inventory for Children.* Los Angeles: Western Psychological Services.

Lachar, D., Gdowski, C. L., & Snyder, D. K. (1982). Broadband dimensions of psychopathology: Factor scales for the Personality Inventory for Children. *Journal of Consulting and Clinical Psychology, 50,* 634–642.

McCarthy, J. M. & Paraskevopoulus, J. (1969). Behavior patterns of learning disabled, emotionally disturbed and average children. *Exceptional Children 36* (2), 69–74.

Mowder, B. A. (1980). Pre-intervention assessment of behavior disordered children: Where does the school psychologist stand? *School Psychology Review,* 9 (1), 5–13.

Wirt, R. D. & Broen, W. E., Jr. (1956). The relation of the children's manifest anxiety scale to the concept of anxiety as used in the clinic. *Journal of Consulting and Clinical Psychology, 20,* 482.

Wirt, R. D., Lachar, D., Klinedinst, J. K., & Seat, P. D. (1977). *Multidimensional description of child personality: A manual for the Personality Inventory for Children,* Los Angeles: Western Psychological Services.

Validation of Discriminant Analysis in Marketing Research

MELVIN R. CRASK
WILLIAM D. PERREAULT, JR.

Introduction

Since marketing researchers first were introduced to discriminant analysis nearly 20 years ago [10, 16], it has become a widely used analytical tool

"Validation of Discriminant Analysis in Marketing Research," Melvin R. Crask and William D. Perreault, Jr., Vol 14 (February 1977), pp. 60–68. Reprinted from the *Journal of Marketing Research,* published by the American Marketing Association. Melvin R. Crask is professor of Marketing at the University of Georgia. William D. Perrault, Jr. is professor of Business Administration at the University of North Carolina, Chapel Hill.

[4–6, 18, 21, 23, 28, 31–33, 35, 36, 38]. The adoption of discriminant function analysis (DFA) techniques has been rapid because marketing researchers and managers alike frequently are concerned with the nature and strength of the relationship between group membership (for example, different market segments) and individual characteristics (such as life style measures). Furthermore, the widespread availability of easy-to-use discriminant analysis computer programs has facilitated implementation of these techniques [7, 8, 26]. The use of DFA in marketing research has proved most beneficial for three major purposes: (1) developing predictive models to *classify* individuals into groups [5, 28, 31, 32], (2) "*profiling*" characteristics of groups which are most dominant in terms of discrimination [4, 6, 21, 27, 33], and/or (3) identifying the major underlying *dimensions* (i.e., discriminant functions) which differentiate among groups [1, 10, 16, 18, 23].

The general applicability of DFA and the ease of computation, however, pose a question which all too often is ignored by marketers who use DFA in sample-based research. How valid are sample-based discriminant analysis results with respect to the broader population of interest? The issue of validity can be raised for each of the purposes of DFA. First, is actual classification potential as high as sample estimates indicate? Second, are the true population profiles what they appear to be from the sample results? Third, are the underlying sample-based dimensions generalizable to the population? The implications of a negative response to any of these questions are evident; no marketing manager wants to commit the resources of his firm on the basis of possibly inaccurate results, and scholars do not wish to misdirect the body of knowledge in an area by reporting sample-specific findings. Although validation problems are not restricted to small-sample research, the issue becomes critical as the sample size is decreased. However, marketing researchers often are forced to use small samples which severely limit their ability to answer the aforementioned questions affirmatively. Thus, the purpose of this report is to discuss the validation of small-sample discriminant analysis results in marketing research.

Toward this objective, a brief review of previous work in this area is provided. Two recently proposed improved alternatives are discussed, and their advantages and limitations examined. The proposed methods of validation are illustrated in the context of a salesman-selection problem. Finally, the operational necessities for using these methods are discussed and other areas of marketing research where the techniques could be applied are considered.

Approaches to Validation

As with other optimizing procedures, the ability of DFA to "find" strong relationships is the cause of the concern over the results obtained because the procedure may simply be capitalizing on relationships that exist as an artifact of the sample. Obviously, the smaller the sample the greater the chance that such spurious relationships will influence the generalizability of the results.

The potential for bias in the use of small samples in discriminant analysis long has been realized in the marketing literature [11, 14, 23, 24]. Several discussions of alternative approaches to validating results have been presented [11, 23, 24], although most have focused on the bias in the error rates of classification. Montgomery [23] recently provided an excellent and concise critique of these approaches. They are reviewed only briefly here to examine their shortcomings and thus give a clearer concept of what a good validation technique should provide.

The most frequently suggested validation approach is the holdout method, in which the sample is randomly split. One of the subsamples is used to develop estimates of the discriminant coefficients, and these coefficients are applied to the observations in the other subsample for classification purposes. With a large data base this approach may be appropriate. In small-sample research its use is impractical because splitting an already small sample makes the derived coefficients even less reliable; the error rates in classification may not be representative of the function which would be derived with the total sample. Furthermore, as typically applied this approach is only useful in considering classification and does not help in determining the validity of the profiles or the underlying dimensions.

Monté Carlo simulations also have been suggested as a mechanism for evaluating discriminant results [11]. Synthetic data are generated and discriminant functions are derived with the same degrees of freedom as the original data. This approach is very useful when the predictors are independent, but this is usually not the case with marketing data. When the predictors are not independent the problem is straightforward: in generating synthetic data it is impossible to model the covariance structure between the predictor variables [23, p. 259]. This method holds great potential if the problems of generating data can be overcome, but until that time results from this approach do not adequately address the problems of marketers.[1]

In a recent study, Montgomery [23] used another method of evaluating the classification results of discriminant analysis. He randomly assigned observations to groups and computed discriminant scores. By repeating this procedure several times, one can compare the results of these classifications with the true group results. By use of the actual data instead of synthetic data, the interrelationships among the sample data are preserved. This approach is interesting and the writers have found it a useful method for evaluating classification results. However, one must question the comparisons being made. Researchers typically are interested in how well the DFA performs in an absolute sense rather than how it performs in comparison with a similar random chance model. Furthermore, like the holdout method, this procedure does not provide a means of evaluating the validity of the DFA results used to define differences between groups.

At a basic level, the validity of DFA results resides in the stability of the coefficients derived. Classifying, profiling, and evaluating the underlying discriminant dimensions all are based directly or indirectly on these coefficients. Considering this fact in light of the foregoing review, marketing researchers would like a validation procedure which provides a mechanism that uses all of the information in the sample data for evaluating the stability of parameter estimates while allowing unbiased estimation of error rates.

Recently, independent research efforts directed toward these criteria have generated two very similar alternatives for validation of DFA. One of these, known as the U-method [19], focuses on the issue of classification errors. The other, generally referred to as "jackknife analysis" [25, 34], focuses on coefficient stability. When the two methods are applied simultaneously, a very useful validation procedure evolves. In fact, the researcher is able to evaluate not only classification

---

[1] If the predicator variables are factor scores or similar independent, normally distributed measures, the Monté Carlo method remains a useful alternative.

rates and coefficient stability, but also the expected classification error rates regardless of sample size. A marketing research example is used to illustrate these approaches after a review of the logic upon which they are based.

Validation: The Jackknife and the U-Method

The jackknife statistic is a general method for reducing the bias in an estimator while providing a measure of the variance of the resulting estimator by sample reuse. The result of the procedure is an unbiased, or nearly unbiased, estimator and its associated approximate confidence interval. Because of the versatility of the technique, it is named after another tool of many uses, the Boy Scout's jackknife [34].

The essence of the jackknife approach is to partition out the impact or effect of a particular subset of the data (e.g., a single case) on an estimate derived from the total sample. Before considering the jackknife as applied in discriminant analysis, it is useful to review its general form.

Suppose that a random sample of size N is under consideration, and that there is an observed variable X for each of the N sampling units. Let the sample be partitioned into k subsets of size M (i.e., $kM = N$), so that a new random sample can be formed by arbitrarily deleting one of the subsets from the original sample. Let θ' be defined as an estimator using *all* the sample values of X, and θ'_i an estimator defined on only those values of X which remain after the "i^{th}" subset of size M has been deleted from the total sample.

The first step in computing the jackknifed estimator is to compute k different *pseudovalues* [29], which are weighted combinations of the θ' and θ'_i values. Specifically, the pseudovalues are:

(1) $J_i(\theta') = k\theta' - (k - 1)\theta'_i, i = 1, \ldots, k.$

Then the jackknife statistic is:

(2) $J(\theta') = \left[\sum_{i=1}^{k} J_i(\theta') \right] \Big/ k = k\theta' - (k - 1)\bar{\theta}'_i;$

thus, the jacknife is simply the average of the pseudovalues.

The logic of the partitioning process by which the pseudovalues are computed (equation 1) is clarified by considering a simplified example. Suppose that a sample of size 5 provides values of 3, 2, 1, 5, and 4. By use of the pseudovalue equation, the impact of the first observation on an estimator, the sample mean, can be eliminated by substituting into equation 1 and computing

$$5[(3 + 2 + 1 + 5 + 4)/5] - 4[(2 + 1 + 5 + 4)/4] = 3.$$

If this process is repeated, with systematic deletion of a different subset (observation) each time, and the computed pseudovalues are averaged, the resulting statistic is the jackknifed mean on the sample.

Computing the jackknife for a mean as in the example illustrates the logic of the pseudovalue derivation but is of little consequence. However, when applied to more complex estimators (e.g., discriminant function coefficients), the pseudovalues and subsequent jackknife estimators have important properties. In particular, the pseudovalues can be treated as independent, identically distributed random variables, and hence can be used to obtain approximate confidence intervals for the jackknife estimate. The confidence intervals can be tested with Student's t having $k - 1$ degrees of freedom [25, 34]. Moreover, when applied to linear estimators, the jackknife is important in itself because the bias in the jackknife estimate has been shown to be less than the bias in the original sample estimate, θ', and frequently approaches zero [14].[2] Furthermore, little efficiency is lost by structuring the subsamples to be very small. Thus the jackknife is very useful in determining the stability of the estimates in many situations where marketing researchers use statistics and particularly in discriminant analysis.

The generalized forms for the computation of pseudovalues (equation 1) and for the jackknife statistic (equation 2) are directly applicable to each of the discriminant coefficients of a discriminant analysis. The procedure is briefly here outlined and clarified further by an illustration.

Step 1. To use jackknife procedures in discriminant analysis, the researcher partitions his sample into k subsamples. The standard discrimi-

nant function is computed by combining all of the subsamples.

Step 2. Then, a discriminant function is computed by using $k - 1$ of the subsamples (i.e., holding out one subsample). This step provides estimates corresponding to the θ'_i and is repeated k times, with a different subsample omitted each time. The pseudovalues, as indicated in equation 1, are derived by weighting and subtracting these coefficients from the estimates of step 1.

Step 3. The jackknifed coefficients are computed from the averaged pseudovalues as indicated in equation 2. The use and interpretation of the jackknifed coefficients are discussed in the next section. The main benefit of this type of analysis is that it provides a foundation for evaluating the *stability* of the coefficients.

If, however, estimation of *error rates* in classification is the principal concern, Lachenbruch and Mickey [19] have proposed a similar sample reuse procedure, the U-method. To apply the U-method, one observation is omitted from the sample and a discriminant function is computed by using the remaining observations. This function is used to determine the group membership of the omitted observation. By repeating this procedure for all observations in the sample, an estimate of misclassification can be obtained for each group by totalling the number of misclassifications in the group and dividing by the total number of cases in the group. Although it might appear that a complete new discriminant analysis would be required for each omitted observation, Bartlett [2] describes a procedure in which only one explicit matrix inversion is required; in fact the U-method is named for this inversion procedure.

By the U-method, there is no effective way to determine the stability of the coefficients of the discriminant function because the U-method does not have the bias reduction properties of the jackknife.[3] However, the procedures used for the jackknife and for the U-method are very similar in that both involve an efficient partitioning of the sample. It is therefore possible to combine the procedures simultaneously to achieve both estimates of the error rates of classification and stability of the coefficients, as shown in the following example.

---

[2] More specifically, the jackknife *completely* eliminates biases which are inversely proportioned to sample size [20, p. 570]. Miller [22], however, has pointed out that in some nonlinear estimation situations the jackknife does not reduce bias; his caveats are useful to researchers who wish to apply the jackknife in situations dissimilar to those discussed in this article.

[3] The bias reduction properties of the jackknife are a function of the weighting procedure used in its calculation. These weights are not applied in the U-method.

The Research Problem

The data for this example were provided by a major manufacturer of consumer goods and concern the selection of salesmen for intermediate markets. Through an outside agency, the company administers personnel tests to new salesmen. After acquiring on-the-job experience, each salesmen is classified by his sales manager as a high or low performer. This illustration is based on a two-group (high versus low performers) discriminant analysis using six personnel test scores as predictors. Data for 24 salesmen are used in the analysis. This situation was chosen as an illustration for several reasons. First, a two-group discriminant analysis, in contrast to a k-group case, allows easier understanding of the techniques presented here; this is not to imply that these procedures are not useful with the k-group case. Instead, with two groups of equal size the issue of classification errors can be evaluated directly by examining discriminant function scores, rather than Mahalanobis' D-square. Second, the use of few cases (salesmen) in relation to the number of variables results in the type of bias which is of concern in discriminant analysis and allows a rather complete presentation of results to facilitate understanding. Finally, the situation is typical of the type of application where the sample is not large enough for traditional holdout methods but where there is very real managerial and legal [15] concern for validation of results.[4]

Analysis

When a standard discriminant analysis is computed with the sample, *all* of the salesmen are classified correctly. Most researchers would question the validity of such results and their applicability to the broader population of sales applicants. Thus, let us evaluate the validity of the discriminant results in three stages. First, the U-method is used to calssify salesmen. Second, the jackknifed discriminant functions are computed and their stability evaluated. Third, the two approaches are integrated to evaluate coefficient stability, classification error rates, and expected classification results if an infinite sample were available.

U-Method Results

As mentioned, developing a discriminant function using all of the subjects and then classifying the same salesmen with this function yielded perfect classification. This result can be expected because, with a small number of observations, each subject has significant impact upon the coefficients of the function. With the U-method, any given observation has no effect upon the coefficients of the function used to classify that observation.

To apply the U-method, the salesmen were grouped randomly into 12 pairs, each pair containing a high and a low performer. By holding out one pair, a discriminant function was computed with the remaining 11 pairs and the pair held out was classified by use of this function. Repeating this process for each pair allowed each salesman to be classified by a discriminant function in which his test scores had no influence upon the coefficients. Table 1 presents the results. Three salesmen (12.5%) are misclassified by the U-method whereas none was misclassified by a standard discriminant analysis. Thus, in this case, the U-method provides a more conservative (higher) estimate of classification error than that observed by use of the total sample. In a Monté Carlo study, Lachenbruch and Mickey [19] found a similar pattern of classification errors, and suggest that the more conservative (U-method) estimates are a better estimate of the error than can be achieved by use of the total sample.

Jackknife Results

The U-method results have provided a basis for evaluating the classification model derived from the discriminant analysis. If the researcher is concerned instead with characterizing between-group profiles or the underlying dimensions which discriminate between groups, the jackknife method would be used. For the salesman-selection problem, the first step is to compute the pseudovalues for the discriminant coefficients. These values are computed according to equation 1, by use of each of the 12 discriminant equations derived from the U-method (Table 2) and the equation which results

---

[4] With regard to the legal requirements of the salesman selection problem, it should be noted that this analysis is based only on salesmen who actually were hired; for strict predictive validity, all *applicants* should be hired, and *then* the relationship between test scores and performance evaluated.

TABLE 1 Classification of Salesmen by Standard Discriminant Analysis, *U*-Method, and Jackknife

Salesman Pair "i"	Total Sample Results		*U*-Method Results		Jackknife Results	
	High Performers	Low Performers	High Performers	Low Performers	High Performers	Low Performers
1	0.8601	0.4895	0.6467	0.4364	1.0595	0.6533[a]
2	1.1621	0.2152	1.2877	0.2071	1.2048	0.3818
3	1.0689	0.0772	1.0853	0.0773	0.9929	0.1488
4	1.2396	0.1209	1.4907	0.1712	1.1281	0.2961
5	1.2199	0.0649	1.2904	0.0568	1.1539	0.1425
6	0.5657	0.0722	0.4566[a]	0.0761	0.6671	0.0692
7	0.9454	-0.0143	0.9001	-0.0138	0.9530	0.1168
8	0.8295	0.0988	0.7885	0.0795	0.8325	0.1619
9	1.0819	0.0499	1.2948	0.0452	0.9122	0.1090
10	0.5137	0.1702	0.4119[a]	0.1489	1.5883	0.1744
11	0.5126	0.0343	0.4541[a]	-0.0073	0.4988[a]	0.1492
12	0.6683	-0.0522	0.6363	0.0662	0.6643	0.0479
Discriminant score average	0.8898	0.1106	0.8936	0.1009	0.8880	0.2042

[a] These scores result in misclassification of the associated salesman.

from the analysis on the total sample. The jack-knifed coefficients (which yield the jackknifed discriminant function) are computed as the averages of the pseudovalues; the jackknifed values are presented in Table 2.

The stability of the jackknifed discriminant coefficients is evaluated directly by computing the traditional standard errors for each set of pseudovalue coefficients (computation of the jackknife standard error follows the traditional calculation of the standard error of the mean). Because the jackknifed coefficients approximate the t distribution [25], each coefficient can be divided by its associated standard error (shown in parentheses under the jackknifed equation in Table 2) to give a t-value; the degrees of freedom for the t-value are based on the number of partitions of the sample (pseudovalues) minus 1, or 11 in this case. In this analysis, only the t-value for the coefficient of test 6 exceeds the critical value, and thus it may be the only true discriminator between the salesman performance groups.

Although the jackknife approach is used mainly for evaluating stability of coefficients, the resulting equations can be used to compute discriminant scores on which the salesmen can be classified. The jackknife classification in this problem results in two misclassifications (see Table 1).

The jackknife provides a basis on which the strength of classification, not just the number classified, can be evaluated. Specifically, confidence intervals about the jackknife discriminant score for a salesman can be computed from the standard errors of the scores obtained from the pseudovalue equations. The larger the confidence interval which is possible without containing the cutpoint between groups, the more securely a salesman is classified. Six salesmen were found to have discriminant scores more than two deviations from the cutpoint, and four other salesmen have values more than 1.6 deviations from the cutpoint. The rest were not so securely classified.

A Combined U-Method/Jackknife Approach

The foregoing approaches are valid if the researcher is interested in either stability or error classification. If it is desirable to evaluate both simultaneously, a combined approach must be used because of bias introduced when both issues are evaluated simultaneously by only one of the methods. A combined approach, suggested in [25], is illustrated and an extension which may be very important in pretest research is discussed.

Suppose each of the 12 salesmen in one group is paired with a salesman in the other group. Holding out one of these pairs and performing a complete jackknife analysis on the remaining 11 pairs yields 11 pseudovalue equations and the resulting jackknifed equation. The pair held out then can be classified by use of each of the 11 pseudovalue equations as well as by the jackknifed equation. Twelve legitimate cross-validations thus are provided for each member of this pair and the pair *was not influential* in the determination of any of the equations. Repeating this procedure for each of the 12 pairs provides a total of 288 cross-validations, and thus yields a good measure of the performance of the variables. In the salesman-selection problem, an average error rate of 13.6% was obtained for these 288 cross-validations, although the error rates for individual salesmen ranged from 4.5 to 45%. These extremes indicate a tremendous influence by the individual salesmen upon the discriminant function.

The pseudovalues also can be used to calculate confidence intervals on the coefficients, as in the foregoing simple jackknife situation. The average coefficient values and their associated confidence intervals yield the equation:

$$D'_j = -4.728 + 0.069T_1 + 0.006T_2 - 0.008T_3 \\ + 0.002T_4 - 0.004T_5 - 0.015T_6.$$
$$(1.162)\ (.016)\ (.002)\ (.005)\ (.003)\ (.004)\ (.001)$$

Tests 1, 2, and 6 all can be considered discriminators from this result, whereas in the original jackknifed equation, D_j (Table 2), only test 6 yielded a significant coefficient. However, D'_j should provide much more accurate coefficient estimation because this estimation is based on more than six times the number of observations used in the first equation.

The literature in the area of error rates in classification suggests yet another possible extension when these two methods are used simultaneously [25]. In general, this extension seeks to answer the question, "What error rate in classification might be expected if the researcher had available an infinite population rather than a limited sample?" Even if these six tests were perfect predictors of high and low performers, dispersion around the population centroids would occur because of the variances of these population means. Depending upon the magnitude of these variances, some

TABLE 2 Adjusted Discriminant and Jackknife Coefficients for Salesman Discriminant Analysis

Salesman Pair Omitted for Pseudovalues	Discriminant Coefficients						
	Test 1	Test 2	Test 3	Test 4	Test 5	Test 6	Constant
(none)	0.03420	0.00317	−0.00915	−0.00295	−0.00194	−0.01413	−0.83552
1	0.04023	0.00337	−0.00665	−0.00225	−0.00355	−0.01370	−1.63210
2	0.02726	0.00310	−0.00535	−0.00311	−0.00051	−0.01245	−0.58232
3	0.03515	0.00304	−0.00946	−0.00322	−0.00191	−0.01420	−0.85930
4	0.05743	0.00511	−0.01741	0.00073	−0.00713	−0.01467	−2.31195
5	0.03845	0.00385	−0.01061	−0.00288	−0.00255	−0.01445	−1.08386
6	0.04356	0.00574	−0.01179	−0.00290	−0.00270	−0.01279	−1.72382
7	0.03457	0.00367	−0.00952	−0.00330	−0.00153	−0.01369	−0.90551
8	0.03321	0.00246	−0.00971	−0.00365	−0.00198	−0.01380	−0.60924
9	0.03452	0.00285	−0.00860	−0.00679	−0.00138	−0.01358	−0.63029
10	0.03059	0.00083	−0.00604	−0.00344	−0.00177	−0.01488	−0.55750
11	0.02813	0.00209	−0.00866	−0.00382	0.00004	−0.01443	−0.30358
12	0.03452	0.00296	−0.01052	−0.00256	−0.00266	−0.01395	−0.73444
Jackknife coefficients	.009	.002	−0.005	−0.001	−0.001	−0.014	.913
Standard error of coefficient	(.026)	(.004)	(.010)	(.005)	(.006)	(.002)	(1.890)

error in classification may occur regardless of the sample size. Finding this error rate is analogous to a determination of the probability of making a Type 1 error. In the foregoing analyses, the researcher knows from the sample results that approximately a 13% error rate is expected. Although this rate is much better than the 50% error rate one would expect if there were no discriminatory power, management might desire a function which would perform at some specified level (e.g., yield no more than 10% error). The researcher or manager would like to know, in effect, whether or not these six tests would meet this criterion if the sample size were increased. The following discussion illustrates how this problem can be attacked.

The population centroids are known to be zero and one, and variation around these centroids can be attributed to two sources: the population standard error and the sample size used. The larger the sample size, the smaller this effect will become. Total variation as well as the variation due to the sample size can be calculated and used to estimate the population standard error. This standard error then can be used to estimate the lower limit of the error rates one might expect.

The total variation is measured by the variation in the mean estimates (the jackknifed estimates). Because grouped data were used, a pooled estimate of the total variation must be obtained by (1) calculating the variance of each group of salesmen (the average of the sum of squared deviations of the mean estimates of the salesmen from their group centroid), (2) adding the variances of both groups, and (3) averaging this sum to obtain a pooled estimate of the total variation.

The variation caused by the sample size is captured by the dispersion of the pseudovalues around the mean estimates. This variation can be derived by (1) calculating the variance of the mean estimate of each salesman (the average of the sum of squared deviations of the salesman's pseudovalues from his mean estimate, or jackknifed value), (2) summing the variances of all salesmen, and (3) averaging this sum to obtain a measure of the variation due to the sample size. The population standard error then can be calculated by subtracting the variation due to the sample from the total variation.

By this procedure the population standard error obtained is 0.2369. Though not exact, this estimate indicates that both population means are approximately two standard deviations from the

cutpoint. Depending upon the assumed distribution, one can estimate the error rate expected regardless of the sample size. For example, the expected error rate for this problem is approximately 5% if a normal distribution is assumed.

Thus, this extension uses small-sample results to develop an estimate of the classification rates that might be expected with much larger samples. This information would be most useful in market pretest research where data collection is very expensive. The researcher can evaluate the results of a small-sample analysis and have a more complete basis for determining whether the expense of collecting data on a large sample would be warranted. It should be emphasized that such "infinite sample" estimates are, in a practical sense, limited to the two-group discriminant case. Though there has been some work to generalize this type of analysis to the multiple-group situation [12], operational solutions have not been developed.

Discussion and Implications

Stability

The researcher interested in validating the stability of discriminant coefficients quickly confronts the problem that "canned" computer programs are not available to implement the jackknife method. To compute the jackknife equations with present software requires many successive computer runs, each with selective deletion of subsamples from the analysis. This process becomes cumbersome (and expensive) when many subsamples are used.

This is not to suggest that coefficient stability should be ignored. Because the stability of the jackknife can be computed with relative ease and accuracy from relatively few pseudovalues, the researcher can divide his sample randomly into a few *nonoverlapping* subgroups. Pseudovalues and the jackknifed equation can be computed by systematically holding out one sample subgroup and computing the pseudo-equations. The confidence intervals on the coefficients (or the discriminant scores) then would be computed as described. Although this type of partitioning of the data is not as "efficient" as holding out individual observations, it is a relatively simple way to evaluate the stability of the coefficients and the general validity of the results of the research.

Classification

In an ideal sense the best estimates of classification accuracy of sample results would be based on the combined jackknife/U-method. Here again good computer software is not available. However, in the salesman performance example, the U-method provided nearly the same error rate estimation alone as it did in combination with the jackknife. The writers have found this similarity of performance to be typical.

Computationally, the U-method can be implemented very easily on the computer. Very flexible discriminant analysis computer programs recently have become available which provide this analysis as an option [8, 9]; one of these [8] is widely available as a program in the BMDP series. Because the classification is based on a jackknifed Mahalanobis' D-square, the researcher can (1) prespecify the prior probabilities of group membership (as when the researcher knows that his sample is not stratified in the same proportion as the total market), (2) use more than two groups, and (3) use different sample sizes for each group.

When the estimate of error classification rates is computed as suggested heretofore, cost is not prohibitive. For example, to provide cost data which might be illustrative of a typical marketing research situation, a four group/six predictor variable discriminant analysis was computed for 100 cases by use of the program described in [8, p. 411–51]. The analysis, including classification based on sample reuse, cost less than three dollars when run on an IBM 360 computer. In short, the incremental cost of the extended analysis is minor.

Other Jackknife Applications

The jackknife approach has been discussed in the context of bias estimation in discriminant analysis. Several other areas where the jackknife has potential for marketing researchers are worth noting briefly.

In general, if an unbiased estimator exists, the jackknife will yield such an estimator. For example, Brillinger [3] has advocated the jackknife in developing estimates of the mean and variance of response variables in market surveys. More specifically, the jackknife is appropriate in any research situation where the analysis procedures (and test statistics) assume normality but the data seriously violate this assumption. This situation

is common in marketing research applications of (multivariate) linear statistics, especially when stratified sampling procedures have been used. In this regard, Ireland and Uselton [17] demonstrate the superiority of the jackknifed F-statistic over the conventional F-statistics under conditions of nonnormal data in regression analysis; the jackknife was more robust to violations of the normality assumptions. Similarly, Mantel [20] has discussed the applicability of the jackknife in other estimation situations where it is not advantageous to assume a form of the distribution for a parameter.

It also has been suggested that the jackknife may be very useful with ratio estimators. For example, consider the case of a company which would like to introduce a product in select test markets and, on the basis of test market sales, to estimate sales levels which might be expected if the product received national distribution. One approach might be to evaluate the ratio of sales to population in the test markets, and then to extrapolate to the total population. There is a major problem, however, because the ratio will vary substantially, introducing bias, depending on which test markets are chosen and any spurious forces in those markets. The jackknife is well applied in such situations to reduce bias in estimation. Raj [30] and Gray and Schucany [13] provide detailed discussion and illustration of the use of jackknife approaches to ratio estimators of this type.

It also has been suggested that the jackknife may be useful in research models based on stochastic processes. There is typically substantial bias in the estimates derived by stochastic models in marketing and consumer behavior research, and in many circumstances this bias can be reduced with the jackknife [37].

The foregoing discussion is not intended to serve as an exhaustive listing of jackknife applications, but rather to emphasize that marketing researchers may find it a useful tool in many situations where bias in results can lead to costly mistakes in marketing strategy.

Summary

This report has addressed the issue of validity in sample-based marketing research, with particular emphasis on problems which arise in applications of discriminant analysis. Two approaches to the validation problem, the U-method and the jackknife, are discussed and illustrated in the context of a marketing research problem. Operational procedures for implementing these methods are suggested, as well as other areas of marketing research where these methods may be valuable aids in analysis and decision making.

References

1. Banks, Seymour. "Why People Buy Particular Brands," in Robert Ferber and Hugh G. Wales, eds., *Motivation and Market Research.* Homewood, Illinois: Richard D. Irwin, Inc., 1958, 277–93.
2. Bartlett, Maurice S. "An Inverse Matrix Adjustment Arising in Discriminant Analysis," *Annals of Mathematical Statistics,* 22 (March 1952), 167.
3. Brillinger, David R. "The Application of the Jackknife to the Analysis of Sample Surveys," *Journal of the Market Research Society,* 8 (April 1966), 74–80.
4. Brody, Robert P. and Scott M. Cunningham. "Personality Variables and the Consumer Decision Process," *Journal of Marketing Research,* 5 (February 1968), 50–7.
5. Churchill, Gilbert A., Jr., Neil M. Ford and Urban B. Ozanne. "An Analysis of Price Aggressiveness in Gasoline Marketing," *Journal of Marketing Research,* 7 (February 1970), 36–42.
6. Claycamp, Henry J. "Characteristics of Thrift Deposit Owners," *Journal of Marketing Research,* 2 (May 1965), 163–70.
7. Dixon, Wilfred J., ed. *BMD: Biomedical Computer Programs.* Berkeley: University of California Press, 1973, 211–54.
8. _____. *BMDP: Biomedical Computer Programs.* Berkeley: University of California Press, 1975, 411–52.
9. Eisenbeis, Robert A. and Robert B. Avery. *Discriminant Analysis and Classification Procedures.* Lexington, Massachusetts: Heath, 1972.
10. Evans, Franklin B. "Psychological and Objective Factors in the Prediction of Brand Choice: Ford versus Chevrolet," *Journal of Business,* 32 (October 1959), 340–69.
11. Frank, Ronald E., William F. Massey and Donald G. Morrison. "Bias in Multiple Discriminant Analysis," *Journal of Marketing Research,* 2 (August 1965), 250–8.
12. Glick, Ned. "Sample-Based Classification Procedures Derived from Density Estimates," *Journal of the American Statistical Association,* 67 (March 1972), 116–22.
13. Gray, Henry L. and W. R. Schucany. *The Generalized Jackknife Statistic.* New York: Marcel Dekker, Inc., 1972.
14. Green, Paul E. "Bayesian Classification Procedures in Analyzing Customer Characteristics," *Journal of Marketing Research,* 1 (May 1964), 44–50.
15. *Guidelines on Employment Selection Procedures.*

Washington, D.C.: Equal Employment Opportunity Commission, August 1970.
16. Harvey, John R. "What Makes a Best Seller," in Robert Ferber and Hugh G. Wales, eds., *Motivation and Market Research.* Homewood, Illinois: Richard D. Irwin, Inc., 1958, 361–81.
17. Ireland, M. Edward and Gene C. Uselton. "A Distribution Free Statistic for Management Scientists," in W. W. Menke and C. H. Whitehurst, eds., *Proceedings,* Eleventh Annual Meeting of the Southeastern Chapter, The Institute of Management Science, October 1975, 25–6.
18. King, William R. "Marketing Expansion—A Statistical Analysis," *Management Science,* 9 (July 1963), 563–73.
19. Lachenbruch, Peter A. and M. Ray Mickey. "Estimation of Error Rates in Discriminant Analysis," *Technometrics,* 10 (February 1968), 1–11.
20. Mantel, Nathan. "Assumption-Free Estimators Using U-Statistics and a Relationship to the Jackknife Method," *Biometrics,* 23 (September 1967), 567–71.
21. Massy, William F. "Discriminant Analysis of Audience Characteristics," *Journal of Advertising Research,* 5 (March 1965), 39–48.
22. Miller, Rupert G. "A Trustworthy Jackknife," *Annals of Mathematical Statistics,* 35 (December 1964), 1594–605.
23. Montgomery, David B. "New Product Distribution: An Analysis of Supermarket Buyer Decisions," *Journal of Marketing Research,* 12 (August 1975), 255–64.
24. Morrison, Donald G. "On the Interpretation of Discriminant Analysis," *Journal of Marketing Research,* 6 (May 1969), 156–63.
25. Mosteller, Frederick and John W. Tukey. "Data Analysis, Including Statistics," in G. Lindsey and E. Aronson, eds., *The Handbook of Social Psychology,* Vol. 2. Reading, Massachusetts: Addison-Wesley, 1968, 80–203.
26. Nie, Norman H., D. Hadlai Hull, Jean G. Jenkins, Karin Steinbrenner and Dale H. Brent, eds. *SPSS: Statistical Package for the Social Sciences.* Englewood Cliffs, New Jersey: McGraw-Hill Book Co., Inc., 1975.
27. Perreault, William D., Jr. and William R. Darden. "GRAFIT: Computer Based Graphics for the Interpretation of Multivariate Analysis," *Journal of Marketing Research,* 12 (August 1975), 343–5.
28. Pessemier, Edgar A., Philip C. Burger and Douglas J. Tigert. "Can New Product Buyers Be Identified?" *Journal of Marketing Research,* 4 (November 1967), 349–54.
29. Quenouille, M. H. "Notes on Bias in Estimation," *Biometrika,* 43 (December 1956), 353–60.
30. Raj, Deg. *Sampling Theory.* New York: McGraw-Hill Book Co., Inc., 1968.
31. Rao, Tanniru R. "Is Brand Loyalty a Criterion for Marketing Segmentation: Discriminant Analysis," *Decision Sciences,* 4 (July 1973), 395–404.
32. Robertson, Thomas S. and John N. Kennedy. "Prediction of Consumer Innovations: Application of Multiple Discriminant Analysis," *Journal of Marketing Research,* 5 (February 1968), 64–9.
33. Shuchman, Abe and Peter C. Riesz. "Correlates of

Persuasibility: The Crest Case," *Journal of Marketing Research*, 12 (February 1975), 7–11.

34. Tukey, John W. "Bias and Confidence in Not-Quite Large Samples" (abstract), *Annals of Mathematical Statistics*, 29 (June 1958), 614.

35. Uhl, Kenneth, Roman Andrus and Lance Poulsen. "How Are Laggards Different? An Empirical Inquiry," *Journal of Marketing Research*, 7 (February 1970), 51–4.

36. Utterback, James M. "Successful Industrial Innovations: A Multivariate Analysis," *Decision Sciences*, 6 (January 1975), 65–77.

37. Watkins, T. A. "Jackknifing Stochastic Process," unpublished Ph.D. dissertation, Texas Tech University, 1971.

38. Wind, Yoram. "Industrial Source Loyalty," *Journal of Marketing Research*, 7 (November 1970), 450–7.

CHAPTER FOUR

Multivariate Analysis of Variance

CHAPTER PREVIEW

As a theoretical construct, multivariate analysis of variance (MANOVA) has been known for several decades since Wilks' original formulation [20]. However, it was not until the development of appropriate test statistics with tabled distributions and the advent of high-speed computers with widely available programs to compute these statistics that MANOVA became a practical tool for applied researchers. Before reading introduction to MANOVA, please review the Definitions of Key Terms.

An understanding of the most important concepts of MANOVA should enable you to:

- Explain the difference between the univariate null hypothesis of ANOVA and the multivariate null hypothesis of MANOVA.
- Discuss the advantages of a multivariate approach to significance testing compared to the more traditional univariate approaches.
- State the assumptions for the use of MANOVA.
- Discuss the different types of test statistics that are available for significance testing in MANOVA.
- Describe the purpose of post hoc tests in ANOVA and MANOVA.
- Identify the techniques that can be used to conduct post hoc tests in MANOVA.
- Describe the purpose of multivariate analysis of covariance (MANCOVA).

KEY TERMS

Analysis of variance A statistical technique used to determine if samples came from populations with equal means. Univariate analysis of variance employs one dependent measure, while multivariate analysis of variance employs two or more to compare populations.

Covariate analysis Use of regression-like procedures to remove extraneous (nuisance) variation in the dependent variables due to one or more uncontrolled metric independent variables (covariates). The covariates are generally assumed to be linearly related to the dependent variables. After adjusting for the influence of the covariates, a standard ANOVA (or MANOVA) is carried out. This adjustment process usually allows for more sensitive tests of treatment effects.

Factor   A nonmetric independent variable, also referred to as a *treatment* or *experimental variable.*

Factorial design   A design with more than one factor (treatment). In factorial designs, we examine the effects of several factors simultaneously by forming groups based on all possible combinations of the levels of the various treatment variables.

Greatest characteristic root   A statistic for testing the null hypothesis in MANOVA. It is based on finding a linear combination of the dependent variables that maximizes group differences.

Hotelling's T^2   Used to assess the statistical significance of the difference between two sets of sample means. A special case of MANOVA for two groups/levels of a treatment variable.

Interaction Effect   In factorial designs, the joint effects of treatment variables in addition to the individual main effects.

Main effects   In factorial designs, the individual effect of each treatment variable on the dependent variable.

Multivariate normal distribution   A generalization of the univariate normal distribution to the case of p variables. A multivariate normal distribution of sample groups is a basic assumption required for the validity of the significance tests in MANOVA.

Null hypothesis   The hypothesis that samples do come from populations with equal means on either a dependent variable (univariate test) or a set of dependent variables (multivariate test). The null hypothesis can be accepted or rejected depending on the results of a test of statistical significance.

Scheffe's contrast method   A procedure for investigating *specific* group differences of interest in conjunction with ANOVA and MANOVA, for example, comparing group mean differences for all pairs of groups.

T-test   Used to assess the statistical significance of the difference between two sample means. A special case of ANOVA for two groups/levels of a treatment variable.

Treatment   The independent variable that a researcher manipulates to see the effect (if any) on the dependent variable(s). The treatment variable can have several levels (e.g., different intensities of advertising appeals might be manipulated to see the effect on consumer believability).

Type 1 error   The probability of rejecting the null hypothesis when it should be accepted, that is, concluding that two means are significantly different when in fact they are the same. Small values of alpha, also denoted as α (e.g., .05, .01) lead to rejection of the null hypothesis as untenable and acceptance of the alternative hypothesis that population means are unequal.

Type 2 error   The probability of accepting the null hypothesis when it should be rejected, that is, concluding that two population means are not significantly different when in fact they are.

Wilks' lambda   An alternative statistic for testing the null hypothesis in MANOVA. It is also referred to as the *maximum likelihood criterion.*

Vector   A vector consists of a set of real numbers, e.g., $X_1 \ldots X_N$.

Vectors can be written as either column vectors or row vectors. Column vectors are considered conventional and row vectors are considered transpose. Column vectors and row vectors are shown as follows:

$$X = \begin{bmatrix} X_1 \\ X_2 \\ \\ X_N \end{bmatrix} \qquad X^T - [X_1 \, X_2 \, . \, . \, . X_N]$$

Column vector   Row vector

The T on the row vector indicates that it is the transpose of the column vector.

What Is Multivariate Analysis of Variance?

Like analysis of variance (ANOVA), multivariate analysis of variance (MANOVA) is concerned with differences between groups (or experimental treatments). However, ANOVA is termed a *univariate* procedure, since we use it to assess group differences on a single metric dependent variable. MANOVA is termed a *multivariate* procedure, since we use it to assess group differences across multiple metric dependent variables simultaneously (i.e., in MANOVA, each treatment group is observed on two or more dependent variables).

As statistical inference procedures, both ANOVA and MANOVA are used to assess the statistical significance of differences between groups. In ANOVA, the null hypothesis tested is the equality of dependent variable means across groups. In MANOVA, the null hypothesis tested is the equality of vectors of means on multiple dependent variables across groups. The distinction between the hypotheses tested in ANOVA and MANOVA is illustrated in Figure 4.1.

Both ANOVA and MANOVA are particularly useful when used in conjunction with experimental designs, that is, research designs in which the researcher directly controls and manipulates one or more independent variables to determine the effect on one (ANOVA) or more (MANOVA) dependent variables. ANOVA and MANOVA provide the tools to judge the reliability of any observed effects (i.e., whether an observed difference is due to a treatment effect or to random sampling variability.)

Case 1: Difference Between Two Independent Groups

To introduce the practical benefits of a multivariate analysis of group differences, we begin our discussion with one of the best-known experimental designs: the two-group randomized design. In this design, each respondent is randomly assigned to only one of the two levels (groups) of the independent variable. In the univariate case, a single metric dependent variable is measured and the null hypothesis is that the two groups have means that are equal. In the multivariate case, multiple metric dependent variables are measured and the null hypothesis is that the two groups have vectors of means that are equal. For a two-group

FIGURE 4.1 Null hypothesis testing of ANOVA and MANOVA.

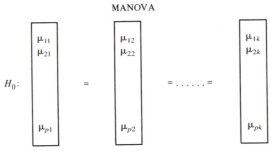

ANOVA

$$H_0 : \mu_1 = \mu_2 = \ldots \mu_k$$

Null hypothesis (H_0) = all the group means are equal, that is, they come from the same population.

MANOVA

$H_0:$

Null hypothesis (H_0) = all the group mean vectors are equal, that is, they come from the same population.

μ_{pk} = mean of variable p, group k

univariate analysis, the appropriate test statistic is the t-statistic (a special case of ANOVA); for a multivariate analysis, the appropriate test statistic is called *Hotelling's T^2* (a special case of MONOVA).

Univariate Approach: The T-Test

The t-test assesses the statistical significance of the difference between two independent sample means. For example, a researcher may expose two groups of respondents to different advertising messages and subsequently ask them to rate their likelihood of purchasing the advertised product on a 10-point scale. To determine whether the two messages lead to reliable differences in purchase intent, a t-statistic is calculated [the t-statistic is the ratio of the difference between sample means $(\overline{X}_1 - \overline{X}_2)$ to its *standard* error (an estimate of the degree of fluctuation between means to be expected due to sampling error rather than real differences between means)]. Large absolute values of the t-statistic lead to rejection of the null hypothesis of no difference in mean purchase intent between groups. Specifically, we determine a *critical value* ($t_{\text{crit.}}$) for our t-statistic by:

1. Specifying a type 1 error rate (denoted as α or *significance level* and equal to the probability of incorrectly rejecting the null hypothesis that $\mu_1 = \mu_2$), that is, concluding that the group means are different when in fact they are not

and

2. Referring to the *t*-distribution with $N_1 + N_2 - 2$ degrees of freedom and a specified α.

   If the value of the *t*-statistic exceeds $t_{\text{crit.}}$, we conclude that the two advertising messages led to different levels of purchase intent (i.e., $\mu_1 \neq \mu_2$), with a type 1 error probability of α.

Example of the Univariate Approach

HATCO has decided to send some of their salespersons to a seminar on effective selling techniques. To evaluate the usefulness of the seminar, HATCO decides to test its effectiveness. The company randomly selects 16 salespersons and sends them to the seminar. At the same time, another random group of 16 salespersons is designated as a control group. Assume that variable X_7 from our data bank is an overall measure of knowledge about selling techniques that is applied to all 32 salespersons the week after the seminar. The null hypothesis HATCO wishes to test is that the course has had no impact on knowledge about selling techniques (i.e., H_0: $\mu_1 = \mu_2$). The alternative hypotheses is that the seminar had positive impact on knowledge (i.e., H_A: $\mu_1 > \mu_2$). The scores obtained by the 32 salespersons on the selling techniques test are graded on a scale from 1 to 15 carried out to three decimal places. These scores are displayed in Table 4.1.

To conduct the test, we decide to specify .05 as the maximum allowable type 1 error rate. Thus, before we conduct the study, we know that 5 times out of 100, we might conclude that the seminar had an impact on the salespersons' knowledge level when in fact it did not. To determine the value for $t_{\text{crit.}}$, we refer to the *t* distribution with $16 + 16 - 2 = 30$ degrees of freedom and $\alpha = .05$. We find that $t_{\text{crit.}} = +1.70$. Next, we compute the value of our *t*-statistic. As shown at the bottom of Table 4.1, $t = 3.82$. Since this exceeds $t_{\text{crit.}}$, we conclude that the seminar was effective in increasing the salespersons' knowledge of selling techniques ($\overline{X}$ for seminar participants = 9.774; $\overline{X}$ for control group = 7.844).

Multivariate Approach: Hotelling's T^2

It is probably unrealistic to assume that a difference between any two experimental treatments will be manifested only in a single measured dependent variable. For example, two advertising messages may not only produce different levels of purchase intent but may affect a number of other (potentially correlated) aspects of the response to advertising (e.g., overall product evaluation, message credibility, interest, attention). Many researchers handle this multiple-criteria situation by repeated application of individual univariate *t*-tests until all of the dependent variables have been analyzed. This approach has serious deficiencies. First, consider what might happen to the type 1 error rate. Assume that we evaluated a series of five dependent variables by separate *t*-tests, each time using .05 as the significant level. Given no real differences in message impact, we would expect to observe a

TABLE 4.1	Treatment 1 = Attended Seminar		Treatment 2 = Control (did not Attend Seminar)	
Salesperson Knowledge Test Scores by Seminar Attendance (Basic Data for Univariate T-Test)	Test Scores:	Knowledge X_7	Test Scores:	Knowledge X_7
	Salesperson	Score	Salesperson	Score
	1	8.599	17	7.952
	2	10.222	18	10.723
	3	9.134	19	8.911
	4	9.766	20	7.113
	5	11.608	21	9.652
	6	10.547	22	8.583
	7	7.532	23	7.688
	8	12.910	24	8.812
	9	9.421	25	5.865
	10	11.869	26	8.149
	11	10.787	27	8.030
	12	7.771	28	6.937
	13	10.041	29	6.362
	14	9.148	30	7.525
	15	8.826	31	5.581
	16	8.210	32	7.625

Means: $X_1 = 9.774$ $\qquad\qquad\qquad$ $X_2 = 7.844$

Variances: 2.278 $\qquad\qquad\qquad\qquad\qquad$ 1.806

Sample Sizes: $N_1 = 16$ $\qquad\qquad\qquad\qquad$ $N_2 = 16$

Standard Error = $\sqrt{\dfrac{1.806}{16} + \dfrac{2.278}{16}} = .505^*$

T-statistic $= \dfrac{9.774 - 7.844}{.505} = 3.82$

* This formula for standard error is appropriate for equal cell sizes. For situations with unequal cell sizes, see [21].

significant effect on any given dependent variable 5 percent of the time. However, across our five separate tests, the probability of a type 1 error will lie somewhere between 5 percent (if all dependent variables are perfectly correlated) and $1 - .95^5 = 23$ percent (if all dependent variables are uncorrelated). Thus, our series of t-tests leaves us without control of our effective type 1 error rate.

A series of t-tests also ignores the possibility that some composite (linear combination) of the dependent variables may provide evidence of an overall group difference that may go undetected by examining each dependent variable separately. Individual t-tests ignore the correlations among the dependent variables and thus use less than the total information available for assessing overall group differences. Hotelling's T^2 in this case (and MANOVA in general) provides a solution to these problems. Hotelling's T^2 solves the type 1 error rate problem by providing a single overall test of group differences across all dependent variables at a specified α level. It solves the composite variable problem by ensuring that the linear combination of the dependent variables that produces the most reliable evidence of group difference is implicitly considered in this test.

How does Hotelling's T^2 achieve these goals? Consider the following equation for a linear combination of the dependent variables:

$$C = W_1Y_1 + W_2Y_2 + \ldots + W_nY_n$$

where

C = composite score for a respondent
W = weights
Y = dependent variables

For *any* set of weights, we could compute composite scores for each respondent and then calculate an ordinary *t*-statistic for the difference between groups on the composite scores. However, if we can find a set of weights that gives the maximum value for the *t*-statistic for this set of data, these weights would be the same as the discriminant function between the two groups (which should not be surprising to those who have read Chapter 3). The maximum *t*-statistic that results from the composite scores produced by the discriminant function can be squared to produce the value of Hotelling's T^2 [8]. The computational formulas for Hotelling's T^2 represent the results of mathematical derivations used to solve for a maximum *t*-statistic (and, implicitly, the most discriminating linear combination of the dependent variables). This is equivalent to saying that if we can find a discriminant function for the two groups that produces a significant T^2, the two groups are considered different across the mean vectors.

How does Hotelling's T^2 provide a test of the hypothesis of no group difference on the vectors of mean scores? Just as the *t*-statistic follows a known distribution under the null hypothesis of no treatment effect on a single dependent variable, Hotelling's T^2 follows a known distribution under the null hypothesis of no treatment effect on any of a set of dependent measures. This distribution turns out to be an F-distribution with $p N_1 + N_2 - 2 - 1$ degrees of freedom after adjustment (where p = the number of dependent variables). That is, we find a tabled value for $F_{crit.}$ at a specified α level and compute $T^2_{crit.}$ as $p(N_1 + N_2 - 2)/(N_1 + N_2 - p - 1) \times F_{crit.}$.

Post Hoc Tests[1]

Given that the T^2-statistic exceeds $T^2_{crit.}$ for a specified α level, we can conclude that the vectors of the mean scores are different. The discriminant function (if computed) tells us what linear combination of the

[1] The follow-up procedure described here is recommended for what are termed *post hoc* tests—tests that are decided on after examining the pattern of the data or "shotgun" approaches in which a large number of tests are made. Another procedure is available for what are termed *a priori* tests—tests that are planned prior to looking at the data from a theoretical or practical decision-making viewpoint. From a pragmatic standpoint,

dependent variables produces the most reliable group difference, but other group comparisons may also be of interest. If we wished to test the group differences individually for each of dependent variables, we could compute a standard t-statistic and compare it to the square root of $T^2_{crit.}$ ($T_{crit.}$) to judge its significance. This procedure would ensure that the probability of *any* type 1 error across all of the tests would be held to α (where α was specified in the calculation of $T^2_{crit.}$) [8].

Example of the Multivariate Approach

In our univariate case 1 example, HATCO evaluated seminar effectiveness by measuring two groups of salespersons (attendees and a control group) on a single dependent variable (variable X_7 from our data bank: overall knowledge of selling techniques). To convert this example into a multivariate case 1 example, we require at least two dependent variables. Let us assume that HATCO also hoped that the seminar would increase motivation to achieve sales goals. Therefore, a motivation test (variable X_8 from our data bank) is also graded on a scale from 1 to 15. Test scores for the 32 salespersons on knowledge and motivation by seminar attendance are presented in Table 4.2. The null hypothesis HATCO wishes to test is that the two vectors of the mean scores are equivalent (i.e., that seminar attendance has no effect on either of the two measures).

To conduct the test, we again specify .05 as the maximum allowable type 1 error. To determine the value for $T^2_{crit.}$, we refer to the F-distribution with 2 and 29 degrees of freedom. We find that $F_{crit.} = 3.33$. For $F_{crit.}$ we could also transform to $T^2_{crit.} = 6.89$ (see the earlier discussion). As shown at the bottom of Table 4.2, the computed value of Hotelling's T^2 (provided by the computer) is 36.17. Since this exceeds $T^2_{crit.}$, we reject the null hypothesis and conclude that the seminar has had some impact on the set of dependent measures.

Two post hoc tests of obvious interest are whether the seminar had an impact on knowledge (X_7) or on motivation (X_8), each considered separately. As shown in Table 4.2, the group means were as follows:

	Knowledge	Motivation
Attended seminar	9.774	10.684
Control group	7.884	8.000
Difference	+1.890	+2.684

situations arise wherein a key dependent variable must be isolated and tested with maximum power. We recommend that an a priori test (in this case, an ordinary t-test is an a priori test for a given dependent variable) be performed in such a situation. However, researchers should be aware that as the number of a priori tests increases, one of the major benefits of a multivariate approach to significance testing—control of the type 1 error rate—is negated.

TABLE 4.2 Salesperson' Test Scores by Seminar Attendance (Basic Data for Hotelling's T^2)

Treatment 1 = Attended Seminar			Treatment 2 = Control (Did Not Attend Seminar)		
Test Scores:	Knowledge X_7	Motivation X_8	Test Scores:	Knowledge X_7	Motivation X_8
Salesperson	Score		Salesperson	Score	
1	8.599	10.364	17	7.952	7.473
2	10.222	9.836	18	10.723	10.724
3	9.134	11.190	19	8.911	8.817
4	9.766	9.868	20	7.113	7.321
5	11.608	13.504	21	9.652	10.369
6	10.547	11.792	22	8.583	6.654
7	7.532	9.988	23	7.688	6.434
8	12.910	11.911	24	8.812	8.527
9	9.421	9.945	25	5.865	7.816
10	11.869	12.032	26	8.149	8.245
11	10.787	11.680	27	8.030	6.820
12	7.771	10.959	28	6.937	8.112
13	10.041	9.914	29	6.362	7.752
14	9.148	10.157	30	7.525	7.433
15	8.826	9.653	31	5.581	6.613
16	8.210	8.169	32	7.625	8.838
Means:	9.774	10.684		7.844	8.000
Sample sizes: $N_1 = 16$				$N_2 = 16$	

Hotelling's $T^2 = 36.17$
Univariate *t*-statistics:
Knowledge variable: $t = 3.82$
Motivation variable: $t = 6.01$

The standard *t*-statistic was previously computed as 3.82 for the difference in knowledge, and as shown in Table 4.2, the *t*-statistic for the difference in motivation is 6.01. Both of these *t*-statistics exceed $\sqrt{T^2_{\text{crit.}}} = \sqrt{6.89} = 2.62$. Thus, we can conclude that the seminar had a positive impact on both motivation and knowledge. We also are confident that the probability of a type 1 error is held to 5 percent across both of these post hoc tests.

Case 2: Difference Between *K* Independent Groups

The two-group randomized design (case 1) is a special case of the more general *k*-group randomized design. In the general case, each respondent is randomly assigned to one of *k* levels (groups) of the independent variable. In the univariate case, a single metric dependent variable is measured and the null hypothesis is that all group means are equal (i.e., $\mu_1 = \mu_2 = \ldots = \mu_k$). In the multivariate case, multiple metric dependent variables are measured and the null hypothesis is that all gorup vectors of mean scores are equal (i.e., $\mu_1 = \mu_2 = \mu_3 = \ldots \mu_k$ where μ refers to a vector or set of mean scores). For a univariate analysis, the appropriate test statistic is the *F*-statistic resulting from ANOVA. For a multivariate analysis, appropriate test statistics include the *greatest characteristic root* (gcr) statistic and Wilks' lambda (also referred to as *Wilks' likelihood ratio criterion* or the *U-statistic*) resulting from MANOVA.

Univariate Approach: ANOVA

In our earlier case 1 discussion, a researcher exposed two groups of respondents to different advertising messages and subsequently asked them to rate their likelihood of purchasing the advertised product on a 10-point scale. Suppose we were interested in evaluating three advertising messages rather than two (i.e., $k = 3$). Respondents would be randomly assigned to one of three groups, and we would have three sample means to compare. To analyze these data, one might be tempted to conduct separate t-tests for the difference between each pair of means (i.e., X_1 versus X_2; X_1 versus X_3; and X_2 versus X_3).

However, we have already seen that multiple tests inflate the overall type 1 error rate. ANOVA avoids type 1 error inflation across comparisons of a number of treatment groups by determining whether the entire set of sample means suggests that the samples were drawn from the same general population. That is, ANOVA is used to determine the probability that differences in means across several groups are due solely to sampling error.

The logic of an ANOVA test is fairly straightforward. As the name *analysis of variance* implies, two independent estimates of the variance in the dependent variable are compared, one of which is sensitive to treatment effects and one of which is not:

1. Within-groups estimate (MS_w: mean square within groups): One estimate of variance of scores on the dependent variable is based on deviations of individual scores from their respective group means. This estimate is influenced by random respondent variability but not by differences between group means. MS_w is sometimes referred to as *error variance*.

2. Between-groups estimate (MS_B: mean square between groups): The second estimate of variance of scores on the dependent variable is based on deviations of group *means* from the overall grand mean of all scores. Under the null hypothesis of no treatment effects (i.e., $u_1 = u_2 = \ldots = u_k$), this variance, like MS_w, is a simple estimate of the sampling variance of scores. However, it can be shown that this variance estimate, unlike MS_w, is also influenced by any treatment effects that exist. That is, population differences in treatment means *increase* the expected value of MS_B. Given that the null hypothesis of no group differences is true, MS_w and MS_B represent independent estimates of population variance. The ratio of MS_B to MS_w gives us a value for an F-statistic. Since group differences tend to inflate MS_B, large values of the F-statistic lead to rejection of the null hypothesis of no difference in means across groups. Specifically, we determine a *critical value* for our F-statistic ($F_{crit.}$) by referring to the F-distribution with $(K - 1)$ and $(N - k)$ degrees of freedom

for a specified level of α (where $N = N_1 + \ldots + N_k$ and k = number of groups). If the value of our F-statistic exceeds $F_{crit.}$, we conclude that the means across all groups are not all equal.

Post Hoc Tests

Although the F-test in ANOVA may allow us to reject the null hypothesis that the k independent sample means are all equal, it does not pinpoint exactly where the significant differences lie. There are many procedures available for further investigation of specific group mean differences of interest. All the procedures attempt to control type 1 error rates across multiple tests. An outline of the many options available is beyond the scope of this chapter. Excellent discussions and explanations of these procedures can be found in Winer [21] and Kirk [9]. We will use one procedure generalizable to MANOVA called *Scheffé's contrast method*. Scheffés contrasts are of the form

$$C = W_1X_1 + W_2X_2 + \ldots + W_kX_k$$

where

C = contrast value
W = weights ($W = 0$)
X = group means

For example, assume that we have three group means. To test for a difference between X_1 and X_2, $C = (1)X_1 + (-1)X_2 + (0)X_3$. To test whether the average of X_1 and X_2 differs from X_3, $C = (.5)X_1 + (.5)X_2 + (-1)X_3$. A separate F-statistic is computed for each Scheffé contrast.

Scheffé's contrast method is very general in that any comparison of interest can be examined. It also holds the probability of *any* type 1 error across all post hoc tests to α (no matter how many tests are done).[2] However, Scheffé's contrast method has been criticized as being extremely conservative. That is, depending on the comparison of interest (e.g., only pairwise comparisons), other post hoc procedures are more powerful (will show significance when Sheffé does not).

Example of the Univariate Approach

Let us assume that HATCO had learned that the seminar on effective selling techniques was offered in both a 3-day and a 5-day version. HATCO takes 48 salespersons and randomly selects 16 to attend the 3-day course, 16 to attend the 5-day course, and 16 to serve as a control group. Again, assume that variable X_7 from our data bank is an overall measure of knowledge about selling techniques that is applied to all 48 salespersons the week after the seminar. The overall null hypothesis

---

[2] Scheffé's contrast method can also handle a priori comparisons. However, the probability of a type 1 error for each a priori comparison is equal to α.

TABLE 4.3 Salespersons' Knowledge Test Scores by Seminar Attendance (Basic Data for ANOVA)

Treatment 1 = Attended 5-Day Seminar		Treatment 2 = Control (Did Not attend Seminar)		Treatment 3 = Attended 3-Day Seminar	
Test Scores:	Knowledge X_7	Test Scores:	Knowledge X_7	Test Scores:	Knowledge X_7
Salesperson	Score	Salesperson	Score	Salesperson	Score
1	8.599	17	7.952	33	6.144
2	10.222	18	10.723	34	8.294
3	9.134	19	8.911	35	11.337
4	9.766	20	7.113	36	8.118
5	11.608	21	9.652	37	10.879
6	10.547	22	8.583	38	11.924
7	7.532	23	7.688	39	10.618
8	12.910	24	8.812	40	7.724
9	9.421	25	5.865	41	12.066
10	11.869	26	8.149	42	10.865
11	10.787	27	8.030	43	9.402
12	7.771	28	6.937	44	6.359
13	10.041	29	6.362	45	10.257
14	9.148	30	7.525	46	7.457
15	8.826	31	5.581	47	9.489
16	8.210	32	7.625	48	10.568
Means:	$\overline{X}_1 = 9.774$		$\overline{X}_2 = 7.844$		$\overline{X}_3 = 9.469$
Variances:	2.278		1.806		3.625
Sample Sizes:	$N_1 = 16$		$N_2 = 16$		$N_3 = 16$

HATCO now wishes to test is that $\mu_1 = \mu_2 = \mu_3$ (i.e., all three groups are equivalent on the knowledge test). The test scores for the 48 salespeople by group are displayed in Table 4.3.

To conduct the test, we specify .05 as the type 1 error rate. To determine the value for $F_{crit.}$, we refer to the F-distribution with $(3 - 1) = 2$ and $(48 - 3) = 45$ degrees of freedom with $\alpha = .05$. We find that $F_{crit.} = 3.20$. The calculation of the F-statistic from ANOVA is usually summarized in an ANOVA table, as shown in Table 4.4. Any mean square (a variance estimate) is a sum of squares (sum of squared deviations) divided by appropriate degrees of freedom. As shown in Table 4.4, the F-statistic = 6.70. Since this exceeds $F_{crit.}$, we can conclude that all group means are not equal.

TABLE 4.4 ANOVA Table for Data from Table 4.3	Source of Variance	Sum of Squares	Mean Square	Degrees of Freedom	F-Ratio
	Between groups	34.443	17.222	2	6.70
	Within groups (error)	115.635	2.570	45	

Example of Post Hoc Testing

As shown in Table 4.3, the group means are as follows:

	Knowledge
Attended 5-day seminar	9.774
Attended 3-day seminar	9.469
Control group	7.844

Examining these means, it appears that attendance at the seminar (for either 3 or 5 days) increased the knowledge level. It also appears that which seminar was attended had little impact on knowledge level. These two hypotheses can be tested with Scheffé's contrast procedure.

For the first hypothesis of an overall effect for seminar attendance (i.e., 9.774 + 9.469)/2 versus 7.844), the Scheffé contrast is significant (assume $\alpha = .05$ as the criterion; formulas for calculation can be found in texts oriented more to the statistician). Thus, we can conclude that seminar attendance increased the knowledge level.

For the second hypothesis of no difference between the 3-day and 5-day seminars (i.e., 9.774 versus 9.469), the Scheffé contrast is not significant. Thus, we can conclude that the length of the seminar had no differential impact on knowledge level.

In summary, univariate ANOVA suggests seminar attendance leads to higher knowledge levels. This effect is the same regardless of whether the 3-day or the 5-day seminar is attended.

Multivariate Approach: MANOVA

In case 2 designs where multiple dependent variables are measured, many researchers proceed with a series of individual F-tests (ANOVAs) until all the dependent variables have been analyzed. As the reader should suspect, this approach suffers from the same deficiencies as a series of t-tests across multiple dependent variables in case 1 designs. That is, a series of F-tests from ANOVA (1) results in an inflated type 1 error rate and (2) ignores the possibility that some composite of the dependent variables may provide reliable evidence of overall group differences. Since individual F-tests ignore the correlations among the dependent variables, they use less than the total information available for assessing overall group differences.

MANOVA again provides a solution to these problems. MANOVA solves our type 1 error rate problem by providing a single overall test of group differences at a specified α level. It solves our composite variable problem by implicitly testing the linear combination of the dependent variables that provides the strongest evidence of overall group differences.

MANOVA can be considered a simple extension of Hotelling's T^2 procedure, with which we are already familiar. That is, we could devise dependent variable weights to produce a composite score for each respondent, as described earlier. In MANOVA we now want to find the set of weights that maximizes the ANOVA F-value computed on the composite scores for all of the groups. The set of weights that maximizes this F-value is called the *first discriminant function. The maximum F-value* itself allows us to compute directly what is called the *greatest characteristic root* statistic [gcr = $(k - 1) F_{max}/(N - k)$] [8].

To obtain a single test of the hypothesis of no group differences on the vectors of mean scores, we could refer to tables of the gcr distribution. Just as the F-distribution follows a known distribution under the null hypothesis of equivalent group means on a single dependent variable, the gcr statistic follows a known distribution under the null hypothesis of equivalent group mean vectors (i.e., group means are equivalent on a set of dependent measures). A comparison of the observed gcr to $gcr_{crit.}$ gives us a basis for rejecting the overall null hypothesis of equivalent group mean vectors.

Other Criteria for Significance Testing in MANOVA

Readers who have some familiarity with MANOVA from other texts or computer programs probably have encountered a more commonly used test statistic for overall significance in MANOVA called *Wilks' lambda* (or the *U-statistic*). We have referred to the greatest characteristic root and the first discriminant function. This implies that there may be additional characteristic roots and discriminant functions. Actually, there are P or $(k - 1)$ (whichever is smaller) characteristic roots and discriminant functions. Basically, unlike the gcr-statistic, which is based on the first (greatest) characteristic root, Wilks' lambda considers all of the characteristic roots. That is, Wilks' lambda examines whether groups are somehow different without being concerned with whether they differ on at least one linear combination of the dependent variables. As it turns out, Wilks' lambda is much easier to calculate than the gcr-statistic. Its formulation is $/W/ \div /W + A/$ where $/W/$ is the determinant (a single number) of the within-groups multivariate dispersion matrix and $/W + A/$ is the determinant of the sum of W and A where A is the between-groups multivariate dispersion matrix [recall our earlier discussion of within-groups and between-groups variance (dispersion) in ANOVA]. The larger the between-groups dispersion, the smaller the value of Wilks' lambda and the greater the implied significance. While the distribution of Wilks' lambda is complex, good approximations for significance testing are available by transforming Wilks' lambda into an F-statistic [14].

Which statistic is preferred? There is some controversy over whether Wilks' lambda or gcr is the preferred test statistic for MANOVA. Most computer packages provide both. Either test is defensible. Other commonly encountered test statistics for MANOVA are Pillai's criterion and Hotelling's trace, both of which are similar to Wilks' lambda, since they consider all of the characteristic roots and can be approximated by an F-statistic.

Post Hoc Tests

After concluding that the group mean vectors are not equivalent, other more specific comparisons may be of interest. For example, we might be interested in whether there are any group differences on a specific dependent or composite dependent variable. A standard ANOVA F-statistic can be calculated and compared to $F_{\text{crit.}} = (N - k)\text{gcr}_{\text{crit.}}/(k - 1)$, where the value of $\text{gcr}_{\text{crit.}}$ is taken from the gcr distribution with appropriate degrees of freedom. Scheffé-type contrasts between groups on any dependent variable can also be tested. These procedures ensure that the probability of *any* type 1 error across all comparisons will be held to α [8]. Note that the above procedures require the use of the gcr distribution. This is one argument in favor of using the gcr-statistic as the overall test in MANOVA.

Other follow-up procedures are available after finding the overall significance from MANOVA. First, a procedure known as *step-down analysis* [18] may be applied. This procedure involves computing a univariate F-statistic for a dependent variable after eliminating the effects of other dependent variables preceding it in the analysis. The procedure is somewhat similar to stepwise regression, but here we examine whether a particular dependent variable contributes unique (uncorrelated) information on group differences. This technique is relatively popular, probably because it is an option in the well-known SPSS computer package. Another follow-up procedure is a further analysis of the discriminant functions, in particular the first discriminant function, to gain additional information about which variables best differentiate between the groups.

Example of the Multivariate Approach

In our univariate case 2 example, HATCO evaluated seminar effectiveness by measuring three groups of salespersons (attended the 5-day seminar, attended the 3-day seminar, or did not attend the seminar) on a single dependent variable (variable X_7 from our data bank: overall knowledge of selling techniques). To convert this example to a multivariate case 2 example, we require at least two dependent variables. As in our earlier multivariate extension of case 1, let us assume that HATCO also hoped that seminar attendance would increase motivation to achieve sales goals. The test scores for the three groups of 16 salespersons on knowledge and motivation (variable X_8 from our data bank) are presented in Table 4.5. The null hypothesis HATCO now wishes to test is that the three sample vectors of the mean scores are equivalent.

Data analysts can use one of several available computer programs to carry out MANOVA. Two of the more popular ones are the BMDP [4] and SAS [15,16] packages. Either can be used for both univariate

TABLE 4.5 Salespersons' Test Scores by Seminar Attendance (Basic Data for MANOVA)

Treatment 1 = Attended 5-Day Seminar			Treatment 2 = Control (Did Not Attend Seminar)			Treatment 3 = Attended 3-Day Seminar		
Test Scores:	Knowledge X_7	Motivation X_8	Test Scores:	Knowledge X_7	Motivation X_8	Test Scores:	Knowledge X_7	Motivation X_8
Sales-person	Score		Sales-person	Score		Sales-person	Score	
1	8.599	10.364	17	7.952	7.473	33	6.144	7.999
2	10.222	9.836	18	10.723	10.724	34	8.294	9.996
3	9.134	11.190	19	8.911	8.817	35	11.337	12.120
4	9.766	9.868	20	7.113	7.321	36	8.118	8.948
5	11.608	13.504	21	9.652	10.369	37	10.879	9.604
6	10.547	11.792	22	8.583	6.654	38	11.924	12.120
7	7.532	9.988	23	7.688	6.434	39	10.618	7.500
8	12.910	11.911	24	8.812	8.527	40	7.724	7.316
9	9.421	9.945	25	5.865	7.816	41	12.066	10.198
10	11.869	12.032	26	8.149	8.245	42	10.865	11.841
11	10.787	11.680	27	8.030	6.820	43	9.402	10.503
12	7.771	10.959	28	6.937	8.112	44	6.359	8.501
13	10.041	9.914	29	6.362	7.752	45	10.257	11.014
14	9.148	10.157	30	7.525	7.433	46	7.457	8.854
15	8.826	9.653	31	5.581	6.613	47	9.489	8.474
16	8.210	8.169	32	7.625	8.838	48	10.568	10.017
Means:	9.774	10.684		7.844	8.000		9.469	9.689

and multivariate analysis of variance and covariance. Table 4.6 provides summary output from the MANOVA performed on the data of Table 4.5. The value of the greatest characteristic root is .697. Referring to the gcr distribution with appropriate degrees of freedom and setting $\alpha = .05$, $gcr_{crit.} = .310$. Since .697 exceeds this value, we can conclude that the mean vectors of the three groups are not equal. The value of Wilks' lambda, also shown in Table 4.6, is .574. An approximate F-statistic associated with this value of Wilks' lambda is 7.05. With 4 and 88 degrees of freedom, $F_{crit.} = 2.40$ ($\alpha = .05$). Since 7.05 exceeds this value, we reach the same conclusion (the mean vectors of the three groups are not equal), as with the gcr-statistic. Again, the difference

TABLE 4.6 MANOVA Summary Table*	gcr = .697
	Wilks' lambda = .574
	Approximate F-statistic = 7.05
	Degrees of freedom = 4 and 88
	$gcr_{crit.}$ ($\alpha = .05$) = .310
	$F_{crit.}$ ($\alpha = .05$) = 2.40

* MANOVA Results based on data of Table 4.5.

is that gcr considers only the first (or best) characteristic root, while Wilks' lambda considers all roots.

Post Hoc Testing

As shown in Table 4.5, the group means are as follows:

	Knowledge	Motivation
Attended 5-day seminar	9.774	10.684
Attended 3-day seminar	9.469	9.689
Control group	7.844	8.000

A number of post hoc tests may be of interest. For example, are group mean differences statistically significant for each dependent variable considered alone? Did attendance at the 5-day seminar result in higher levels of motivation than attendance at the 3-day seminar? Did attendance at the 3-day seminar result in higher levels of motivation than not attending any seminar (control group)? All such questions can be answered with the multivariate extension of Scheffé's contrast procedure outlined earlier. Examination of discriminant functions (which have to be obtained by using a discriminant analysis program, as described in Chapter three) and step-down analyses described earlier are generally more useful first steps in post hoc analysis as the number of dependent variables increases.

Factorial Designs

In ANOVA, independent variables are termed *factors*. A factor may have two or more levels. A design with more than two factors is called a *factorial design* (in general, a design with n factors is called an *n-way factorial design*).

As an example, assume that a cereal manufacturer wishes to examine the impact of three different color possibilities (red, blue, and green) and three different shapes (stars, cubes, and balls) on the overall consumer evaluation of a new cereal. We could examine the impact of both of these independent variables simultaneously by employing a 3 × 3 factorial design. Respondents would be randomly assigned to evaluate (e.g., a 10-point overall evaluation scale) one of the nine possible combinations of color and shape. In analyzing this design, three different overall effects can be tested with ANOVA:

1. The main effect of color: Are there any differences between the mean ratings given to red (i.e, including all ratings of red stars, red cubes, and red balls), blue, and green?
2. The main effect of shape: Are there any differences between the mean ratings given to stars (i.e., including all ratings of red stars, blue stars, and green stars), cubes, and balls?
3. The interaction effect of color and shape: Looking at the overall difference between colors, is this difference the same when we exam-

TABLE 4.7
Example of an
ANOVA Summary
Table for a
Factorial Design

Source of Variance	Mean Square	Degrees of Freedom	F-Statistic
Color	4.5	2	3.0*
Shape	1.5	2	1.0
Color × shape	1.6	4	1.1
Error	1.5	63	

* Significant; $\alpha = .05$.

ine it separately for stars, cubes, and balls? For example, if red was rated very high overall but received a very low rating when it was rated as a ball (relative to blue and green), this would be evidence of an interaction effect. That is, the effect of color depends on what shape we are considering. We could ask this interaction question in an equivalent fashion by asking whether the effect of shape depends on what color we are considering.

In ANOVA for factorial designs, each of these three effects would be tested with an F-statistic. A hypothetical example is presented in Table 4.7. The F-tests suggest that color has a significant impact on overall evaluation but shape does not. The effect of color is consistent for each shape (i.e., the interaction is not significant).

The MANOVA for factorial designs is a straightforward extension of ANOVA. That is, for every F-statistic in ANOVA that evaluates an effect on a single dependent variable, there is a corresponding multivariate statistic (e.g., gcr or Wilks' lambda) that evaluates the same effect on a set (vector) of dependent variable means.

MANOVA Counterparts of Other ANOVA Designs

As the reader may be aware, there are many types of ANOVA designs—including factorial designs, repeated measures designs, split plot designs—discussed in standard experimental design texts [9,12,21]. For every ANOVA design, there is a multivariate counterpart. That is, any ANOVA on a single dependent variable can be extended to MANOVA extensions of other types of ANOVA designs, we would have to discuss each ANOVA design in detail. Clearly, this is not possible in a single chapter, since entire books are devoted to the subject of ANOVA designs. For more information, the reader is referred to more statistically oriented texts [1,2,3,5,6,7,8,13,19].

ANCOVA and MANCOVA

In any univariate ANOVA design, metric independent variables, referred to as *covariates*, can be included. The design is then termed an *analysis of covariance* (ANCOVA) design. Metric covariates are typically included in an experimental design to remove extraneous influences from the dependent variable, thus increasing measurement precision. Procedures similar to linear regression are employed to remove variation in the dependent variable associated with one or more covariates. Then a conventional ANOVA is carried out on the adjusted dependent variable.

An effective covariate in ANCOVA is one that is highly correlated with the dependent variable but not correlated with the independent variables. That is, variance in the dependent variable forms the basis of our error term in ANOVA. If we have a covariate that is correlated with the dependent variable, we can explain some of the variance (through linear regression) and we are left with only residual variance in the dependent variable. This residual variance provides a smaller error term for the F-statistic and thus a more efficient test of treatment effects.

Multivariate analysis of covariance (MANCOVA) is a simple extension of the principles of ANCOVA to multivariate (multiple dependent variables) analysis. That is, MANCOVA can be viewed as MANOVA of regression residuals—variance in the dependent variables not explained by the covariate(s).

Assumptions of ANOVA and MANOVA

The univariate test procedures of ANOVA described in this chapter are only valid (in a formal sense) if it is assumed that the dependent variable is normally distributed and that variances are equal for all treatment groups. There is evidence, however, [12,21] that F-tests in ANOVA are robust with regard to these assumptions. Thus, the F-statistic is relatively insensitive to violations of these assumptions except in extreme cases.

For the multivariate test procedures of MANOVA to be valid, the set of p dependent variables must follow a mulitvariate normal distribution (i.e., any linear combination of the dependent variables must follow a normal distribution [8] and variance-covariance matrices must be equal for all treatment groups). Very little is known about violations of these assumptions [8].

Summary

It may be unrealistic to assume that a difference between experimental treatments will be manifested only in a single measured dependent variable. Unfortunately, many researchers handle multiple-criteria situations by repeated application of individual univariate tests until all of the dependent variables have been analyzed. This approach can seriously inflate type 1 error rates and ignores the possibility that some composite of the dependent variables may provide the strongest evidence of reliable group differences. Appropriate use of MANOVA provides solutions to both of these problems.

We have not covered all of the types of experimental designs for MANOVA in this chapter, nor would such an undertaking be practical in a book oriented to the nonstatistician. It is hoped that the reader has obtained sufficient stimulation, understanding, and confidence to tackle some of the more sophisticated designs of MANOVA described in more statistically oriented texts.

QUESTIONS

1. Design a two-way factorial MANOVA experiment. What are the different sources of variance in your experiment? What would the interaction test tell you?

2. Locate and compare at least two "canned" MANOVA computer programs. What are the essential differences between them, especially with regard to the printout of results?

3. After finding the overall or global significance, there are at least three approaches to doing follow-up tests: (a) the use of Scheffé contrast procedures; (b) step-down analysis, which is similar to stepwise regression in that each successive F-statistic is computed after eliminating the effects of the previous dependent variables; and (c) examination of the discriminant function(s). Try to name the practical advantages and disadvantages of each of these approaches.

4. Describe some data analysis situations in which MANOVA and MANCOVA would be appropriate in your areas of interest. What types of uncontrolled variables or covariates might be operating in each of these situations?

REFERENCES

1. Anderson, T. W., *Introduction to Multivariate Statistical Analysis.* New York: Wiley, 1958.
2. Cattell, R. B. (ed.), *Handbook of Multivariate Experimental Psychology.* Chicago: Rand McNally, 1966.
3. Cooley, W. W., and P. R. Lohnes, *Multivariate Data Analysis.* New York: Wiley, 1971.
4. Dixon, W. J., (ed.), *Biomedical Computer Programs.* Los Angeles: Health Science Computing Faculty, School of Medicine, University of California, 1978, pp. 751–64.
5. Green, P. E., *Analyzing Multivariate Data.* Hinsdale, Ill.: Holt, Rinehart & Winston, 1978.
6. Green, P. E., and J. Douglas Carroll, *Mathematical Tools for Applied Multivariate Analysis.* New York: Academic Press, 1978.
7. Green, P. E., and D. S. Tull, *Research for Marketing Decisions,* 3rd ed. Englewood Cliffs, N.J.: Prentice-Hall, 1979.
8. Harris, R. J., *A Primer of Multivariate Statistics.* New York: Academic Press, 1975.
9. Kirk, R. E., *Experimental Design: Procedures for the Behavioral Sciences.* Belmont, Calif. Brooks/Cole, 1968.
10. McGraw-Hill, Inc., *SPSS-X Users's Guide,* 2nd ed. Chicago, 1986.
11. McGraw-Hill, Inc., *SPSS-X Advanced Statistics Guide.* Chicago, 1986.
12. Meyers, J. L., *Fundamentals of Experimental Design.* Boston: Allyn & Bacon, 1975.
13. Morrison, D. F., *Multivariate Statistical Methods.* New York: McGraw-Hill, 1967.
14. Rao, C. R., *Linear Statistical Inference and Its Application,* 2nd ed. New York: Wiley, 1978.
15. SAS Institute, *SAS User's Guide: Basics,* Version 5 ed. Cary, N.C., 1985.
16. SAS Institute, *SAS User's Guide: Statistics,* Version 5 ed. Cary, N.C., 1985.
17. Sheth, J. N., *Multivariate Methods for Market and Survey Research.* Chicago: American Marketing Association, 1977, pp. 83–96.
18. Stevens, J. P., "Four Methods of Analyzing Between Variations for the *k*-Group MANOVA Problem," *Multivariate Behavioral Research,* vol. 7 (October, 1972), pp. 442–454.
19. Tatsuoka, M. M., *Multivariate Analysis: Techniques for Education and Psychological Research.* New York: Wiley, 1971.
20. Wilks, S. S., "Certain Generalizations in the Analysis of Variance," *Biometrika,* vol. 24 (1932), pp. 471–94.
21. Winer, B. J., *Statistical Principles in Experimental Design.* New York: McGraw-Hill, 1962.

SELECTED READINGS

The Effects of Social Skills Training and Peer Involvement on the Social Adjustment of Preadolescents

KAREN LINN BIERMAN
WYNDOL FURMAN

Children who are unaccepted by their peers may be deprived of a number of important experiences provided through positive peer contact. Peer interactions play significant and unique roles in facilitating the development of appropriate assertiveness, altruistic behavior, moral reasoning, and other social competencies involving reciprocal give-and-take relationships (Hartup, 1978). Moreover, continued isolation from positive peer contact has been linked with a number of serious adjustment problems in later adolescence and adulthood (Cowen, Pederson, Babigian, Izzo, & Trost, 1973; Roff, Sells, & Golden, 1972).

Several investigators have suggested that some children are unable to gain peer acceptance because they have inadequate social skills (Asher, Oden, & Gottman, 1977; Combs & Slaby, 1977). For example, deficits in communication skills have been associated with ratings of poor social competence and low sociometric status (Gottman, Gonso, & Rasmussen, 1975; Ladd, 1981; Minkin, Braukmann, Minkin, Timbers, Fixsen, Phillips, & Wolf, 1976). Recently, coaching programs have

"The Effects of Social Skills Training and Peer Involvement on the Social Adjustment of Preadolescents," Karen Linn Bierman and Wyndol Furman, Vol. 55 (1984), pp. 151–162. Reprinted from *Child Development*, published by the Society for Research in Child Development, Inc. Karen Linn Bierman is a professor of psychology at Pennsylvania State University. Wyndol Furman is a professor of psychology at the University of Denver.

been designed to teach unaccepted grade school children new social skills. Research indicates that coaching can improve children's social behavior (Ladd, 1981; La Greca & Santogrossi, 1980), but the effects on peer acceptance have been mixed. While coaching has produced increases in sociometric status for younger children (Oden & Asher, 1977; Ladd, 1981), it has not had an effect on the peer acceptance of preadolescents (La Greca & Santogrossi, 1980; Hymel & Asher, Note 1).

One factor moderating the impact of social skills training programs may be the reactions of peers to unaccepted children. Peers often develop derisive stereotypes and negative expectations for children of low sociometric status (Koslin, Haarlow, Karlins, & Pargament, 1968; Sherif, Harvey, White, Hood, & Sherif, 1961). In fact, peers can inadvertently foster inappropriate social behavior (Patterson, Littman, & Bricker, 1967). They may converse less with children they dislike and may even respond less positively to appropriate social overtures made by such children.

Given the potential negative influence of peers, some investigators have focused on changing peer attitudes and responses rather than working directly with unaccepted children. One means of altering peer attitudes involves the use of superordinate goals—that is, goals that can be attained only if all children in a group work together and cooperate. In their well-known Robber's Cave experiment, Sherif et al. (1961) demonstrated the effectiveness of superordinate goals in reducing the hostility and negative stereotypes that had emerged between two groups of campers during a series of competitive activities. Other investigators have also found that children behave more cooperatively and like each other better after working together toward a superordinate goal (Bryan, 1975).

Regardless of whether the focus has been on

training social skills or restructuring the peer environment, the generalization and maintenance of treatment gains has been a major concern. For example, skill training programs with preadolescents have resulted in behavior changes but not changes in peer acceptance (La Greca & Santogrossi, 1981; Hymel & Asher, Note 1). Similarly, programs using superordinate goals produce short-term gains in peer acceptance, but such gains often fade quickly (Rucker & Vincenzo, 1970). Additionally, investigators have not measured the effects of superordinate goal programs on children's social behavior or social skills. The poor generalization that has been observed may reflect the limited focus of the interventions employed. Even when their social skills have been enhanced by coaching, unpopular preadolescents may find it difficult to change the negative attitude and responses of their peers. Alternatively, the children who have experienced positive, cooperative exchanges with peers may not have the social skills needed to sustain peer acceptance once the environmental support of these programs is removed. Perhaps programs that focus on both enhancing skills and changing peer responses may be more effective.

The purpose of this study was to compare the effects of conversation skills training and positive peer involvement under superordinate goals on various dimensions of preadolescents' social competence. These two intervention strategies were compared using a factorial design. Treatments that did or did not provide skill training were crossed with treatments that did or did not involve peer interaction under superordinate goals. Unaccepted fifth and sixth graders with poor social skills were identified on the basis of sociometric ratings and behavioral observations and randomly assigned to one of the four resulting treatment conditions: (1) skill training only (individual coaching), (2) positive peer involvement only (group experience), (3) a combined treatment including skills training and positive peer involvement (group experience with coaching), or (4) a no-treatment control. In the coaching conditions, children were trained on three conversational skills related to peer acceptance. Children in the peer involvement conditions were given opportunities to interact with peers under a superordinate goal—making video films together. Children in the group experience with coaching condition received skill training within the context of positive peer involvement under superordinate goals,

while children in the no-treatment condition received neither. The immediate and sustained effects of skill training and peer involvement were evaluated on measures of conversational skills, peer acceptance, rates of peer interaction, and self-perceptions. This range of measures was included to provide a comprehensive picture of the treatment effects and because previous research suggested that each variable may be a distinct aspect of social competence (Gottman, 1977).

It was anticipated that skill training and peer involvement would have different effects. Coaching is designed to promote skill acquisition, so children who received skill training were expected to show immediate and sustained improvement on measures of conversational skills. Their perceptions of their social efficacy were also expected to increase as a result of acquiring the new skills. Consistent with previous findings, skill training was not expected to have a direct impact on the acceptance of these preadolescents. On the other hand, it was anticipated that positive peer group experiences would facilitate peer acceptance. The experience of more positive interactions with peers was also expected to enhance children's perceptions of their social competence and social efficacy. Without skill training, however, these children were not expected to show changes on measures of conversation skills. Furthermore, because of continuing skill deficits, the gains that these children made in peer acceptance and peer interaction were not expected to maintain at follow-up. Only those children who received both treatments were expected to show sustained improvements in peer acceptance at follow-up.

Method

Subjects

Subjects were 28 male and 28 female fifth and sixth graders, selected from a total sample of 396 children in six schools within a large metropolitan area. The sample predominantly consisted of Caucasian children from middle- to upper-class socioeconomic backgrounds. Subjects were selected by a two-step process. First, children who ranked in the lower third in sociometric status on the Roster and Rating Scale were identified as low in peer acceptance. These children were then observed in peer group interactions, and those who had the lowest ratings in conversational skill performance were selected as subjects. Thus, identified children were both low in peer acceptance and deficient in conversational skills. For the treatment conditions including peers, an additional 28 male and 28 female peers were randomly selected from the group of classmates who scored in the upper two-thirds of their class in sociometric status.

Dependent Measures

Differential treatment effects were expected in the areas of conversational skills, peer acceptance, rates of peer interaction, and self-perception. Multiple measures were included to represent each area.[1]

CONVERSATIONAL SKILLS. Conversational skills were selected as a focus for intervention because they have been linked repeatedly to ratings of poor social competence and low sociometric status (Gottman et al., 1975; Minkin et al., 1976) and training in conversational skills has been found to enhance social functioning (Kelly, Furman, Phillips, Hathorn, & Wilson, 1979; Ladd, 1981). Based on previous intervention studies, three conversational skills were selected for assessment and training: (1) self-expression—sharing information about oneself; (2) questioning—asking others about themselves; and (3) leadership bids—giving help, suggestions, invitations, and advice.

Three measures assessed children's conversational skills. A structured dyadic conversation tapped skill acquisition, observations of conversational behavior in a peer group setting assessed the use of these skills in peer interactions, and a written questionnaire measured children's cognitive representations of the three skills.

To assess skill acquisition, a modified version of the Kelly et al. (1979) procedure was used. A structured dyadic interview was conducted by a young-looking college student who was presented to children as an older peer. The interview consisted of 24 stimulus statements of three types, designed to prompt the three identified conversational skills. Children's responses were taped, transcribed, and coded as described below.

[1] Copies of all experimental measures and the treatment protocol are available from the first author.

To assess conversational skills during peer interactions, a modified version of Ladd's (1981) procedure was employed. In particular, children were observed during two 15-min sessions with eight same-sex peers from their classroom. In one session they engaged in a group art activity; in the other they had an opportunity for unstructured conversation. Observers watched each child for 1 min at a time, alternating between 6 sec of observing and 6 sec of recording behaviors that occurred in the prior 6-sec interval. The exact coding system is described below.

A third measure, the Conversational Skill Concept Scale, was designed by the present investigators to assess children's knowledge of conversational skills. Children were presented with three problem situations, such as talking to peers at lunch, and were asked to list alternative verbal responses in these situations.

For each of these three measures, statements that contained personal references or information about oneself were coded as "self-expressions"; all questions and statements functioning as questions were coded as "questions"; suggestions, advice, invitations, or directives were coded as "leadership bids." Other coding categories were "social noise," which included screaming, hooting, and singing, and "talk," which included all other verbal statements not coded elsewhere. Additionally, "no information" was coded during the dyadic conversation when a child answered "I don't know," and "silence" was coded during the group observations when a child made no comment during a 6-sec interval. For analyses, a total conversational skill performance score was computed for each of the three measures by summing their self-expressions, questions, and leadership bids.

Three undergraduates unaware of the exact purpose of the study served as coders. All coders received 8 weeks of training that included reviewing the written manual and practice coding of videotaped and live interactions. They also participated in booster sessions involving videotaped reviews during the course of the study. To assess interrater agreement, multiple raters coded approximately 25% of the observations or protocols. The κ coefficients were calculated using the formula $(P_0 - P_c)/(1 - P_c)$, where P_0 was the percentage agreement and P_c was the percentage of chance agreement. For the dyadic interview, κ's ranged from .80 to .95, with a mean of .90. For the group observation, κ's ranged from .76 to .92, with a mean of .85. The κ's for the written Conversational Skill Concept Scale ranged from .99 to .53, with a mean of .86.

PEER ACCEPTANCE.   Two peer rating scales were employed to assess peer acceptance. The Roster and Rating Scale (Hymel & Asher, Note 1) was used to assess children's sociometric status. On this scale, each child used a five-point Likert scale to rate their feelings of friendship for each classmate. Sociometric scores were derived by averaging the ratings given to each child by same-sex classmates, excluding those peers involved as partners in the treatment. Same-sex ratings were used as they tend to provide a more valid and reliable estimate of peer acceptance at this age (Hymel & Asher, Note 1). Additionally, for the subjects in the group experience condition or the group experience with coaching condition, a partner sociometric score was derived by calculating the average rating given to the subject by the peers actually in treatment with the subject.

Additional information concerning the peer reputations of subjects was attained from the withdrawal scale of the Pupil Evaluation Inventory (PEI) (Pekarik, Prinz, Liebert, Weintraub, & Neale, 1976). On this scale, children indicate whether each of nine descriptions of withdrawn behavior apply to each same-sex classmate. Scores consisted of the mean number of nominations received.

RATES OF PEER INTERACTIONS.   In addition to the observations of children's conversational skills previously mentioned, it was desirable to attain a nonobtrusive measure of children's interaction rates in naturalistic settings. Two measures were collected in an attempt to estimate unobtrusively the naturalistic interaction levels—behavioral observations of interaction rates during lunch and teacher ratings of children's classroom interactions. In the lunchroom observations, observers stood at a distance from the children to minimize their obtrusiveness. They used a 6-sec time sampling technique to record the occurrence or nonoccurrence of verbal peer interactions. Children were observed for a total of 50 intervals interspersed over at least two lunch periods. Interrater agreement (κ) was .90.

Finally, teachers were asked to rate each subject on six seven-point Likert items concerning the child's level of peer interaction in the classroom (e.g., "How often does this child talk with others in the class?"). The internal consistency for the scale was .86 (Cronbach's α).

SELF-PERCEPTIONS.   The Social Self-Efficacy Scale was designed by the present investigators to measure children's perceived efficacy on the three targeted skills as well as their more general feelings of social efficacy. Fifteen nine-point Likert items asked children about their perceived ability to express themselves, ask questions, or make leadership bids (e.g., "In a group of kids from your class, how good are you at expressing your opinion and saying what you think?"), and five other Likert items asked about their perceived ability to become accepted. The internal consistency of the scale was .94.

Additionally, the social subscale of Harter's (1982) Perceived Competence Scale for Children was administered. Using a series of seven four-point Likert items, children were asked to indicate how socially competent they perceived themselves to be.

Procedure

PRETREATMENT ASSESSMENT.   Initially, classrooms of children were administered a questionnaire containing measures of peer acceptance, conversational skill knowledge, and self-perceptions of social competence. In each classroom, the six children who received the lowest sociometric scores were observed during two 15-min sessions within groups of eight same-sex classmates. Of the unaccepted children, those showing the greatest deficits in conversational skills in peer group interactions were selected as subjects. These identified subjects then participated in a structured dyadic conversation and were observed during two or three nonconsecutive lunch periods. Finally, teacher ratings of their peer interactions were collected.

TREATMENT.   The identified subjects were randomly assigned to one of the four treatment conditions: (1) individual coaching, (2) group experience, (3) group experience with coaching, or (4) no treatment. All treatments consisted of 10 half-hour sessions over a 6-week period. Two women, the first author and a trained college graduate, served as coaches following a specific treatment manual. The two coaches were counterbalanced across the treatment conditions.

In the two conditions involving skills training (individual coaching and group experience with coaching), three conversational skills (self-expressions, questions, and leadership bids) were coached using instruction, rehearsal, and performance feedback techniques. In the treatments involving peer interactions with superordinate goals (group experience and group experience with coaching), children were given the goal of making videotapes together, an approach similar to Chandler's (1973) program. The specific procedures for the four treatment conditions were as follows.

The individual coaching condition was designed to provide skill training, but without peer involvement. Here the children met individually with an adult trainer. The children were told that their help was needed in making videotapes to teach college students what children their age talked about and did together. They were also told they needed to practice for the films. Five half-hour "practice sessions" were then held, during which the coach presented, discussed, and practiced the three conversational skills with the child. In the remaining five sessions, the coach and child alternated between making a film of their conversations in one session and then watching it and reviewing skill performance the following session.

The group experience condition was designed to provide a peer group experience with superordinate goals but without skill training. Here, children met in groups of three—one identified child and two same-sex peers randomly chosen from classmates scoring in the upper two-thirds in sociometric status. These groups were also told that their help was needed in making friendly interaction films for the university. In their five practice sessions, the adult coach facilitated their group planning but gave no specific training in conversational skills. In the next five sessions, these children also made and reviewed films. While the coach provided comments about the children's films, the coach did not prompt, cue, or reward specific skill performance.

In the group experience with coaching condition, children also met in triads (one identified child and two same-sex classroom peers). They were given the rationale about making friendly interaction films and spent five sessions practicing for the films as the children in the other conditions did. During these group practice sessions, however, the coach introduced, discussed, and had the children specifically practice the three conver-

sational skills. In the remaining five sessions when the group alternated making and reviewing their films, the coach used the review sessions to incorporate skill performance feedback. Skill performance by each of the members of the triad was prompted, noted and reinforced.

POSTTREATMENT AND FOLLOW-UP ASSESSMENTS. After the treatments were completed, all of the pretreatment measures were repeated. A similar follow-up assessment was administered 6 weeks later.

Results

Pretreatment Scores

To determine whether the measures differentiated between the targeted children and their classmates, univariate analyses of variance were conducted on the pretreatment scales that all children in the classrooms had completed. Highly significant differences were found in the peer group and lunchroom interactions and on the Roster and Rating Scale, PEI withdrawal scale, and social subscale of the Perceived Competence Scale for Children, all F's > 15, p's $< .001$, and nearly significant differences were found on the Conversational Skill Concept Scale, $F(1,108) = 3.75$, $p < .06$. The difference on the other available measure, the Social Self-Efficacy Scale, was in the expected direction but not significant, $F(1,108) = 2.39$, $p < .15$. Additionally, main effects for sex were found on the Conversational Skill Concept Measure, $F(1,108) = 14.76$, $p < .001$, and the Roster and Rating Scale, $F(1,108) = 4.69$, $p < .05$. Sex × subject status interactions were found on the Roster and Rating Scale, $F(1,108) = 3.93$, $p < .05$, and PEI withdrawal scale, $F(1,108) = 6.57$, $p < .01$. Both of the sex effects reflected higher scores by girls, but on the measure of peer acceptance the difference was only significant for the identified children (Duncan multiple range test, $p < .05$). Finally, on the PEI withdrawal scale, identified boys were rated as more withdrawn than identified girls, $p < .05$.

Since identified children were randomly assigned to the various conditions, no pretreatment differences among the conditions were expected. To rule out this possibility, however, the targeted children's pretreatment scores on each measure were examined in 2 (skill training) × 2 (peer involvement) × 2 (sex) ANOVAs. No significant condition effects were found on any measure except for the social subscale of the Perceived Competence Scale for Children, where subjects who were to receive skill training had higher scores, $F(1,48) = 6.74$, $p < .01$. These analyses also yielded a series of sex differences that paralleled the ones reported in the previous paragraph.

In summary, analyses of pretreatment scores indicated that the targeted children were less socially competent than their classmates on almost all dependent measures. Additionally, with one exception, no preexisting condition differences were found.

Treatment Effects

PLAN.   Treatment effects were assessed by subjecting posttreatment or follow-up scores to 2 (skill training) × 2 (peer involvement) × 2 (sex) multivariate analyses of covariances (MANCOVAs) in which corresponding pretreatment scores were used as covariates. Preliminary analyses to test the assumption of homogeneity of regression coefficients found only a chance number of significant interactions between a covariate and another effect, two of 20 univariate cases, all other p's > .10. Separate MANCOVAs were run for each of the four conceptual sets of variables: conversational skills, peer acceptance, rates of interaction, and self-perceptions. If a multivariate effect was significant, univariate analyses of covariance were conducted to determine the nature of the effect.

CONVERSATIONAL SKILLS.   It was hypothesized that skill training would promote children's acquisition of conversational skills. A MANCOVA of the posttreatment scores on the three measures of conversational skills (skill performance during a dyadic conversation, during peer group interactions, and on the Conversational Skill Concept Scale) yielded a significant main effect for skill training, $F(3,43) = 9.84$, $p < .001$, and a peer involvement × sex interaction effect, $F(3,43) = 5.76$, $p < .01$. Main effects for skill training at posttreatment were found on the ANCOVAs of skill performance during a dyadic conversation, $F(1,47) = 28.12$, $p < .001$, and during peer group interactions, $F(1,47) = 13.12$, $p < .01$. Examination of the means revealed that children who received skill training showed significantly higher levels of skill performance on both measures than those who did not (see Table 1). The posttreatment ANCOVA on the written measure of conversational skills, the Conversational Skill Concept Scale, revealed a significant peer involvement × sex interaction, $F(1,47) = 10.69$, $p < .01$. Duncan multiple range follow-up tests showed that the girls experiencing peer involvement had significantly higher posttreatment scores on this measure than did girls not involved with peers or boys in any condition, p's < .05.

The MANCOVA on follow-up scores revealed a continuing main effect for skill training, $F(3,43) = 6.28$, $p < .001$, and no other significant effects. Univariate ANCOVAs showed sustained main effects for skill training on skill performance in the dyadic conversation, $F(1,47) = 10.58$, $p < .01$,

TABLE 1   Mean Conversational Skills Scores

Measures	Time of Assessment		
	Pretreatment	Posttreatment	Follow-up
Dyadic interview:			
Skill training conditions	31.32	52.29**	50.32**
No skill training conditions	39.29	33.54**	36.32**
Peer group interaction:			
Skill training conditions	15.57	21.29**	19.00**
No skill training conditions	14.64	11.93**	12.43**
Written skill concept:			
Skill training conditions	11.21	15.64	14.21*
No skill training conditions	11.36	14.93	9.92*

Note. Scores indicate the mean frequencies that children engaged in a skill (self-expression, questions, or leadership bids) during the dyadic or peer group interaction or mentioned a skill on the written concept measure.

* $p < .10$.
** $p < .05$.

and in peer group interactions, $F(1,47) = 8.44$, $p < .01$, and a near significant effect for skill training on the written Conversational Skill Concept Scale, $F(1,47) = 3.87$, $p < .06$.

Thus, skill training did produce strong and sustained improvements in children's conversational skill performance. These effects were well documented on the two behavioral measures of conversational skills and suggested on the written measure. No significant interactions with group involvement were found, suggesting that the main effect of skill training on conversational skills was not affected by the individual or group context of training.

PEER ACCEPTANCE. The second hypothesis was that treatments providing opportunities for positive peer interactions would promote peer acceptance. A MANCOVA on the posttreatment sociometric and PEI withdrawal ratings yielded a significant main effect for peer involvement, $F(2,45) = 3.73$, $p < .05$, and a significant peer involvement × sex interaction effect, $F(2,45) = 3.54$, $p < .05$. As shown in Table 2, univariate ANCOVAs revealed that children who participated in the peer group treatments received significantly higher sociometric ratings after treatment than children who were not involved with peers, $F(1,47) = 4.85$, $p < .05$. This main effect of peer involvement was not modified by any significant interaction with the inclusion or exclusion of associated skill training. No significant effects were found on the PEI withdrawal scale. The MANCOVA of follow-up scores revealed no continuing effects, p's > .10.

The primary measure of peer acceptance did not include the ratings by the partners of children in the two peer involvement conditions. The partner ratings were examined in separate ANCOVAs in which group condition and sex were factors. The analysis of both posttreatment and follow-up scores revealed significant condition effects, $F(1,26) = 4.73$, $p < .05$; $F(1,26) = 5.97$, $p < .05$, respectively. As shown in Table 2, children who received group coaching were subsequently liked more by their partners than children who received only group experience.

Thus, peer involvement produced significant but temporary improvements in children's general peer acceptance. On the other hand, the children who received group coaching continued to be liked more by those peers actively engaged in treatment with them.

RATES OF PEER INTERACTION. Two measures focused on children's rates of peer interaction in naturalistic settings—observations of lunchtime interaction rates and teacher ratings of classroom interactions. A MANCOVA on posttreatment scores yielded a main effect for peer involvement, $F(2,45) = 5.12$, $p < .01$. Subsequent ANCOVAs revealed that children who had received peer involvement had significantly higher lunchroom interaction rates than children not treated with peers, $F(1,47) = 6.96$, $p < .01$. No effects were observed on the teacher ratings.

At follow-up, the MANCOVA on the two measures of interaction rates showed a near significant main effect for skill training, $F(2,45) = 2.45$, $p < .10$. Again, the univariate ANCOVA on teacher

TABLE 2 Mean Peer Acceptance Scores

Measures	Time of Assessment		
	Pretreatment	Posttreatment	Follow-up
Roster and Rating Scale:			
Peer involvement conditions	2.59	2.83**	2.74
No peer involvement conditions	2.68	2.65**	2.70
PEI withdrawal scale:			
Peer involvement conditions	2.99	3.29	3.32
No peer involvement conditions	2.83	2.84	2.80
Partner sociometric:			
Group coaching condition	2.68	3.29**	3.32**
Group experience condition	2.18	2.43**	2.29**

Note. Higher scores indicate greater acceptance on the Roster and Rating Scale and on the partner sociometric and greater withdrawal on the PEI. Partner sociometrics were only available in the two peer involvement conditions.
** $p < .05$.

ratings showed no significant effects, but the AN-COVA on children's rates of peer interaction at lunchtime revealed a significant main effect for skill training, $F(1,47) = 4.66$, $p < .05$. Six weeks following treatment, children who had received skill training had significantly higher lunchtime interaction rates than children who had not. While peer involvement alone produced a significant increase in children's naturalistic interactions immediately following treatment, skill training was apparently necessary to produce sustained improvements.

SELF-PERCEPTIONS. The MANCOVA on the post-treatment scores on the Social Self-Efficacy Scale and the social subscale of the Perceived Competence Scale for Children revealed a significant main effect for peer involvement, $F(2,45) = 3.79$, $p < .05$. Univariate ANCOVAs showed this effect to result predominantly from a near significant effect of peer involvement on the Social Self-Efficacy Scale, $F(1,47) = 3.57$, $p < .07$. Children who received group treatments tended to rate themselves as more socially competent than those who did not. At follow-up, the MANCOVA on the two self-perception measures revealed no significant main effects or interactions.

Treatment Effects on Peers

Although the peer involvement treatment was not designed to affect the peers, it was of interest to determine whether the experience had any positive or negative effects on them. The partners' scores on the sociometric measure and the PEI Withdrawal Scale were subjected to a 2 (group condition) × 2 (sex) MANCOVA; no significant effects were found, all p's > .10. Similarly, no effects were found on a MANCOVA on the partners' posttreatment or follow-up scores on the Social Self-Efficacy Scale and the social subscale of the Perceived Competence Scale for Children.

The only available measure of the partners' conversational skills was their written responses on the Conversational Skill Concept Scale. Univariate ANCOVAs revealed a near significant condition effect for posttreatment scores, $F(1,51) = 2.66$, $p < .10$, and a significant condition effect for follow-up scores, $F(1,51) = 5.86$, $p < .05$. Peers who participated in the group coaching experience displayed higher scores than peers who only received group experience. Apparently, skill training did have a positive effect on the acquisition of skill concepts for the peers involved.

Discussion

The two interventions examined here had strong, positive, and differential effects on the social competencies of the targeted children. Skill training promoted sustained increases in conversational skills observed both during a dyadic conversation and during small group interactions with peers. Additionally, 6 weeks after treatment, coached children were talking more with peers during lunch and received higher scores on the written measure of conversational skills. In contrast, peer involvement under superordinate goals did not have a major impact on conversational skills. Instead, peer involvement produced significant, though temporary, improvements in classroom sociometric status, lunchtime rates of peer interaction, and feelings of social efficacy. No significant interactions were observed between the two treatments, yet the group experience with coaching condition had an additive advantage. Only children in this combined condition shared general and sustained improvements in peer partner acceptance as well as social skills and peer interaction rates.

Mechanisms Responsible for Treatment Effects

In any intervention study, one must consider the possibility that placebo or other nonspecific treatment effects are responsible for the demonstrated changes. While this study did not include an attention-placebo control group, such an explanation can be safely ruled out for several reasons. First, previous coaching studies that have included attention-placebo control groups have not found such procedures to have any effects (Ladd, 1981; Oden & Asher, 1977). Second, if a placebo effect were present, one would have expected to find general differences among the three treatment conditions and the no-treatment condition. On the contrary, this study revealed predicted and specific effects of the different treatments on various dimensions of social competency.

Consider the effects produced by skill training. Coaching programs are based on the theoretical premise that unaccepted children lack the social skills necessary to make friends. When children are given instruction and practice in these skills, they are expected to use these skills increasingly in their peer interactions and acquire greater peer approval. In the present study, a series of measures provided an opportunity to examine the effects of coaching along a gradient of generalization. The dyadic conversation and the Conversational Skill Concept Scale assessed children's behavioral repertoire and skill concepts. In this study, changes were observed on skill performance in the dyadic conversation and to some degree on the Conversational Skill Concept Scale. Thus, the coaching program did appear to teach children new conversational skills.

A third measure of conversational skills, observations in small groups of classroom peers, provided an estimate of the extent to which children generalized their skill performance to peer interactions. Positive effects for coaching were found on this measure both at the end of treatment and in the follow-up assessment. Additionally, at the time of follow-up, coached children began to show increased rates of peer interaction during lunchtime.

Consistent with previous research, this study found that children receiving coaching did not gain greater peer acceptance. Our findings help clarify the reasons for this consistent failure. In particular, one might have argued that at least some of the programs failed to change peer acceptance because they were ineffective in teaching the children new social skills. In this study, however, the children learned the targeted skills and applied them in their interactions with peers, yet they remained unaccepted by their peers. Thus, these results raise questions about the relationship between social skills and peer acceptance at this age. Perhaps conversational skills are not a primary cause of peer rejection at this age. Alternatively, peer attitudes may be controlled by factors other than the targeted children's behavior.

In fact, the effects of the peer involvement treatments used here do suggest that peer acceptance is controlled by other factors. While coaching procedures did not affect peer acceptance, peer involvement did. Children who were involved with peers received higher sociometric ratings immediately afterward, although these improvements did not last. More enduring improvements in peer acceptance were documented, however, for children who received group experience with coaching. When peer involvement and coaching were

combined, peer partners liked the identified children more, even at the time of the follow-up. Apparently, coaching is necessary but not sufficient for fostering greater peer acceptance.

What mechanisms may account for the effects of the peer involvement? First, it is important to note that the changes in the partners' ratings do not seem to be artifactual. If the partners were simply trying to please the experimenter, one would have expected the two peer group treatments to have similar effects, but they did not. Instead, the provision of a superordinate goal may have contributed directly to the changes in the peer partners' attitudes. Previous research has shown that interactions in the context of a superordinate goal is an effective means for reducing intergroup conflict and facilitating cooperation and liking among group members (Bryan, 1975; Sherif et al., 1961). When a superordinate goal is present, environmental conditions are created that are hypothesized to promote interdependence among group members (Hyman & Singer, 1968). To attain the goal, group members must increase their positive and cooperative interactions, which in turn may lead to increased liking among group members. Alternatively, the status of low accepted children may have increased because they made integral contributions to the groups' goals.

The presence of a superordinate goal, however, merely maximizes the probability that the group will try to use the skills of all members. If certain children actually do not have skills that can facilitate group goal attainment, they will continue to be rejected. Since they had been trained in relevant skills, the coached children may have been more capable of contributing to the group task than the noncoached children. Additionally, the group experience with coaching condition may have enhanced communication among group members; improved communication should facilitate the group's ability to work together and produce more successful and rewarding films. Either the hypothesized differences in the targeted children's skills or group communication could account for the finding that the partners' ratings changed more in the group experience with coaching condition than in the group experience alone condition.

Other factors may also account for the changes in peer partner ratings. While the peer partners may previously have avoided the disliked child out of habit, the group interactions may have led them to realize that the targeted child can be a rewarding playmate. In this case, involving the peers in the coaching process per se or providing superordinate goals may not be essential components of the treatment. Rather, it may be enough to provide coached children with opportunities for positive peer interaction so that their improved behavior would become salient to their peers. Clearly, further work is required to delineate the mechanisms underlying the effects of group experience with coaching.

In addition to the effects on peer acceptance, peer involvement had some effect on feelings of social efficacy, while skill training did not. In the social domain one only has the reaction of the recipient as the criteria to evaluate one's social competence. Thus, children's perceptions of social self-efficacy may not be based primarily on their actual behavioral abilities but instead they may be based on their perceptions of peer responses. If so, changes in self-efficacy may be more likely to occur in peer involvement programs where peers' responses have been modified than in other treatments where only the identified children's behavior has been altered.

Developmental Considerations in Treatment Programs

Research on normal social-developmental processes provided the foundation for the design, predictions, and interpretation of the present clinical intervention. In the past, clinical research with adults has often served as the basis for deciding which skills to teach children (Furman, 1980; Furman & Drabman, 1981). In contrast, our selection of conversational skills was based on the developmental literature indicating that these skills were important components of social competence in childhood (Gottman et al., 1975). Similarly, we decided to assess the effects of the treatment on four dimensions of social competence (conversational skills, social interaction, peer acceptance, and self-perceptions) because previous developmental literature had indicated that they were distinct aspects of childhood social competence.

The selection of intervention strategies provides the nicest illustration of the importance of considering developmental factors. Previous social skills programs have focused principally on changing the behavioral or cognitive capabilities of the child; when peer responses were considered, they were usually viewed as criteria by which to evaluate the adequacy of change in the targeted children's behavior. Peer expectations and attributions may actively influence how peers interpret and respond to a child's behavior. For example, even when they engage in similar behaviors, children of different sociometric status may receive differential peer responses (Putallaz & Gottman, 1981). These data point out the importance of considering how peers tend to respond to target children's behavior.

Moreover, the influence of sociometric status on peer responses seems to increase during the period of preadolescence. Developmental studies have shown that the stability and structure of peer group norms begin to change during preadolescence. During these years, children become more peer-oriented and the norms and standards of the peer group become more clear, consensually valid, and differentiated (Bowerman & Kinch, 1959). There is greater group consensus about the reputations of group members, and sociometric status shows less fluctuation (Horrocks & Buker, 1951). In other words, the social structure begins to crystallize.

Thus, by preadolescence, peer acceptance may be affected increasingly by peer group norms. If so, changes in children's behavior may have less impact on their acceptance by the group. Consequently, it may be important to supplement skill training for preadolescents with environmental manipulations that maximize the probability that peers will recognize and accept the new competencies of the coached children. Consistent with this hypothesis, this study found that coaching alone did not improve the peer acceptance of preadolescents, although similar interventions have been effective with younger children. It would be interesting to compare directly the differential effects of coaching and peer involvement programs with preadolescents and younger elementary school children. If the preceding hypothesis is correct, the contribution of peer involvement programs beyond that obtained by skill training should be greater for the older children.

Although the peer involvement programs had a larger impact on peer acceptance than the coaching programs did, the changes in general sociometric status were only temporary ones. To obtain lasting effects on the crystallized group structure of preadolescents, it may be necessary to have a more extended peer involvement program or have more peers participate in some aspects of the program. In this study only two peers were involved in the program. Perhaps if more children had been involved, sustained and generalized effects on classroom sociometric status may have been observed. At the same time, we believe that it is important that during any particular session the group size be small enough that the identified child is an intergral part of the group.

While the importance of considering developmental factors in designing intervention programs has been emphasized in the preceding paragraphs, it is also important to note that these clinical interventions have implications for normal developmental research. For example, the children who were having difficulties in peer relations did show deficits in conversational skills, supporting the hypothesis that such skills are a central facet of peer interaction at this age. Similarly, the differential treatment effects on the four dimensions of social competence suggest that these dimensions are distinct facets of social adjustment. Finally, the fact that the present coaching program did not have the same effect as coaching programs with younger children provides indirect support for the developmental hypothesis that peer group processes change and intensify during preadolescence.

Summary

In summary, this study found coaching to be effective in teaching preadolescents new behavioral skills. Coaching alone, however, was not sufficient to enable children at this age to modify the attitudes of their peers without structured support. Changes in peer attitudes and in children's self-perceptions of social efficacy were found only when opportunities for positive peer interaction under superordinate goals were provided. By including a series of different measures of social competencies, it was possible to identify some of the mechanisms that may be responsible for some of these differential effects.

This study illustrates the value of developmental considerations in designing intervention programs (Furman, 1980). The fields of developmental and child clinical psychology have been isolated from each other. Research and interventions conducted by child clinical psychologists are often not well funded in developmental theory. Conversely, developmental researchers often do not consider the implications of their work for clinical applications. We hope that this article provides an illustration of how both research on childhood psychopathology and on normal development can be enhanced through integrative efforts.

Reference Note

Hymel, S., & Asher, S. R. *Assessment and training of isolated children's social skills.* Paper presented at the biennial meeting of the Society for Research in Child Development, New Orleans, March 1977.

References

Asher, S. R., Oden, S., & Gottman, J. M. Children's friendships in school settings. In L. G. Katz (Ed.), *Current topics in early childhood education* (Vol. 1). Norwood, N.J.: Ablex, 1977.

Bowerman, C. E., & Kinch, J. W. Changes in family and peer orientation of children between the fourth and tenth grades. *Social Forces,* 1959, *37,* 206–211.

Bryan, J. H. Children's cooperation and helping behavior. In M. Hetherington (Ed.), *Review of child development research* (Vol. 5). Chicago: University of Chicago Press, 1975.

Chandler, M. J. Egocentrism and social behavior: The assessment and training of social perspective taking skills. *Developmental Psychology,* 1973, *9,* 326–332.

Combs, M. L., & Slaby, D. A. Social skills training with children. In B. Lahey & A. Kazdin (Eds.), *Advances in clinical child psychology* (Vol. 1). New York: Plenum, 1977.

Cowen, E. L., Pederson, A., Babigian, H., Izzo, L. D., & Trost, M. A. Long-term follow-up of early detected vulnerable children. *Journal of Consulting and Clinical Psychology,* 1973, *41,* 438–446.

Furman, W. Promoting social development: Developmental implications for treatment. In B. Lahey & A. Kazdin (Eds.), *Advances in clinical child psychology* (Vol. 3). New York: Plenum, 1980.

Furman, W., & Drabman, R. Methodological issues in child behavior therapy. In M. Hersen, R. M. Eisler, & P. M. Miller (Eds.), *Progress in behavior modification* (Vol. 13). New York: Academic Press, 1981.

Gottman, J. M. Toward a definition of social isolation in children. *Child Development,* 1977, *48,* 513–517.

Gottman, J. M., Gonso, J., & Rasmussen, B. Social interaction, social competence and friendship in children. *Child Development,* 1975, *46,* 709–718.

Harter, S. The Perceived Competence Scale for Children. *Child Development,* 1982, *53,* 87–97.

Hartup, W. W. Children and their friends. In H. McGurk (Ed.), *Issues in childhood social development.* London: Methuen, 1978.

Horrocks, J. E., & Buker, M. E. A study of the friendship fluctuations of preadolescents. *Journal of Genetic Psychology,* 1951, *78,* 131–144.

Hyman, H. H., & Singer, E. (Eds.) *Readings in reference group theory and research.* New York: Free Press, 1968.

Kelly, J. A., Furman, W., Phillips, J., Hathorn, S., & Wilson, T. Teaching conversational skills to retarded adolescents. *Child Behavior Therapy*, 1979, *1*, 36–43.

Koslin, B. L., Haarlow, R. N., Karlins, M., & Pargament, R. Predicting group status from members' cognitions. *Sociometry*, 1968, *31*, 64–75.

Ladd, G. Effectiveness of a social learning method for enhancing children's social interaction and peer acceptance. *Child Development*, 1981, *52*, 171–178.

La Greca, A. M., & Santogrossi, D. A. Social skills training with elementary school students: A behavioral group approach. *Journal of Consulting and Clinical Psychology*, 1980, *48*, 220–227.

Minken, N., Braukmann, C. J., Minken, B. L., Timbers, G. D., Fixsen, D. L., Phillips, E. L., & Wolf, M. M. The social validation and training of conversation skills. *Journal of Applied Behavior Analysis*, 1976, *9*, 127–140.

Oden, S., & Asher, S. R. Coaching children in social skills for friendship making. *Child Development*, 1977, *48*, 495–506.

Patterson, G. R., Littman, R. A., & Bricker, W. Assertive behavior in children: A step toward a theory of aggression. *Monographs of the Society for Research in Child Development*, 1967, *32*(5, Serial No. 113).

Pekarik, E. G., Prinz, R. J., Liebert, D. E., Weintraub, S., & Neale, J. M., The Pupil Evaluation Inventory: A sociometric technique for assessing children's social behavior. *Journal of Abnormal Child Psychology*, 1976, *4*, 83–97.

Putallaz, M., & Gottman, J. M. An interactional model of children's entry into peer groups. *Child Development*, 1981, *52*, 986–994.

Roff, M., Sells, S. B., & Golden, M. M. *Social adjustment and personality development in children.* Minneapolis: University of Minnesota Press, 1972.

Rucker, C. N., & Vincenzo, F. M. Maintaining social acceptance gains made by mentally retarded children. *Exceptional Children*, 1970, *36*, 679–680.

Sherif, M., Harvey, O. J., White, B. J., Hood, W. R., & Sherif, C. W. *Intergroup conflict and cooperation: The robber's cave experiment.* Norman: University of Oklahoma, 1961.

Assessment of Stress-Related Psychophysiological Reactions in Chronic Back Pain Patients

HERTA FLOR
DENNIS C. TURK
NIELS BIRBAUMER

The classification *chronic back pain* (CB) commonly refers to back pain presumably originating in the spine or the surrounding tissue that is not related to a specific acute disease process. Numer-

ous physiological factors have been proposed as being of etiological significance for CBP, namely, inflammatory processes, degenerative changes, structural deformities, traumatic incidents, and muscular or ligamentous strain (e.g., Cailliet, 1981; Loeser, 1980). There is, however, virtually no empirical evidence to support an isomorphic relationship between any of these factors and CBP (Flor & Turk, 1984; Nachemson, 1979).

The lack of empirical evidence for an organic pathology of CBP has resulted in the consideration of the interaction among psychological and physiological factors (Fordyce, 1976; Sternbach, 1974). Gentry and Bernal (1977) have proposed a respondent model of CBP in which they assume that classical conditioning of pain and tension may occur in acute pain states, leading to a pain-tension circle and subsequent chronic back pain. The pain may be exacerbated by conditioned fear of movement leading to avoidance of activity and immobilization, (e.g., Lethem, Slade, Troup, & Bentley, 1983). Some preliminary tests of the assumptions of the respondent model have been conducted; however, the results are mixed. For example, several investigators have reported that in comparison to normal subjects. CBP patients demonstrated higher electromyogram (EMG) levels in various body positions or during differential relaxation (e.g., Grabel, 1973; Hoyt et al., 1981; Kravitz, Moore, & Glaros, 1981). In contrast, Collins, Cohen, Naliboff, and Schandler (1982) found lower back EMG levels in certain body positions.

The EMG levels in the lower back of CBP patients following exposure to various stressors have received some attention (e.g., Collins et al., 1982; Grabel, 1973). The results do not support the assumption that stressors produced high levels of EMG activity specifically in the back muscles of CBP patients. Rather, elevated frontalis muscle tension levels have been observed in CBP subjects during resting baseline as well as in response to different laboratory stressors (Collins et al., 1982).

Several limitations of the studies conducted prohibit drawing any valid conclusions regarding the adequacy of the respondent model. For example, in some studies EMG measurements were uni-

"Assessment of Stress-Related Psychophysiological Reactions in Chronic Back Pain Patients," Herta Flor, Dennis C. Turk, and Niels Birbaumer, Vol. 53 (3) 1985, pp. 354–364. Reprinted from the *Journal of Consulting and Clinical Psychology*, published by the American Psychological Association, Inc. Herta Flor is a professor of psychology at the University of Bonn. Dennis C. Turk is on the faculty at Yale University. Niels Birbaumer is on the faculty of the Universitat Tubingen, the Federal Republic of Germany.

lateral (e.g., Grabel, 1973), and some physiological differences may not have been detected. When exposure to stressful stimuli has been tested, investigators have relied on laboratory stressor tasks that may have had little personal relevance for the subjects (e.g., Collins et al., 1982). Additionally, investigators have failed to include appropriate control groups. For example, no comparisons with other chronic pain syndromes have been performed, nor have several muscle sites been examined to test for the specificity of abnormal reactions in the paraspinal muscles of the CBP patients (e.g., Grabel, 1973; Hoyt et al., 1981). Finally, most of the studies on the psychophysiology of back pain have relied exclusively on EMG elevations, although the delay in return to baseline levels may be an equally important parameter (Philips, 1977).

Recently, Flor and her collaborators (Flor, Birbaumer, & Turk, 1984; Turk & Flor, 1984) proposed a diathesis–stress model of CPB that integrates many of the disparate findings noted above. The model postulates that CBP may result from the interaction of personally relevant stressful events with a predisposing organic or psychological condition. Intense or recurrent, potentially aversive stimulation in a predisposed individual with maladaptive or inadequate coping abilities is hypothesized to lead to extensive and sustained reactions of the back muscles. Further, the diathesis–stress model suggests that increases in muscle tension might lead to ischemia, reflex muscle spasms, oxygen depletion, and the release of pain-eliciting substances (e.g., histamine, substance p). The subsequent pain might then act as a new stressor feeding into a vicious circle. The development of movement-related anticipatory anxiety and subsequent immobility and the reinforcement of pain behaviors may exacerbate the pain problem. Moreover, the anticipatory anxiety may produce hyperactivity (muscle tension) and may further exacerbate nociception and, subsequently, pain.

The diathesis–stress model (Flor et al., 1984; Turk & Flor, 1984) predicts that CBP patients will exhibit elevated and prolonged reactions specifically of the back musculature to stress compared with healthy controls or other pain patients. Further, it predicts that muscular reactions to stress will only occur in response to personally relevant stressors. Additionally, the model would hypothesize that these muscular hyperreactions will be best predicted by psychological variables (e.g., anxiety, helplessness) as compared to organic variables (e.g., amount of degenerative damage, number of surgeries, duration of the pain problem) and that anxiety levels and immobility are related to CBP patients. The purpose of this study is to examine each of these predictions.

Method

SUBJECTS

Seventeen patients who had been suffering from CBP for more than 6 months and had been referred to the West Haven, Connecticut, Veterans Administration Medical Center Pain Management Program for evaluation made up the CBP group. Patients with neurological complications (e.g., nerve impingement) or inflammatory disorders (e.g., arthritis) were excluded. All patients' diagnoses were included in their medical records by their primary physicians (e.g., neurosurgeon). The patients' medical records were reviewed by the investigators, and the diagnoses were categorized according to the classification of the American Rheumatism Association (Blumberg, Bunim, Calkins, Pirani, & Zwaifler, 1964). The first two investigators independently classified the patients and agreed on the classification of all of them.

All patients who were currently in psychiatric treatment (an exclusion criterion of the Pain Management Program) and all non-CBP patients with a documented history of back pain were excluded from the study. Of the 65 patients who had originally volunteered for the study, 4 were excluded due to equipment malfunction and 7 nonCBP subjects were excluded because they had a significant CBP history or were suffering from acute back pain. Two subjects were excluded because they did not comply with the experimental instructions (e.g., excessive movement during the monitoring period), and 1 CBP patient was excluded on the suspicion of surgery-related denervation potentials in the back musculature.

The general pain (GP) group consisted of 17 patients who had been suffering from pain syndromes unrelated to their back for more than 6 months. The healthy control (H) group comprised 17 hospital patient volunteers who did not experience back pain but who were in treatment for other medical problems (e.g., diabetes).

The three experimental groups were matched according to sex, age, marital status, education (years of school), and employment level (categorized as employed, unemployed, disabled, retired, or in training). Chi-square analyses and F tests revealed no significant group differences (all $ps > .29$). Seventy-eight percent of the subjects were male. The average age of the subjects was

47 years (ranging from 23 to 73 years); 72% were married, 14% were single, and 14% were widowed or divorced. The mean number of years of education was 13 (ranging from 5 to 22 years). Only 45% were currently employed or in training, whereas 27% were retired, 20% (in all three groups) were totally disabled, and 8% were unemployed.

The two pain groups did not differ in pain duration, pain intensity, number of surgeries, medication, or pain-related compensation (all $ps > .24$). The mean duration of the pain in the CBP group was 10.79 years ($SD = 13.09$; range = 8 months to 38 years), and in the general pain group, the mean duration was 8.93 years ($SD = 8.37$; range = 1 year to 30 years; $t[28] = .74$, ns). Eight patients in the CBP group and 6 patients in the GP group had undergone at least one pain-related surgery. Examination of the medical records revealed that the amount and type of analgesic medication was comparable in the two pain groups. Nine patients in the CBP group showed moderate to severe degenerative changes of their spine on X rays, and 8 displayed minimal or no identifiable change. Nine patients in the CBP and 8 patients in the GP group received pain-related compensation.

The medical diagnosis in the CBP group ranged from degenerative disc disease/spondylosis (7 patients) to low back pain of "unknown origin" (4 patients), disc hernia (3 patients), sciatica (1 patient), chronic unstable low back syndrome (1 patient), and spondylolisthesis (1 patient). In the GP group, a wide range of pain locations and etiologies was present. Seven patients suffered from rheumatic disease and 3 from headaches; 2 had chest or abdominal pain; and 1 each complained of cancer-related pain, hip pain, peripheral vascular disease, postherpetic neuralgia, and stump pain.

INSTRUMENTATION

All physiological variables were recorded by a Cyborg Biolab 11, an eight-channel computer-operated (Apple II) physiograph. Surface electromyograms were recorded from the frontalis and the lumbosacral erector spinae muscles (low back pain patients, $n = 14$) or the trapezuis muscle (upper back pain patients, $n = 3$) using standardized electrode placement (Lippold, 1967). The trapezius and erector spinae measurements were taken bilaterally because side asymmetries are believed to be important in CBP (Cram & Steger, 1983; Wolf et al., 1982).

To obtain measurements, 15-mm silver silver chloride (AgAgCl) sensors filled with Spectra 360 electrode gel were attached to the relevant muscle sites after the skin had been cleansed with 70% alcohol and a mild abrasive (Redux) had been applied to the skin. Heart rate (HR) was measured by a photopletysmorgraphic sensor attached to the second finger of the subject's right hand. Skin resistance level (SRL) was measured via 4-mm AgAgCl sensors that were filled with Surgilube surgical jelly and attached to the basal and palmar surface of the subject's left hand after it had been cleaned with soap and water.

The EMG signals were filtered by a 100- to 250-Hz bandpass filter, a preset range for the Biolab.[1] All signals were fed into a remote differential amplifier. Electrode resistance was kept below 10,000 kiloohms. The signals were read at a sampling rate of 10 per s, and the digitized signal was averaged for 2-s intervals and stored for further analysis.

DEPENDENT VARIABLES

Dependent measures consisted of physiological, motor, and subjective measures (Birbaumer, 1977; Sanders, 1979). The mean value of the initial baseline phase of the paraverteberal EMGs, frontalis EMG, HR, and SRL were used for group comparisons as indicators of the tonic activity in the subjects. The reactivity to the experimental tasks was measured as change from the pretrial baseline to the trial level and as time to return to baseline in s in the poststress phase. The SRL and HR measurements served as manipulation checks because they are regarded as indicators of general arousal (Lang. 1971). Spinal mobility was assessed by asking patients to bend at the waist and to reach down with their fingers as if to touch the floor. The distance from finger tips to the floor was then measured and used as an indicant of spinal mobility (e.g., Moll & Wright, 1980).

Patients rated the amount of present pain, tension, stress, and effort required by the task before and after the baseline and each experimental condition on 11-point scales ranging from none to very much. The State form of the State-Trait Anxiety Inventory (STAI; Spielberger, Gorsuch, & Lushene, 1970) was administered before the baseline and after the last experimental trial. The ratings of perceived tension and stress and the STAI scores served as manipulation checks for the experimental trials and as indication of the subjects' anxiety levels.

---

[1] The limitation of the measurement of the EMG signal to a 100- to 250-Hz range may have led to a loss of measurement sensitivity. This might account for the relatively low paraspinal EMG baseline levels obtained. However, it does not invalidate our conclusions on the reactivity of the paraspinal EMG. A lack of sensitivity should, if anything, have limited our ability to observe significant changes in EMG activity.

In addition, daily pain intensity ratings (Turk, Meichenbaum, & Genest, 1983), the Pain Rating Index–Total from the McGill Pain Questionnaire (Melzack, 1975), and the pain ratings during the experimental session were used as indicators of the patients' pain intensity. Depression was assessed using the Depression Adjective Check List (DACL; Lubin, 1965) and the Beck Depression Inventory (BDI; Beck, Ward, Mendelson, Mock, & Erbaugh, 1961). Finally, the Pain Experience Scale (PES; Turk, 1981), which assesses the patient's emotional and cognitive reactions to the pain experience, was used.

EXPERIMENTAL CONDITIONS AND PROCEDURES

Patients were informed that they were participating in an experiment that assessed the relationship of stress, muscle tension, and chronic back pain. Patients completed the consent form and the STAI and answered questions on demographic data and disease characteristics prior to the experimental trials. Next, spinal mobility was assessed. The subjects washed their hands and were seated in a straight-back arm chair; electrodes were attached; and the initial pain, tension, stress, and effort ratings were recorded.

Following a 10-min resting baseline, during which the subjects were to sit still with their eyes open, four experimental trials were presented in a predetermined and counterbalanced order. Each trial lasted 1 min and was preceded by a 1-min baseline phase and followed by a 5-min return to baseline phase.

The four experimental conditions included two personally relevant stressors, a general stressor, and a neutral condition. During the personally relevant stress trials, subjects were asked to recall a recent stressful event or a recent pain episode and describe it for 1 min. During the general stress trial, subjects were asked to count backwards by 7s from 758 as quickly as possible. During the neutral trial, subjects were asked to recite the alphabet for 1 min. After each trial, pain, tension, stress, and effort ratings were recorded. Subjects were instructed to move as little as possible, and extent of movement was monitored by the experimenter on a graphic screen display. Any movement-related activity (which appeared as a spike on the screen) was noted together with the sample number during which it occurred. The data were then edited for movement-related outliers. Excessive movement was rarely noted except in the case of 2 subjects who, as noted above, were excluded from the study.

The recitation of the alphabet served as a control for speaking and experimental participation. The choice of the stress situation discussed was

left to subjects, because it was hypothesized that only those situations that were experienced as personally relevant by the subjects would affect muscular reactivity. The arithmetic task served as a contrast for the personally relevant tasks, and it has frequently been employed as a general source of stress in other studies (Collins et al., 1982).

After the presentation of the fourth trial, the electrodes were removed and the STAI was readministered, debriefing was conducted, and the experimental session was terminated. The entire procedure required approximately 90 min.

DATA REDUCTION AND ANALYSIS

All physiological data were averaged for the baseline period (10 min), the pretrial baseline (1 min), the trial (1 min), and the posttrial baseline (5 min). Return to baseline was defined as the point in time in the posttrial baseline when the physiological values first reached pretrial baseline values and was calculated in s. A time of 300 s was recorded when the values did not return to pretrial baseline within the 5-min posttrial interval. A 3×4 mixed design with one between (groups) and one within (trials) variable was used to test the major hypotheses of the study.

Multivariate analyses of variance (MANOVAs) and Geisser-Greenhouse corrected mixed model analyses of variance (ANOVAs) were computed for the pretrial-trial differences because of heterogeneous variance-covariance matrices (Muller, Otto, & Benignus, 1983; Reinsel, 1982). In addition to these analyses, univariate ANOVAs were computed for between-groups comparisons in baseline physiological values and other data. Alternatively, Kruskal-Wallis ANOVAs were calculated when distributions deviated from normality or unstable variances were observed (Hays, 1973). T tests and chi-square analyses were computed for single-group comparisons. Mann-Whitney U and Wilcoxon tests were calculated for a posteriori group comparisons for the physiological variables be-

TABLE 1 Left and Right Mean Paravertebral EMG (Sitting) and Standard Deviations in Microvolts for the Experimental Groups

Group	Left Back		Right Back	
	M	SD	M	SD
Chronic back pain	3.63	6.80	2.48	2.21
General pain	1.33	1.21	1.57	0.98
Healthy controls	1.23	0.63	1.47	0.88

Note. EMG = electromyogram.

cause they had large variance differences. In these cases, Bonferroni-corrected probability levels were used (Kirk, 1968). To analyze the relationship among various physiological and psychological variables within and between groups, correlation and regression analyses were performed.

Results

Resting Baseline Levels

Table 1 shows the subjects' mean resting EMG values at the right and left paraspinal muscles.

Kruskal-Wallis ANOVAs were calculated with regard to the large variance differences in the groups. Although the EMG levels at the left back differ significantly, $\chi^2(2) = 6.327$, $p = .042$, with the CBP group displaying significantly higher levels than the two control groups, the right-back EMG levels are not significantly different, $\chi^2(2) = 2.783$, $p = .249$. Both sides of the back display higher mean resting tension levels and higher variances in the CBP group as compared to the two other groups.

Within the group of CBP patients, only 4 displayed extreme resting baseline levels, the remainder ($n = 13$) displayed levels quite similar to the GP and H groups, which did not differ from each other. The CBP patients with high or average EMG resting levels did not differ from one another in organic variables such as surgery, degenerative change, duration of pain, or amount of pain (all $ps > .30$), but they differed in State Anxiety and depressed mood ($p < .05$). No significant lateral asymmetries were identified in the CBP patients. Only three patients in the CBP group had lateral differences of more than two microvolts.

Reactivity to Stress in the Back Musculature

Table 2 includes the means and standard deviations for the left-back EMG values during the pretrial, trial, and posttrial phases for all conditions and all groups.

The MANOVA for the left-back EMG reactivity yielded a highly significant effect for groups, $F(2, 48) = 8.84$, $p = .001$; a marginally significant effect for trials, Wilks's lambda = 0.86, average $F(3, 46) = 2.53$, $p = .07$; and a significant Group × Trial interaction, Wilks's lambda = .74, average $F(6, 92) = 2.54$, $p = .026$. A posteriori Mann-Whitney U tests (adjusted alpha = .008) reveal significant differences between the CBP group and

TABLE 2 Means and Standard Deviations for the Left Paraspinal EMG (Sitting) During the Experimental Conditions in Microvolts

Condition	Chronic Back Pain	General Pain	Healthy Controls
Alphabet Pretrial			
M	2.74	1.18	1.21
SD	4.89	0.91	0.67
Trial			
M	3.09	1.59	1.41
SD	4.03	1.29	0.59
Posttrial			
M	2.82	1.30	1.26
SD	4.63	1.25	0.80
Arithmetic Pretrial			
M	2.93	1.18	1.30
SD	4.93	0.78	0.69
Trial			
M	3.70	1.64	1.70
SD	5.17	1.20	0.79
Posttrial			
M	3.26	1.35	1.18
SD	5.59	1.11	0.65
Pain Pretrial			
M	3.08	1.28	1.15
SD	5.70	1.07	0.63
Trial			
M	5.28	1.55	1.40
SD	5.51	1.12	0.62
Posttrial			
M	3.67	1.32	1.24
SD	5.32	1.18	0.82
Stress Pretrial			
M	2.61	1.49	1.13
SD	4.80	1.39	0.62
Trial			
M	4.50	1.83	1.44
SD	5.06	1.49	0.64
Posttrial			
M	2.97	1.67	1.16
SD	4.72	1.40	0.66

Note. EMG = electromyogram.

the two other groups only in the pain and personal-stress conditions. Wilcoxon tests (adjusted alpha = .006) were significant only in the CBP group for the alphabet versus the pain and personal-stress conditions. Figure 1 displays the change in EMG values for the left back during the experimental conditions.[2] As Table 2 and Figure 1 show,

[2] The right back showed an equivalent response pattern, which is therefore not displayed.

FIGURE 1 Left paraspinal electromyogram reactivity to the stress condition. (CBP = chronic back pain; GP = general pain; H = healthy controls.)

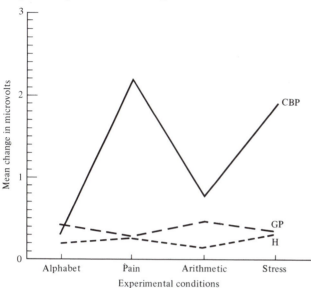

the left-back EMG displays greater reactivity only during the personally relevant stressful situations (the pain and personal-stress trials).

The right paravertebral muslces display a similar reaction. The MANOVA reveals a significant groups effect, $F(2, 43) = 11.62$, $p < .001$; a significant trials effect, Wilks's lambda = .78, average $F(3, 46) = 4.28$, $p = .010$; and a significant Groups × Trials interaction, Wilks's lambda = .65, average $F(6, 92) = 3.67$, $p = .003$. A posteriori Mann-Whitney U and Wilcoxon tests were significant for the CBP versus the other two groups (adjusted alpha = .008) and the alphabet versus the pain and personal-stress trial. In addition, only the CBP group demonstrated a significant arithmetic versus personal-stress trial difference (adjusted alpha = .006). To summarize, only the CBP group displays abnormal EMG elevations when exposed to the pain or personal-stress conditions. This holds for both the right and left sides of the back.

Return to Baseline

An additional prediction based on the diathesis–stress model was the delayed return to pretrial baseline values in the CBP patients' paravertebral EMGs. Figures 2 and 3 show the mean time to return to baseline for both the left and right back muscles.

Repeated measures ANOVAs with a priori contrasts (CBP vs. GP and H) revealed a marginal trials effect, $F(1, 48) = 3.34$, $p = .07$; significant group effect, $F(1, 48) = 12.31$, $p = .001$; and significant interaction, $F(1, 48) = 7.56$, $p = .008$, for the left back. Additionally, there was a trend for significant trials, $F(1, 48) = 3.32$, $p = .08$, and a Groups × Trials interaction, $F(1, 48) = 3.55$, $p = .07$, for the right back. A posteriori comparisons demonstrate that for the left back, the CBP group and the healthy controls differ significantly from each other in the pain and stress condition ($p = .008$). Furthermore, the alphabet condition is significantly different from the pain and personal-stress condition in the CBP group ($p < .006$), and the alphabet differs significantly from the arithmetic and personal-stress condition in the GP group ($p < .006$). For the right back, a posteriori tests again show a significant difference between the CBP group and the healthy controls only in the pain and personal-stress conditions ($p < .008$). Within-group comparisons reveal a significant difference between the alphabet and two stress trials in the CBP group (alpha < .006).

To summarize the results for the return to baseline measures, the delay in return to baseline of the back EMGs is similar, although the effects are somewhat less dramatic than the increases in EMG level. The left back displays a clear prolongation of the return to baseline for the CBP versus the

FIGURE 2 Delay in return to baseline in the left paraspinal electromyogram response. (CBP = chronic back pain; GP = general pain; H = healthy controls.)

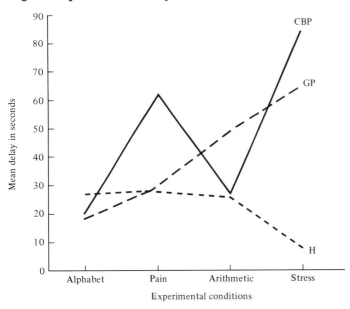

FIGURE 3 Delay in return to baseline in the right paraspinal electromyogram response. (CBP = chronic back pain; GP = general pain; H = healthy controls.)

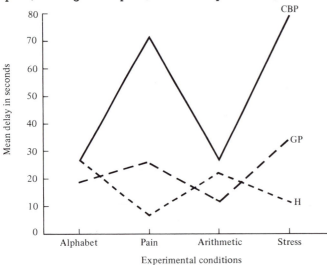

other groups in the personally relevant stress situations. The right back shows similar but only marginal effects.

Specificity Versus Generality of Muscular Reactivity

The subjects' frontalis EMG values were recorded to allow conclusions about the specificity versus generality of the back muscle response. The analysis of the frontalis EMG data revealed no group differences in resting baseline levels ($F < 1$) or in reactivity, Group × Trial interaction; Wilks's lambda = 0.87, average $F(6, 92) = 1.09$, $p = .17$. There is only a significant trials effect, average $F(3, 46) = 8.50$, $p < .001$. All groups showed stronger frontalis responses in the stress conditions versus the neutral condition, $F(1, 48) = 9.00$, $p = .004$.

For the delay in return to baseline there is again only a significant trials effect, neutral versus stress; $F(1, 48) = 5.26$, $p = .023$. No stimulus-specific response peak in the frontalis during mental arithmetic was found, nor was there a stronger reaction in the pain groups as compared to the healthy controls. Thus, paraspinal EMG reactivity was only observed for the CBP group and only for the personally relevant stress conditions. This supports a response specificity model.

Manipulation Checks

As noted above, the frontalis activity increased with the stressfulness of the experimental condition regardless of the experimental group, suggesting successful manipulation of stress in all groups. Additionally, HR and SRL had been assessed as manipulation checks for the stressful versus neutral conditions. Again, there were no differences in the groups' mean baseline HR levels ($F < 1$) nor in the change in HR during personal stress conditions. There was, however, a trend toward a significant conditions effect, average $F(3, 45) = 2.28$, $p = .09$. Although the two pain groups responded with undifferentiated HR increase, the healthy controls showed an HR increase only during mental arithmetic (Wilcoxon test: $p < .006$) but a decrease in the other conditions. Time to return to baseline did not reveal any significant effects for the heartrate.

Analysis of the SRL reactivity and return to

baseline yielded only a significant trials effect, average $F(3, 39) = 4.38$, $p = .009$, and average $F(3, 40) = 5.32$, $p = .004$, respectively. The stress conditions showed lowered resistance levels (indicating higher arousal) than the neutral trial, contrast $F(1, 41) = 5.40$, $p = .025$.

In summary, frontalis EMG, HR, and SRL only show significant main effects for trials without any significant group differences or interactions. Frontalis EMG, HR, and SRL only show stronger responses in the stressful conditions, indicating successful manipulation of the stress versus nonstress conditions. This is confirmed by the subjects' ratings of the stressfulness of the experimental conditions. The stress rating of the alphabet in contrast to the three stress conditions is highly significant, $F(1, 48) = 9.09$, $p = .004$, with means of 4.77, 4.73, 3.62, and 1.63 for the stress, pain, arithmetic, and alphabet conditions, respectively. The same is true for the effort ratings, contrast alphabet versus stress conditions: $F(1, 48) = 66.64$, $p < .001$.

Relationship of Back EMG Reactivity and Psychological Variables

A further question of interest was how psychological and physical variables are related to this muscular hyperreactivity of the back in CBP patients. The diathesis–stress model (Flor et al., 1984; Turk & Flor, 1984) predicts that dysphoric mood and the ability to cope with stressful or painful stimulation are important mediators of the psychophysiological response.

Stepwise regression analysis reveals that the best predictors for the EMG reactivity are the patients' Beck Depression Inventory scores ($\beta = 1.009$) and Pain Experience Scale scores ($\beta = 0.582$), which explain almost 76% of the variance; sample size adjusted $R^2 = .76$, $R = .91$, $F(2, 5) = 11.86$, $p = .013$. These data suggest that the patients with high paraspinal EMG reactivity are those who are most depressed, most worried, and emotionally affected by their pain problem (and possibly also by other stressful events). In contrast, stepwise regression with the organic variables (i.e., amount of degenerative change, surgery, pain duration) as independent variables did not yield a significant effect, nor did they significantly improve the prediction of the psychological variables. A median split of the CBP subjects into high versus low responders and cross tabulation of the data with the amount of degenerative change, number of surgeries, and pain duration yielded no significant relationship between the reactivity and these organic variables (all $ps > .60$ for chi-square analyses) but a significant relationship between the reactivity and the STAI and DACL scores (t tests between responders and nonresponders: $p = .05$). These data confirm the relationship between muscular hyperactivity and dysphoric mood predicted by the diathesis–stress model.

Spinal Mobility

The diathesis–stress model predicts high immobility in the CBP patients related to high levels of anxiety. The mean finger-to-floor distance in the CBP group is 27.88 cm ($SD = 25.39$ cm); the average distances for the GP group and the healthy controls are 9 cm ($SD = 9.91$ cm) and 1.35 cm ($SD = 3.43$ cm), respectively. An ANOVA and subsequent Tukey tests revealed a highly significant difference between the CBP and the two control groups, $F(2, 48) = 12.61$, $p < .001$, Tukey's HSD $p < .01$. As predicted, a significant relationship between State Anxiety and immobility can be found ($r = .59$, $p < .01$). Interestingly, immobility was significantly related to the amount of pain (pain card index: $R = .71$, $p < .001$) and the number of back surgeries ($r = .47$, $p = .029$) but not to the presence of surgery per se ($p > .60$).

Discussion

This study investigated the psychophysiological responses of CBP patients and examined specific hypotheses based on the diathesis–stress model (Flor et al., 1984; Flor & Turk, 1984; Turk & Flor, 1984). The data presented indicate that only CBP patients demonstrate abnormal back muscle reactivity to personally relevant stressors and prolonged delay in return to baseline levels. These results suggest that the abnormal paravertebral muscle response is characteristic of the CBP patients and is not a phenomenon of chronic pain in general (because the GP group does not overreact), nor is it part of a general stress reaction. The general pain and healthy controls indicated comparable levels of arousal, SRL, HR, and frontal EMG but no back muscle response.

It was also hypothesized that abnormal back

muscle reactions would only occur in stress situations that were of particular significance to the patient, such as the pain and personal-stress conditions. This hypothesis was confirmed and replicates the findings of Collins et al. (1982). The results of this study clearly indicate that the amount of physiological responding is related to depression and the negative evaluation of painful and stressful experiences but not to organic variables.

The findings suggest that abnormal reactions to stress and delayed return to baseline of the paraspinal muscle tension levels may be implicated in the development and maintenance of CBP. Similar conclusions regarding EMG reactivity and latency were, in regard to tension headaches, drawn by Philips (1977). In addition, it was confirmed that CBP patients had reduced spinal mobility, and it was shown that State Anxiety, pain levels, and spinal immobility were related in the CBP patients. In sum, all of the predictions based on the diathesis–stress model were supported.

This study incorporates several improvements over previous ones. Healthy controls as well as a general pain group were included to permit differentiation between back-pain-related effects and those attributable to chronic pain in general. Furthermore, bilateral EMG recordings from several sites were included, as was assessment of HR and SRL, to allow conclusions regarding the specificity versus generality of the back muscle response. The relationship among organic factors, psychological factors, and muscular stress repsonse was examined. Perhaps most important, patient-selected stress stimuli were included, in contrast to experimenter-selected stressors.

There are, however, several potential limitations of our study. One problem was the possibility of carry-over effects among the experimental conditions due to the repeated measures design. To avoid this effect, the presentation of the stimuli was counterbalanced, each condition was followed by a 5-min posttrial baseline, and additional 2- to 7-min intertrial intervals were interspersed. Two ANOVAs using weights to simulate carry-over effects between the pretrial baselines and the conditions were computed to examine carry-over effects statistically. Both ANOVAs yielded nonsignificant effects.

Another problem was the presence of surgery in some back pain patients. Comparison between the patients with and without surgery showed no differences in their resting EMG levels, reactivity of the back muscles, or delayed return to baseline (all $ps > .30$). The possible existence of denervation potentials was controlled by excluding patients whose baselines were very variable (more than 500% difference across the trials). Although the amount and type of analgesic medication was similar in the CBP and GP groups, the failure to control for the effect of various types of medication on EMG activity may be a limitation of this study. Future research should examine this question in more detail.

An additional problem was possible auto-correlation in the data. Time series analysis could not be computed due to the unavailability of a main frame transfer and the amount of time required to use ARIMA models with large numbers of subjects. Thus, repeated measures analyses were computed instead, and MANOVA was used as a conservative estimate if heterogeneous variance-covariance matrices were present.

Yet, another issue that needs to be addressed is that of excessive mobility and, hence, higher EMG levels of the CBP patients in contrast to the other groups. Care was taken to avoid the introduction of movement-related error in the data. All subjects were instructed about the great importance of sitting immobile during data acquisition. If movements occurred, they were recorded by the experimenter, who could observe both the subject and a continuous graphic display of the data. Any movement-related outliers in the data were noted together with the sample number during which they occurred. Before further analysis, all movement-related outliers were removed. Moreover, movements would be expected to occur in all four conditions equally so that the effects of the stress and pain situation could not be attributed solely to movement-related error.

The clinical relevance of these findings must also be considered. Although the microvolt changes observed were quite small, it is possible that chronic, low-level EMG hyperactivity may serve to maintain, exacerbate, or potentiate pain. Moreover, it should be noted that the experimental manipulation was quite weak (discussion of a stressful or painful event for 1 min) and conducted in a static, seated position. Research on the utility and validity of the diathesis–stress model of CBP should include replication of these results with more potent sources of stress. Additionally, the study of stress-related hyperactivity should be extended to dynamic as well as static body positions.

The results of this study may have important implications for both the assessment and treatment of CBP. Identification of muscular hyperactivity to personally relevant stressors may serve as a useful criterion for the inclusion of a stress-management component within a comprehensive treatment program of CBP.

References

Beck, A. T., Ward, C. H., Mendelson, M., Mock, J., & Erbaugh, J. (1961). An inventory for measuring depression. *Archives of General Psychiatry, 4,* 561–571.

Birbaumer, N. (1977). Angst als Forschungsgegenstand der experimentellen Psychologie. In N. Birbaumer (Ed.), *Psychophysiologie der Angst* (pp. 1–21). München, West Germany: Urban & Schwarzenberg.

Blumberg, B. S., Bunim, J. J., Calkins, E., Pirani, C. L., & Zwaifler, N. J. (1964). Nomenclature and classification of arthritis and rheumatism (tentative) accepted by the American Rheumatism Association. *Bulletin of Rheumatic Diseases, 14,* 31–33.

Cailliet, R. (1981). *Low back pain syndrome.* Philadelphia, PA: Davis.

Collins, G. A., Cohen, M. M., Naliboff, B. D., & Schandler, S. L. (1982). Comparative analysis of paraspinal and frontalis EMG, heartrate and skin conductance in chronic low back pain patients and normals to various postures and stress. *Scandinavian Journal of Rehabilitation Medicine, 14,* 39–46.

Cram, J., & Steger, J. C. (1983). EMG scanning in the diagnosis of chronic pain. *Biofeedback and Self-Regulation, 8,* 229–242.

Flor, H., Birbaumer, N., & Turk, D. C. (1984, June). *Ein Diathese-Stress Modell chronischer Rückenschmerzen: Empirische Überprüfung und therapeutische Implikationen* [A diathesis-stress model of chronic back pain: Empirical evaluation and therapeutic implications]. Paper presented at the third symposium on Research in Clinical Psychology, Bad Liebenzell, West Germany.

Flor, H., & Turk, D. C. (1984). Etiological theories and treatments for chronic back pain: I. Somatic models and interventions. *Pain, 19,* 105–121.

Fordyce, W. E. (1976). *Behavioral methods for chronic pain and illness.* St. Louis, MO: Mosby.

Gentry, W. D., & Bernal, A. (1977). Chronic pain. In R. B. Williams & W. D. Gentry (Eds.), *Behavioral approaches to medical treatment* (pp. 173–191). New York: Ballinger.

Grabel, J. A. (1973). Electromyographic study of low back pain muscle tension in subjects with and without chronic low back pain. *Dissertation Abstracts International, 34,* 2929B–2930B.

Hays, W. (1973). *Statistics for the social sciences.* London: Holt, Rinehart & Winston.

Hoyt, W. H., Hunt, H. H., DePauw, M. A., Bard, D., Shaffer, F., Passias, J. N., Robbins, D. H., Reunyon, D. G., Semrad, S. E., Symonds, J. T., & Watt, K. C. (1981). Electromyographic assessment of chronic low back pain syndrome. *Journal of the American Osteopathic Association, 80*, 722–730.

Kirk, R. E. (1968). *Experimental design: Procedures for the behavioral sciences.* Belmont, CA: Brooks-Cole.

Kravitz, E., Moore, M. E., & Glaros, A. (1981). Paralumbar muscle activity in chronic low back pain. *Archives of Physical Medicine and Rehabilitation, 62,* 172–176.

Lang, P. J. (1971). The application of psychophysiological methods to the study of psychotherapy and behavior modification. In A. E. Bergin & S. L. Garfield (Eds.), *Handbook of psychotherapy and behavior change* (pp. 75–125). New York: Wiley.

Lethem, J., Slade, P. D., Troup, J. D. G., & Bentley, G. (1983). Outline of a fear-avoidance model of exaggerated pain perception. *Behavior Research and Therapy, 21,* 401–408.

Lippold, O. C. J. (1967). Electromyography. In P. H. Venables & J. Martin (Eds.), *A manual of psychophysiological methods* (pp. 38–69). Amsterdam: North-Holland.

Loeser, J. D. (1980). Low back pain. In J. J. Bonica (Ed.), *Research publications: Association for research on nervous and mental disease, Vol. 58, Pain* (pp. 363–377). New York: Raven Press.

Lubin, B. (1965). Adjective checklist for measurement of depression. *Archives of General Psychiatry, 12,* 57–67.

Melzack, R. (1975). The McGill Pain Questionnaire: Major properties and scoring methods. *Pain, 1,* 277–299.

Moll, J., & Wright, V. (1980). The measurement of spinal movement. In M. I. V. Jayson (Ed.), *The lumbar spine and back pain* (pp. 157–184). Turnbridge Wells, Kent, England: Pitman Medical.

Muller, K. E., Otto, D. A., & Benignus, V. A. (1983).

Design and analysis issues and strategies in psychophysiological research. *Psychophysiology, 20,* 212–218.

Nachemson, A. A. (1979). A critical look at the treatment for low back pain. *Scandinavian Journal of Rehabilitation Medicine, 11,* 143–149.

Philips, C. (1977). A psychological analysis of tension headache. In S. Rachman (Ed.), *Contributions to medical psychology* (pp. 91–114). Oxford, England: Pergamon Press.

Reinsel, G. (1982). Multivariate repeated-measurement of growth curve models with multivariate random effects covariate structure. *Journal of the American Statistical Association, 77,* 190–195.

Sanders, S. H. (1979). Behavioral assessment and treatment of clinical pain: Appraisal of current status. In R. M. Hersen, P. Eisler, & P. M. Miller (Eds.), *Progress in behavior modification* (Vol. 8, pp. 249–291). New York: Academic Press.

Spielberger, C., Gorsuch, R., & Lushene, N. (1970). *Manual for the State-Trait Anxiety Inventory.* Palo Alto, CA: Consulting Psychologists Press.

Sternbach, R. A. (1974). *Pain patients: Traits and treatment.* New York: Academic Press.

Turk, D. C. (1981). *The Pain Experience Scale (PES).* Unpublished questionnaire, Yale University.

Turk, D. C., & Flor, H. (1984). Etiological theories and treatments for chronic back pain. II. Psychological factors and interventions. *Pain, 19,* 209–233.

Turk, D. C., Meichenbaum, D. H., & Genest, M. (1983). *Pain and behavioral medicine: A cognitive-behavior perspective.* New York: Guilford Press.

Wolf, S. L., Nacht, M., & Kelly, J. R. (1982). EMG biofeedback training during dynamic movement for low back pain patients. *Behavior Therapy, 13,* 395–406.

CHAPTER FIVE

Canonical Correlation Analysis

CHAPTER PREVIEW

Until recent years, canonical correlation analysis was a relatively unknown statistical technique. The availability of canned computer programs has witnessed its increased application to research problems. This chapter introduces the data analyst to the multivariate statistical technique of canonical correlation analysis. Specifically, we (1) describe the nature of canonical correlation analysis; (2) illustrate its application; and (3) discuss its potential advantages and limitations. Before reading the chapter, you should familiarize yourself with the key terms.

An understanding of the most important concepts in canonical correlation analysis should enable you to:

- State the similarities and differences between multiple regression, factor analysis, discriminant analysis, and canonical correlation.
- Summarize the conditions that must be met for application of canonical correlation analysis.
- State what the canonical root measures and point out its limitations.
- State how many independent canonical functions can be defined between the two sets of original variables.
- Compare the advantages and disadvantages of the three methods for interpreting the nature of canonical functions.
- Define redundancy and compare it with multiple regression's R^2.

Key Terms

Canonical correlation Measures the strength of the overall relationships between the linear composites of the predictor and criterion sets of variables. In effect, it represents the bivariate correlation between the two linear composites.

Canonical loadings Referred to in SAS as *canonical structure matrices*, they measure the simple linear correlation between the independent variables and their respective linear composites, and can be interpreted like factor loadings.

Canonical roots Squared canonical correlations. They provide an estimate of the amount of shared variance between the respective optimally weighted linear composites of criterion and predictor variables.

Criterion variables Dependent variables.

Linear composites Also referred to as *linear combinations, linear* *compounds,* and *canonical variates;* they represent the weighted sum of two or more variables. In canonical analysis, each canonical function has two separate linear composites (canonical variates), one for the set of criterion variables and one for the set of predictor variables.

Orthogonal   A mathematical constraint specifying that the canonical functions are independent of each other. In other words, the canonical functions are derived so that each is at a right angle to all other functions when plotted in multivariate space.

Predictor variables   Independent variables.

Redundancy index   The amount of variance in one set of variables explained by a linear composite of the other set of variables. It can be computed for both the dependent and the independent sets of variables.

What Is Canonical Correlation Analysis?

In Chapter Two you studied multiple regression analysis, which can be used to predict the value of a single (metric) criterion variable from a linear function of a set of predictor (independent) variables. For some research problems, interest may not center on a single criterion (dependent) variable. Rather, the analyst may be interested in relationships between sets of multiple criterion and multiple predictor variables. Canonical correlation analysis is a multivariate statistical model that facilitates the study of interrelationships among sets of multiple criterion (dependent) variables and multiple predictor (independent) variables. That is, whereas multiple regression predicts a single dependent variable from a set of multiple independent variables, canonical correlation predicts multiple dependent variables from multiple independent variables.

Canonical correlation is considered to be the general model on which many other multivariate techniques are based. As you can see from Table 5.1, canonical correlation places the fewest restrictions on the types of data on which it operates. Since the other techniques impose more rigid restrictions, it is generally believed that the information obtained from such techniques is of higher quality and may be presented in a more interpretable manner. For this reason, many researchers view canonical correlation as a last-ditch effort to be used when all other higher-level techniques have been exhausted.

Hypothetical Example of Canonical Correlation

To clarify further the nature of canonical correlation, let us consider an extension of the example used in Chapter 2. Recall that the HATCO survey results used family size and income as predictors of the number of credit cards a family would hold. The problem involved examining the relationship between two independent variables and a single dependent variable.

Suppose HATCO was interested in the broader concept of credit usage by consumers. To measure this concept, it seems logical that

TABLE 5.1
The Relationship Between Canonical Correlation and Other Multivariate Techniques

Canonical correlation
$$Y_1 + Y_2 + Y_3 + \ldots Y_n = X_1 + X_2 + X_3 + \ldots X_n$$
(metric or nonmetric)   (metric or nonmetric)

Multivariate analysis of variance
$$Y_1 + Y_2 + Y_3 + \ldots Y_n = X_1 + X_2 + X_3 + \ldots X_n$$
(metric)   (nonmetric)

Analysis of Variance
$$Y_1 = X_1 + X_2 + X_3 + \ldots X_n$$
(metric)   (nonmetric)

Multiple Discriminant Analysis
$$Y_1 = X_1 + X_2 + X_3 + \ldots X_n$$
(nonmetric)   (metric)

Multiple Regression Analysis
$$Y_1 = X_1 + X_2 + X_3 + \ldots X_n$$
(metric)   (metric)

HATCO should consider not only the number of credit cards held by the family but also the family's average monthly dollar charges on all credit cards. The problem involves predicting two dependent measures simultaneously (number of credit cards and average dollar charges), and multiple regression is capable of handling only a single dependent variable. Thus, canonical correlation would be used because it is appropriate when multiple dependent variables are involved.

The problem of predicting credit usage is illustrated in Table 5.2. The two dependent variables used to measure credit usage—number of credit cards held by the family and average monthly dollar expenditures on all credit cards—are listed on the left side. The two independent variables selected to predict credit usage—family size and family income—are shown on the right side. By using canonical correlation analysis, HATCO can predict the composite measure of credit usage, consisting of both dependent variables, rather than having to compute two separate regression equations, one for each of the dependent variables. The result of applying canonical correlation is a measure of the strength of the relationship between two sets of multiple variables. This measure is expressed as a canonical correlation coefficient (R) between the two sets.

TABLE 5.2
Prediction of Credit Usage

Composite of Dependent (Criterion) Variables		Composite of Independent (Predictor) Variables
Number of credit cards held by the family		Family size
Average monthly dollar expenditures on all credit cards		Family income
Multiple dependent variables	}R{	Multiple independent variables

Objectives of Canonical Analysis

Canonical correlation analysis is the most generalized member of the family of multivariate statistical techniques (which includes multiple correlation, regression, and discriminant analysis), and is directly related to principal components-type factor analytic models. The goal of canonical correlation is to determine the primary independent dimensions that relate one set of variables to another. In particular, the objectives may be any or all of the following:

1. Determining whether two sets of variables (measurements made on the same objects) are independent of one another or, conversely, determining the magnitude of the relationships that may exist between the two sets.
2. Deriving a set of weights for each set of criterion and predictor variables such that the linear combinations themselves are maximally correlated.
3. Deriving additional linear functions that maximize the remaining correlation, subject to being independent of the preceding set (or sets) of linear compounds.
4. Explaining the nature of whatever relationships exist between the sets of criterion and predictor variables, generally by measuring the relative contribution of each variable to the canonical functions (relationships) that are extracted.

As noted from the preceding description, canonical analysis is a method for dealing mainly with composite association between sets of multiple criterion and predictor variables. By using this technique, it is possible to develop a number of independent canonical functions that maximize the correlation between the linear composites of sets of criterion and predictor variables.

Application of Canonical Correlation

Discussion of the application of canonical correlation analysis is organized around two topics. The first focuses on deriving the canonical functions and the second on the output information.

Deriving the Canonical Functions

The basic input data for canonical correlation analysis are two sets of variables. We assume that each set can be given some theoretical meaning, at least to the extent that one set could be defined as the independent variable and the other as the dependent variable. The underlying logic of canonical correlation involves the derivation of a linear combination of variables from each of the two sets of variables so that the correlation between the two linear combinations is maximized.

The application of canonical correlation does not stop with the derivation of a single relationship between the sets of variables. Instead, a number of pairs of linear combinations—referred to as *canonical variates*—may be derived. The maximum number of canonical variates (functions) that can be extracted from the sets of variables equals the number of variables in the smallest data set, independent or dependent. For example, when the research problem involves five independent (predictor) variables and three dependent (criterion) variables, the maximum number of canonical functions that can be extracted is three.

The derivation of successive canonical variates is similar to the procedure used with unrotated factor analysis (see Chapter 6). That is, the first factor extracted accounts for the maximum amount of variance in the set of variables. Then the second factor is computed so that it accounts for as much as possible of the variance not accounted for by the first factor, and so forth, until all factors are extracted. Therefore, successive factors are derived from residual or leftover variance from earlier factors. Canonical correlation analysis follows a similar procedure but focuses on accounting for the maximum amount of the relationship *between* the two sets of variables rather than within a single set of variables. The result is that the first pair of canonical variates is derived so as to have the highest intercorrelation possible between the two sets of variables. The second pair of canonical variates is then derived so that it exhibits the maximum relationship between the two sets of variables that was not accounted for by the first pair of variates. In short, successive pairs of canonical variates are based on residual variance, and their respective canonical correlations (which reflect the interrelationships between the variates) become smaller as each additional function is extracted. That is, the first pair of canonical variates exhibits the highest intercorrelation, the next pair the second largest correlation, and so forth.

One additional point about the derivation of canonical variates: As has been noted, successive pairs of canonical variates are based on residual variance. Therefore, each of the pairs of variates is orthogonally independent of all other variates derived from the same set of data.

Output Information from Canonical Analysis

The four most important types of output information derived through canonical correlation analysis are (1) the canonical variates, (2) the canonical correlations between the variates, (3) the statistical significance of the canonical correlations, and (4) the redundancy measure of shared variance for the canonical functions.

Each canonical function consists of a pair of variates, one for each of the subsets of variables entered into the analysis. In other words, each canonical function has two variates, one representing the independent variables and the other the dependent variables. The canonical variates are interpreted on the basis of a set of correlation coefficients, usually referred to as *canonical loadings* or *structure correlations*.[1] Just as with factor analysis, these coefficients reflect the importance of the original variables in deriving the canonical variates. Thus, the

---

[1] Some canonical analyses do not compute correlations between the variables and the variates. In such cases the canonical weights are considered comparable but not equivalent for purposes of our discussion.

larger the coefficient, the more important it is in deriving the canonical variate. Also, the criteria for determining the significance of canonical structure correlations are the same as with factor loadings (see Chapter Six).

Two other types of information provided by a canonical analysis are the canonical correlations and their respective levels of statistical significance. The strength of the relationship between the pairs of variates is reflected by the canonical correlation. When squared, the canonical correlation represents the amount of variance in one canonical variate that is accounted for by the other canonical variate. This also may be referred to as the *amount of shared variance between the two canonical variates.* Squared canonical correlations are referred to as *canonical roots* or *eigenvalues.* As with all correlation coefficients, canonical or otherwise, various statistics can be utilized to assess their level of significance—usually expected to be at or beyond the .05 level to be considered significant.

The last type of information of concern to us at this point is the redundancy measure of shared variance. At this point, a detailed explanation of this measure may be confusing. Just remember that using the canonical root as the only measure of shared variance may lead to some misinterpretation. As a result, a redundancy measure can be computed to provide additional information concerning the variance shared by the two sets of variables [12].

An Illustrative Example

To illustrate the application of canonical correlation, we shall use variables drawn from the data bank introduced in Chapter One. Recall that the data consisted of a series of measures obtained on a sample of 50 HATCO salespersons. The variables included six social-psychological measures of attitudes, two measures assessing salespersons' motivation and knowledge, and two different methods of training.

In demonstrating the application of canonical correlation, we use the first eight variables as input data. The six measures of social-psychological attitudes (variables X_1 through X_6) are designated as the set of multiple independent variables or the predictor variables. The measures of knowledge and motivation (variables X_7 and X_8) are identified as the set of multiple dependent variables or the criterion variables. The statistical problem involves identifying any latent relationships between a salesperson's social-psychological attitudes and his or her level of knowledge and motivation.

A canonical correlation analysis was performed on the set of two criterion variables and the six predictor variables using the CANCORR procedure in the SAS statistical package. CANCORR first finds linear combinations of the two sets of variables and creates two canonical variables, so that the correlation between these variables is maximized. The correlation between them is referred to as the *canonical correlation.* The procedure then finds another set of linear functions that produces a second set of canonical variables. The second set of variables has the second highest canonical correlation coefficient. The procedure continues until the number of pairs of canonical functions equals the number of variables in the smaller group, which in our example is two since the criterion set includes only the two measures of knowledge and motivation.

Standard output for the CANCORR procedure includes (1) the corrected sums-of-squares and cross-products matrix for the group 1 or criterion variables; (2) the corrected sums-of-squares and cross-products matrix for group 2 or predictor variables; (3) the corrected between-groups cross-products matrix; (4) means of canonical variables for groups 1 and 2; (5) canonical correlations; (6) Wilks' lambda, Pillai's trace, Hotelling-Lawley's trace, and Roy's greatest root; (7) error degrees of freedom for each Rao's approximate F-statistic; (8) canonical loadings; (9) canonical cross-loadings; (10) canonical weights; and (11) canonical R-square.

In addition, the CANCORR procedure enables the user to obtain a canonical redundancy analysis and univariate multiple regression statistics for predicting each of the criterion variables from the predictor set of independent variables. By using the OUT= option, the user may generate a data set containing the original variables plus the canonical scores for each set of canonical functions. This data set can then be printed, or the PLOT procedure can be used to plot each canonical variable against its counterpart in the other group. The CANCORR procedure also allows the creation of a second data set through the use of the OUTSTAT= option. This second data set contains the various statistics generated by CANCORR, including the canonical correlations and coefficients. The package then allows the user to rotate the canonical coefficients by using the FACTOR procedure. Such rotations allow the user to identify the underlying relationships more clearly when more than one canonical function is significant.

The canonical analysis reported in Table 5.3 is based on the SAS CANCORR program. The table is designed to resemble a CANCORR printout. A similar printout is provided by the SPSS package.

By now you should be familiar with what canonical correlation analysis is and how it can be applied. Subsequent discussion focuses on the following topics: (1) which canonical functions should be inter-

TABLE 5.3 HATCO Canonical Analysis Relating Knowledge and Motivation of Sales Trainees to Social-Psychological Attitudes		Canonical Correlation	Canonical Correlation Analysis				
		Canonical Correlation	Adjusted CANCORR	Approx. Std. Err.	Canonical R-Square	F-Stat.	Prob. > F
	1	.9367	.9244	.0175	.8774	15.7930	0.0000
	2	.4800	.3543	.1099	.2304	2.5748	0.0401

Multivariate Test Statistics			
Statistic	Value	Approx. F	Prob > F
Wilks' lambda	0.0943	15.793	7.55E-17
Pillai's trace	1.1078	8.899	8.24E-11
Hotelling-Lawley trace	7.4589	25.485	9.29E-23
Roy's greatest root	7.1595	51.31	5.17E-18

preted? (2) which methods should be used to interpret the nature of canonical function relationships? (3) what are the limitations of canonical correlation analysis as a research technique?

Which Canonical Functions Should Be Interpreted?

As with research using other statistical techniques, the most common practice is to analyze those functions whose canonical correlation coefficients are statistically significant beyond some level, typically .05 or above. Thus, variables in each set that contribute heavily to shared variance for these functions are considered to be related to each other. The other independent functions are deemed insignificant, and the relationships among the variables are not interpreted.

The authors believe that the use of a single criterion such as the level of significance is too superficial. Instead, it is recommended that three criteria be used in conjunction with each other to decide which canonical functions should be interpreted. By interpretation we mean examining the canonical loadings to determine how the original variables from the two data sets are related. The three criteria are (1) the level of statistical significance of the function, (2) the magnitude of the canonical correlation, and (3) the redundancy measure for the percentage of variance accounted for from the two data sets.

LEVEL OF SIGNIFICANCE.   The level of significance of a canonical correlation that is generally considered to be the minimum for interpretation is the .05 level. The .05 level (along with the .01 level) has become the generally accepted level for considering a correlation coefficient statistically significant. This is largely due to the availability of tables for these levels, borrowed from other disciplines where higher confidence levels are desired. These levels are not necessarily required in all situations, however, and researchers from various disciplines frequently must rely on results based on lower levels of significance.

Several statistics can be used for evaluating the significance of canonical roots. The most widely used test, and the one normally printed out in computer packages, is the F-statistic based on Rao's approximation.

MAGNITUDE OF THE CANONICAL RELATIONSHIPS.   The size of the canonical correlations also should be considered in deciding which functions to interpret. No generally accepted guidelines have been established regarding acceptable sizes for canonical correlations. Rather, the decision is usually made based on the contribution of the findings to better understanding of the research problem being studied. It seems logical that the guidelines suggested for significant factor loadings (see Chapter Six) might be useful with canonical correlations. This is particularly true when one considers the fact that canonical correlations refer to the variance explained in the canonical variates (linear composites), not the original variables.

REDUNDANCY MEASURE OF SHARED VARIANCE.   Recall that squared canonical correlations (roots) provide an estimate of the shared variance between the canonical variates. While this is a simple and appealing measure of the shared variance, it may lead to some misinterpretation. This is because the squared canonical correlations represent the variance shared by the linear composites of the sets of criterion and predictor variables, and not the variance extracted from the sets of variables [1]. Thus, a relatively strong canonical correlation may be obtained between two linear composites (canonical variates) even though these linear composites may not extract significant portions of variance from their respective sets of variables [10].

Since canonical correlations may be obtained that are considerably larger than previously reported bivariate and multiple correlation coefficients, there may be a temptation to assume that canonical analysis has uncovered substantial relationships of conceptual and practical significance. Before such conclusions are warranted, however, further analysis involving measures other than canonical correlations must be undertaken to determine the amount of the dependent variable variance that is accounted for or shared with the independent variables [7].

To overcome the inherent bias and uncertainty in using canonical roots (squared canonical correlations) as a measure of shared variance, a redundancy index has been proposed [12]. The redundancy index is the equivalent of computing the squared multiple correlation coefficient between the total predictor set and each variable in the criterion set, and then averaging these squared coefficients to arrive at an average R^2. It provides a summary measure of the ability of a set of predictor variables (taken as a set) to explain variation in the criterion variables (taken one at a time). As such, the redundancy measure is perfectly analogous to multiple regression's R^2 statistic, and its value as an index is similar.

The Stewart-Love index of redundancy is a measure that tries to calculate the amount of variance in one set of variables that can be explained by the variance in the other set. This index serves as a measure of accounted-for variance, similar to the R^2 calulation used in multiple regression. The R^2 measures the amount of variance in the dependent (criterion) variable explained by the regression function of the independent (predictor) variables. In the regression cases, the total variance in the dependent variable is equal to 1, or 100%. Remember that canonical correlation is different from multiple regression in that it does not deal with a single criterion variable. It has a criterion set that is a composite of several variables, and this composite has only a portion of each independent variable's total variance. For this reason, we cannot assume that 100% of the variance in the criterion set is available to be explained by the predictor set. The predictor set of variables can only be expected to account for the variance in the canonical variate of the criterion set. Remember that the variance in a criterion canonical variate represents the portion of the amount of shared variance in the criterion set. For this reason, the calculation of the redundancy index is a two-step process. The first step involves calculating the amount of variance from the criterion set of variables that is included in the criterion canonical variate. The second step involves calculating the amount of variance in the criterion canonical variate that can be explained by the predictor set canonical variate. The redundancy index is then found by multiplying these two components.

To calculate the amount of shared variance in the criterion set that is included in the criterion canonical variate, let us first consider how the regression R^2 statistic is calculated. The R^2 is simply the square of the correlation coefficient r, which represents the correlation between the actual dependent variable and the predicted value. In the canonical case, we are concerned with the correlation between the criterion canonical variate and each of the criterion variables. Such information can be obtained from the canonical structure. Remember that the canonical structure includes the canonical loadings, which represent the correlation between each input variable and its own canonical variate. By squaring each of the criterion loadings, one may obtain a measure of the amount of variation in each of the criterion variables that is explained by the criterion canonical variate. To calculate the amount of shared variance that is explained by the canonical variate, a simple average of the squared loadings is used. Table 5.4 shows the calculation of the amount of shared variance explained by the first canonical variate.

The second step of the redundancy process involves the percentage of variance in the criterion canonical variate that can be explained by the predictor canonical variate. This is simply the squared correlation between the predictor canonical variate and the criterion canonical variate, which is otherwise known as the *canonical correlation*. The squared canonical correlation is commonly called the *canonical R^2.* This information can be taken directly from the SAS/CANCORR printout shown in Table 5.3.

As mentioned previously, the redundancy index is derived by multiplying the two components to find the amount of shared variance that can be explained by each canonical function. It is important to note that in order to have a high redundancy index, one must have a high

TABLE 5.4		L_1	L_1^2
Calculation of the	X_7 (Knowledge)	.9126	.8328
Redundancy Index	X_8 (Motivation)	.9299	.8647
for the First			1.6975

Calculation of the Redundancy Index for the First Canonical Function

Shared Variance in the Criterion Set Explained
By the Criterion Canonical Variate
$(1.6975)/2 = .8488$

Canonical R-Square
.8774

Redundancy Index
$.8488 \times .8774 = .7447$

canonical correlation and a high degree of shared variance explained by the criterion variate. A high canonical correlation alone does not ensure a valuable canonical function. Using the figures given in the HATCO example, you can calculate the redundancy index as shown in Table 5.4. Table 5.5 shows the canonical redundancy analysis as seen on the SAS/CANCORR printout. The proportion of variance of the criterion variables explained by the opposite canonical variable is what is commonly refered to as the *redundancy index*. This is .7447, as can be seen on the top right-hand side of the table.

As noted earlier, the redundancy index of the shared variance of the HATCO canonical functions (see Table 5.6) is illustrated in Table 5.5. Results are shown for both the dependent and independent sets, although in most instances the researcher is only concerned with the variance extracted from the dependent variable set. The redundancy index for the example problem indicates that 74 percent of the variance in the dependent variables has been explained by the canonical variate for the independent variable set. While the redundancy index lowers the shared variance estimate obtained from the canonical roots, it still represents a substantial amount of shared variance. Moreover, it provides the analyst with a much more realistic measure of the predictive ability of canonical relationships.

What is the minimum acceptable redundancy index needed to justify the interpretation of canonical functions? Just as with canonical correlations, no generally accepted guidelines have been established. The analyst must judge each canonical function in light of its theoretical and practical significance to the research problem being investigated to determine if the redundancy index is sufficient to justify interpretation. Also, a test for the significance of the redundancy index has been developed [2], although it has not been widely utilized.

TABLE 5.5 SAS /CANCORR Redundancy Analysis		Standardized Variance of the *Criterion* Variables Explained By:				
		Their Own Canonical Variable			The Opposite Canonical Variable	
		Proportion	Cumulative Proportion	Canonical R-Square	Proportion	Cumulative Proportion
	1	.8487	.8487	.8774	.7447	.7447
	2	.1513	1.0000	.2304	.0349	.7796
		Standardized Variance of the *Predictor* Variables Explained By:				
		Their Own Canonical Variable			The Opposite Canonical Variable	
		Proportion	Cumulative Proportion	Canonical R-Square	Proportion	Cumulative Proportion
	1	.2942	.2942	.8774	.2582	.2582
	2	.1054	.3996	.2304	.0243	.2824

TABLE 5.6
Canonical
Structure for the
HATCO Example

Correlations Between the Predictor Variables and Their Canonical Variables
(Canonical Loadings)

	P1	P2
X_1 (self-esteem)	.7562	−.1259
X_2 (locus of control)	.0135	−.1707
X_3 (alienation)	.6415	−.1128
X_4 (social responsibility)	.3799	.6835
X_5 (Machiavellianism)	.7391	−.2595
X_6 (political opinion)	.3019	.2003

Correlations Between the Criterion Variables and Their Canonical Variables
(Canonical Loadings)

	C1	C2
X_7 (knowledge)	.9126	−.4089
X_8 (motivation)	.9299	.3678

Correlations Between the Predictor Variables and the Canonical Variables of
the Criterion Set (Canonical Cross-Loadings)

	C1	C2
X_1 (self-esteem)	.7083	−.0605
X_2 (locus of control)	.0126	−.0819
X_3 (alienation)	.6009	−.0542
X_4 (social responsibility)	.3559	.3281
X_5 (Machiavellianism)	.6923	−.1245
X_6 (political opinion)	.2828	.0961

Correlations Between the Criterion Variables and the Canonical Variables of
the Predictor Set (Canonical Cross-Loadings)

	P1	P2
X_7 (knowledge)	.8548	−.1963
X_8 (motivation)	.8711	.1766

* P1 and C1 stand for the variates of the first canonical function; P1 = predictor variate
of the first function, C1 = criterion variate of the first function.

Interpretation Methods for Canonical Functions

If the canonical relationship is statistically significant and the magnitude of the canonical root and the redundancy index is acceptable,
the analyst still needs to make substantive interpretations of the results.
This involves examining the canonical functions to determine the relative importance of each of the original variables in deriving the canonical relationships. Three methods have been proposed: (1) canonical
weights, (2) canonical loadings (structure correlations), and (3) canonical cross-loadings. The results of these methods are shown for the
HATCO data set in Tables 5.6 and 5.7.

CANONICAL WEIGHTS.   The traditional approach to interpreting canonical
functions involves examining the sign and magnitude of the canonical
weight assigned to each variable in computing the canonical functions.
Variables with relatively larger weights contribute more to the functions,
and vice versa. Similarly, variables whose weights have opposite signs
exhibit an inverse relationship with each other and those with the

---

TABLE 5.7 Canonical Weights for the HATCO Example

Standardized Canonical Coefficients for the Predictor Set

	Function1	Function2
X_1 (self-esteem)	.1564	.7679
X_2 (locus of control)	.0567	.3901
X_3 (alienation)	.6034	−.0921
X_4 (social responsiblity)	.3436	1.5137
X_5 (Machiavellianism)	.5047	−1.1328
X_6 (political opinion)	−.0320	−.8768

Standardized Canonical Coefficients for the Criterion Set

X_7 (knowledge)	.5138	−1.2989
X_8 (motivation)	.5712	1.2747

same sign exhibit a direct relationship. However, interpreting the relative importance or contribution of a variable by its canonical weight is subject to the same criticisms associated with the interpretation of beta weights in regression techniques. For example, a small weight may mean either that its corresponding variable is irrelevant in determining a relationship or that it has been partialed out of the relationship because of a high degree of multicollinearity. Another problem with the use of canonical weights is that these weights are subject to considerable instability (variability) from one sample to another. This is because the computational procedure for canonical analysis yields weights that maximize the canonical correlations for a particular sample of observed dependent and independent variable sets [7]. These problems suggest considerable caution in using canonical weights to interpret the results of canonical analysis.

CANONICAL LOADINGS. In recent years canonical loadings have been increasingly used as a basis for interpretation because of the deficiencies in utilizing weights. Canonical loadings, referred to sometimes as *structure correlations*, measure the simple linear correlation between an original observed variable in the dependent or independent set and the set's canonical variate. The methodology considers each independent canonical function separately and computes the within-set variable–variate correlation [1]. That is, for each set of variables, dependent and independent, the correlation is computed between each original observed variable and its respective canonical variate. Thus, the canonical loading reflects the variance that the observed variable shares with the canonical variate and can be interpreted like a factor loading in assessing the relative contribution of each variable to each canonical function.

Canonical loadings like weights may be subject to considerable variability from one sample to another. This variability suggests that loadings, and hence the relationships ascribed to them, are sample specific, due to change or the result of extraneous factors [7]. Canonical loadings are considered relatively more valid than weights as a means of interpreting the nature of canonical relationships. But the analyst still must

be cautious when using loadings for interpreting canonical relationships, particularly with regard to the external validity of the findings.

CANONICAL CROSS-LOADINGS.   The computation of canonical cross-loadings has been suggested as an alternative to conventional loadings
[4]. This procedure involves correlating each of the original observed
dependent variables directly with the independent canonical variate.
Recall that conventional loadings correlate the original observed variables with their respective variates after the two canonical variates (dependent and independent) are maximally correlated with each other.
Thus, cross-loadings do provide a more direct measure of the dependent–independent variable relationships by eliminating an intermediate step involved in conventional loadings.

How to Interpret Canonical Functions

Several different methods for interpreting the nature of canonical relationships were discussed in the last section. The question remains,
however: Which method should the analyst use? Since most canonical
problems necessitate the use of a computer, the analyst frequently
must use whichever method is available in the standard statistical
packages. The use of cross-loadings is the preferred approach and can
be applied using the SAS statistical package. If the SAS package is
not available, the analyst is forced either to compute the cross-loadings
by hand or to select another method of interpretation, since none of
the other popular statistical software packages provide this information.
The widely used SPSS package does provide canonical loadings, while
the BMD package provides canonical weights. The canonical loadings
approach is somewhat more valid than the use of weights. Therefore,
whenever possible, it is recommended that the loadings approach be
a second alternative to the canonical cross-loadings method.

Since Table 5.6 includes the cross-loadings obtained from the SAS
printout, let's use the information to interpret our results. In studying
the first canonical function, we see that variables X_7 and X_8 show high
correlations with the predictor canonical variate (P_1): .8548 and .8711,
respectively. By squaring these terms, we may find the percentage of
the variance for each of the variables explained by P1. The results
show that 73 percent of the variance in X_7 and 76 percent of the variance
in X_8 is explained by P1. Looking at the predictor variables' cross-
loadings, we see that variables X_1 and X_5 both have high correlations
of roughly .70 with the criterion canonical variate. From this information, we see that approximately 50 percent of the variance in these
two variables is explained by the criterion variate (the 50 percent is
obtained by squaring the correlation coefficient; $.7 \times .7 = .49 = \approx$
50%). One should note the correlation of X_3 (.6009). While this may
appear high, one must realize that after squaring this correlation, only
36 percent of the variation is included in the canonical variate.

In evaluating the second function, one should first consider the
redundancy. It is .0349, which says that less than 4 percent of criterion
variance is explained by the predictor variate. With such a small per-

centage, one must question the value of the function. This is an excellent example of a statistically significant canonical function that does not significantly explain a large proportion of the criterion variance.

A final question in relation to interpretation is, how do we use the signs? The answer is that like signs indicate a direct relationship and opposite signs an inverse one. For example, on function 1 all the signs are positive. Therefore all the variables are directly related. But on function 2 some variables are positive and others are negative. Thus, some of the relationships are direct and others are inverse.

Limitations of Canonical Analysis

Applications of canonical correlation analysis have expanded substantially in recent years. Like other multivariate techniques, however, it is not without certain limitations. When selecting a statistical technique for data analysis and interpreting the results of canonical correlation, it is important to keep the following limitations in mind: (1) the canonical correlation reflects the variance shared by the linear composites of the sets of variables, not the variance extracted from the variables; (2) canonical weights derived in computing canonical functions are subject to a great deal of instability; (3) canonical weights are derived to maximize the correlation between linear composites, not the variance extracted; and (4) it is difficult to identify meaningful relationships between the subsets of independent and dependent variables because precise statistics have not yet been developed to interpret canonical analysis and we must rely on inadequate measures such as loadings or cross-loadings [7]. These limitations are not meant to discourage the use of canonical correlation. Rather, they are pointed out so that the effectiveness of canonical correlation as a research tool will be enhanced.

Summary

Canonical correlation analysis is a useful and powerful technique for exploring the relationships among multiple criterion and predictor variables. The technique is primarily descriptive, although it may be used for predictive purposes. Results obtained from a canonical analysis should suggest answers to questions concerning the number of ways in which the two sets of multiple variables are related, the strengths of the relationships, and the nature of the relationships so defined.

Canonical analysis enables the data analyst to combine into a composite measure what otherwise might be an unmanageably large number of bivariate correlations between sets of variables. It is useful for identifying overall relationships between multiple independent and dependent variables, particularly when the data analyst has little a priori knowledge about relationships among the sets of variables. Essentially, the analyst can apply canonical correlation analysis to a set of variables, select those variables (both independent and dependent) that appear to be significantly related, and run subsequent canonical correlations with the more significant variables remaining, or individual regressions.

QUESTIONS

1. Under what circumstances would you select canonical correlation analysis instead of multiple regression as the appropriate statistical technique?

2. What three criteria should you use in deciding which canonical functions should be interpreted? Explain the role of each.

3. How would you interpret a canonical correlation analysis?

4. What is the relationship between the canonical root, the redundancy index, and multiple regression's R^2?

5. What are the limitations associated with canonical correlation analysis?

6. Why has canonical correlation analysis been used much less frequently than the other multivariate techniques?

REFERENCES

1. Alpert, Mark I., and Robert A. Peterson, "On the Interpretation of Canonical Analysis." *Journal of Marketing Research*, vol. IX, (May 1972), p. 187.
2. Alpert, Mark I., Robert A. Peterson, and Warren S. Martin. "Testing the Significance of Canonical Correlations," *Proceedings*, American Marketing Association, vol. 37, 1975, pp. 117–119.
3. Bartlett, M. S. "The Statistical Significance of Canonical Correlations," *Biometrika*, vol. 32 (1941), p. 29.
4. Dillon, W.R. and M. Goldstein, *Multivariate Analysis: Methods and Applications.* New York: Wiley, Inc., 1984
5. Green, P. E., *Analyzing Multivariate Data.* Hinsdale, Ill.: Holt, Rinehart & Winston, 1978.
6. Green, P. E., and J. Douglas Carroll, *Mathematical Tools for Applied Multivariate Analysis.* New York: Academic Press, 1978.
7. Lambert, Z., and R. Durand, "Some Precautions in Using Canonical Analysis," *Journal of Marketing Research*, vol. XII (November 1975), pp. 468–475.
8. McGraw-Hill, *SPSS-X Users's Guide*, 2nd ed., Chicago, 1986.
9. McGraw-Hill, *SPSS-X Advanced Statistics Guide*, Chicago, 1986.
10. SAS Institute, *SAS User's Guide: Basics*, Version 5 ed. Cary, N.C., 1985.
11. SAS Institute, *SAS User's Guide: Statistics*, Version 5 ed. Cary, N.C., 1985.
12. Stewart, Douglas, and William Love, "A General Canonical Correlation Index." *Psychological Bulletin*, vol. 70 (1968), pp. 160–163.

SELECTED READINGS

Using Canonical Correlation to Construct Product Spaces for Objects with Known Feature Structures

MORRIS B. HOLBROOK
WILLIAM L. MOORE

Marketing researchers often have the problem of constructing product spaces based on subjective attribute ratings or other structured judgmental responses. Various *compositional* techniques can be used to address this problem. For example, considerable attention has been given to the use of *factor analysis* (Howard and Sheth 1969; Urban 1975) and *discriminant analysis* (Johnson 1970, 1971; Pessemier 1977) to derive spatial representations of multiattribute data. These two compositional approaches have been discussed and compared extensively in the literature (e.g., Hauser and Koppelman 1979; Huber and Holbrook 1979; Moore 1981). In contrast, with the exception of work by Green (1978; Green, Rao, and DeSarbo 1978), a third possible compositional approach via *canonical correlation analysis* (CCA) appears to have been relatively neglected by marketing researchers.

When product stimuli are coded as zero-one dummy variables (e.g., Green, Rao, and DeSarbo 1978), this CCA approach is equivalent to multiple discriminant analysis (MDA; Cooley and Lohnes

"Using Canonical Correlation to Construct Product Spaces for Objects With Known Feature Structures," Morris B. Holbrook and William L. Moore, Vol. 19 (February 1982), pp. 87–98. Reprinted from the *Journal of Marketing Research*, published by the American Marketing Association. Morris B. Holbrook and William L. Moore are Associate Professors, Graduate School of Business, Columbia University.

1971; Green 1978; McKeon (1966). In some cases, however, products exhibit "known" feature structures in the sense that their characteristics profiles are manipulated experimentally or measured objectively. Here, *other CCA methods* can be used for constructing product spaces. These differ from MDA procedures in ways that render CCA very useful in some situations where MDA cannot be applied. Because these potentially extended applications of CCA have not (to our knowledge) appeared elsewhere in the literature and therefore may not be appreciated by marketing researchers, we explore some important questions about the use of canonical correlation to form product spaces based on attribute ratings of objects with known feature structures.

Using Canonical Correlation to Construct Product Spaces

In general, canonical correlation derives a series of linear combinations (called *canonical variates*) formed from two sets of variables so as to maximize correlations within each pair of canonical variates subject to a constraint of zero correlation between successive pairs. General discussions appear, for example, in the texts by Anderson (1958) and Cooley and Lohnes (1971). Levine (1977) and Green (1978) have recently provided good overviews oriented toward the social sciences and marketing.

The constrained correlation maximization is accomplished by solving the following eigenstructure problem.

$$(1) \qquad (\mathbf{S}_{12}\mathbf{S}_{22}^{-1}\mathbf{S}_{21} - \lambda_k \mathbf{S}_{11})\mathbf{w}_k = \mathbf{O}$$

where $\mathbf{S}_{ij}$ is the cross-products matrix between the ith and j^{th} sets of variables $(i,j = 1,2)$. Here,

$\mathbf{w}_k$ is the kth set of weights associated with the first set of variables and

$$(2) \qquad \mathbf{v}_k = (1/\lambda_k^{1/2})\mathbf{S}_{22}^{-1}\mathbf{S}_{21}\mathbf{w}_k$$

is the corresponding set of weights associated with the second.

In the marketing context, one can distinguish three ways in which canonical correlation can be used to construct product spaces. For all three, the first set of canonical variables consists of judgmental responses to each object on a number of subjective attribute-rating scales. These subjective responses could consist of ratings on a continuous or multicategory graded scale, as in our study, or categorical judgments coded as discrete dummy variables, as in Carroll's (1969, 1973) categorical conjoint measurement followed in applications by Green and Wind (1975) and by Green, Rao, and DeSarbo (1978). In either case, the first set of attribute scores is related to a second canonical set consisting of dummy variables coded according to one of three schemes. Differences in the manner of coding this second set of dummy variables result in the distinctions among what we call the *objects-*, *features-*, and *interactions*-based CCA methods.

In each type of CCA method, the product space is constructed by using the first K attribute-based canonical variates as coordinates in a K-dimensional spatial representation. In other words, the $\mathbf{w}_k$ weights associated with the first set of canonical variables (equation 1) are multiplied by the appropriate attribute ratings and then summed to create the coordinate of interest on each spatial dimension.

Further, each CCA approach is related to a corresponding direct or indirect extension of categorical conjoint measurement (Carroll 1969, 1973). This technique is a form of conjoint analysis in which objectively manipulated stimulus profiles (coded as dummy variables) are related to a dependent variable that is also categorical. Carroll has shown that this approach is equivalent to canonical correlation between two sets of dummy variables. In our context, its value lies in the interpretation of $\mathbf{w}_k$ weights obtained for categorically coded attributes as "optimal" scale values. These applications as well as possible extensions of categorical conjoint measurement are noted in the following discussion of objects-, features,- and interactions-based canonical correlation.

Objects-Based Method

In the objects-based method, each object (j) is represented by a separate zero-one dummy variable (D_j, $j = 1, \ldots, J$) coded for object j and zero otherwise. These object-specific dummy variables are then regarded as a second set of canonical variables to be related to a first set consisting of attribute ratings for each object. The attribute-based canonical variates (i.e., linear combinations of the attribute ratings) may therefore be regarded as the coordinates of objects on dimensions that best account for the judged differences among objects.

This objects-based CCA approach seems to be the only one of the three that has been described in the literature. For example, Green, Rao, and DeSarbo (1978) used canonical correlation between categorical vacation attributes (coded as zero-one dummy variables) and dummy variables representing vacation sites to derive a perceptual space for holiday locations. As support for this procedure, they cited work by Maxwell (1961) on dichotomous variables. Further support comes from McKeon's (1966) general comparison of canonical and discriminant analyses and from Green's (1978) own extension of this comparison to marketing.

Specifically, when objects-based canonical correlation is performed across individuals, $\mathbf{S}_{12}\mathbf{S}_{22}^{-1}\mathbf{S}_{21}$ in equation 1 is equivalent to the between-objects cross-products matrix and $\mathbf{S}_{11}$ is equivalent to the total cross-products matrix. Hence, the attribute-weight vectors ($\mathbf{w}_k$) that solve equation 1 also maximize the ratio of between-to-total variance. These are the same vectors that maximize the between-to-within-objects variance in discriminant analysis (Cooley and Lohnes 1971, p. 249; Green 1978; McKeon 1966; Moore and Holbrook 1981).[1]

As noted by Green, Rao, and DeSarbo (1978), when categorically coded attributes are used in objects-based CCA, this approach is "in the spirit of categorical conjoint measurement" (p. 190). It

---

[1] Because the Jth object-identifying dummy variable (D_J) is uniquely determined by the other $J - 1$ dummy variables, the $\mathbf{S}_{22}$ matrix is singular and cannot be inverted to satisfy equations 1 and 2. This difficulty can be resolved by computing the generalized inverse (McKeon 1966) or by simply omitting the Jth dummy variable in the second canonical set (Green 1978). Either approach produces results equivalent to MDA (Green 1978). For convenience, the latter was used in the analyses that follow.

differs from the strict categorical conjoint approach in that objects are coded as separate dummy variables rather than in accord with their profiles of objective characteristics or features. However, it resembles the categorical conjoint technique in that "optimal" scale values are derived by relating attribute categories to an objective set of dummy variables.

When analysis is performed at the individual level and there is only one observation per object, the within-objects covariance matrix is undefined in discriminant analysis and $S_{12}S_{22}^{-1}S_{21} = S_{11}$ in objects-based canonical correlation so that both techniques are infeasible (for a proof, see Moore and Holbrook 1981). In some situations, however, one may wish to construct a product space at the individual level of analysis. One approach to this problem, developed by Pessemier (1973), is to ask each respondent for a subjective estimate of within-objects covariance on each pair of attributes. Another, proposed here, is to narrow one's focus to a restricted set of feature effects contained implicitly in the attribute data. This approach requires the use of objects with "known" feature structures (i.e., characteristics profiles that have been either manipulated experimentally or measured objectively) and is embodied in the features- and interactions-based CCA methods described hereafter.[2]

Features-Based Method

The features-based CCA method defines a set of canonical variables consisting of zero-one dummy variables (F_i, $i = 1, \ldots, I$) representing the I features in the stimulus-design matrix (coded one if an object has the ith feature and zero otherwise). This set of dummy variables is then related to the set of attribute ratings. A maximum of I pairs of canonical variates can be extracted by this procedure. Then, as in the objects-based case, the attribute-based variates may be interpreted as coordinates of the objects on I dimensions that do the best possible job of reproducing the brand features, subject to the noncorrelation constraints.

If attributes are coded categorically, features-based CCA extends one form of categorical con-

joint measurement. Green and Wind (1975) provide an illustration in which the dependent variable (type of end use) consists of exhaustive and mutually exclusive categories to be related to the features underlying a factorially designed product (bars of soap). In contrast, features-based CCA uses a number of attribute scales rather than just one dependent variable. Thus, if multiple n-point attribute ratings are coded as $n - 1$ (or fewer) categorical dummy variables, features-based CCA may be regarded as a form of categorical conjoint measurement with multiple sorts on the dependent variable.

This features-based CCA method has the advantage of applicability at the individual level where MDA or objects-based CCA is infeasible. A disadvantage of features-based CCA, however, is that it derives a product space based only on the main effects of features and therefore does not take account of feature interactions that may reflect important cue-configural determinants of attribute ratings. By definition, MDA and objects-based CCA construct spaces based on all such interactions underlying the design of stimulus objects. As recently demonstrated by Carmone and Green (1981), debated by Neslin (1981), and reviewed at length by Holbrook and Moore (1981a, b), where cue configurality occurs in evaluative judgments, such interactions may account for a significant portion of ratings variance. Therefore, at the individual level of analysis, a potentially useful compromise between the objects-based and features-based CCA methods is to include those feature interactions that appear to be important—either on a priori grounds or on the basis of preliminary data analysis. Such a selective use of feature interactions in canonical correlation give rise to the third approach.

Interactions-Based Method

The interactions-based method defines a set of canonical variables consisting of (1) the aforementioned dummy variables representing each design feature (F_i, $i = 1, \ldots, I$) and (2) zero-one dummy variables associated with some subset of the feature interactions (F_{ij}, F_{ijk}, F_{ijkl}, . . .). For example, the analyst might choose to focus on all two- and three-way interactions, as represented by dummy variables defined as follows:

[2] In experimental studies, these stimuli may be presented as verbal descriptions, pictures, or physical objects. In design, they resemble those typically used in conjoint analysis (Green and Wind 1975).

(3)
$$F_{ij} = F_i \cdot F_j$$
$$F_{ijk} = F_i \cdot F_j \cdot F_k$$

where i, j, and k are taken over the I features of interest. The set of dummy variables thus defined (F_i, F_{ij}, F_{ijk}) is then related via canonical correlation to the set of attribute ratings. The resulting attribute-based canonical variates may again be regarded as object coordinates that do the best possible job of reflecting the underlying feature structure, which now includes various important feature interactions of interest.[3]

The extension to categorical conjoint measurement is straightforward in the case of attributes coded as categories. Such an extension is indicated by Green and Wind's (1975) discussion of the potential use of interaction terms (p. 148), but does not appear to have been applied elsewhere in the literature.

If the objects' feature structure has been designed orthogonally and if *all* main effects and interactions $(F_i, F_{ij}, F_{ijk}, F_{ijk...})$ are included in the dummy variable set, the full interactions-based method will produce a product space equivalent to that obtained by the objects-based CCA solution because the same attribute weights ($\mathbf{w}_k$ in equation 1) are recovered in both methods. However, the second set of weights ($\mathbf{v}_k$ in equation 2) will differ between the two approaches, requiring different interpretations of the spaces. In other words, the objects- and *full* interactions-based CCA methods have different interpretations but do generate the same spatial representation. A proof of this important conclusion is given by Moore and Holbrook (1981).

Here, we emphasize the conceptual link between objects-, interactions-, and features-based methods. Specifically, if one begins with objects-based CCA, the restricted interactions-based approach may be regarded as an embedded model that omits unimportant (usually higher level) interactions, thereby gaining both parsimony and applicability at the individual level of analysis. As increasing numbers of interactions are left out, the solution approaches that derived by features-

based CCA, in which interaction terms are omitted altogether. Interactions-based CCA is therefore the most general formulation, producing spaces equivalent to those obtained by the objects-based (features-based) method when all feature interactions are included (omitted). In this generalized approach, the number of interactions to retain hinges delicately on questions of *a priori* conceptualization (deductive) and/or statistical testing (inductive).[4]

Interpretive Considerations

The degree of differences in fit among the three canonical correlation methods will depend on the extent to which cue configurality occurs in the relationships between stimulus features and subjective attribute ratings. In situations where a major portion of variance in judgmental responses is accounted for by interactions among stimulus features, objects-based CCA (or MDA) should construct spaces that represent these cue configuralities better than do spaces constructed by features-based CCA. To the extent that important interactions are included in the appropriate set of canonical variables, interactions-based CCA may offer a good approximation to the full objects-based or MDA solution. In contrast, where important feature interactions do not occur, major differences in degree of fit are unlikely to appear in spaces constructed by the three methods. It follows that only an analyst who is concerned primarily with main feature effects or who has compelling reasons to believe that cue configuralities are missing from the judgmental data may safely rely on features-based CCA. Others should adhere to the objects- or interactions-based method.

These considerations suggest that comparisons among results from the various CCA methods may provide interpretative insights about the role and importance of cue configuralities in a set of judgmental data. For example, these interpretive considerations may bear on the relationship of feature interactions to the familiar aggregation problem.

[3] An alternative to the coding scheme reflected by equation 3 is to use effects-type coding of each feature ($-1, 1$) instead of zero-one coding. This coding scheme has the advantage of maintaining orthogonality between main and interaction effects, but the disadvantage of complicating the interpretation of feature-interaction vectors (discussed hereafter). Accordingly, because total explained variance was the central consideration in our study and because this criterion was not affected by the choice of coding scheme, the more easily interpretable zero-one coding was used in the analysis that follows.

[4] It might be added that various *mixed* methods are also available in which *both* objects- *and* interactions-based dummy variables are included in the second canonical set. For example, if objects are identified by name, such a mixed approach might be used to remove the judgmental effects of brand loyalty. Though potentially helpful in some situations (e.g., when working with real brands), this mixed method does not appear warranted in cases where hypothetical stimuli are designed orthogonally, as in the study reported hereafter. Accordingly, we do not discuss it further.

Here, three important generalities are likely to apply in many situations. First, in *aggregate-level* analysis, where *heterogeneity* in cue configurality is likely, feature interactions may tend to cancel out so that objects-, interactions-, and features-based methods may give closely comparable results. Second, if respondent *subgroups* are chosen so as to be relatively *homogeneous* in configurality, objects- and features-based analysis at the group level may show meaningful differences in fit to main feature effects and feature interactions, respectively. Third, if analysis is conducted at the *individual level,* differences between features- and interactions-based solutions should depend on the extent to which cue configurality accounts for variance in any given respondent's attribute ratings.

These generalities indicate that comparisons among contrasting CCA solutions may facilitate interpretations about the role of cue configurality in judgmental data at various levels of aggregation. The purpose of our article is to illustrate (1) the application of objects-, features-, and interactions-based CCA to the construction of product spaces, (2) some considerations that arise in applying these CCA methods at the aggregate, subgroup, and individual levels of analysis, and (3) some tentative conclusions from comparing CCA results at different levels of aggregation. Accordingly, as the purpose is primarily illustrative, the applications described are new analyses of data previously reported by Holbrook and Moore (1981b). They provide concrete examples of using canonical correlation to construct product spaces for objects with known feature structures.

Method

Research Setting

Holbrook and Moore's (1981b) study focused on judgmental responses to contrasting clothing designs and used an experimental manipulation in which factorially arrayed sweaters were presented to subjects in either verbal or pictorial form. Because only one subset of data is needed for an illustrative application and because there is theoretical support for greater cue configurality in the case of pictorial presentations (Holbrook and Moore 1981a), we analyzed data from those subjects who evaluated sweater pictures.

Stimuli

Sweaters were designed to differ on five features, each with two levels.

Feature	Levels
pattern	solid vs. striped
length	short vs. long
fit	loose vs. tight
sleeves	no sleeves vs. sleeves
neck	V-neck vs. turtle-neck

These five features were combined in all 32 possible ways to create the 2^5 array of factorially designed sweaters shown in Figure 1. These stimuli (each about 3×3 inches) were presented in random orders across subjects to minimize artifacts due to stimulus sequencing or respondent fatigue.

Task

Subjects provided attribute ratings of the sweater pictures on the following 20 semantic differential scales (Osgood, Suci, and Tannenbaum 1957).

dull	_:_:_:_:_:_	lively
masculine	_:_:_:_:_:_	feminine
weak	_:_:_:_:_:_	strong
orderly	_:_:_:_:_:_	disorderly
active	_:_:_:_:_:_	passive
dynamic	_:_:_:_:_:_	static
displeasing	_:_:_:_:_:_	pleasing
stylish	_:_:_:_:_:_	unstylish
interesting	_:_:_:_:_:_	uninteresting
heavy	_:_:_:_:_:_	light
sporty	_:_:_:_:_:_	formal
patterned	_:_:_:_:_:_	random
ordinary	_:_:_:_:_:_	unusual
smooth	_:_:_:_:_:_	rough
bad	_:_:_:_:_:_	good
fancy	_:_:_:_:_:_	plain
simple	_:_:_:_:_:_	complex
beautiful	_:_:_:_:_:_	ugly
hard	_:_:_:_:_:_	soft
refined	_:_:_:_:_:_	tacky

Selection of these bipolar adjective pairs were based on the evaluative-activity-potency factor structure reported by Osgood and his colleagues, with additional items suggested by 15 informal interviews in which respondents (drawn from the same student population as used in the main expe-

FIGURE 1   Pictures of 32 Factorially Designed Sweaters

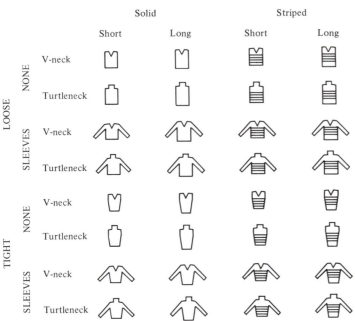

riment) discussed their reactions to 20 samples of real sweaters. Order and directions of the scales were randomized in an effort to minimize yea-saying and response-style biases.

Subjects

Subjects in the pictorial treatment were 30 randomly assigned members of an MBA class at Columbia University who satisfied a course requirement by spending about an hour completing the questionnaire. All participants showed general familiarity with and interest in the test product (Ferber 1977).

Canonical Correlation Analysis

Ratings on each of the 20 adjectival scales were standardized across sweaters within each subject (cf. Bass and Wilkie 1973). Canonical correlations were then performed (1) at the *aggregate* level across both individuals and sweaters ($N = 30 \times 32 = 960$), (2) on a *subgroup* of half the sample selected for homogeneity of cue configurality ($N = 15 \times 32 = 480$), and (3) at the *individual* level on two subjects chosen to differ in their degrees of cue configurality ($N = 32$).

In each canonical correlation, standardized attribute ratings formed the first variable set and some dummy-variable representation of objects or features formed the second. The latter dummy-variable codings can be summarized as follows.

CCA Method	Dummy Variables in Second Canonical Set
(4a) Objects-based[5]	$D_j = \begin{cases} 1 \text{ for the } j^{th} \text{ sweater} \\ 0 \text{ otherwise} \end{cases}$ where $j = 1, \ldots, 31$
(4b) Features-based	$F_i = \begin{cases} 1 \text{ if sweater has } i^{th} \text{ feature} \\ 0 \text{ otherwise} \end{cases}$ where $i = 1, \ldots, 5$
(4c) Interactions-based	$F_i = \begin{cases} 1 \text{ if sweater has } i^{th} \text{ feature} \\ 0 \text{ otherwise} \end{cases}$ where $i = 1, \ldots, 5;$ $F_{ij} = F_i \cdot F_j$

[5] To adapt SPSS to the considerations of perfect linear dependency described in footnote 1, one perfectly dependent dummy variable was omitted from the analysis. This step preserves the correspondence between MDA and objects-based CCA solutions (Green 1978).

where $i = 1, \ldots, 4;$
$$j = 2, \ldots, 5;$$
$$F_{ijk} = F_i \cdot F_j \cdot F_k$$
where $i = 1, \ldots, 3;$
$$j = 2, \ldots, 4;$$
$$k = 3, \ldots, 5;$$
$$F_{ijkl} = F_i \cdot F_j \cdot F_k \cdot F_l$$
where $i = 1, 2; j = 2, 3;$
$$k = 3, 4; l = 4, 5.$$

On the grounds that perfect representation of the stimulus structure would require a minimum of five independent discriminations (one for each feature) whereas the maximum number of canonical variates obtainable in the features-based CCA is five (the number of dummy variables in the smaller canonical set), all canonical correlation solutions were derived and compared in five dimensions. The spatial coordinates of each sweater for each individual (k) on each CCA axis were computed as canonical variates using the appropriate set of canonical weights applied to the relevant vector of attribute ratings by that subject.

$$(5) \qquad \mathbf{C}_k = \mathbf{A}_k \mathbf{W}$$

where $\mathbf{C}_k$ is a 32×5 matrix of spatial coordinates of each sweater for individual k, $\mathbf{A}_k$ is individual k's 32×20 matrix of attribute ratings, and $\mathbf{W}$ is a 20×5 matrix of canonical weights.[6] For purposes of comparing spaces based on different sample sizes ($N = 960, 480, 32$), centroids ($\bar{c}_{ji}$) were computed for each sweater.

$$(6) \qquad \bar{c}_{ji} = \left(\sum_{k=1}^{K} c_{jik} \right) / K$$

where $\bar{c}_{ji}$ is the coordinate of centroid j on axis i, c_{jik} is individual k's coordinate for sweater j on axis i, and $j = 1, \ldots, 32$, $i = 1, \ldots, 5$, and $k = 1, \ldots, K$ (with $K = 30, 15,$ or 1).[7]

[6] Note that the solutions for spatial coordinates ($\mathbf{C}_k$) could be rotated according to the varimax or some other criterion. However, because each pair of CCA variates has a clear interpretation in terms of maximum shared residual variance, such rotations were not utilized in our analysis.

[7] The spatial representations thus derived are based on attribute ratings that represent both perceptions (e.g., patterned/random) and preferences (e.g., beautiful/ugly). An important task for marketing researchers is to separate (Holbrook and Huber 1979) and relate (Holbrook 1981) the two types of jugement. However, this issue does not appear to be central to the present methodological comparison. Accordingly, all 20 adjectival scales were used in the CCA analyses. One should therefore keep in mind that the resulting sweater spaces may be regarded as *both* cognitive *and* affective.

For each CCA solution, the proportion of variance in the first canonical set (attributes) accounted for by the second canonical set (dummy variables) was determined by computing Stewart and Love's (1968) index of *redundancy*. This index weights each canonical r^2 by the corresponding explained variance in the dependent set (i.e., the average of the squared attribute loadings discussed hereafter) and sums across canonical pairs. The redundancy index has been described at length by Alpert and Peterson (1972), Cooley and Lohnes (1971), Fornell (1978), Green (1978), Lambert and Durand (1975), and Levine (1977).

Interpretive Analysis

Interpretation of the sweater spaces was based on three types of supplementary correlational analysis.

ATTRIBUTE LOADINGS. First, canonical variates were correlated with adjectival ratings to obtain the relevant attribute loadings (cf. Cooley and Lohnes 1971; Lambert and Durand 1975; Levine 1977). The resulting loadings or "structure" matrix facilitated interpretation of the CCA dimensions.

FEATURE VECTORS. Second, where interest focused on the geometric relationship of a CCA solution to the underlying stimulus characteristics, feature and feature-interaction vectors were positioned in the space by means of a regression procedure like that incorporated as Phase IV of PREFMAP (Carroll 1972). Such vectors help to explain the CCA space because objects whose projections are farther toward the head of a vector tend more to have the feature or interaction the vector represents. They therefore assist in the important task described by Green and Wind (1975) as "developing relationships (where possible) between . . . psychological spaces and objective spaces involving 'controllable' variables" (p. 151).

FIT. Third, fits of the five spatial dimensions to the underlying structure of objective sweater characteristics were assessed by regressing spatial coordinates on each axis against the relevant subset of features and feature interactions. The squared multiple correlations of these five separate regressions were then averaged to obtain *mean R^2* as an index of overall fit between a CCA space and

FIGURE 2 Aggregate objects-based CCA space for 30 subjects rating
32 sweaters.

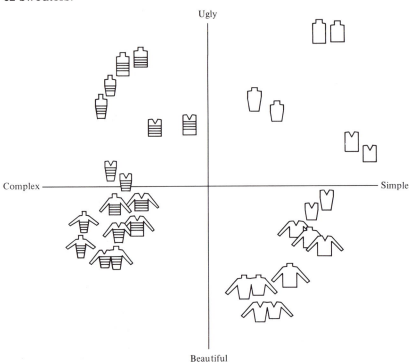

sweater features.[8] Comparisons of fit between CCA solutions made possible conclusions about the role of cue configurality in subjective judgments.

Results

Aggregate-Level Analysis

CANONICAL CORRELATION.   For the aggregate sample ($N = 960$), objects- and features-based canonical correlations yield redundancies for the first five variate pairs of .204 and .186, respectively. Both fits are rather low, undoubtedly because of the considerable heterogeneity that one might expect

in the attribute ratings of any given sweater across individuals (particularly where affective components are involved).[9]

   Results for the objects-based canonical correlation were validated against a conventional multiple discriminant analysis with the anticipated equivalence of CCA and MDA solutions. To illustrate the meaning of these solutions, the first two attribute-specific canonical variates are plotted in Figure 2, with each sweater's position determined by its centroids as given by equations 5 and 6. The redundancy of this two-dimensional objects-based solution is .180, indicating that the three remaining canonical dimensions add little incremental explanation of variance in the attribute ratings (.204 − .180 = .024).

ATTRIBUTE LOADINGS.   Interpretation of the objects-based CCA space is illustrated by examining the correlations across respondents between attribute ratings and canonical variates. These attribute loadings suggest that the horizontal dimension

---

[8] This regression-analytic measure of fit (mean R^2) differs from that implied by the redundancy index discussed before. Redundancy measures the degree to which variance in all the attributes is explained by the features and interactions included in a given canonical correlation. If some attribute(s) is (are) strongly associated with some feature interaction(s)—but only weakly related to other attributes—then redundancy could actually decrease when interactions are added to the canonical analysis. In contrast, the regression measure of fit assesses the degree to which object positions in a CCA space are accounted for by the underlying feature structure. By definition, mean R^2 cannot decrease when interactions are included in the regressions predicting object coordinates.

[9] The increase in redundancy due to the inclusion of all feature interactions was very small (.204 − .186 = .018). Here and elsewhere in the analysis, this accords with the considerations raised in footnote 8.

FIGURE 3 Aggregate objects-based CCA space with Vectors indicating features and feature interactions fitting better than $R = .62$.

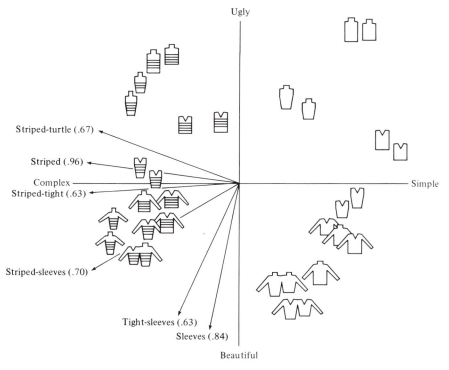

represents *simplicity*, with high loadings on simple/complex (−.83), fancy/plain (.77), and active/passive (.45), whereas the vertical axis indicates aggregate *evaluation* with high loadings on beautiful/ugly (.84), refined/tacky (.80), and displeasing/pleasing (−.80).[10]

FEATURE VECTORS.   One can easily see that the horizontal axis tends to distinguish between striped and solid sweaters whereas the vertical dimension separates those with and without sleeves. If sweater positions were represented by linguistic or numerical symbols, however, such contrasts would not be so obvious to the unaided eye— especially if the focus on feature interactions, not just simple main effects. Suppose, for example, that the sweaters in Figure 2 were represented by the numbers 1 through 32 instead of little pictures. This format would correspond to the common practice of using names, letters, or numbers to indicate object positions in a product space.

In such situations, interpretation of the CCA space may be greatly facilitated by the introduction of vectors to represent product features and feature interactions. These vectors show the directions of maximal increase in the tendencies of objects to have the features and interactions of interest.

This point is illustrated by Figure 3, which contains vectors for main and interaction effects with regression fits better than $R = .62$ ($p < .001$) in the objects-based CCA space. This figure indicates the strong contrasts due to stripes ($R = .96$) and sleeves ($R = .84$). Moreover, it reveals some interactions that might not have been noticed on casual inspection. For example, the striped-tight interaction ($R = .63$) is closely associated with the complexity dimension; further, if one wishes to build a beautiful sweater, the tight-sleeves combination comes into play ($R = .63$). Thus, feature and interaction vectors may provide clues useful in interpreting the CCA space and linking it to "actionable" or "controllable" variables.[11]

[10] Note that the familiar *potency* dimension appeared as the third canonical variate, with high loadings on heavy/light (−.74), masculine/feminine (−.41), smooth/rough (.40) and hard/soft (−.39). However, because it contributed little to shared variance or redundancy (.017), this dimension does not appear in Figure 2.

[11] As an alternative guide to interpreting the CCA axes, features or interactions could be represented as points in the space via the addition of a quadratic term such as that found in Phase III of PREFMAP (Carroll 1972). Though an interesting possibility for future study, this somewhat tangential methodological extension is not pursued in our illustrative analysis.

FIT. The type of pictorial representation shown in Figure 3 deepens one's understanding of the relationships between features and rated attributes at the aggregate level. With respect to the question of cue configurality, further insights emerge from a comparison of the results obtained by objects- and features-based CCA.

Recall that, where feature interactions do not strongly affect attribute ratings, objects- and features-based solutions should be closely comparable in their degrees of fit to the underlying feature structure. These fits were assessed by regressing sweater coordinates along each axis on the design features, with hierarchic inclusion of their two- to four-way interactions. Such regressions were performed for each axis across both sweaters and respondents (N = 960) and across sweater centroids (N = 32). The two approaches produce identical regression coefficients, but the latter gener-

ally tends to give stronger fits at the expense of a considerable loss in degrees of freedom.

Results of these regressions are reported in Table 1, in which mean R^2 may be taken as an overall index of fit between features (with and without interactions) and the competing five-dimensional CCA solutions. This table supports our aforementioned expectations that, at the aggregate level of analysis, features- and objects-based spaces should perform about equally well in fitting main feature effects (.226 versus .221; .770 versus .688) and feature interactions (.256 versus .265; .996 versus .992). Such differences in fit are extremely small in magnitude, attesting to the relative unimportance of feature interactions at the aggregate level of analysis where cue configuralities might be expected to be very heterogeneous and therefore rather weak in their overall impact. This result raises the question of whether more dramatic dif-

TABLE 1 Regression Fits (R^2) Between Sweater Features and Objects- versus Features-Based CCA Dimensions in Aggregate Analysis

Observations	Type of CCA	Dimension	Main Effects	2-Way	3-Way	4-Way
Subjects and sweaters (N = 906)	Objects-based	1	.496[c]	.501[c]	.505[c]	.508[c]
		2	.370[c]	.424[c]	.427[c]	.428[c]
		3	.140[c]	.171[c]	.185[c]	.186[c]
		4	.088[c]	.107[c]	.116[c]	.119[c]
		5	.011	.041[c]	.080[c]	.083[c]
		Mean	.221	.249	.263	.265
	Features-based	1	.496[c]	.500[c]	.504[c]	.507[c]
		2	.378[c]	.417[c]	.420[c]	.421[c]
		3	.145[c]	.170[c]	.179[c]	.181[c]
		4	.093[c]	.108[c]	.113[c]	.116[c]
		5	.020[b]	.031[a]	.048[b]	.053[b]
		Mean	.226	.245	.253	.256
Sweater centroids (N = 32)	Objects-based	1	.976[c]	.986[c]	.994[c]	.999
		2	.865[c]	.989[c]	.997[c]	1.000[c]
		3	.747[c]	.911[c]	.986[b]	.992
		4	.725[c]	.881[c]	.951[a]	.981
		5	.125	.490	.955[a]	.987
		Mean	.688	.851	.977	.992
	Features-based	1	.977[c]	.986[c]	.994[c]	.999
		2	.896[c]	.990[c]	.997[c]	1.000[a]
		3	.800[c]	.939[c]	.988[c]	.997
		4	.792[c]	.916[c]	.966[a]	.987
		5	.385[a]	.585	.910	.997
		Mean	.770	.883	.971	.996

[a] $p < .05$.
[b] $p < .01$.
[c] $p < .001$.

ferences in fit might appear in subgroups chosen to be more homogeneous in response to cue configuralities. This issue was explored in the subgroup analysis.

SUBGROUP ANALYSIS. Differences in the fits of objects- and features-based spaces were examined for a subgroup selected to be relatively homogeneous in their responsiveness to two-way interactions. Somewhat arbitrarily, this "configural" subgroup selection was based on the pattern of two-way effects found by Holbrook and Moore (1981b) in conjoint analyses of their first judgmental factor. Specifically, the 15 subjects showing the most positive responses to the most common sleeves/turtleneck interaction were chosen to compose the configural subgroup. Note, however, that feature interactions in this subgroup were heterogeneous in other respects so that the configural subgroup can

be considered more homogeneous than the remaining subjects only in their shared tendency to show the most common interaction effect.

For this configural subgroup, objects- and features-based canonical correlations produced redundancies of .235 and .191, which suggest that heterogeneity is still too great to permit the explanation of a large amount of variance in the attribute ratings.

Comparisons between objects-based and features-based regression fits for the configural subgroup appear in Table 2 (which is based on analyses analogous to those underlying Table 1). Like the aggregate results, these comparisons continue to show only small differences in fit between features- and objects-based spaces in representing main feature effects (.251 versus .239; .707 versus .595) and feature interactions (.312 versus .331; .990 versus .965).

TABLE 2 Regression Fits (R^2) Between Sweater Features and Objects- Versus Features-Based CCA Dimensions for the Configural Subgroup

Observations	Type of CCA	Dimension	Main Effects	2-Way	3-Way	4-Way
				Highest Level of Interactions Included		
Subjects and sweaters ($N = 480$)	Objects-based	1	.450[c]	.542[c]	.556[c]	.560[c]
		2	.421[c]	.471[c]	.478[c]	.481[c]
		3	.203[c]	.222[c]	.251[c]	.273[c]
		4	.111[c]	.174[c]	.194[c]	.202[c]
		5	.010	.099[c]	.133[c]	.140[c]
		Mean	.239	.302	.322	.331
	Features-based	1	.501[c]	.517[c]	.526[c]	.530[c]
		2	.392[c]	.477[c]	.486[c]	.488[c]
		3	.208[c]	.243[c]	.260[c]	.277[c]
		4	.125[c]	.161[c]	.176[c]	.188[c]
		5	.030[a]	.049	.068	.076
		Mean	.251	.289	.303	.312
Sweater centroids ($N = 32$)	Objects-based	1	.802[c]	.966[c]	.992[c]	1.000
		2	.875[c]	.978[c]	.994[c]	.999
		3	.698[c]	.765[a]	.862	.940
		4	.536[c]	.846[c]	.940	.981
		5	.065	.639	.861	.906
		Mean	.595	.839	.930	.965
	Features-based	1	.944[c]	.975[c]	.991[c]	1.000[b]
		2	.804[c]	.978[c]	.996[c]	.999
		3	.724[c]	.844[c]	.903	.962
		4	.661[c]	.852[c]	.931	.991
		5	.401[a]	.642	.897	.999
		Mean	.707	.858	.944	.990

[a] $p < .05$.
[b] $p < .01$.
[c] $p < .001$.

These findings suggest that the so-called "configural" subgroup was not sufficiently homogeneous in its interaction effects to elicit the comparative differences that we had anticipated. One way to circumvent this problem would be to search for still more homogeneous subgroups. Another is to conduct analyses and comparisons at the individual level of analysis.

Individual-Level Analysis

To draw out the contrast between configural and nonconfigural subjects, the results of Holbrook and Moore (1981b) were again consulted to select individuals with different levels of feature interactivity. Specifically, total number of two-way feature interaction (significant at $p < .05$ in conjoint analyses across four attribute factors) was counted and taken as an index on which to base a selection of the most and least configural subjects (10 and zero two-way interactions, respectively). Expectations about the comparative results for interactions- and features-based CCA methods were then examined separately for each subject. In this analysis, interactions-based canonical correlations were performed hierarchically for each subject for two-way to four-way interactions. However, the inclusion of three- and four-way effects failed to improve noticeably on the variance explained with two-way interactions. This finding is itself potentially important, and its implications are noted in our concluding discussion. On the basis of this result, the two-way effects were the highest level included in the interactions-based CCA results reported hereafter.

Configural Subject

For the configural subject, interactions- and features-based canonical correlations produced redundancies of .517 and .491—dramatically stronger than those obtained in aggregate and subgroup analyses. This improvement in fit may be attributed to two factors: (1) loss of degrees of freedom in the CCA analysis and (2) homogeneity of response at the individual level.

Regression analyses comparable to those described before gave the results reported in the top half of Table 3. These results support our expectations about comparative fits for a configural subject at the individual level of analysis in that

(1) features-based CCA strongly out-performs the interactions-based method in fitting main feature effects (.706 versus .380) and (2) the interactions-based approach provides noticeably better fits when interactions are included as part of the fit criterion (.804 versus .992). These results indicate that, on a subject chosen for his tendency toward two-way configurality, interactions-based CCA spaces can provide extremely good fits to the underlying structure of features and feature interactions.

Nonconfigural Subject

The conconfigural subject also shows high redundancies of .517 and .498 in the interactions- and features-based analyses, respectively. Moreover, this subject continues to exhibit a difference between features- and interactions-based solutions in fitting main effects (.825 versus .521) and feature interactions (.885 versus .991). Therefore the "nonconfigural" subject appears to be more interactive in his responses than we had anticipated, and this unexpected degree of configurality is well represented by differences between the features- and interactions-based fits.

Conclusions About Individuals

In sum, comparisons of features- and interactions-based fits at the individual level of analysis suggest that even respondents not originally viewed as "configural" may show strong effects of feature interactions on product spaces obtained by CCA analysis. The fact that such effects of cue configurality may disappear when one aggregates across individuals emphasizes that feature interactions tend to be heterogeneous and, therefore, to "wash out" when viewed cross-sectionally.

Discussion

Limitations

The methods we propose and illustrate do have limitations. Foremost among them is the reliance on stimulus objects with a feature structure that is known, discrete, and orthogonal. An important issue is the degree to which CCA methods for constructing spaces can be extended to cases

TABLE 3 Regression Fits (R^2) Between Sweater Features and Interactions- Versus Features-Based CCA Dimensions for Two Subjects

Subject	Type of CCA	Dimension	Main Effects Only	2-Way Interactions Included
Configural subject	Interactions-based (N = 32)	1	.391[a]	1.000[c]
		2	.225	1.000[c]
		3	.482[b]	1.000[c]
		4	.569[c]	1.000[c]
		5	.233	.962[c]
		Mean	.380	.992
	Features-based (N = 32)	1	.954[c]	.968[c]
		2	.879[c]	.926[c]
		3	.609[c]	.717[a]
		4	.594[c]	.686
		5	.496[b]	.724[a]
		Mean	.706	.804
Nonconfigural subject	Interactions-based (N = 32)	1	.654[c]	1.000[c]
		2	.883[c]	1.000[c]
		3	.144	1.000[c]
		4	.274	1.000[c]
		5	.648[c]	.955[c]
		Mean	.521	.991
	Features-based (N = 32)	1	.969[c]	.979[c]
		2	.940[c]	.965[c]
		3	.829[c]	.874[c]
		4	.745[c]	.874[c]
		5	.640[c]	.735[a]
		Mean	.825	.885

[a] $p < .05$.
[b] $p < .01$.
[c] $p < .001$.

where object characteristics are continuous, correlated, and "known" in the sense of being objectively measured (rather than experimentally manipulated). Such a situation might arise, for example, when the research focuses on real brands instead of hypothetical objects constructed by the researcher.[12] Whether spaces derived by interactions-based canonical correlation will remain equally useful in studying judgments of these more realistic stimuli is a question that suggests important directions for future research.

Conclusions

Subject to these limitations, several conclusions might be drawn from the illustrative analyses. First, canonical correlation affords a method for constructing spatial representations of attribute data that may be extremely useful in cases such

[12] For a comment on this kind of analysis, see footnote 4.

as individual-level analysis where MDA is inappropriate but where interactions-based CCA remains feasible.[13]

Second, even where conventional MDA can be applied to subgroup or aggregate data, interactions-based CCA may encourage insights into which feature interactions are most important in discriminating judgmentally among objects. For example, if a set of hypothesized feature interac-

[13] As mentioned by one reviewer, categorical conjoint measurement could also be applied at the *individual* level of analysis. For example, if each seven-point attribute scale were coded as six (or fewer) category-specific dummy variables and then used in canonical correlation against selected features and feature interactions, "optimal" attribute-scale values could be derived in the manner discussed previously. Such an application, though an interesting avenue for potential future research, is beyond the scope of our study. Here, because of limited degrees of freedom at the individual level of analysis, categorical conjoint measurement would permit considering no more than a few multicategorical attributes among the 20 available. Instead, in the manner of frequent marketing research practice (cf. Bass and Wilkie 1973), we have treated the attribute ratings as continuous scales standardized within subjects across objects to avoid scale-response biases.

tions can be deduced *a priori*, one may test them systematically by including the appropriate features and interaction terms in an interactions-based CCA. Alternatively, in adopting a more inductive approach, one might pursue the following steps of analysis.

1. Begin with canonical correlation using all available interaction terms.
2. Apply regression measures of fit and examine results for the strength and significance of configural effects.
3. Drop weak configural effects from a subsequent round of more selective interactions-based CCA.
4. Iterate until only important interaction effects remain.

Certainly, the results of such a search procedure should be tested against a validation sample. The important point, however, is that—whether one's investigation is exploratory or hypothetico-deductive—interactions-based canonical correlation makes possible a selective kind of analysis that would be infeasible with the conventional discriminant approach.

Third, the overall pattern of results shown in Tables 1 to 3 suggests that, in this experimental setting, one need not go beyond the level of two- or at most three-way interactions to understand the structure of attribute judgments. In general, fits attained with four-way interactions were only marginally better than three-way fits, often failing to justify the additional degrees of freedom consumed. For many products, it appears doubtful that most consumers could or would process information at the level of four- or higher-way interactions. Indeed, we suspect that, given limits to cognitive capacity, two-way interactions probably constitue a realistic ceiling in the kinds of judgments we collected. The results for individual subjects in Table 3 appear to support the existence of such a ceiling.

Future Directions

When CCA methods are applied to real brands whose objective characteristics are measured rather than manipulated experimentally, the range of real-world stimuli is likely to be characterized by many more than the five features chosen for investigation in our study. As the number of fea-

tures or characteristics (n) increases, however, the number of potential feature interactions increases exponentially (e.g., $n^2 - n$ and $n^3 - 3n^2 + 2n$ for two- and three-way interactions, respectively). It follows that the use of real brands will force the consideration of highly restricted sets of feature interactions via appropriate selective interactions-based CCA methods. Here, considerable interest will focus on whether two-way interactions are sufficient to account for variance in subjective attribute ratings or whether three- and higher-way configuralities must also be included.

One possible way to examine this issue might be to focus on products such as menus, wardrobes, decors, or music groups that are expected to generate numerous two-way interactions. If higher order interactions do not appear for these product classes, one might safely assume that they are also unimportant in more conventional product categories. Thus, the proposed use of interactions-based CCA should help to address this question in future research on constructing product spaces for objects with known feature structures.

References

Alpert, Mark I. and Robert A. Peterson (1972), "On the Interpretation of Canonical Analysis," *Journal of Marketing Research*, 9 (May), 187–92.
Anderson, T. W. (1958), *An Introduction to Multivariate Statistical Analysis*. New York: John Wiley & Sons, Inc.
Bass, Frank M. and William L. Wilkie (1973), "A Comparative Analysis of Attitudinal Predictions of Brand Preference," *Journal of Marketing Research*, 10 (August), 262–9.
Carmone, Frank J. and Paul E. Green (1981), "Model Misspecification in Multiattribute Parameter Estimation," *Journal of Marketing Research*, 18 (February), 87–93.
Carroll, J. Douglas (1969), "Categorical Conjoint Measurement," paper presented at Mathematical Psychology Meeting, Ann Arbor, MI.
——— (1972), "Individual Differences and Multidimensional Scaling," in *Multidimensional Scaling: Theory and Applications in the Behavioral Sciences*, Vol. 1, Roger N. Shepard, A. Kimball Romney, and Sara Beth Nerlove, eds. New York: Seminar Press.
——— (1973), "Categorical Conjoint Measurement," in *Multiattribute Decisions in Marketing: A Measurement Approach*, Paul E. Green and Yoram Wind, eds. Hinsdale, IL: Dryden Press.
Cooley, W. W. and P. R. Lohnes (1971), *Multivariate Data Analysis*, New York: John Wiley & Sons, Inc.
Ferber, Robert (1977), "What Is the JCR Editorial Policy on Samples and Sample Requirements," *Newsletter*, Association for Consumer Research, 7 (December), 16.

Fornell, Claes (1978), "Problems in the Interpretation of Canonical Analysis: The Case of Power in Distributive Channels," *Journal of Marketing Research*, 15 (August), 489–91.

Green, Paul E. (1978), *Analyzing Multivariate Data.* Hinsdale, IL: Dryden Press.

———, Vithala R. Rao, and Wayne S. DeSarbo (1978), "Incorporating Group-Level Similarity Judgments in Conjoint Analysis," *Journal of Consumer Research*, 5 (December), 187–93.

——— and Yoram Wind (1975), "Recent Approaches to the Modeling of Individuals' Subjective Evaluations," in *Attitude Research Bridges the Atlantic*, Philip Levine, ed. Chicago: American Marketing Association.

Hauser, John R. and Frank S. Koppelman (1979), "Alternative Perceptual Mapping Techniques: Relative Accuracy and Usefulness," *Journal of Marketing Research*, 16 (November), 495–506.

Holbrook, Morris B. (1981), "Integrating Compositional and Decompositional Analyses to Represent the Intervening Role of Perceptions in Evaluative Judgments," *Journal of Marketing Research*, 18 (February), 13–28.

——— and Joel Huber (1979), "Separating Perceptual Dimensions from Affective Overtones: An Application to Consumer Aesthetics," *Journal of Consumer Research*, 5 (March), 272–83.

——— and William L. Moore (1981a), "Cue Configurality in Esthetic Responses," in *Symbolic Consumer Behavior*, Elizabeth C. Hirschman and Morris B. Holbrook, eds. Ann Arbor, MI: Association for Consumer Research.

——— and ——— (1981b), "Feature Interactions in Consumer Judgments of Verbal Versus Pictorial Presentations," *Journal of Consumer Research*, 8 (June), 103–13.

Howard, John A. and Jagdish N. Sheth (1969), *The Theory of Buyer Behavior*, New York: John Wiley & Sons, Inc.

Huber, Joel and Morris B. Holbrook (1979), "Using Attribute Ratings for Product Positioning: Some Distinctions Among Compositional Approaches," *Journal of Marketing Research*, 16 (November 507–16.

Johnson, Richard M. (1970), "Multiple Discriminant Analysis: Applications to Marketing Research," working paper, Market Facts, Inc.

——— (1971), "Market Segmentation: A Strategic Management Tool," *Journal of Marketing Research*, 8 (February), 13–18.

Lambert, Zarrel V. and Richard M. Durand (1975), "Some Precautions in Using Canonical Analysis," *Journal of Marketing Research*, 12 (November), 468–75.

Levine, M. S. (1977), *Canonical Analysis and Factor Comparison*. Beverly Hills, CA: Sage Publications.

Maxwell, A. E. (1961), "Canonical Variate Analysis When the Variables Are Dichotomous," *Educational and Psychological Measurement*, 21, 259–71.

McKeon, James J. (1966), "Canonical Analysis: Some Relations Between Canonical Correlation, Factor Analysis, Discriminant Function Analysis, and Scaling Theory," *Psychometric Monographs*, No. 13.

Moore, William L. (1981), "Predictive Power of Joint Space Models Constructed with Composition Techniques," *Journal of Business Research*, forthcoming.

——— and Morris B. Holbrook (1981), "Equivalences Among Several Types of Canonical Correlation Analysis," working paper, Columbia University.

Neslin, Scott A. (1981), "Linking Product Features to Perceptions: Self-Stated Versus Statistically Revealed Importance Weights," *Journal of Marketing Research*, 18 (February), 80–6.

Osgood, C. E., G. J. Suci, and P. H. Tannenbaum (1957), *The Measurement of Meaning.* Urbana: University of Illinois Press.

Pessemier, Edgar A. (1973), "Single Subject Discriminant Configurations," Working Paper 406, Purdue University.

——— (1977), *Product Management.* Santa Barbara, CA: John Wiley & Sons, Inc.

Stewart, D. K. and W. A. Love (1968), "General Canonical Correlation Index," *Psychological Bulletin*, 70, 160–3.

Urban, Glen L. (1975), "PERCEPTOR: A Model for Product Positioning," *Management Science*, 8 (April), 858–71.

Content and Response-Style in the Construct Validation of Self-Report Inventories: A Canonical Analysis[1]

RICHARD L. GREENBLATT
GERALD J. MOZDZIERZ
THOMAS J. MURPHY

A common problem in establishing the validity of self-report inventories is the confounding of the purported meaning of test scores with response styles such as social desirability, deviation responses, and acquiescence (Wiggins, 1973, pp. 415–439). For example, Block (1965) asserted that socially desirable responses reflect the co-occurrence of social desirability and lack of psychopathology. In contrast, Edwards (1967) maintained that social desirability reflects a non-deliberate

"Content and Response-Style in the Construct Validation of Self-Report Inventories: A Canonical Analysis," Richard L. Greenblatt, Gerald J. Mozdzierz, and Thomas J. Murphy, Vol. 40 (6) 1984, pp. 1414–1420. Reprinted from the *Journal of Clinical Psychology*, published by the Clinical Psychology Publishing Co., Inc. Richard L. Greenblatt is affiliated with the Veterans Administration Hospital, Hines Illinois and the University of the Health Sciences/The Chicago Medical School. Gerald J. Mozdzierz and Thomas J. Murphy are affiliated with the Veterans Administration Hospital, Hines Illinois and Loyola University, Stritch School of Medicine. This study was supported in part by basic institutional research funds from the Veterans Administration.

tendency of Ss to describe themselves favorably and that test scores may be better interpreted in terms of their socially desirable features rather than their purported content. Accordingly, in using relatively well-established scales or behavioral measures as criteria in the validation of a self-report inventory, investigators must determine whether the newly determined relationships result from the purported content of the new inventory or a response style such as social desirability. The issue is still alive, as witnessed by the recent research on the relationship between hopelessness, suicidality, and social desirability (Linehan & Nielson, 1981, 1983; Nevid, 1983); problems in social desirability also have emerged in the behavioral literature (Rock, 1982).

In order to establish the validity of measures while holding constant the possible confounding effects of a third variable, researchers have used partial correlation (McNemar, 1969, pp. 182–185). For example, an investigator may correlate two measures of anxiety, while holding constant their social desirability via partial correlation.

The present paper demonstrates the advantages of using canonical correlation, an increasingly popular and flexible statistical procedure (Darlington, Weinberg, & Walberg, 1975; Holland, Levi, & Watson, 1980), for removing the effects of a confounding variable. More specifically, we wish to demonstrate that canonical correlation analysis can challenge the results derived from a series of partial correlations and can describe relationships between variables that otherwise would go undetected. This goal is consistent with the tenets of establishing the construct validity of measures. Accordingly, the present study attempted to determine whether a general measure of mental health was related to inventories that assess psychopathology beyond what one could expect by Ss responding in a socially desirable manner.

Method

Subjects

Ss were 64 hospitalized, male, alcoholic veterans who were participating in a 4-week treatment program. Their mean age was 41.52 years (SD = 12.82); mean level of education was 11.66 years (SD = 2.40).

Procedure

Scale scores on two personality inventories, the MMPI and the Psychological Screening Inventory (PSI) (Lanyon, 1970), were used as measures of psychopathology; the Sulliman Scale of Social Interest, a 50-item, true-false, self-report inventory (SSSI) (Sulliman, 1973) was used as a global measure of adjustment/mental health (Crandall, 1980). Social interest was Adler's (Ansbacher & Ansbacher, 1956) term for empathic humanistic identification with one's fellow man, which allows one to form ties with others and is reflected in other-directedness. The Marlowe-Crowne Scale (SDS)(Crowne & Marlowe, 1960) defined social desirability.

The patients completed (under informed consent) the MMPI, PSI, SDS, and SSSI as part of a battery of tests used for counseling and research purposes. Pearson product-moment correlations were computed between the SSSI and (1) each MMPI scale; and (2) each PSI scale, followed by partial correlations with SDS partialled out. The scale scores of these instruments were subjected to two canonical correlation analyses. The two sets of variables for the first canonical analysis were (1) SSSI and SDS; and (2) the 13 standard MMPI scales. The two sets of variables for the second canonical analysis were (1) SSSI and SDS; and (2) the 5 PSI scales. The canonical loadings, i.e., the correlation of each original scale score with the canonical scores, are reported.

Results

Table 1 reveals that most MMPI scales correlated significantly with SSSI, with inconsequential changes after the partialing out of SDS. Table 2 displays a similar pattern for the PSI scales.

Table 3 displays the two significant variates that result from the first canonical analysis of SSSI, SDS, and the MMPI. Examination of the first variate reveals that it was influenced strongly by SSSI and that SSSI was related inversely to nearly all MMPI scales. Note further that the only exception to this pattern was the positive loading of the K scale (.61). SDS also attained a moderate correlation on the first variate (.40). Thus, the first canonical variate indicates the following: Ss with high social interest tend to obtain low scores on measure of clinical maladjustment; moreover, these Ss tend to respond to items in a socially desirable

TABLE 1 Correlations of SSSI with MMPI Scales and SDS and Partial Correlations with SDS Held Constant

MMPI Scale	Correlation Coefficient	Partial Correlation Coefficient with SDS Held Constant
L	−.06	−.08
F	−.70***	−.67***
K	.49***	.45***
Hs	−.28*	−.22
D	−.39**	−.37**
Hy	−.21	−.16
Pd	−.37**	−.32**
Mf	−.08	−.07
Pa	−.55***	−.59***
Pt	−.51***	−.44***
Sc	−.67***	−.64***
Ma	−.35**	−.26*
Si	−.44***	−.42***
SDS	.25*	

* $p < .05$.
** $p < .01$.
*** $p < .001$.

TABLE 3 Canonical Loadings of SSSI, SDS, and MMPI Scales

	Variates	
	Variate 1	Variate 2
First set		
SSSI	.99	−.16
SDS	.40	.92
Second set		
L	−.02	.49
F	−.84	−.07
K	.61	.24
Hs	−.35	−.21
D	−.47	−.01
Hy	−.26	.16
Pd	−.48	−.31
Mf	−.12	−.16
Pa	−.66	−.01
Pt	−.65	−.40
Sc	−.83	−.26
Ma	−.46	−.38
Si	−.55	−.24
Canonical correlation	.85	.60
Significance level based on χ^2	.001	.02

TABLE 2 Correlations of SSSI with PSI Scales and SDS and Partial Correlations with SDS Held Constant

PSI Scale	Correlation Coefficient	Partial Correlation with SDS Held Constant
AL	−.65***	−.63***
SN	−.54***	−.50***
DI	−.53***	−.49***
EX	.23	.25*
DE	.50***	.45***
SDS	.25*	

* $p < .05$.
** $p < .01$.
*** $p < .001$.

manner. However, the tendency of Ss with high social interest to respond in a socially desirable manner (.40 loading) is not as marked as the tendency of those same Ss with high social interest to score low measures of clinical maladjustment (e.g., .99 loading for SSSI and −.83 loading on Sc).

In contrast to the first variate, the second variate was influenced strongly by SDS (.92 loading) with low to modest negative loadings for most MMPI scales. Note that the highest loading for the MMPI was the L scale (.49 loading); the reader will recall

that this scale is comprised of items that express culturally sanctioned behaviors. The −.16 loading for SSSI indicates its negligible influence on the variate. Thus this second variate, dominated by SDS, is composed of social desirability as a response style. In turn, the content-oriented implications of high social interest and low psychopathology are absent from this variate. In summary, the first variate demonstrates a content-oriented dimension in that social interest shows a strong inverse relationship to measures of psychopathology, whereas social desirability shows a moderate inverse relationship to psychopathology. The second variate demonstrates a response style dimension insofar as social desirability is related to endorsement of culturally sanctioned items.

Comparison of the canonical analysis with the partial correlations reveals that both analyses support the inverse relationship of psychopathology and social interest. However, the unique contributions of the content dimension and the stylistic dimension are masked if only partial correlations are examined. Table 1 displays the small changes in the correlations of the SSSI and MMPI scales when SDS is partialed out. However, when the squared canonical correlations in Table 3 are examined (.72 for variate 1 and .36 for variate 2),

TABLE 4 Canonical Loadings of SSSI, SDS, and
PSI Scales

	Variate 1	Variate 2
First set		
SSSI	.86	−.52
SDS	.71	.70
Second set		
AL	−.76	.54
SN	.75	−.02
DI	−.75	−.08
EX	.15	−.63
DE	.76	.27
Canonical		
correlation	.85	.36
Significance level		
based on χ^2	.001	.083

the hitherto undetected influence of social desirability can be seen more clearly; that is, the proportion of explainable variance (approximately one-third) attributed to social desirability is larger than one would expect from examination of the partial correlations.

Table 4, which contains the PSI canonical analysis, shows a contrasting but equally interesting picture. The one significant variate contains high loadings for both SSSI and SDS. If the second variate reached significance, given the large magnitude of the loadings for SDS and SSSI, one must still untangle their respective influences on PSI. Hence, the data do *not* support a relationship between social interest and the psychopathology measures of the PSI independent of the relationship between social desirability and the PSI.

Note further that the partial correlations (Table 2) support a statistically significant relationship between SSSI and PSI scales. These results contrast with the canonical analysis, where one is unable to distinguish whether Ss are responding to content-oriented characteristics of the PSI or to the response style characteristics. In this set of data, content and response style are indistinguishable.[2]

Discussion

The findings support the use of canonical correlation analysis rather than partial correlations in

[2] Results similar to those of the PSI were obtained with a battery of MMPI research scales. These tables are available from the authors.

evaluating the relationships between multi-scale inventories and a new scale when the investigator wishes to remove the effects of a confounding variable such as social desirability

The canonical analyses demonstrated that an examination of only partial correlations may yield spurious conclusions. That is, the canonical correlation analyses showed that the MMPI scales discriminated between social interest and social desirability, whereas the PSI scales failed to discriminate between these variables. Furthermore, the MMPI canonical analysis revealed the co-occurrence of high social interest, low psychopathology, and a moderate tendency to select socially desirable responses. An investigator who examined the partial correlations (Table 1 and 2) would have found support for the relationship between both established inventories and social interest even with social desirability partialled out. We maintain that from the perspective of construct validation (Cronbach & Meehl, 1955; Nevid, 1983) the latter interpretation is potentially misleading because only the multivariate procedures parsimoniously assess the extent to which the sets of variables converge (or produce high correlations) and diverge (or produce low correlations). From another perspective, weighted MMPI scale scores could predict Ss' social interest and social desirability; PSI-based predictions of social interest and social desirability are confounded. Moreover, the canonical correlation analysis takes into account the intercorrelations of a set of predictors and does not require multiple tests of significance.

Application of the canonical method of analysis may shed further light on the continuing controversy over response styles and content of new scales by separating their characterological and stylistic components. Moreover, investigators can use the present methodology in relating any two sets of variables (e.g., observations, external criteria, or other non-test behaviors) while controlling for the effects of a confounding variable. In that way, the number, magnitude, and content of the dimensions that underlie such relationships can be determined in a more parsimonious and potentially more revealing manner than partial correlation allows.

Finally, comments on issues relevant to the proposed methodology are in order. Canonical correlation is appropriate with two sets of variables, although it can be applied with more sets of variables in conjunction with other multivariate procedures (Skinner, 1978). Statisticians question the appropriateness of commonly used significance tests and the interpretation of canonical variates (Harris, 1975; Isaacs & Milligan, 1983). The present study focused on the establishment of the construct validity of self-report inventories. Once construct validity has been established, application of these inventories for clinical prediction still requires attention to base rates and selection ratios (Meehl & Rosen, 1955).

References

Ansbacher, H. L., & Ansbacher, R. R. (Eds.). (1956). *The individual psychology of Alfred Adler: A systematic presentation in selections from his writings.* New York: Basic Books.

Block, J. (1965). *The challenge of response sets.* New York: Appleton-Century-Crofts.

Crandall, J. E. (1980). Adler's concept of Social Interest: Theory, measurement, and implications for adjustment. *Journal of Personality and Social Psychology, 39,* 481–495.

Cronbach, L. J., & Meehl, P. E. (1955). Construct validity in psychological tests. *Psychological Bulletin, 52,* 281–302.

Crowne, D. P., & Marlowe, D. A. (1960). A new scale of social desirability independent of psychopathology, *Journal of Consulting Psychology, 24,* 349–354.

Darlington, R. B., Weinberg, S. L., & Walberg, H. J. (1975). Canonical variate analysis and related techniques. In D. J. Amick & H. J. Walberg (Eds.), *Introductory multivariate analysis.* Berkeley: McCutchan.

Edwards, A. E. (1967). The social desirability variable: A review of the evidence. In I. A. Berg (Ed.), *Response set in personality assessment.* Chicago: Aldine.

Harris, R. J. (1975). *A primer of multivariate statistics.* New York: Academic Press.

Holland, T. R., Levi, M., & Watson, C. G. (1980). Canonical correlation in the analysis of a contingency table. *Psychological Bulletin, 87,* 334–336.

Isaacs, P. D., & Milligan, G. W. (1983). A comment on the use of canonical correlations in the analysis of contingency tables. *Psychological Bulletin, 9,* 378–381.

Lanyon, R. I. (1970). Development and validation of a psychological screening inventory. *Journal of Consulting and Clinical Psychology Monograph, 35* (1, Part 2).

Linehan, M. M., & Nielson, S. L. (1981). Assessment of suicide ideation and para-suicide: Hopelessness and social desirability. *Journal of Consulting and Clinical Psychology, 49,* 773–775.

Linehan, M. M., & Nielson, S. L. (1983). Social desirability: Its relevance to the measurement of hopelessness and suicidal behavior. *Journal of Consulting and Clinical Psychology, 51,* 141–143.

McNemar, A. (1969). *Psychological statistics.* New York: John Wiley.

Meehl, P. E., & Rosen, A. (1955). Antecedent probability and the efficiency of psychometric signs, patterns, and cutting scores. *Psychological Bulletin, 52,* 194–216.

Nevid, J. S. (1983). Hopelessness, social desirability, and construct validity. *Journal of Consulting and Clinical Psychology, 51,* 139–140.

Rock, D. L. (1982). The confounding of two self-report assertion measures with the tendency to give socially desirable responses in self-deception. *Journal of Consulting and Clinical Psychology, 49,* 743–744.

Skinner, H. A. (1978). The art of exploring predictor-criterion relationships. *Psychological Bulletin, 85,* 327–337.

Sulliman, J. R. (1973). The development of a scale for the measurement of Social Interest. (Doctoral dissertation, Florida State University). *Dissertation Abstracts, 34,* 6. (University Microfilms No. 73-31, 567).

Wiggins, J. S. (1973). *Personality and prediction: Principles of personality assessment.* Reading, MA: Addison-Wesley.

Factor Structure and Correlates of Ratings of Inattention, Hyperactivity, and Antisocial Behavior in a Large Sample of 9-Year-Old Children From the General Population

ROB McGEE
SHEILA WILLIAMS
PHIL A. SILVA

The *Diagnostic and Statistical Manual of Mental Disorders* (DSM-III; American Psychiatric Association, 1980) has proposed an attention deficit disorder (ADD) to replace the concept of hyperactivity. ADD may or may not be accompanied by hyperactive behavior, but the focus of the disorder has become attentional dysfunction rather than excessive motor activity. The relations among attention deficit, hyperactive behavior, and conduct problems or aggressive-antisocial behavior, however, are yet to be clearly described. In Rutter's (1982) review of the concepts of minimal brain

"Factor Structure and Correlates of Ratings of Inattention, Hyperactivity, and Antisocial Behavior in a Large Sample of 9-Year-Old Children From the General Population," Rob McGee, Sheila Williams, and Phil A. Silva, Vol. 53 (4) 1985, pp. 480–490. Reprinted from the *Journal of Consulting and Clinical Psychology,* published by the American Psychological Association, Inc. The authors are affiliated with the Department of Pediatrics and Child Health, University of Otago, Dunedin, New Zealand.

dysfunction and the hyperkinetic syndrome, he concluded that there is no good empirical support for a broad concept of hyperactivity as a separate disorder. Furthermore, he argued that even though emphasis has shifted from activity to attention deficits, the problem of the validity of the disorder remains. For example, the features of one disorder may be present in others, such as inattentiveness associated with conduct disorder, depression, or anxiety. In addition, lack of agreement among measures may be just as much a problem for inattentiveness as it is for hyperactivity. If attention deficit, hyperactivity, and conduct problems represent meaningful clinical conditions, then the issue still remaining is whether they may be differentiated on the basis of variables other than the symptoms that define the disorders, for example, treatment or etiological variables (Achenbach, 1981; Loney, 1980; Rutter, 1982).

Evidence from factor analytic studies of behavioral descriptions has shown that inattention, hyperactivity in the sense of motor excess, and antisocial or aggressive behaviors often emerge as separate factors (Aman, Werry, Fitzpatrick, Lowe, & Waters, 1983; Conners, 1969; Lahey, Stempniak, Robinson, & Tyroler, 1978; Loney, Langhorne, & Paternite, 1978; O'Leary & Steen, 1982; Trites & Laprade, 1983; Werry & Hawthorne, 1976). However, behavioral measures based on these factors may be highly correlated. Unfortunately, there are relatively few studies examining the validity of these behavioral dimensions in terms of differential associations with other variables.

Lahey, Green, and Forehand (1980) examined the interrelations among hyperactivity, conduct disorder, and attention deficit using a combination of direct observation, teacher ratings, peer evaluation, and academic performance. They concluded that although inattentiveness and hyperactivity may be separate behavioral dimensions, conduct disorder and hyperactivity are not. Furthermore, they argued that the results do not confirm the existence of hyperactivity as a separate diagnostic category.

On the other hand, Loney et al. (1978) found differential associations between factor scores for aggressiveness and hyperactivity and a number of background variables. More recently, Milich, Loney, and Landau (1982) confirmed and extended these findings with a sample of 90 boys referred to a psychiatric outpatient service. Taken together, these two studies suggest the desirability of distinguishing between hyperactive and aggressive dimensions of behavior. These two studies,

however, did not make a distinction between inattentive and hyperactive behaviors.

A slightly different approach to the problem of the distinctiveness of childhood disorders has been to use behavioral dimensions derived from factor analytic studies to identify various groups of children scoring at the extremes of the dimensions. In a descriptive sense, it is clearly possible to identify children with pure and mixed forms of disorder, such as conduct disorder with or without hyperactive behavior or pure hyperactivity without conduct disorder (August & Stewart, 1982; Lynn, Mirkin, Lanese, Schmidt, & Arnold, 1983; McGee, Williams, & Silva, 1984a; Stewart, Cummings, Singer, & deBlois, 1981; Trites & Laprade, 1983). The question of differential associations between these disorders and various background variables is more difficult to answer. There is some evidence to suggest that hyperactivity is associated with a more general cognitive impairment, whereas aggressive behavior is not (August & Stewart, 1982; McGee et al., 1984a). However, variables that might be considered to be of etiological significance, for example, family background, do not appear to differentiate strongly among children with these disorders (McGee, Williams, & Silva, 1984b; Sandberg, Wieselberg, & Shaffer, 1980).

The present study examines further the issue of the independence of hyperactive and antisocial behavior in a large sample of 9-year-old New Zealand children. It reports the factor structure of inattentive, impulsive, hyperactive, and antisocial behavior ratings and examines the relations between the derived factors and other developmental and background variables. The developmental variables were chosen on the basis of the research literature, particularly their relation with hyperactive behavior, and included cognitive measures (Rosenthal & Allen, 1978), speech articulation measures (Thorley, 1984), and both fine and gross motor development (Denckla & Rudel, 1978). The background characteristics of the children included socioeconomic, maternal, and family variables.

Method

Subjects

The subjects were 926 non-clinic-referred children aged 9 from the Dunedin Multidisciplinary Child Development Study, which is a longitudinal investigation of their health, development, and behavior. The history of this study and characteristics of the sample have been described by McGee and Silva (1982). To summarize, the children were part of a cohort born at Dunedin's only obstetric hospital (Queen Mary Hospital), between April 1, 1972 and March 31, 1973. The sample was first traced at age 3, and 1,139 children were eligible for inclusion because they were living within the province of Otago, in which Dunedin is situated. Of these 1,139 children, 1,037 (91.0%) were assessed; the remaining children either were traced too late or the parents were unable to cooperate. Subsequently, the sample was assessed at age 5 ($N = 991$), at age 7 ($N = 954$), and at age 9 ($N = 955$).

In terms of cross-cultural perspective, New Zealand enjoys an overall high standard of living, and when compared with other countries on various indexes of living standards, New Zealand ranks in the first 10 (New Zealand Department of Statistics, 1976). When compared with New Zealand as a whole on a measure of socioeconomic status (SES; Elley & Irving, 1972), the Dunedin sample tended to be socioeconomically advantaged, with overrepresentation in the two highest SES levels and underrepresentation in the two lowest levels. In addition, the sample was underrepresentative of Maori and Polynesian races, with about 2% being more than half Maori and Polynesian compared with about 10% for New Zealand (New Zealand Department of Statistics, 1976). The predominant European background of the children suggests that the sample would be comparable to other English-speaking, Western cultures.

MEASURES

Teacher Ratings. The Rutter Scale B for teachers consists of 26 items covering a variety of behaviors to which the teacher responds *does not apply* (0), *applies somewhat* (1), and *certainly applies* (2). A factor analysis of the results of the 7-year-olds with this scale identified three factors: Aggressiveness, Hyperactivity, and Anxiety–Worry (McGee et al., in press). The questionnaire, however, has only three items relating to hyperactive behaviors: poor concentration, restless–hardly ever still, and squirmy–fidgety. Consequently, at the 9-year assessment the Rutter scale for teachers was administered again, but additional items to assess inattention, impulsivity, and hyperactivity in more detail were also given. These new items were taken from DSM-III criteria for ADD.[1] Three of these additional items related to lack of self-

confidence, quick mood changes, and assertiveness mentioned in DSM-III as possible associated features of ADD. These three were included in the present study because they were found to be associated with hyperactive behavior in an analysis of the 7-year results from the Dunedin project (McGee et al., 1984a). The new items were rated in the same manner as the original scale items, and the full set of 31 behaviors selected for analysis is shown in Table 1.

There are three important points to note about the items chosen for analysis. In the first instance, about half of the items were constructed ad hoc, although based on behavioral descriptions from DSM-III. These items were piloted on a trial sample of children, and several teachers' made suggestions regarding the appropriate wording of items. However, the new items are essentially unvalidated. Second, the items were chosen to give a broad range of behaviors thought to relate to ADD/hyperactivity/antisocial behavior and included some items that were not strictly part of DSM-III diagnostic criteria. Third, five items relating to anxiety-unhappiness were included in the factor analysis. In McGee et al.'s (in press) factor analysis of the Rutter teacher scale at age 7, they identified a factor made up of worries, fearful, miserable, fussy, and solitary, and these were included in the present study to see whether this factor would be replicated. Completed ratings were available for 926 children, representing a follow-up of 81.3% of the eligible sample at age 3. There were 480 boys and 446 girls.

Parent Ratings. At the 9-year assessment, the parent completed the Rutter Child Scale A, a 31-item questionnaire covering the same range of behaviors as the teacher scale. In addition, the parents were given the extra questions relating to inattention, impulsivity, and hyperactivity except for Question 11, calling out in class. Consequently, it was possible to compare teacher and parent ratings of these behaviors.

Background Measures. Three kinds of background variables were studied in relation to the behavioral ratings, namely, cognitive measures, speech–motor measures, and family background. These measures were restricted to the 7-year and 9-year assessments. Cognitive development was assessed by verbal and performance IQs (VIQ and PIQ) obtained from the Wechsler Intelligence Scale for Children–Revised (WISC–R) administered when the children were aged 7 and 9 (Wechsler, 1974).[2] Reading at both ages was measured by the

---

[1] Unfortunately, the DSM-III item, "Always on the go. Acts as if driven by a motor," was not included. However, for the purposes of the present article (i.e., factor analysis), the ratings appeared to be sufficiently diverse to cover the problem areas of inattention, impulsivity, and hyperactivity.

[2] The WISC–R subscales of Comprehension from the Verbal scale and Picture Arrangement from the Performance scale were omitted at both ages because of time constraints on the testing schedule.

TABLE 1 Distribution of Responses "Applies Somewhat" and "Certainly Applies" for the Teacher Ratings and Sorted Factor Loadings

Item[a]	Applies Somewhat	Certainly Applies	Factor			
			I	II	III	IV
1. Fails to complete tasks	26.3	9.1*	0.81	0.20	0.08	0.12
2. Poor concentration	25.8	9.5*	0.79	0.23	0.23	0.14
3. Easily distracted	31.4	11.9*	0.77	0.23	0.27	0.10
4. Not listening to instructions	31.6	9.4*	0.74	0.25	0.19	0.17
5. Difficulty organizing work	29.3	7.1*	0.71	0.15	0.17	0.27
6. Needs help/attention	20.6	7.3*	0.70	0.25	0.12	0.25
7. Shifts activities	15.4	5.5*	0.64	0.29	0.36	0.06
8. Difficulty staying with play	13.6	2.2*	0.60	0.27	0.17	0.16
9. Acts before thinking	34.9	6.3*	0.49	0.23	0.25	0.12
10. Lacks self-confidence	33.3	5.1	0.49	0.00	−0.05	0.58
11. Frequently fights	14.6	2.8*	0.19	0.76	0.17	0.12
12. Bullies other children	9.2	2.7*	0.21	0.73	0.17	−0.01
13. Often tells lies	11.0	2.2*	0.29	0.68	0.03	0.03
14. Irritable	14.6	2.9*	0.09	0.66	0.38	0.23
15. Often disobedient	13.4	2.5*	0.31	0.65	0.39	−0.03
16. Quick changes in mood	11.9	3.2	0.11	0.64	0.28	0.24
17. Impatient waiting turns	15.7	3.2*	0.18	0.62	0.41	0.07
18. Destructive	5.2	0.6*	0.24	0.61	0.14	0.02
19. Not much liked	16.7	3.1	0.32	0.55	0.05	0.31
20. Assertive approach	17.0	3.2*	0.03	0.49	0.33	0.00
21. Stolen things	4.0	1.6	0.28	0.48	−0.01	0.00
22. Trouble sitting still	20.4	4.0*	0.44	0.27	0.68	−0.01
23. Very restless	19.4	3.7*	0.34	0.34	0.67	0.08
24. Squirmy, fidgety	19.5	4.5*	0.42	0.26	0.63	0.08
25. Calls out in class	13.7	4.5*	0.23	0.35	0.55	0.01
26. Often worried	31.7	5.6	0.08	0.04	0.10	0.71
27. Tends to be fearful	23.8	2.8	0.27	−0.03	−0.02	0.68
28. Often appears miserable	15.1	2.2	0.16	0.30	0.13	0.53
29. Rather solitary	29.3	4.9	0.23	0.19	−0.08	0.47
30. Fussy	13.2	1.0*	−0.02	−0.01	0.02	0.37
31. Truants from school	1.2	0.4	—	—	—	—

Note. Factor I = Inattention; Factor II = Antisocial Behavior; Factor III = Hyperactivity; and Factor IV = Worry–Fearful. Factor loadings above 0.45 are underscored. A loading of 0.45 indicates that the factor accounts for 20% of the variance on the item.

[a] Items 2, 11 to 15, 18, 19, 21, 23, 24, and 26 to 31 are taken from the Rutter Child Scale B for teachers. A full description of the behavioral items is available from the authors.

* $p < .05$; indicates item showed significant sex difference in the distribution of ratings.

Burt Word Reading Test–1974 Revision (Scottish Council for Research in Education, 1976), and spelling ability was assessed at age 9 by a 25-item scale adapted from a test described by Smith and Pearce (1966). These tests were administered by trained psychometrists who did not have access to the results of the teacher and parent behavior ratings.

Speech articulation at both ages was measured by the Dunedin Articulation Check (DAC), a 20-item scale scored for the correct articulation of 6 phonemes and 14 groups of sounds (Justin,

Lawn, & Silva, 1983). Motor ability at ages 7 and 9 was assessed by the nine subtests of the Arnheim and Sinclair (1974) Basic Motor Abilities Test. For the present study, the two subscales of Tapping Ability and Standing Long Jump were chosen for analysis. A factor analysis of the motor test (Wilson, Silva, & Williams, 1981) indicated that these two subscales had the highest loadings on a factor of Fine Motor Control (tapping ability) and a second factor of Large Muscle Control (long jump).

The family background measures included the

following variables: (a) SES as indicated by the Elley and Irving (1976) index for New Zealand that ranks occupations from highest (Rank 1) to lowest (Rank 6); (b) whether the child came from a solo-parent family; (c) whether the child's natural parents had separated any time from birth to age 9; (d) mother's report of her mental health, using the Rutter, Tizard, and Whitmore (1970) Malaise Inventory, a 24-item symptom checklist administered at both ages; and (e) quality of family relationships when the child was 7 and 9, based on maternal report using the Cohesiveness, Expressiveness, and Conflict subscales of the Family Environment Scale (FES; Holahan & Moos, 1983).

PROCEDURE

A similar general procedure has been used at each assessment phase of the study. Prior to the child attending the unit, the parents received details concerning the nature of the assessments, and written consent was sought for the child's participation in the assessment program. Parents were also asked for permission to seek behavioral reports and details of progress from the child's school. Children enrolled in the study were assessed at the Dunedin unit within 1 month of their ninth birthday on a variety of measures regarding their health, cognitive, motor, and behavioral status. Children living in other parts of New Zealand who could not attend in Dunedin were assessed by the Psychological Service of the Department of Education. The child's parent, usually the mother, completed extensive questionnaires to gather background information, including the Rutter Scale A and other behavioral items. The child's teacher was also asked to complete the Rutter Scale B and additional questions relating to attentiveness and behavior during class.

Results

The distribution of responses "applies somewhat" and "certainly applies" for the 31 items in the teacher questionnaire is presented in Table 1. Sex differences were examined using chi-square ($2df$) tests, with the level of significance for each χ^2 being adjusted ($p < .0016$ or $.05/31$) using the Bonferroni inequality to control for inflation of the overall Type I error rate beyond $p < .05$ (Grove & Andreasen, 1982). Twenty-two items had a significant sex difference in the distribution of ratings as shown in Table 1. On most items relating to inattention, impulsivity, hyperactivity, and antisocial behavior, the boys were rated as showing the behavior more often than the girls. The only

behavior shown more often by girls than by boys was being fussy or overparticular.

To examine the factor structure of the teacher ratings, factor analyses using principal factoring with iteration followed by varimax rotation (Nie, Hull, Jenkins, Steinbrenner, & Bent, 1975) were carried out separately for boys and girls. As there was little difference between these two analyses in terms of the relative sizes of factor loadings of the various items, the data were reanalyzed for the total sample, and those results are reported here. For this analysis, Item 31 (truants from school) was not included because of the low frequency of ratings 1 and 2 (1.6%). all other items had combined frequencies for these ratings over 5.5%. The two least frequent behaviors were stealing (Item 21, 5.6%) and being destructive (Item 18, 5.8%). However, these items were included because of their importance in the group of antisocial behaviors, and both subsequently showed communalities comparable with the other items.

The factor analysis of the 30 remaining items resulted in four factors with eigenvalues greater than one. The eigenvalues prior to rotation were 11.76, 3.06, 2.09, and 1.38; the eigenvalue of the fifth factor extracted was 0.93. The first four factors accounted for 60.9% of the variance prior to iteration. Presented in Table 1 are the sorted factor loadings after rotation for these four factors. Factor I, Inattention, showed highest loadings for items relating to poor concentration, distractability, difficulties in organizing and completing work, shifting from activities, not listening, jumping into activities without thinking, needing help and attention, and lacking self-confidence. Factor II. Antisocial Behavior, replicates and extends the Aggressiveness factor identified by McGee et al. (in press). This factor loaded most highly for ratings of fighting, bullying, disobedience, destructiveness, irritability, impatience, lying and stealing, and not being liked. Factor III, Hyperactivity, had high loadings for restlessness, squirminess, trouble sitting still, and calling out in class. Finally, Factor IV, Worry–Fearful, loaded on worries, fearful, lacks self-confidence, unhappy, and solitary. The item measuring fussy behavior had a loading of 0.37 on this last factor and had loadings below 0.05 on the first three factors.

To study further the properties of the behavioral dimensions identified, four measures were calculated from the unweighted sums of items shown in Table 1. As Item 10 (lacks self-confidence) had loadings greater than 0.45 on two factors, it was not included in either the Inattention (Factor I) or Worry–Fearful (Factor IV) measure. Consequently, Inattention was based on Items 1–9 in Table 1; Antisocial Behavior was based on Items 11–21; Hyperactivity, on Items 22–25; and Worry–Fearful, on Items 26–29. The reliability of these four measures was examined using coefficient alpha, an index of internal consistency (Nunnally, 1967). This measure is an estimate of the expected correlation of a test with all alternate forms of the test having the same number of items. Coefficient alpha was .93 for Inattention, .91 for antisocial Behavior, .87 for Hyperactivity, and .71 for Worry–Fearful. An examination of the corrected item-total correlations for each of the four measures indicated that all correlations were significant ($p < .05$). Consequently, no items were deleted from the subscales. These results showed that the measures based on the factor analysis of the teacher ratings had a high level of reliability based on internal consistency.

The parent ratings were also factor analyzed, and a four-factor solution accounted for 46.0% of the variance prior to iteration. The four factors identified were very similar in terms of item composition to those based on the teacher ratings. For example, in the case of the Inattention factor, the following items all had loadings above 0.45: being easily distracted, poor concentration, difficulty staying with play, not listening to instructions, shifting activities, failing to complete tasks, difficulty organizing work, and needing help and attention. Acts before thinking loaded 0.38 on this factor and had low loadings on the other three factors. The main difference between the teacher

and parent factor structure was that in the latter, being miserable loaded more highly on the Antisocial Behavior factor (0.46) than on the Worry–Fearful factor (0.39). However, given the overall similarity in factor structures, the results of the teacher factor analysis were used to form the measures of inattention, antisocial behavior, hyperactivity, and worry–fearful for the parent ratings. This procedure resulted in comparable measures for the teacher and parent ratings and appeared preferable to constructing a slightly different set of measures for each set of ratings. Coefficient alphas for the parent measures were as follows: .86 for inattention, .81 for antisocial behavior, .74 for hyperactivity, and .59 for worry–fearful. As was the case with the teacher measures, all item-total correlations were significant ($p < .05$).

The product-moment intercorrelations between the various derived scales are presented in Table 2. Most correlations were significant, although many indicated relatively weak associations between some of the measures. This is at least partially due to the power of the statistical tests, given the large sample size. Inspection of Table 2 indicates that correlations within a particular rating source (teacher or parent) were higher than those between the two sources. Of particular interest are the correlations among the measures of inattention, antisocial behavior, and hyperactivity. For both the parent and teacher ratings, these scales showed moderately high intercorrelations. To examine the pattern of relations among these latter three scales more closely, partial correlations between pairs of measures were calculated while controlling for the remaining two mea-

TABLE 2 Correlations Among the Teacher and Parent Measures of Inattention, Antisocial Behavior, Hyperactivity, and Worry–Fearful at Age 9

	Teacher				Parent			
Measure	1	2	3	4	1	2	3	4
Teacher								
1. Inattention		.58*	.65*	.43*	.35*	.29*	.24*	.02
2. Antisocial Behavior			.66*	.33*	.14*	.25*	.15*	−.01
3. Hyperactivity				.24*	.24*	.28*	.26*	−.03
4. Worry–Fearful					.16*	.08	.10	.16*
Parent								
1. Inattention						.57*	.46*	.41*
2. Antisocial Behavior							.45*	.41*
3. Hyperactivity								.20*
4. Worry–Fearful								

* $p < .05$ (adjusted $p < .001$ for each correlation), otherwise correlation is nonsignificant.

sures. These partial correlations were as follows: inattention–antisocial behavior (controlling for hyperactivity and worry–fearful), .21 and .37 for teachers and parents, respectively; inattention–hyperactivity, .45 and .27; and antisocial behavior–hyperactivity, .46 and .28. All partial correlations were significant ($p < .05$). Those for the teacher ratings suggest a weaker relation between inattention and antisocial behavior than is shown in Table 2. Hyperactivity still correlated moderately with both of these behaviors. The results for the parent measures suggest weaker relations between all three behaviors.

The final analyses to be reported investigate the relations between the behavior measures derived from the factor analysis and the background variables. These relations are reported in terms of canonical correlations between linear combinations of predictor variables (cognitive, speech–motor, and family characteristics) and the four behavioral measures. This form of analysis is equivalent to a multiple regression with k independent and m dependent variables (Kerlinger

TABLE 3 Results of Canonical Correlation Analysis for the Cognitive and Behavioral Measures Showing Unconstrained and Constrained Solutions

Measure	Unconstrained		Constrained	
	Teacher	Parent	Teacher	Parent
Cognitive				
VIQ				
7	0.14	0.18	0	0
9	0.01	0.13	0	0
PIQ				
7	−0.17	−0.10	0	0
9	0.47	0.42	1	1
Reading				
7	0.35	0.19	1	1
9	0.06	0.12	0	1
Spelling				
9	0.33	0.29	1	1
Behavioral				
Inattention	−1.14	−0.86	−1	−1
Antisocial Behavior	−0.03	−0.08	0	0
Hyperactivity	0.26	−0.24	0	0
Worry–Fearful	0.01	0.11	0	0
Rc	0.51*	0.38*	0.48*	0.36*

Note. V = verbal; P = performance. The numbers 7 and 9 refer to ages. Values shown are canonical coefficients. Rc represents canonical correlation.
* $p < .05$.

& Pedhazur, 1973). The aim was to investigate whether differential relations existed between the background variables and the behavior measures. For each canonical analysis two solutions are reported. The first is the traditional canonical correlation; the second solution is based on linear functions where the values of the canonical weights are constrained or limited to +1, 0, and −1 (De Sarbo, Hausman, Lin, & Thompson, 1982). One difficulty with the traditional canonical correlation concerns the somewhat arbitrary inclusion–exclusion of variates resulting from any subjective evaluation of whether coefficients are large enough to be considered significantly different from zero (Share, 1984). Constrained canonical correlation (see De Sarbo et al., 1982, for a fuller discussion of this topic) provides a method of avoiding the problem of subjective interpretation of coefficient size by deriving a solution in which variates can have only positive (+1), negative (−1), or null (0) effects. Given these constraints on the values of the canonical correlation coefficients, the method derives a globally optimum correlation between two linear combinations of variables.

Presented in Table 3 are the unconstrained and constrained solutions for the canonical analysis of the cognitive variables and the behavioral measures for the teacher and parent ratings.[3] The first point to note is that the unconstrained and constrained solutions provide very similar values of the canonical correlations. Second, it is clear from the table that the cognitive variables were related only to the behavioral measure of inattention, and this was true for both teacher and parent ratings. Third, the same three cognitive measures (reading at age 7, PIQ at age 9, and spelling at age 9) were predictive (in inverse fashion) of poorer inattention ratings by teachers and parents. Reading at age 9 was also predictive of parent ratings of inattention.

The results of the analysis for the speech–motor variates and the behavioral measures are presented in Table 4. Once again there was little difference between the two solutions in terms of the canonical correlations. The speech–motor measures were related to the measure of inattention

[3] These analyses are reported for the whole sample in the interests of a more parsimonious presentation of results. Canonical correlations were obtained for boys and girls separately, and in general, the values of the canonical correlation and the pattern and size of the canonical coefficients did not differ markedly between the sexes.

TABLE 4 Results of the Canonical Correlation Analysis for the Speech–Motor Measures and the Behavioral Measures

Measure	Unconstrained		Constrained	
	Teacher	Parent	Teacher	Parent
Speech				
7	0.33	0.45	1	1
9	0.56	0.49	1	1
Tapping				
7	0.19	0.33	1	1
9	0.34	0.12	1	0
Long jump				
7	−0.08	−0.21	−1	0
9	−0.10	−0.06	0	0
Behavioral				
Inattention	−1.20	−0.79	−1	−1
Antisocial				
Behavior	0.08	−0.29	0	0
Hyperactivity	0.24	−0.13	0	0
Worry–Fearful	0.08	0.15	0	0
Rc	0.38*	0.31*	0.36*	0.29*

Note. The numbers 7 and 9 refer to ages. Rc represents canonical correlation.
* $p < .05$.

TABLE 5 Results of the Canonical Correlation Analysis for the Family Background and Behavioral Measures

Measure	Unconstrained		Constrained	
	Teacher	Parent	Teacher	Parent
Family				
background				
Low SES	0.42	0.14	1	1
Solo parent	0.63	0.20	1	1
Parent				
separation	0.31	0.07	1	0
Malaise				
7	0.17	0.35	0	1
9	0.12	0.29	1	1
FRI				
7	−0.04	−0.22	0	−1
9	−0.09	−0.39	0	−1
Behavioral				
Inattention	0.37	0.09	1	0
Antisocial				
Behavior	0.64	0.69	1	1
Hyperactivity	0.01	0.24	0	1
Worry–Fearful	0.19	0.23	0	1
Rc	0.23*	0.39*	0.21*	0.37*

Note. SES = socioeconomic status. Low SES indicates semi-skilled or unskilled; Malaise represents the Malaise Inventory (Rutter et al., 1970); the numbers 7 and 9 refer to ages; FRI is the Family Relations Index (Holahan & Moos, 1983); Rc represents canonical correlation.
* $p < .05$.

only, and again there was a reasonable degree of similarity between the solutions for the teacher and parent ratings. Speech articulation at ages 7 and 9 and tapping ability at age 7 were associated with parent and teacher ratings of inattention. In addition, tapping ability at age 9 was predictive of teacher ratings of inattention, as was long jump (in inverse fashion) at age 7. The interpretation of this latter finding, however, is unclear.

The final set of results are shown in Table 5, which presents the canonical analysis for the family background measures and the behavior ratings. Low SES indicated an SES Level V or VI (semi-skilled or unskilled) according to the Elley and Irving (1976) index. Reported quality of family relationships was indicated by the Family Relations Index (FRI; Holahan & Moos, 1983). The FRI combined scores on the FES subscales of Cohesion, Expressiveness, and Conflict; the higher the score, the greater the level of social support within the family. In contrast with the previous canonical analysis, the canonical correlations for the parent ratings were somewhat higher than were those for the teachers. In addition, there was rather less similarity between the parent and teacher solutions for the background measures than was the case with the cognitive and motor measures. Teacher ratings of inattention and anti-

social behavior were predicted primarily by the more visible aspects of the family (SES, solo parenting, and separated parents). Parental ratings of antisocial, hyperactive, and fearful behaviors were predicted by low SES and solo parenting in combination with maternal psychological symptoms and poor relationships within the family.

Discussion

An examination of factor analytic studies indicates several studies that have identified a Hyperactivity factor distinct from an Antisocial or Aggressiveness factor (e.g, Lahey et al., 1978; Loney et al., 1978; Milich et al., 1982; Trites & Laprade, 1983). However, the Hyperactivity factor in these studies had elements of both inattention and restlessness–overactivity. Other analyses (Aman et al., 1983; Conners, 1969) have reported separate factors of Inattention and Hyperactivity. There is some evidence to suggest that a general Hyperactivity factor and an Antisocial factor are differentially related to other background variables (Loney et al., 1978; Milich et al., 1982). On the other hand, Lahey et al. (1980) argued that inattentiveness represents a disorder distinct from both hyperactivity and antisocial behavior and that the latter two are probably not separate disorders. Others have similarly stressed the importance of attention difficulties as the core disorder in the hyperactive syndrome (Rosenthal & Allen, 1978).

The present study examined the factor structure of ratings of inattentive, impulsive, hyperactive, antisocial, and anxious behaviors. The emergence of a Worry–Fearful factor replicated the findings of the earlier factor analysis on the sample at age 7 (McGee et al., in press) and is in agreement with numerous other studies (e.g., Quay, 1979). Of more interest was the fact that separate factors of Inattention, Hyperactivity, and Antisocial Behavior were identified. The Inattention factor described a behavioral dimension relating to the planning, organization, and execution of tasks or activities. In addition, it included some of the items that appear under the heading "Impulsivity" in DSM-III (acts before thinking, difficulty organizing work, and shifts activities), and it is of some interest that there was no impulsivity grouping, as such. The Hyperactivity factor described a dimension of restlessness and overactivity. The Antisocial Behavior factor was similar to other factors so-named in the literature (Quay, 1979). There was relatively little agreement between the teacher and parent ratings on these dimensions. For example, the highest correlation was only .35 for the inattention measure. This relatively low agreement between parent and teacher ratings is not new (e.g, Rutter et al., 1970; Touliatos & Lindholm, 1981) and is part of a more general phenomenon of information tending to cluster together according to its source (Loney et al., 1978). Arguments concerning the situational specificity of behaviors have been advanced to account for previous findings of low agreement between parent and teacher ratings of hyperactive and antisocial behavior. However, the hypothesis that there is a generalized attention deficit in individuals classified as having ADD (which is implicit in DSM-III criteria) is not supported in the present study by the low correlation between parent and teacher ratings of inattention. Brown and Wynne (1984) reported low correlations among different measures of attention during task performance. It would appear, therefore, that lack of agreement among measures is

likely to be as much a problem in the assessment of attention deficit as it has been for hyperactivity.

As far as the independence of these ratings is concerned, the results support Lahey et al.'s (1980) argument for considering attention deficits as separate from antisocial behavior. The two were only moderately correlated (.21 for teachers and .37 for parents) when their relations with hyperactivity and worry were controlled. Hyperactivity, on the other hand, correlated a little more highly with both inattention and antisocial behavior. These present results suggest that the Inattentive, Hyperactive, and Antisocial factors represent dimensions describing at least partially different aspects of behavior. In this regard, the emergence of Inattention and Hyperactivity as separate factors is in agreement with some previous factor analyses (Aman et al., 1983; Conners, 1969) and corresponds to the DSM-III distinction between attention deficit and hyperactivity. However, other studies on both whole population and clinic-referred samples have not found distinct Inattentive and Hyperactive factors. The question, then, as to the importance of this distinction and the further differentiation between these behaviors and antisocial behaviors cannot be decided on the results of factor analyses, per se. The results of any factor analysis will inevitably depend on the nature of the item pool and the characteristics of the sample.

Although it is possible to distinguish between inattentive, hyperactive, and antisocial behaviors in a phenomenological or descriptive sense, whether such distinctions have clinical importance depends on whether or not these behaviors are differentially related to other variables (Achenbach, 1981). The research of Loney et al. (1978) and Milich et al. (1982), based on this approach, provides strong support for the distinction between aggressive and inattentive-hyperactive behaviors. In the present study, the results of the canonical correlation analyses provide further evidence for the continued distinction between inattentive, hyperactive, and antisocial behaviors.

For both parent and teacher ratings, the cognitive and speech–motor variables predicted inattention but did not predict hyperactivity or antisocial behavior. Reading, spelling, performance IQ, speech articulation, and fine motor ability were the best predictors of inattention. Milich et al. (1982) found that full-scale IQ was unrelated to their Hyperactive and Aggressive factors. The results of this study, however, suggest that verbal and performance IQ may need to be considered separately. More important, the findings indicate the need for a theory of attention dysfunction that predicts the kinds of performance decrements associated with attention deficit disorder (Rosenthal & Allen, 1978). The findings do not support the hypotheses that deficits in verbal abilities (Prentice & Kelly, 1963) or reading disability (Rutter et al., 1970) are associated with antisocial behavior. The relation between inattention and speech articulation and fine motor ability is of interest given previous reports of an association between motor abnormalities and hyperactivity (Denckla & Rudel, 1978). Earlier research has suggested that hyperactive children are slower than nonhyperactive controls on motor tasks (Rosenthal & Allen, 1978), and the results from the tapping test support this. Speech articulation problems have also been associated with the hyperactive syndrome (Thorley, 1984). The findings of our study suggest slow motor performance and articulation problems are predictive of the inattentive, rather than the hyperactive, component of behavior.

Although cognitive and motor variables were associated with ratings of inattention, family background measures were related to the other behaviors. In the case of the teacher ratings, low SES, solo parenting, separations, and maternal symptoms predicted both inattention and antisocial behavior. For the parent ratings, low SES, solo parenting, family conflict, and mother's psychological health predicted antisocial behavior, hyperactivity, and worry. These results partially confirm the Loney et al. (1978) findings that family background variables are associated more with aggressive behaviors than with hyperactive behaviors. However, the source of the ratings appears to be an important determinant of both predictor variables and the behaviors predicted, with the cognitive measures showing more consistency between parents and teachers.

Overall, the findings of this article suggest that a group of inattentive behaviors closely resembling the DSM-III category of attention deficit disorder can be identified. Furthermore, these behaviors have cognitive and speech–fine motor correlates that distinguish them from both hyperactivity and antisocial behavior. As such, inattention may represent the core dysfunction in the hyperactive syndrome. One question arising from this is whether hyperactivity in the sense of overactivity–restlessness–fidgetiness represents a disorder at all. Lahey et al. (1980) argued that it does

not. However, it is clear that children who achieve a high score on hyperactivity alone can be found (Lynn et al., 1983), and Klein and Young (1979) identified low problem hyperactives who showed neither inattentiveness nor conduct disorder. We would therefore argue that for the present, the behavioral dimension of hyperactivity should continue to be differentiated from both inattention and antisocial behavior.

References

Achenbach, T. M. (1981). The role of taxonomy in developmental psychopathology. In M. E. Lamb & A. L. Brown (Eds.), *Advances in developmental psychology* (Vol. 1, pp. 159–198). Hillsdale, NJ: Erlbaum.

Aman, M. G., Werry, J. S., Fitzpatrick, J., Lowe, M., & Waters, J. (1983). Factor structure and norms for the revised behaviour problem checklist in New Zealand children. *Australian and New Zealand Journal of Psychiatry, 17,* 354–360.

American Psychiatric Association. (1980). *Diagnostic and statistical manual of mental disorders* (3rd ed.). Washington, DC: Author.

Arnheim, D. D., & Sinclair, W. A. (1974). *The clumsy child.* St. Louis: Mosby.

August, G. J., & Stewart, M. A. (1982). Is there a syndrome of pure hyperactivity? *British Journal of Psychiatry, 140,* 305–311.

Brown, R. T., & Wynne, M. E. (1984). An analysis of attentional components in hyperactive and normal boys. *Journal of Learning Disabilities, 17,* 162–166.

Conners, C. K. (1969). A teacher rating scale for use in drug studies with children. *American Journal of Psychiatry, 126,* 884–888.

Denckla, M. B., & Rudel, R. G. (1978). Anomalies of motor development in hyperactive boys. *Annals of Neurology, 3,* 231–233.

De Sarbo, W. S., Hausman, R. E., Lin, S., & Thompson, W. (1982). Constrained canonical correlation. *Psychometrika, 47,* 389–516.

Elley, W. B., & Irving, J. C. (1972). A socio-economic index for New Zealand based on levels of education and income from the 1966 census. *New Zealand Journal of Educational Studies, 7,* 155–167.

Elley, W. B., & Irving, J. C. (1976). Revised socioeconomic index for New Zealand. *New Zealand Journal of Educational Studies, 11,* 25–36.

Grove, W. M., & Andreasen, N. C. (1982). Simultaneous tests of many hypotheses in exploratory research. *Journal of Nervous and Mental Disease, 170,* 3–8.

Holahan, C. J., & Moos, R. H. (1983). The quality of social support: Measures of family and work relationships. *British Journal of Clinical Psychology, 22,* 157–162.

Justin, C., Lawn, L., & Silva, P. A. (1983). *The Dunedin articulation check.* Dunedin, New Zealand: Otago Speech Therapy Association.

Kerlinger, F. N., & Pedhazur, E. J. (1973). *Multiple regression in behavioral research.* New York: Holt, Rinehart & Winston.

Klein, A. R., & Young, R. D. (1979): Hyperactive boys in their classroom: Assessment of teacher and peer perceptions, interactions and classroom behaviors. *Journal of Abnormal Child Psychology, 7,* 425–442.

Lahey, B. B., Green, K. D., & Forehand, R. (1980). On the independence of ratings of hyperactivity, conduct problems, and attention deficits in children: A multiple regression analysis. *Journal of Consulting and Clinical Psychology, 48,* 566–574.

Lahey, B. B., Stempniak, M., Robinson, E. J., & Tyroler, M. J. (1978). Hyperactivity and learning disabilities as independent dimensions of child behavior problems. *Journal of Abnormal Psychology, 87,* 333–340.

Loney, J. (1980). Hyperkinesis come of age: What do we know and where should we go? *American Journal of Orthopsychiatry, 50,* 28–41.

Loney, J., Langhorne, J. E., & Paternite, C. E. (1978). An empirical basis for subgrouping the hyperkinetic/minimal brain dysfunction syndrome. *Journal of Abnormal Psychology, 87,* 431–441.

Lynn, D. J., Mirkin, I. R., Lanese, D. M., Schmidt, H. S., & Arnold, L. E. (1983). Correspondence between DSM-II hyperkinetic reaction and DSM-III attention deficit disorder. *Journal of the American Academy of Child Psychiatry, 22,* 349–350.

McGee, R., & Silva, P. A. (1982). *A thousand New Zealand children: Their health and development from birth to seven* (Special Rep. Series No. 8). Auckland: Medical Research Council of New Zealand.

McGee, R., Williams, S., Bradshaw, J., Chapel, J. L., Robins, A., & Silva, P. A. (in press). The Rutter scale for completion by teachers: Factor structure and relationships with cognitive abilities and family adversity for a sample of New Zealand children. *Journal of Child Psychology and Psychiatry.*

McGee, R., Williams, S., & Silva, P. A. (1984a). Behavioral and developmental characteristics of aggressive, hyperactive and aggressive-hyperactive boys. *Journal of the American Academy of Child Psychiatry, 23,* 270–279.

McGee, R., Williams, S., & Silva, P. A. (1984b). Background characteristics of aggressive, hyperactive and aggressive–hyperactive boys. *Journal of the American Academy of Child Psychiatry, 23,* 280–284.

Milich, R., Loney, J., & Landau, S. (1982). Independent dimensions of hyperactivity and aggression: A validation with playroom observation data. *Journal of Abnormal Psychology, 91,* 183–198.

New Zealand Department of Statistics. (1976). *New Zealand yearbook.* Wellington: Government Printer.

Nie, N. H., Hull, C. H., Jenkins, J. G., Steinbrenner, K., & Bent, D. H. (1975). *Statistical package for the social sciences* (2nd ed.). New York: McGraw-Hill.

Nunnally, J. C. (1967). *Psychometric theory.* New York: McGraw-Hill.

O'Leary, S. G., & Steen, P. L. (1982). Subcategorizing hyperactivity: The Stony Brook Scale. *Journal of Consulting and Clinical Psychology, 50,* 426–432.

Prentice, N. M., & Kelly, F. J. (1963). Intelligence and delinquency: A reconsideration. *Journal of Social Psychology, 60,* 327–337.

Quay, H. C. (1979). Classification. In H. C. Quay & J. S. Werry (Eds.), *Psychopathological disorders of childhood* (2nd ed., pp. 1–42). New York: Wiley.

Rosenthal, R. H., & Allen, T. W. (1978). An examination of attention, arousal, and learning dysfunctions of hyperkinetic children. *Psychological Bulletin, 85,* 689–715.

Rutter, M. (1982). Syndromes attributed to "minimal brain dysfunction" in childhood. *American Journal of Psychiatry, 139,* 21–33.

Rutter, M., Tizard, J., & Whitmore, K. (1970). *Education, health and behaviour.* London: Longmans.

Sandberg, S. T., Wieselberg, M., & Shaffer, D. (1980). Hyperkinetic and conduct problem children in a primary school population: Some epidemiological considerations. *Journal of Child Psychology and Psychiatry, 21,* 293–311.

Scottish Council for Research in Education. (1976). *The Burt Word Reading Test–1974 revision.* London: Hodder & Stroughton.

Share, D. L. (1984). Interpreting the output of multivariate analyses: A discussion of current approaches. *British Journal of Psychology, 75,* 349–362.

Smith, C. T. W., & Pearce, D. W. (1966). Testing spelling: Attainment norms and comparison for pupils from 9–13 years. *National Education, 1,* 117–120.

Stewart, M. A., Cummings, C., Singer, S., & deBlois, C. S. (1981). The overlap between hyperactive and unsocialized aggressive children. *Journal of Child Psychology and Psychiatry, 22,* 34–45.

Thorley, G. (1984). Hyperkinetic syndrome of childhood: Clinical characteristics. *British Journal of Psychiatry, 144,* 16–24.

Touliatos, J., & Lindholm, B. W. (1981). Congruence of parents' and teachers' ratings of children's behavior problems. *Journal of Abnormal Child Psychology, 9,* 347–354.

Trites, R. L., & Laprade, K. (1983). Evidence for an independent syndrome of hyperactivity. *Journal of Child Psychology and Psychiatry, 24,* 573–586.

Wechsler, D. (1974). *The Wechsler Intelligence Scale for Children–Revised.* New York: Psychological Corp.

Werry, J. S., & Hawthorne, D. (1976). Conners teacher questionnaire: Norms and validity. *Australian and New Zealand Journal of Psychiatry, 10,* 257–262.

Wilson, J. G., Silva, P. A., & Williams, S. (1981). An assessment of motor ability in seven year olds. *Journal of Human Movement Studies, 7,* 221–232.

CHAPTER SIX

Factor Analysis

CHAPTER PREVIEW

The multivariate statistical technique of factor analysis has found increased use during the past decade in the various fields of business-related research, especially marketing and personnel management. The purpose of this chapter is to describe factor analysis, a technique particularly suitable for analyzing the complex, multidimensional problems encountered by researchers and businesspeople. This chapter defines and explains in broad conceptual terms the fundamental aspects of factor analytic techniques. Basic guidelines for presenting and interpreting the results of these techniques also are included to clarify further the methodological concepts. Before proceeding further, it is helpful to review the key terms.

Factor analysis can be utilized to examine the underlying patterns or relationships for a large number of variables and determine if the information can be condensed or summarized in a smaller set of factors or components. An understanding of the most important factor analysis concepts should enable you to:

- Differentiate factor analytic techniques from other multivariate techniques.
- State the major purposes of factor analytic techniques.
- Identify the difference between component analysis and common factor analysis models.
- Tell when component analysis and common factor analysis should be utilized.
- Identify the difference between R and Q factor analysis.
- Explain the concept of rotation of factors.
- Tell how to determine the number of factors to extract.
- Explain the purpose of factor scores and how they can be used.
- Explain how to select surrogate variables for subsequent analysis.
- State the major limitations of factor analytic techniques.

KEY TERMS

Common factor analysis A factor model in which the factors are based upon a reduced correlation matrix. That is, communalities are inserted in the diagonal of the correlation matrix, and the extracted factors are based only on the common variance, with specific and error variance excluded.

Communality The amount of variance an original variable shares with all other variables included in the analysis.

Component analysis A factor model in which the factors are based upon the total variance. With component analysis, unities (1's) are used in the diagonal of the correlation matrix, which computationally implies that all of the variance is common or shared.

Correlation matrix A table showing the intercorrelations among all variables.

Eigenvalue The column sum of squares for a factor; also referred to as the *latent root*. It represents the amount of variance accounted for by a factor.

Factor A linear combination of the original variables. Factors also represent the underlying dimensions (constructs) that summarize or account for the original set of observed variables.

Factor loadings The correlation between the original variables and the factors, and the key to understanding the nature of a particular factor. *Squared factor loadings* indicate what percentage of the variance in an original variable is explained by a factor.

Factor matrix A table displaying the factor loadings of all variables on each factor.

Factor rotation The process of manipulating or adjusting the factor axes to achieve a simpler and pragmatically more meaningful factor solution.

Factor score Factor analysis reduces the original set of variables to a new smaller set of variables, or factors. When this new smaller set of variables (factors) is used in subsequent analysis (e.g., discriminant analysis), some measure or score must be included to represent the newly derived variables. This measure (score) is a composite of all of the original variables that were important in making the new factor. The composite measure is referred to as a *factor score*.

Oblique factor solutions A factor solution computed so that the extracted factors are correlated. Rather than arbitrarily constraining the factor solution so the factors are independent of each other, the analysis is conducted to express the relationship between the factors that may or may not be orthogonal.

Orthogonal Refers to the mathematical independence of factor axes to each other (i.e., at right angles or 90 degrees).

Orthogonal factor solutions A factor solution in which the factors are extracted so that the factor axes are maintained at 90 degrees. Thus, each factor is independent of or orthogonal from all other factors. The correlation between factors is arbitrarily determined to be zero.

Trace The sum of the square of the numbers on the diagonal of the correlation matrix used in the factor analysis, it represents the total amount of variance on which the factor solution is based. With component analysis, the trace is equal to the number of variables, based on the assumption that the variance in each variable is equal to 1. With common factor analysis, the trace is equal to the sum of the communalities on the diagonal of the reduced correlation matrix (also equal to the amount of common variance for the variables being analyzed).

What Is Factor Analysis?

Factor analysis is a generic name given to a class of multivariate statistical methods whose primary purpose is data reduction and summarization. Broadly speaking, it addresses itself to the problem of analyzing the interrelationships among a large number of variables (e.g., test scores, test items, questionnaire responses) and then explaining these variables in terms of their common underlying dimensions (factors). For example, a hypothetical survey questionnaire may consist of 100 questions, but since not all of the questions are identical, they do not all measure the basic underlying dimensions to the same extent. By using factor analysis, the analyst can identify the separate dimensions being measured by the survey and determine a factor loading for each variable (test item) on each factor.

Factor analysis (unlike multiple regression, discriminant analysis, or canonical correlation, in which one or more variables is explicitly considered the criterion or dependent variable and all others the predictor or independent variables) is an interdependence technique in which all variables are simultaneously considered. In a sense, each of the observed (original) variables is considered as a dependent variable that is a function of some underlying, latent, and hypothetical set of factors (dimensions). Conversely, one can look at each factor as a dependent variable that is a function of the originally observed variables.

Purposes of Factor Analysis

The general purpose of factor analytic techniques is to find a way of condensing (summarizing) the information contained in a number of original variables into a smaller set of new composite dimensions (factors) with á minimum loss of information; that is, to search for and define the fundamental constructs or dimensions assumed to underlie the original variables. More specifically, four functions factor analysis techniques can perform are as follows:

1. Identify a set of dimensions that are latent (not easily observed) in a large set of variables; this is also referred to as *R factor analysis*.
2. Devise a method of combining or condensing large numbers of people into distinctly different groups within a larger population; this is also referred to as *Q factor analysis*.
3. Identify appropriate variables for subsequent regression, correlation or discriminant analysis from a much larger set of variables (see the section "How to Select Surrogate Variables for Subsequent Analysis").
4. Create an entirely new set of a smaller number of variables to partially or completely replace the original set of variables for inclusion in subsequent regression, correlation or discriminant analysis (see the section "How to Use Factor Scores").

Approaches 1 and 2 take the identification of the underlying dimensions or factors as ends in themselves; the estimates of the factor loadings are all that is required for the analysis. Method 3 also relies on the factor loadings but uses them as the basis for identifying variables for subsequent analysis with other techniques. Method 4 requires that estimates of the factors themselves (factor scores) be obtained; then the factor scores are used as independent variables in a regression, discriminant, or correlation analysis.

Factor Analysis Decision Diagram

Figure 6.1 shows the general steps followed in any application of factor analysis techniques. The starting point in factor analysis, as with other statistical techniques, is the research problem. If the objective of the

FIGURE 6.1   Factor analysis decision diagram.

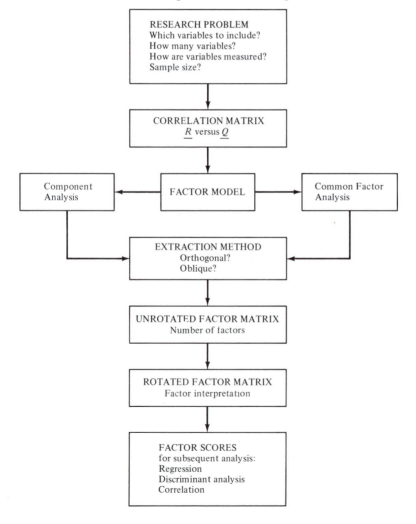

research is data reduction and summarization, factor analysis is the appropriate technique to use. Questions that the analyst needs to answer at this point are: what variables should be included, how many variables should be included, how are the variables measured, and is the sample size large enough? Regarding the question of variables, any variables relevant to the research problem can be included as long as they are appropriately measured. Raw data variables for factor analysis are generally assumed to be of metric measurement. In some cases, dummy variables (coded 0–1), although considered nonmetric, can be used. Regarding the sample size question, the researcher generally would not factor analyze a sample of fewer than 50 observations, and preferably the sample size should be 100 or larger. As a general rule, there should be four or five times as many observations as there are variables to be analyzed. This ratio is somewhat conservative, and in many instances the researcher is forced to factor analyze a set of variables when only a 2:1 ratio of observations to variables is available. When dealing with smaller sample sizes and a lower ratio, the analyst should interpret any findings cautiously.

One of the first decisions in the application of factor analysis involves the calculation of the correlation matrix. Based upon the research problem, the analyst must define the relevant universe for analysis. The alternative would be to examine either the correlations between the variables or the correlations between the respondents. For example, suppose you have data on 100 respondents in terms of 10 characteristics. It is possible to calculate the correlations between each of the 10 characteristics or between each of the individuals. If the objective of the research is to summarize the characteristics, the factor analysis would be applied to a correlation matrix of the variables. This is the most common type of factor analysis and is referred to as R *factor analysis*. It corresponds to the first purpose of factor analysis outlined previously. Factor analysis also may be applied to a correlation matrix of the individual respondents. This type of analysis, called Q *factor analysis*, was previously identified as a second possible purpose of factor analysis. A Q factor analysis approach is not utilized very frequently because of computational difficulties. Instead, most analysts utilize some type of cluster analysis or hierarchical grouping technique to group individual respondents.

Numerous variations of the general factor model are available. The two most frequently employed factor analytic approaches are component analysis[1] and common factor analysis. Selection of the factor model depends upon the analyst's objective. The component model is used when the objective is to summarize most of the original information (variance) in a minimum number of factors for prediction purposes. In contrast, common factor analysis is used primarily to identify underlying factors or dimensions not easily recognized. Both of these factor models will be discussed in more detail in the following sections.

---

[1] Many texts refer to this approach as *principal components*. For our purposes, component analysis is the same as principal components analysis.

In addition to selecting the factor model, the analyst must specify how the factors are to be extracted. Two options are available: orthogonal factors and oblique factors. In an orthogonal solution, the factors are extracted in such a way that the factor axes are maintained at 90 degrees, meaning that each factor is independent of all other factors. Therefore, the correlation between factors is arbitrarily determined to be zero. An oblique factor solution is more complex than an orthogonal one. In fact, an entirely satisfactory analytic procedure has not been devised for oblique solutions. They are still the subject of considerable experimentation and controversy. As the term *oblique* implies, the factor solution is computed so that the extracted factors are correlated. Oblique solutions assume that the original variables or characteristics are correlated to some extent; therefore, the underlying factors must be similarly correlated. To summarize, orthogonal factor solutions are mathematically simpler to handle, while oblique factor solutions are more flexible and more realistic, because the theoretically important underlying dimensions are not assumed to be unrelated to each other.

The choice of an orthogonal or oblique rotation should be made on the basis of the particular needs of a given research problem. If the goal of the research is to reduce the number of original variables, regardless of how meaningful the resulting factors may be, the appropriate solution would be an orthogonal one. Also, if the researcher wants to reduce a larger number of variables to a smaller set of uncorrelated variables for subsequent use in a regression or other prediction technique, an orthogonal solution is best. However, if the ultimate goal of the factor analysis is to obtain several theoretically meaningful factors or constructs, an oblique solution is appropriate. This is because, realistically, very few variables are uncorrelated, as in an orthogonal solution.

When a decision has been made on the correlation matrix, the factor model, and the extraction method, the analyst is ready to extract the initial unrotated factors. By examining the unrotated factor matrix, the analyst can explore the data reduction possibilities for a set of variables and obtain a preliminary estimate of the number of factors to extract. Final determination of the number of factors must wait, however, until the factor matrix is rotated and the factors are interpreted.

Depending upon the reason for applying factor analysis techniques, the research may stop with factor interpretation or proceed to the computation of factor scores and subsequent analysis with other statistical techniques. If the objective is simply to identify logical combinations of variables or respondents (purposes 1 and 2), the analyst will stop with the factor interpretation. If the objective is to identify appropriate variables for subsequent application to other statistical techniques (purpose 3), the analyst would examine the factor matrix and select the variable with the highest factor loading as a surrogate representative for a particular factor dimension. If the objective is to create an entirely new set of a smaller number of variables to replace the original set of variables for inclusion in a subsequent type of statistical analysis (purpose 4), composite factor scores would be computed to represent each of the factors. The factor scores would then be used as the raw data to represent the independent variables in a regression, discriminant, or correlation analysis.

Approaches for Deriving the Correlation Matrix

As noted previously, one of the first decisions in the application of factor analysis focuses on the approach to use in calculating the correlation matrix. The analyst could derive the correlation matrix based on the computation of correlations between the variables. This would be an R-type factor analysis, and the result would be a factor pattern demonstrating the underlying relationships of the variables. The analyst could also elect to derive the correlation matrix based on the correlations between the individual respondents. This is Q-type factor analysis, and the results would be a factor matrix that would identify similar individuals. For example, if the individual respondents are identified by number, the resulting factor pattern might tell you that individuals 1, 5, 7, and 10 are similar. These respondents would be grouped together because they exhibited a high loading on the same factor. Similarly, respondents 2, 3, 4, and 8 would perhaps load together on another factor. We would label these individuals as being similar. From the results of a Q factor analysis, we could identify groups or clusters of individuals demonstrating a similar response pattern on the variables included in the analysis.

A logical question at this point would be, how does Q-type factor analysis differ from cluster analysis? The answer is that both approaches compare a series of responses to a number of variables and place the respondents into several groups. The difference is that the resulting groups for a Q-type factor analysis would be based on the intercorrelations between the means and standard deviations of the respondents. In a typical cluster analysis approach, groupings would be devised based on a distance measure between the respondents' scores on the variables being analyzed [10]. To illustrate this difference, consider Table 6.1. This table contains the scores of four respondents over three

TABLE 6.1 Comparisons of Score Profiles for Q-Type Factor Analysis and Hierarchical Cluster Analysis

Respondent	Variables 1	2	3
A	7	6	7
B	6	7	6
C	4	3	4
D	3	4	3

different variables. A Q-type factor analysis of these four respondents would yield two groups with similar variance structures. The two groups would consist of respondents A and C versus B and D. In contrast, the clustering approach would be sensitive to the distances among the respondents' scores and would lead to a grouping of the closest pairs. Thus, with a cluster analysis approach, respondents A and B would be placed in one group and C and D in the other group.

Common Factor Analysis and Component Analysis

There are two basic models that the analyst can utilize to obtain factor solutions. They are known as *common factor analysis* and *component analysis*. To select the appropriate model, the analyst must understand something about the types of variance. For the purposes of factor analysis, total variance consists of three kinds: (1) common, (2) specific, and (3) error. These types of variance and their relationship to the factor model selection process are illustrated in Figure 6.2. *Common variance* is defined as that variance in a variable that is shared with all other variables in the analysis. *Specific variance* is that variance associated with only a specific variable. *Error variance* is that due to unreliability in the data-gathering process or a random component in the measured phenomenon. When using component analysis, the total variance is considered and hybrid factors are derived that contain small proportions of unique and in some instances error variance, but not enough in the first few factors to distort the overall factor structure. Specifically, with component analysis, unities are inserted in the diagonal of the correlation matrix. Conversely, with common factor analysis, communalities are inserted in the diagonal, and the factors are derived based only on the common variance. From a variance point of view, there is a big difference between inserting unity in the diagonal and using communality estimates. With unity in the diagonal, the full variance is brought into the factor matrix, as shown in Figure 6.2. Common

FIGURE 6.2   Types of variance carried into factor matrix.

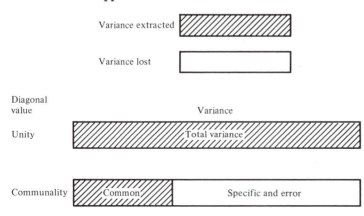

| Variance extracted | |
| Variance lost | |

Diagonal value	Variance
Unity	Total variance
Communality	Common / Specific and error

factor analysis substitutes communality estimates in the diagonal, and the resulting factor solution is based only on common variance.

The common factor and component analysis models are both widely utilized. The selection of one model over the other is based upon two criteria: (1) the objective of the researcher conducting the factor analysis and (2) the amount of prior knowledge about the variance in the variables. When the analyst is primarily concerned about prediction, determining the minimum number of factors needed to account for the maximum portion of the variance represented in the original set of variables, and has prior knowledge suggesting that unique and error variance represent a relatively small proportion of the total variance, the appropriate model to select is the component analysis model. In contrast, when the primary objective is to identify the latent dimensions or constructs represented in the original variables, and the researcher has little knowledge about the amount of unique or error variance and therefore wishes to eliminate this variance, the appropriate model to select is the common factor model. It is important to note at this point that either model can be selected and applied with relative ease. This is made possible by computers that derive good approximations of communalities through repeated calculations.

The Rotation of Factors

An important concept in factor analysis is the rotation of factors. The term *rotation* in factor analysis means exactly what it implies. Specifically, the reference axes of the factors are turned about the origin until some other position has been reached. The simplest case is an orthogonal rotation in which the axes are maintained at 90 degrees. It is possible to rotate the axes and not retain the 90 degree angle between the reference axes. This rotational procedure is referred to as an *oblique rotation*. Orthogonal and oblique factor rotations are demonstrated using a graphic approach in Figures 6.3 and 6.4.

As was pointed out in Figure 6.1, two stages are involved in the derivation of a final factor solution. First, the initial unrotated factor matrix is computed to assist in obtaining a preliminary indication of the number of factors to extract. In computing the unrotated factor matrix, the analyst is simply interested in the best linear combination of variables—best in the sense that the particular combination of original variables would account for more of the variance in the data as a whole than any other linear combination of variables. Therefore, the first factor may be viewed as the single best summary of linear relationships exhibited in the data. The second factor is defined as the second best linear combination of the variables subject to the constraint that it is orthogonal to the first factor. To be orthogonal to the first factor, the second one must be derived from the proportion of the variance remaining after the first factor has been extracted. Thus, the second factor may be defined as the linear combination of variables that accounts for the most residual variance after the effect of the first factor

FIGURE 6.3   Orthogonal factor rotation.

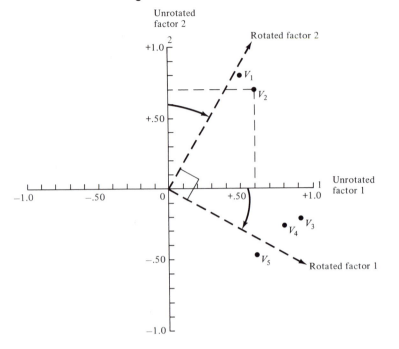

FIGURE 6.4   Oblique factor rotation.

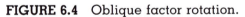

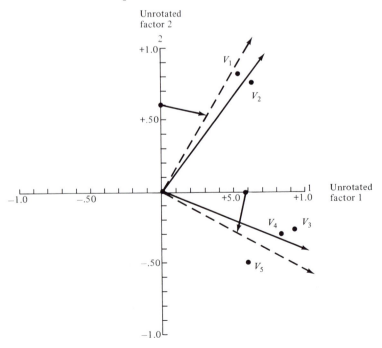

is removed from the data. Subsequent factors are defined similarly until all the variance in the data is exhausted.

Unrotated factor solutions achieve the objective of data reduction, but the analyst must ask if the unrotated factor solution (while fulfilling desirable mathematical requirements) will provide information that offers the most adequate interpretation of the variables under examination. In most instances the answer to this question will be no. Therefore, the basic reason for employing a rotational method is to achieve simpler and theoretically more meaningful factor solutions. Rotation of the factors in most cases improves the interpretation by reducing some of the ambiguities that often accompany initial unrotated factor solutions.

The unrotated factor solution may or may not provide a meaningful patterning of variables. If the unrotated factors are expected to be meaningful, the user may specify that no rotation be performed. Generally, rotation will be desirable because it simplifies the factor structure and because it is usually difficult to determine whether unrotated factors will be meaningful or not.

As indicated earlier, unrotated factor solutions extract factors in the order of their importance. The first factor tends to be a general factor with almost every variable loading significantly, and it accounts for the largest amount of variance. The second and subsequent factors are then based upon the residual amount of variance. Each accounts for successively smaller portions of variance. The ultimate effect of rotating the factor matrix is to redistribute the variance from earlier factors to later ones to achieve a simpler, theoretically more meaningful, factor pattern.

To illustrate the concept of factor rotation, examine Figure 6.3, in which five variables are depicted in a two-dimensional factor diagram. The vertical axis represents the unrotated factor II and the horizontal axis represents unrotated factor I. The axes are labeled with a 0 at the origin and extending outward up to a $+1.0$ or a -1.0. The numbers on the axes represent the factor loadings. The variables are labeled V_1, V_2, V_3, V_4, and V_5. The factor loading for variable 2 on the unrotated factor II would be determined by drawing a dashed line horizontally to the vertical axis for factor II. Similarly, a vertical would be drawn from variable 2 to the horizontal axis of the unrotated factor I in order to determine the loading of variable 2 on factor I. A similar procedure would be followed for the remaining variables until all the loadings are determined for all factor variables. The factor loadings for the unrotated and rotated solutions are displayed in Table 6.2 for comparison purposes. On the unrotated first factor, all of the variables load fairly high. On the unrotated second factor, variables 1 and 2 are very high in the positive direction. Variable 5 is moderately high in the negative direction, while variables 3 and 4 have considerably lower loadings in the negative direction.

From visual inspection of Figure 6.3, it is obvious that there are two clusters of variables. Variables 1 and 2 go together, as well as variables 3, 4, and 5. However, such patterning of variables is not so

		Unrotated Factor Loadings		Rotated Factor Loadings	
TABLE 6.2 Comparison Between Rotated and Unrotated Factor Loadings	**Variable**	I	II	I	II
	V_1	.50	.80	.03	.94
	V_2	.60	.70	.16	.90
	V_3	.90	−.25	.95	.24
	V_4	.80	−.30	.84	.15
	V_5	.60	−.50	.76	−.13

obvious from the unrotated factor loadings. By rotating the original axes clockwise, as indicated in Figure 6.3, we obtain a completely different factor loading pattern. Note that in rotating the factors, the axes are maintained at 90 degrees. This signifies that the factors are mathematically independent and that the rotation has been orthogonal. After rotating the factor axes, variables 3, 4, and 5 load very high on factor I, and variables 1 and 2 load very high on factor II. Thus, the clustering or patterning of these variables into two groups is more obvious after the rotation than before, even though the relative position or configuration of the variables remains unchanged.

The same general principles pertain to oblique as to orthogonal rotations. The oblique rotation method is more flexible because the factor axes need not be orthogonal. It also is more realistic because the theoretically important underlying dimensions are not assumed to be uncorrelated with each other. In Figure 6.4 the two rotational methods are compared. Note that the oblique factor rotation represents the clustering of variables more accurately. This is because each rotated factor axis is now closer to the respective group of variables. Also, the oblique solution provides us with information about the extent to which the factors are actually correlated with each other.

Most factor analysts agree that many direct unrotated solutions are not sufficient. That is, in most cases rotation will improve the interpretation by reducing some of the ambiguities that often accompany the preliminary analysis. The major option available to the analyst in rotation is to choose an orthogonal or an oblique method. The ultimate goal of any rotation is to obtain some theoretically meaningful factors, and, if possible, the simplest factor structure. Orthogonal rotational approaches are more widely used because all computer packages performing factor analysis contain orthogonal rotation options. Only a few computer packages contain the oblique rotational option. Orthogonal rotations are also utilized more frequently because the analytical procedures for performing oblique rotations are not as well developed and are still subject to considerable controversy. When the objective is to utilize the factor results in a subsequent statistical analysis, the analyst should always select an orthogonal rotation procedure. This is because the factors are orthogonal and therefore eliminate collinearity. However, if the analyst is simply interested in obtaining theoretically meaningful constructs or dimensions, the oblique factor rotation is more desirable because it is theoretically and empirically more realistic.

One final topic needs to be discussed regarding the rotation of factors. Several different approaches are available for performing either orthogonal or oblique rotations. Only a limited number of oblique rotational procedures are available and the analyst will probably be forced to accept the one that is accessible. Since none of the oblique solutions have been demonstrated to be analytically superior, no further comment will be made on oblique rotational methods. Rather, the focus will be on orthogonal approaches.

In practice, the objective of all methods of rotation is to simplify the rows and/or columns of the factor matrix to facilitate interpretation. By simplifying the rows, we mean making as many values in each row as close to zero as possible. By simplifying the columns, we mean making as many values in each column as close to zero as possible. Three major orthogonal approaches have been developed. They are QUARTIMAX, VARIMAX and EQUIMAX. The ultimate goal of a QUARTIMAX rotation is to simplify the rows of a factor matrix. That is, it focuses on rotating the initial factor so that a variable loads high on one factor and as low as possible on all other factors. In contrast to QUARTIMAX, the VARIMAX criterion centers on simplifying the columns of the factor matrix. Note that in QUARTIMAX approaches many variables can load high or near high on the same factor because the technique centers on simplifying the rows. With the VARIMAX rotational approach, the maximum possible simplification is reached if there are only 1's and 0's in a single column. The EQUIMAX approach is a compromise between the QUARTIMAX and VARIMAX criteria. Rather than concentrating either on simplification of the rows or on simplification of the columns, it tries to accomplish some of each; thus the name EQUIMAX is used for this approach.

The QUARTIMAX method has not proved very successful in producing simpler structures. Its difficulty is that it tends to produce a general factor in the rotations. Regardless of one's concept of a "simpler" structure, inevitably it involves dealing with clusters of variables; a method that tends to create a large general factor (i.e., QUARTIMAX) is not in line with the goals of rotation.

In contrast to QUARTIMAX, the VARIMAX criterion centers on simplifying the columns of the factor matrix. That is, the VARIMAX method maximizes the sum of variances of required loadings of the factor matrix. Recall that in QUARTIMAX approaches many variables can load high or near high on the same factor because the technique centers on simplifying the rows. With the VARIMAX rotational approach, there tends to be some high loadings (i.e., close to -1 or $+1$) and some loadings near 0 in each column of the matrix. The logic is that interpretation is easiest when the variable-factor correlations are either close to $+1$ or -1, thus indicating a clear association between the variable and the factor, or close to 0, indicating a clear lack of association. This indicates the fundamental aspect of a simple structure.

Although the QUARTIMAX solution is analytically simpler than the VARIMAX solution, VARIMAX seems to give a clearer separation of the factors. In general, Kaiser's experiment [2] indicates that the factor pattern obtained by VARIMAX rotation tends to be more invariant than that obtained by the QUARTIMAX method when different subsets of variables are analyzed.

The VARIMAX method has proved very successful as an analytic approach to obtaining an orthogonal rotation of factors. Even in those cases where the results do not meet the analyst's concept of a simple structure, the solution is close enough to reduce greatly the labor of finding a satisfactory rotation.

No specific rules have been developed to guide the analyst in selecting a particular orthogonal rotational technique. In most instances, the analyst will simply utilize the rotational technique that is a standard output of the computer program used. Most programs have only a single rotational option, usually VARIMAX. Thus, the VARIMAX method of rotation is the most widely utilized. However, there is no compelling analytical reason to favor one rotational method over another. Whenever possible, the choice should be made on the basis of the particular needs of a given research problem.

Criteria for the Number of Factors to be Extracted

How do we decide on the number of factors to extract? When a large set of variables is factored, the analysis will extract the largest and best combinations of variables first, and then proceed to smaller, less understandable combinations. In deciding when to stop factoring (that is, how many factors to extract), the analyst generally will begin with some predetermined criterion, such as the a priori or the latent root criterion, to arrive at a specific number of factors to extract (these two techniques will be discussed in more detail later). After the initial solution has been derived, the analyst will make several additional trial rotations—usually one less factor than the initial number and two or three more factors than were initially derived. Then, on the basis of information contained in the results of these several trial analyses, the factor matrices will be examined and the best representation of the data will be used to assist in determining the number of factors to extract. To use an analogy, choosing the number of factors to be interpreted is something like focusing a microscope. Too high or too low an adjustment will obscure a structure that is obvious when the adjustment is just right. Therefore, by examining a number of different factor structures derived from several trial rotations, the analyst can compare and contrast to arrive at the best representation of the data.

An exact quantitative basis for deciding the number of factors to extract has not been developed. However, the following stopping criteria for the number of factors to extract are currently being utilized.

LATENT ROOT CRITERION.   The most commonly used technique is referred to as the *latent root criterion*. This rule is very simple to apply. But it does differ depending on whether the analyst has chosen either component analysis or common factor analysis as the basic model. Recall that in component analysis 1's are inserted in the diagonal of the correlation matrix and the entire variance is considered in the analysis. In component analysis only the factors having latent roots (eigenvalues) greater than 1 are considered significant; all factors with latent roots less than 1 are considered insignificant and disregarded.

Many factor analysts utilize only the eigenvalue 1 criterion. However, when the common factor model is selected, the eigenvalue 1 criterion should be adjusted slightly downward. With the common factor model, the eigenvalue cutoff level should be lower and approximate either the estimate for the common variance of the set of variables or the average of the communality estimates for all variables. In fact, some analysts contend that any positive eigenvalue obtained in a common factor analysis indicates that the factor qualifies for examination.

The rationale for the eigenvalue criterion is that any individual factor should account for at least the variance of a single variable if it is to be retained for interpretation. The eigenvalue approach is probably most reliable when the number of variables is between 20 and 50. In instances where the number of variables is less than 20, there is somewhat of a tendency for this method to extract a conservative number of factors. When more than 50 variables are involved, however, it is not uncommon for too many factors to be extracted.

A PRIORI CRITERION.   The *a priori criterion* is a simple yet reasonable criterion under certain circumstances. When applying it, the analyst already knows how many factors to extract before undertaking the factor analysis. The analyst simply instructs the computer to stop the analysis when the desired number of factors has been extracted. This approach is useful if the analyst is testing a theory or hypothesis about the number of factors to be extracted. It also can be justified in instances where the analyst is attempting to replicate another researcher's work and extract exactly the same number of factors that was previously found.

PERCENTAGE OF VARIANCE CRITERION.   The *percentage of variance criterion* is another approach. Using it, the cumulative percentages of the variance extracted by successive factors is the criterion. No absolute cutting line has been adopted for all data. However, in the hard sciences the factoring procedure usually should not be stopped until the extracted factors account for at least 95 percent of the variance or the last factor accounts for only a small portion (less than 5 percent). In contrast, in the social sciences, where information is often less precise, it is not uncommon for the analyst to consider a solution that accounts for 60 percent of the total variance (and in some instances even less) as a satisfactory solution.

SCREE TEST CRITERION.   Recall that with the component analysis factor model, the later factors extracted contain both common and unique variance. While all factors contain at least some unique variance, the proportion of unique variance in later factors is substantially higher than in earlier factors. The *scree tail test* is an approach used to identify the optimum number of factors that can be extracted before the amount of unique variance begins to dominate the common variance structure [1]. The scree test is derived by plotting the latent roots against the number of factors in their order of extraction, and the shape of the resulting curve is used to evaluate the cutoff point. Figure 6.5 plots the first 18 factors extracted in a study by the authors. Starting with the first factor, the plot slopes steeply down initially and then slowly becomes an approximately horizontal line. The point at which the curve first begins to straighten out is considered to indicate the maximum number of factors to extract. In the present case, the first 10 factors would qualify. Beyond 10, too large a proportion of unique variance would be included; thus, these factors would not be acceptable. Note that in using the latent root criterion, only eight factors would have been considered. In contrast, using the scree test provides us with two more factors. As a general rule, the scree tail test will result in at least one and sometimes two or three more factors being considered as significant than will the latent root criterion [1].

In practice, most factor analysts seldom use a single criterion in determining how many factors to extract. Instead, they initially use a criterion such as the latent root as a guideline for the first rotation. Then several additional trial rotations are undertaken, and by considering the initial criterion and comparing the factor interpretations for several different trial rotations, the analyst can select the number of factors to extract based upon the initial criterion and the factor structure

FIGURE 6.5 Eigenvalue plot for scree test criterion.

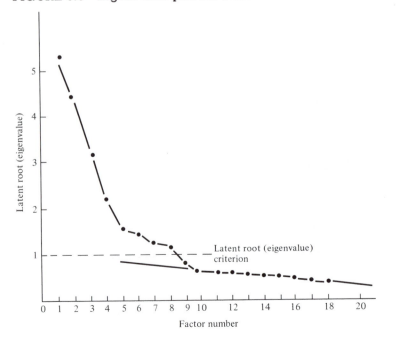

that best represents the underlying relationship of the variables. In short, the ability to assign some meaning to the factors, or to interpret the nature of the variables, becomes an extremely important consideration in determining the number of factors to extract.

Criteria for the Significance of Factor Loadings

In interpreting factors, a decision must be made regarding which factor loadings are worth considering.

1. The first suggestion is not based on any mathematical proposition. It is a rule of thumb that has been used frequently by factor analysts as a means of making a preliminary examination of the factor matrix. In short, factor loadings greater than $\pm$ .30 are considered significant. Loadings $\pm$ .40 are considered more important, and if the loadings are $\pm$ .50 or greater, they are considered very significant. Thus, the larger the absolute size of the factor loading, the more significant the loading is in interpreting the factor matrix. These guidelines are considered useful when the sample size is 50 or larger. This approach may appear too simplistic, yet compared with other criteria it is quite rigorous and acceptable.
2. As pointed out previously, a factor loading represents the correlation between an original variable and its factor. In determining a significance level for interpretation of loadings, an approach could be used that is similar to that of interpreting correlation coefficients. Specifically, loadings of at least $\pm$ .19 and $\pm$ .26 are recommended for the 5 and 1 percent levels, respectively, when the sample size is 100. When the sample size is 200, $\pm$ .14 and $\pm$ .18 are recommended for the 5 and 1 percent levels of significance. Finally when the sample size is at least 300, loadings of $\pm$ .11 and $\pm$ .15 are recommended for the 5 and 1 percent levels, respectively. Since it is difficult to assess the amount of error involved in factor analytic studies, it is probably safer to adopt the 1 percent level as the criterion for significance.
3. A disadvantage of methods 1 and 2 is that the number of variables being analyzed and the specific factor being examined are not considered. It has been shown that as the analyst moves from the first factor to later factors, the acceptable level for a loading to be judged significant should increase. The fact that unique variance and error variance begin to enter in later factors means that some upward adjustment in the level of significance should be included [2].
   The number of variables being analyzed is also important in deciding which loadings are significant. As the number of variables being analyzed increases, the acceptable level for considering a loading significant decreases. Adjustment for the number of variables is particularly true as one moves from the first factor extracted to later factors. Specifically, when the sample size is 50 and the desired significance level is .05, the following guidelines are applicable: (1) a significant loading on the fifth factor with 20 variables would

be ± .292, whereas with 50 variables a significant loading would
be ± .267; (2) a significant loading on the tenth factor with 20 variables would be ± .353, but with 50 variables it would be only ±
.274. Similar guidelines can be given when the sample size is 100
and the significance level is .05: (1) a significant loading on the
fifth factor with 20 variables would be ± .216, but with 50 variables
it would drop to ± .202; (2) a significant loading on the tenth factor
with 20 variables would be ± .261, but only ± .214 with 50 variables.

To summarize the criteria for the significance of factor loadings,
the following guidelines can be stated: (1) the larger the sample size,
the smaller the loading to be considered significant; (2) the larger the
number of variables being analyzed, the smaller the loading to be considered significant; (3) the larger the number of factors, the larger the
size of the loading on later factors to be considered significant for
interpretation.

Interpreting a Factor Matrix

Interpreting the complex interrelationships represented in a factor matrix is no simple matter. By following the procedure outlined in the
following paragraphs, however, the factor interpretation procedure can
be simplified considerably.

1. Examine the factor matrix. Each column of numbers represents
a separate factor. The columns of numbers are the factor loadings for
each variable on each factor. For identification purposes, the computer
printout will usually identify the factors from left to right by the numbers
1, 2, 3, 4, and so forth. It also will identify the variables by number
from top to bottom. To further facilitate interpretation, the analyst
should write the name of each variable in the left margin beside the
variable numbers.

2. To begin the interpretation, the analyst should start with the
first variable on the first factor and move horizontally from left to
right, looking for the *highest loading* for that variable on any factor.
When the highest loading (largest absolute factor loading) is identified,
if it is significant the analyst should underline it. The analyst should
then go to the second variable and, again moving from left to right
horizontally, look for the highest loading for that variable on any factor
and underline it. This procedure should be continued for each variable
until all variables have been underlined once for their highest loading
on a factor. Recall that for sample sizes of less than 100, the lowest
factor loading to be considered as significant would in most instances
be ± .30.

It should be noted that the process of underlining only the single
highest loading as significant for each variable is an ideal that the
analyst should strive for but can seldom achieve. When each variable
has only one loading on one factor that is considered significant, the
interpretation of the meaning of each factor is simplified considerably.

In practice, many variables will have several moderate-sized loadings, all of which are significant, and the job of interpreting the factors is much more difficult. This is because a variable with several significant loadings must be considered in interpreting (labeling) all the factors on which it has a significant loading. Since most factor solutions do not result in a simple structure solution (a single high loading for each variable on only one factor), the analyst will, after underlining the highest loading for a variable, continue to evaluate the factor matrix by underlining all significant loadings for a variable on all the factors. Ultimately, the analyst tries to minimize the number of significant loadings on each row of the factor matrix (that is, the loadings associated with one variable) and to maximize the number of loadings with negligible values.

3. Once all the variables have been underlined on their respective factors, the analyst should examine the factor matrix to identify variables that have not been underlined and therefore do not load on any factor. If there are variables that do not load on any factor, the analyst has two options: (1) interpret the solution as it is and simply ignore those variables without a significant loading or (2) critically evaluate each of the variables that do not load significantly on any factor. This evaluation would be in terms of the variable's overall contribution to the research as well as its communality index. If the variable(s) is of minor importance to the study's objective and/or has a low communality index, the analyst may decide to eliminate the variable or variables and derive a new factor solution with the nonloading variables eliminated.

4. When a factor solution has been obtained in which all significant variables are loading on a factor, the analyst attempts to assign some meaning to the pattern of factor loadings. Variables with higher loadings are considered more important in this stage of factor interpretation. They greatly influence the name or label selected to represent a factor. Thus, the analyst will examine all the underlined variables for a particular factor and, placing greater emphasis on those variables with higher loadings, will attempt to assign a name or label to a factor that accurately reflects to the greatest extent possible what the several variables loading on that factor represent. It is important to note that this label is not derived or assigned by the factor analysis computer program, but rather is intuitively developed by the factor analyst based upon its appropriateness for representing the underlying dimensions of a particular factor. This procedure is followed for each of the extracted factors. The final result will be a name or label that represents each of the derived factors as accurately as possible.

In some instances, it is not possible to assign a name to each of the factors. When such a situation is encountered, the analyst may wish to use the label undefined to represent a particular factor or factors derived by that solution. In such cases, the analyst interprets only those factors that are meaningful and disregards undefined or less meaningful ones. It is important to note, however, that in describing the factor solution, the factor analyst indicates that these factors were derived but were undefinable, and only those factors representing meaningful relationships were interpreted.

An Illustrative Example

In the preceding sections, the major empirical questions concerning the application of factor analysis have been covered. To clarify these topics further, an illustrative example of the application of factor analysis is presented based upon data from the data bank presented in Chapter One. The first six variables from the data bank were selected to conduct the factor analysis. Here we will take the reader through a step-by-step application and interpretation of a component analysis and a common factor analysis of the same six variables. The rotational approach will be orthogonal and VARIMAX for both factor models. First, we will consider component analysis. Then we will look at common factor analysis.

Component Analysis

As noted earlier, factor analysis procedures are based upon the initial computation of a complete table of intercorrelations among the variables (correlation matrix). This correlation matrix is then transformed to obtain a factor matrix. The correlation matrix provides an initial indication of the relationships among the variables. However, interpretation of the correlation matrix still involves the examination of numerous relationships and can be extremely difficult. For example, even 20 variables yield a correlation matrix containing 210 separate entries. Since such a table of numbers is frequently too large to grasp and interpret effectively, the question arises of whether it might be possible to develop an even more condensed arrangement that will represent the underlying order in the data better. In many instances, the answer to this question is yes. Through the application of factor analytic techniques, it is frequently possible to reduce the correlation matrix to a smaller set of relationships—the factor matrix. Of course, this assumes that a certain degree of underlying order exists in the data being analyzed.

Table 6.3 shows the correlation matrix for the six variables drawn from the data bank. From inspection of the correlation matrix, we can see that six variables are related at the .38 level or above (under-

TABLE 6.3		Correlations Among Variables					
Component Analysis	Variable	1	2	3	4	5	6
Correlation Matrix	X_1 Self-esteem	1.00	−.38	.51	.02	.61	.04
	X_2 Locus of control		1.00	−.50	.23	.48	.14
	X_3 Alienation			1.00	−.13	.06	−.05
	X_4 Social responsibility				1.00	.24	.77
	X_5 Machiavellianism					1.00	.16
	X_6 Political opinion						1.00

lined). But it is difficult to derive a complete and clear understanding of their relationships. From a factor analysis of these variables, it should be possible to do so.

The result of the first stage in the computation of factors is shown in Table 6.4, the unrotated component analysis factor matrix. To begin the analysis, let's explain the numbers included in the table. Four columns of numbers are shown. The first three are the results for the three factors that are extracted. The fourth column provides summary statistics covering all of the factors in this particular factor solution. The matrix of factor loadings consists of the three columns of numbers beside the six variables. The numbers at the bottom of each column are the *column sum of squared factor loadings* and the percentage of trace. The column sum of squared factor loadings (eigenvalue) shown at the bottom of each column of factor loadings indicates the relative importance of each factor in accounting for the variance associated with the set of variables being analyzed. Note that the sums of squares for factors 1, 2, and 3 are 2.12, 1.91, and 1.28, respectively. As expected, the unrotated factor solution has extracted the factors in the order of their importance, with factor 1 accounting for the most variance, factor 2 for slightly less, and factor 3 for the least amount of variance. At the far right-hand side of the row of sums of squares is the number 5.31, which represents the total sum of squares. The total sum of squared factor loadings is obtained by adding the individual sums of squares for each of the factors. It represents the total amount of variance extracted by the factor solution.

The *percentage of trace* for each of the three factors is also shown at the bottom of the table. The percentages of trace for factors 1, 2, and 3 are 35.33, 31.83, and 21.33, respectively. The percent of trace is obtained by dividing each factor's sum of squares by the trace for the set of variables being analyzed. For example, if the sum of squares of 2.12 for factor 1 is divided by the trace of 6.0, the result will be the percentage of trace, or 35.33 percent for factor 1. By adding the percentages of trace for each of the three factors together, we obtain the total percentage of trace extracted for the factor solution. The total percentage of trace can be used as an index to determine how well a

TABLE 6.4
Unrotated Component Analysis Factor Matrix

Variables	1	2	3	Communality
X_1 Self-esteem	−.19	.92	.20	.92
X_2 Locus of control	.73	−.39	.54	.91
X_3 Alienation	−.50	.65	−.19	.71
X_4 Social responsibility	.78	.28	−.45	.89
X_5 Machiavellianism	.45	.62	.64	.99
X_6 Political opinion	.70	.31	−.55	.89
Sum of squares (eigenvalue)	2.12	1.91	1.28	5.31
Trace*	35.33	31.83	21.33	88.50

*Trace = 6.0.

particular factor solution accounts for what all the variables together represent. If the variables are all very different from each other, this index will be low. If the variables fall into one or more highly redundant or related groups, and if the extracted factors account for all the groups, the index will approach 100. The index for the present solution shows that 88.50 percent of the total variance is represented by the information contained in the factor matrix. Therefore, the index for this solution is high and the variables are in fact highly related to each other.

The *row sum of squared factor loadings* is shown at the far right side of the table. These figures, referred to in the table as *communalities*, show the amount of variance in a variable that is accounted for by the three factors taken together. The size of the communality is a useful index for assessing how much variance in a particular variable is accounted for by the factor solution. Large communalities indicate that a large amount of the variance in a variable has been extracted by the factor solution. Small communalities show that a substantial portion of the variance in a variable is unaccounted for by the factors. For instance, the communality figure of .71 for variable X_3 indicates that it has less in common with the other variables included in the analysis than does variable X_5, which has a communality of .99.

Having defined the various elements of the unrotated factor matrix, let's examine the factor loading patterns. As anticipated, the first factor accounts for the largest amount of variance and is a general factor, with every variable except X_1 loading significantly. Also, based upon the sample size, the number of variables, and the factor number, variable X_2 loads on all three factors significantly. Variable X_3 loads on factors 1 and 2, variable X_4 loads on factors 1 and 3, variable X_5 loads on all three factors, and variable X_6 loads on all factors. Based on this factor loading pattern, interpretation would be extremely difficult and theoretically less meaningful. Therefore the analyst should proceed to rotate the factor matrix to redistribute the variance from the earlier factors to the later factors. This should result in a simpler and theoretically more meaningful factor pattern.

The VARIMAX rotated component analysis factor matrix is shown in Table 6.5. Note that the total amount of variance extracted is the

TABLE 6.5
VARIMAX Rotated Component Analysis Factor Matrix

Variables	Factor 1	Factor 2	Factor 3	Communality
X_1 Self-esteem	.64	.02	.71	.92
X_2 Locus of control	−.87	.11	.36	.91
X_3 Alienation	.83	−.05	.18	.71
X_4 Social responsibility	−.12	.93	.11	.89
X_5 Machiavellianism	−.13	.13	.98	.99
X_6 Political opinion	−.005	.94	.04	.89
Sum of squares (eigenvalue)	1.89	1.78	1.64	5.31
Trace*	31.50	29.67	27.33	88.50

* Trace = 6.00.

same in the rotated solution as it was in the unrotated one—88.50. Two major differences are obvious, however. First, the variance has been redistributed so that the factor loading pattern is different, and the percentage of variance for each of the factors is different also. Second, factor 1 in the unrotated solution is now factor 2 in the rotated solution because it accounts for less variance, and factor 2 is now factor 1 because it accounts for more variance (recall that factors are extracted so that the first factor accounts for the most variance, the second the next most, and so on). Specifically, in the VARIMAX rotated factor solution, the first factor accounts for 31.50 percent of the variance but the second factor accounts for 29.67 percent. The third factor accounts for 27.33 percent of the variance. Also, recall that in the unrotated factor solution, all variables except X_1 loaded significantly on the first factor. In the rotated factor solution, however, variables X_4 and X_6 load significantly on factor 2, variables X_1, X_2, and X_3 load significantly on factor 1, and variables X_1 and X_5 load significantly on factor 3. The only variable that loads significantly on two factors is X_1, which loads on factors 1 and 3. To some extent, the analyst may consider variable X_2 as also loading significantly on two factors—1 and 3. The loading of .36 for variable X_2 on factor 3 could be considered significant. But since the difference between .36 and the next highest loading of .64 for variable X_1 on factor 1 is so great, the analyst probably would consider this differential as too large and the loading for variable X_2 would not be judged as significant. It should be apparent that factor interpretation has been simplified considerably by rotating the factor matrix.

Common Factor Analysis

Common factor analysis is one of the two major factor analytic models that will be discussed here. The difference between component analysis and common factor analysis is that the latter considers only the common variance associated with a set of variables. This is accomplished by factoring a "reduced" correlation matrix with communalities in the diagonal instead of unities. Comparison of the numbers on the diagonal of the common factor correlation matrix (Table 6.6) with the numbers shown on the diagonal of the component analysis correlation matrix (Table 6.3) will demonstrate the difference between the correlation

TABLE 6.6 Common Factor Analysis Correlation Matrix	Variable	Correlations Among Variables					
		1	2	3	4	5	6
	X_1 Self-esteem	.97	−.35	.51	.04	.59	.03
	X_2 Locus of control		.97	−.48	.25	.46	.13
	X_3 Alienation			.38	−.08	.05	−.07
	X_4 Social responsibility				.63	.22	.48
	X_5 Machiavellianism					.98	.17
	X_6 Political opinion						.61

matrix for the two factor models. The numbers on the diagonal of the component analysis correlation matrix are all 1's, whereas for the common factor model all the numbers on the diagonal are less than 1.

Examination of the sizes of the communalities for each variable will suggest whether the variables' loading pattern in the common factor solution will differ from the component analysis solution. Specifically, the communalities for variables X_1, X_2 and X_5 are all very high. Therefore, the factor loading pattern for these variables in the common factor model should not differ substantially from what it was in the component analysis. In contrast, the communalities for variables X_4, X_6, and particularly X_3 are much lower than with the component analysis correlation matrix. This suggests that the loading pattern for these three variables will probably be different than it was with the component factor model. This difference may result in the variables loading on a different factor, or it may simply mean that they load on the same factor as with the component model, but the size of the loading may differ since their communality is lower.

Turning next to the VARIMAX rotated common factor analysis factor matrix (Table 6.7), let's examine how it compares with the component analysis rotated factor matrix. The information provided for the common factor solution is similar to that for the component analysis solution. Sums of squares, percentage of variance, communalities, total sums of squares, and total variance extracted are all provided, just as with the component analysis solution. The information that differs is the row "Percentage of Trace." Recall that with component analysis the trace is equal to the number of variables (based on the assumption that the variance in each variable is equal to 1). In contrast, the trace for a common factor analysis solution is equal to the sum of the communalities on the diagonal of the correlation matrix. For the present example, the sum of the communalities (trace) is 4.54. The percentage of trace is included to demonstrate the relative importance of each of the factors in accounting for the variance included in the reduced common factor solution. For example, by dividing the sum of squares

TABLE 6.7 VARIMAX Rotated Common Factor Analysis Factor Matrix		Factors			
	Variables	**1**	**2**	**3**	**Communality**
	X_1 Self-esteem	.73	.65	.04	.96
	X_2 Locus of control	−.88	.41	.13	.96
	X_3 Alienation	.60	.10	−.07	.38
	X_4 Social responsibility	−.10	.11	.74	.57
	X_5 Machiavellianism	−.05	.97	.16	.97
	X_6 Political opinion	−.02	.04	.73	.53
	Sum of squares (eigenvalue)	1.68	1.56	1.13	4.37
	Percentage of trace*	37.00	34.36	24.89	96.47
	Percentage of variance	28.00	26.00	18.83	72.83

* Trace = 4.54 = 75.6% of total variation is common.

for factor 1 (1.68) by the trace (4.54), we obtain the percentage of trace for factor 1—37.00 percent. By summing the percentage of trace for the factors, we obtain the total percentage of trace for the factor solution. For the present example, it is 96.47 percent. This percentage is useful because it tells the analyst how much of the common variance (trace) is accounted for by the factor solution.

Another percentage that may be useful to the analyst is the percentage of total variation that is common variance. As can be noted, the amount of variance included in the reduced common factor solution is 4.54 as opposed to 6.0. Thus, by dividing the amount of common variance (4.54) by the amount of total variance (6.0), we obtain the percentage of total variation that is common variance. The result of this computation reveals that 75.6 percent of the total variation is common.

Comparison of the information provided in the common factor analysis factor matrix and the component analysis factor matrix shows several differences. The factor loading pattern is similar in that variables X_1 and X_5 load together, variables X_1, X_2 and X_3 load together, and variables X_4 and X_6 load together. The primary difference, is that with the component analysis factor matrix, the factor with variables X_4 and X_6 loading was factor 2, whereas with the common factor analysis, variables X_4 and X_6 loaded on factor 3. Similarly, variables X_1 and X_5 loaded on factor 3 with the component analysis solution, whereas these same variables loaded on factor 2 with the common factor solution. The logic for this reversal in the order of extraction for the factors is based upon the communalities for the respective variables. Specifically, note that the communalities for variables X_4 and X_6 are relatively low. Therefore, when the factor solution is based only on the common variance, as with a common factor solution, the order of extraction would dictate that factors with variables exhibiting lower communalities would be extracted later than they would be under a component analysis when the total variance is included in the correlation matrix.

Naming of Factors

When a satisfactory factor solution has been derived, the analyst usually will attempt to assign some meaning to it. The process involves substantive interpretation of the pattern of factor loadings for the variables, including their signs, in an effort to name each of the factors. Before interpretation, a minimum acceptable level of significance for a factor loading must be selected. All significant factor loadings typically are used in the interpretation process. But variables with higher loadings will influence to a greater extent the name or label selected to represent a factor. The signs are interpreted just as with any other correlation coefficients. On each factor, like signs mean that the variables are positively related and opposite signs mean that the variables are negatively related. In orthogonal solutions the factors are independent of each other. Therefore, the signs for a factor loading relate only to the factor that they appear on, not to other factors in the solution.

Let's look at the results shown in Table 6.7 to illustrate this procedure.

Our factor solution was derived from a common factor VARIMAX rotation of the six social-psychological variables. Our cutoff point for this solution is all loadings ± .60 or above (underlined in Table 6.7). This relatively high cutoff was possible because many high loadings were obtained. Variable X_2, locus of control, has a loading of .41 on factor 2 and could be considered significant. But it is loading substantially below all the other variables considered significant (.41 versus .60 or higher), is less than half the size of its loading on factor 1, and to include it would violate the guidelines for simple structure factor solutions (only one loading on any factor for each variable). Thus, it was not considered significant.

Substantive interpretation is based on the significant higher loadings. Factor 1 has three significant loadings, factor 2 has two significant loadings, and factor 3 has two significant loadings (significant loadings are underlined). Looking at factor 2, we see that variables X_1, self-esteem, and X_5, Machiavellianism, are positively related to each other. This suggests that persons in our sample who are high in self-esteem are also high in Machiavellianism. A possible name for this factor on the loading pattern is "indiscriminant high confidence achiever." But perhaps you may want to assign your own label.

Turning next to factor 1, we note that variables X_1, self-esteem, and X_3, alienation, are positively related to each other and negatively related to variable X_2, locus of control. From the description of our data bank in Chapter One, we know the following scoring procedures for the variables: X_1, self-esteem = high scores indicate higher self-esteem; X_2, locus of control = high scores indicate external locus of control; and X_3, alienation = high scores indicate a person who is not alienated. From this scoring procedure and the signs of the variable loadings, we can interpret this factor as an individual who is high in self-esteem, exhibits an internal locus of control, and is not alienated (the opposite interpretation for all three variables is also possible). A possible label for this factor is "self-confident achiever." But again, you may wish to develop your own name to represent the factor.

The process of naming factors has been demonstrated. You will note that it is not very scientific and is based on the subjective opinion of the analyst. Different analysts will no doubt assign different names to the same results because of the difference in their background and training. For this reason, the process of labeling factors is subject to considerable criticism. But if a logical name can be assigned that represents the underlying nature of the factors, it usually facilitates the presentation and understanding of the factor solution and therefore is a justifiable procedure.

How to Select Surrogate Variables for Subsequent Analysis

If the researcher's objective is to identify appropriate variables for subsequent application with other statistical techniques (purpose 3), the researcher could examine the factor matrix and select the variable with the highest factor loading as a surrogate representative for a particular factor dimension. If there is one factor loading for a variable that is substantially higher than all other factor loadings, the variable with the obviously higher loading would be selected for subsequent analysis to represent that factor. In some instances, the selection process is more difficult because two or more variables have loadings that are significant and fairly close to each other. In such cases, the analyst would have to examine critically the several factor loadings that are of approximately the same size and select only one as a representative of a particular dimension. This decision would be based on the researcher's a priori knowledge of theory suggesting that a particular variable would be more logically representative of the dimension that has been identified. Also, the analyst may have knowledge suggesting that the raw data for a variable that is loading slightly lower is in fact more reliable than the raw data for the highest loading variable. In such cases, the analyst may choose the variable that is loading slightly lower as the variable to represent a particular factor.

Let's examine the data provided in Table 6.5 to clarify the procedure for selecting surrogate variables. First, recall that surrogate variables would be selected only when the rotation is orthogonal. This is because, when the analyst is interested in using surrogate variables in subsequent analyses, he or she wants to observe to the extent possible the assumption that the independent variables should be uncorrelated with each other. Thus, an orthogonal solution would be selected instead of an oblique one. Focusing on the factor loadings for factor 2, we see that the loading for variable X_4 is .93 and for variable X_6 is .94. The selection of a surrogate is difficult in cases like this because the sizes of the loadings are so close. However, if the analyst has no a priori evidence to suggest that the reliability or validity of the raw data for one of variables is better than for the other, and if neither would be theoretically more meaningful for the factor interpretation, the analyst would select variable X_6 as the surrogate variable. In contrast, the loadings for factor 3 are .98 for variable X_5 and .71 for variable X_1. Therefore, because of its substantially higher loading, the analyst would select variable X_5 as the surrogate variable to represent factor 3 in a subsequent analysis.

How to Use Factor Scores

When the analyst is interested in creating an entirely new set of a smaller number of composite variables to replace the original set of variables, either in part or in whole, he or she computes factor scores (purpose 4). Factor scores are composite measures for each factor representing each subject. The original raw data measurements and the factor analytic results are utilized to compute factor scores for each individual. Using our data bank example, each individual would have had six raw data measurements representing each of the original six variables. After computation of factor scores to represent the factor solution, each individual would be represented by only three composite measures rather than the original six measures. These three composite measures or factor scores would represent each of the three factors that were derived in the factor solution. Conceptually speaking, the factor score represents the degree to which each individual scores high on the group of items that load high on a factor. Thus, an individual who scores high on the several variables that have heavy loadings for a factor surely will obtain a high factor score on that factor. The factor score, therefore, shows that an individual possesses a particular characteristic represented by the factor to a high degree. Most factor analysis computer programs compute scores for each respondent on each factor to be utilized in subsequent analysis. The analyst would merely have to select the factor score option, and these scores would either be printed out or punched out for use by the analyst.

Creating factor scores to represent factor structures is not a difficult task for the analyst. Thus, the question may arise of whether factor scores or surrogate variables should be used by the analyst. Both have advantages and disadvantages, and no clear-cut answer is available for all situations. Factor scores have the advantage of representing a composite of all variables loading on the factor, whereas surrogate variables represent only a single variable. But since factor scores are based on correlations with all of the variables in the factor, and these correlations are likely to be much less than 1.0, the scores are only approximations of the factors and, as such, are error-prone indicators of the underlying factors. The decision rule, therefore, would be that if the scale is a well-constructed, valid, and reliable instrument, the factor scores are probably the best alternative. But if the scale is untested and exploratory, with little or no evidence of reliability and/or validity, surrogate variables should probably be used.

Summary

The multivariate statistical technique of factor analysis has been presented in broad conceptual terms. Basic guidelines for interpreting the results were included to clarify further the methodological concepts. An example of the application of factor analysis was presented based upon the data bank in Chapter 1.

Factor analysis can be a highly useful and powerful multivariate statistical technique for effectively extracting information from large data bases. Factor analysis helps the investigator make sense of large bodies of interrelated data. When it works well, it points to interesting relationships that might not have been obvious from examination of the raw data alone or even a correlation matrix. Potential applications of factor analytic techniques to problem solving and decision making in business research are numerous. The use of these techniques will continue to grow as increased familiarity with the procedures is gained by academicians and practitioners.

Factor analysis is a much more complex and lengthy subject than might be indicated by this brief exposition. Four of the most frequently cited limitations are as follows: First, there are many techniques for performing factor analyses. Controversy exists over which technique is best. Second, the subjective aspects of factor analysis (deciding how many factors to extract, which technique should be used to rotate the factor axes, which factor loadings are significant) are all subject to many differences in opinion. Third, the computational labor involved in conducting factor analysis and any other multivariate technique with large data bases necessitates the use of computers. With the rapid spread of computers, this particular limitation has diminished. Fourth, the problem of reliability is real. Like any other statistical procedure, a factor analysis starts with a set of imperfect data. When the data change because of changes in the sample, the data-gathering process, or the numerous kinds of measurement errors, the results of the analysis also change. The results of any single analysis are therefore less than perfectly dependable. This problem is especially critical because the results of a single-factor analytic solution frequently look plausible. It is important to emphasize that plausibility is no guarantee of validity or even stability.

QUESTIONS

1. What are three problem situations in which factor analysis is the appropriate multivariate statistical technique to apply?

2. What is the difference between an orthogonal and an oblique factor rotation? When would the application of each approach be most appropriate?

3. What guidelines can you use to determine the number of factors to extract? Explain each briefly.

4. How do you use a factor loading matrix to interpret the meaning of factors?

5. How and when should you use factor scores in conjunction with other multivariate statistical techniques?

6. What is the difference between Q-type factor analysis and cluster analysis?

7. When would the analyst use oblique factor analysis instead of orthogonal factor analysis?

8. How can factor analysis be used in conjunction with other "dependence" multivariate techniques discussed in earlier chapters?

REFERENCES

1. Cattell, R. B., "The Scree Test for the Number of Factors," *Multivariate Behavioral Research*, vol. 1 (April 1966), pp. 245–276.
2. Dillion, W. R., and M. Goldstein, *Multivariate Analysis: Methods and Applications.* New York: Wiley, 1984.
3. Green, P. E., *Analyzing Multivariate Data.* Hinsdale, Ill.: Holt, Rinehart, & Winston, 1978.
4. Green, P. E., and J. Douglas Carroll, *Mathematical Tools for Applied Multivariate Analysis.* New York: Academic Press, 1978.
5. McGraw-Hill, *SPSS-X Users's Guide*, 2nd ed. Chicago, 1986.
6. McGraw-Hill, *SPSS-X Advanced Statistics Guide*, Chicago, 1986.
7. SAS Institute, *SAS User's Guide: Basics*, Version 5 ed. Cary, N. C., 1985.
8. SAS Institute, *SAS User's Guide: Statistics*, Version 5 ed. Cary, N. C., 1985.

SELECTED READINGS

The Organizational Context of Market Research Use

ROHIT DESHPANDE

Largely as a function of developments in its environment, marketing is asking introspective questions about its own efficiency. At the beginning of the 1980s we have seen the rapid growth of the marketing function over the past two decades slowed under the impacts of inflation, raw material shortages, unemployment and recession. These economic changes necessitate a reassessment of strategies that had earlier proved successful. The drive now is to become leaner, more efficient in the use of available resources and more oriented toward the future (Wind 1980).

If we are to believe that the U. S. and other postindustrial economies are moving from an "Age of Product Technology" to a "Knowledge-based Society" (Bell 1976), we should be increasingly concerned with our ability to manage our corporate knowledge systems. The growth and even survival of today's business entities will depend on their strategies for handling and processing information. The more current this information, the greater the ability of managers to make policy decisions based upon it. In turn, the effectiveness of those decisions will be measured in terms of market information.

The marketing function is somewhat unique in that the information gathering and analysis processes in firms have been institutionalized as marketing research departments or divisions. Al-

"The Organizational Context of Market Research Use," Rohit Deshpande, Vol. 46 (Fall 1982), pp. 91–101. Reprinted from the *Journal of Marketing*, published by the American Marketing Association. Rohit Deshpande is a professor in the Department of Marketing at the University of Texas at Austin.

though these specialized information processing units have existed for some time, very little examination has been given to the effectiveness of research in providing information at the right place for the right decision. Additionally, it is only very recently that any attention has been paid to the factors that affect the usefulness of marketing research.

The issue of examining marketing's R&D has not gone unnoticed. The critical costs of inadequate utilization of marketing tools and techniques have been mentioned recently by a special AMA/Marketing Science Institute joint commission (Myers, Massy and Greyser 1980). The commission's members were surprised at the relatively low rate of adoption at the line manager level of new marketing knowledge generated over a period encompassing the past 25 years. Their major recommendation was to develop better ways "to bridge the gaps between knowledge-generation and knowledge-utilization" (Myers, Greyser and Massy 1979, p. 27). Both marketing practitioners and academics support these observations and agree that much problem oriented research is not used (Dyer and Shimp 1977, Ernst 1976, Kover 1976, Kunstler 1975). However, little formal research has been conducted in this area (Greenberg, Goldstucker and Bellenger 1977; Krum 1978; Luck and Krum 1981). Most observations about the factors affecting use of marketing research have been limited to introspective, albeit careful, analyses of personal experiences (Hardin 1973, Kunstler 1975, Newman 1962).

The issue of inadequate utilization of available research information is not unique to marketing. Underuse occurs in all areas of applied research activity. Most recently it has received much empirical attention in the policy sciences and has led to the creation of the area of inquiry called Knowledge Utilization (Caplan, Morrison and

Stambugh 1975; Rich 1975; Weiss 1977; Weiss and Bucuvalas 1980). Developments in this area indicate that an understanding of the research use phenomenon lies in examining the organizational contexts in which policy decisions are made. The design of the decision making structures of organizations sometimes provides clues as to why some of them are more efficient at using research than others.

As Day and Wind (1980) have commented, senior management has come to believe that focusing only on a customer-oriented search for competitive advantage may be shortsighted. There is a need to widen the scope of empirical attention in marketing by looking at relationships beyond those of the company and its customers. One set of these relationships deals with managers *within* an organization. Unless the structure of work relationships in a firm has been designed to optimize managerial effectiveness, the company-customer transactions will suffer and, in turn, negatively impact on the firm's long-term profitability. Yet the influence of organizational structure on the marketing function has seldom been studied systematically (Bonoma, Zaltman and Johnston 1977; Silk and Kalwani 1980; Spekman and Stern 1979). This issue is particularly important in the knowledge utilization area since parallel findings in the policy sciences, as mentioned earlier, indicate the importance of organizational design in influencing research use. In the pursuit of marketing effectiveness it may be useful to examine what forms of marketing organization appear best suited to manage the marketing research process efficiently (Wind 1980). This paper looks at the issue by surveying marketing managers in major U. S. business firms.

This paper does not intend to develop or extend a paradigm in organizational theory but attempts to look at why some consumer product companies make more use of marketing research than others. In the process, it is necessary to acquaint the reader with several organizational studies conducted in the past. Although these studies did not explicitly study marketing departments or market research-based decisions, it is possible to transfer the knowledge gained in those studies to the marketing area. This is the task of the following section.

Information Use in an Organizational Setting

Harold Wilensky, in his famous treatise on organizational intelligence, writes of barriers to the use

of information in organizations: "Intelligence failures are rooted in structural problems that cannot be fully solved; they express universal dilemmas of organizational life that can, however, be resolved in various ways at varying costs. In all complex systems, hierarchy, specialization, and centralization are major sources of distortion and blockage of intelligence" (1967, p. 42). Why is this so? According to Wilensky, an organization that has a long hierarchial structure and emphasizes rank is likely to have much distortion occurring as information flows upward from junior through senior managerial levels. Due to the differential selective perception of information by different individuals, new knowledge takes on different shades of meaning as it passes from one person to the next. This distortion is further accentuated by the tendency of lower level managers to show themselves in the most favorable light to their superiors. In the case of the marketing organization, therefore, although senior marketing managers may wish to exert more effective control by centralizing information (and thereby its use), the knowledge with which they are provided may be a far cry from what was initially gathered by junior members of the marketing department.

Yet the attempt at resolving this problem by greatly decentralizing information collection and decision making activities may not serve the need either. What Wilensky refers to as the dilemma of centralization may occur ". . . if intelligence is lodged at the top, too few officials and experts with too little accurate and relevant information are too far out of touch and too overloaded to function effectively; on the other hand, if intelligence is scattered throughout many subordinate units, too many officials and experts with too much specialized information may . . . delay decisions while they warily consult each other. . . . More simply, plans are manageable only if we delegate; plans are coordinated in relation to organizational goals only if we centralize" (1967, p. 58).

It appears therefore that the structure of a marketing organization may impact on the use of research information by its managers in one of two opposing ways. It is not entirely clear whether few rules and procedures and extensive decentralization of decision making authority will help or hinder the organization's use of information.

It is helpful at this point to look at some past studies in the sociology of organizations to see whether an empirical resolution can be found to Wilensky's dilemma. Two sociologists, Michael

Moch and Edward Morse, recently studied the impact of organizational size and centralization on the adoption of innovations in 1,000 U. S. hospitals (Moch and Morse 1977). Their study found that in larger organizations a great deal of task specialization and role differentiation occurred. "Large organizations are in a better position to employ specialists and formally to differentiate responsibilities assigned to personnel in order to accommodate variation in input material. By employing specialists, the organization gains access to knowledge of new ideas, practices, and technical skill . . . [Also] formally differentiating task responsibilities to organizational personnel focuses their interests within specialized areas" (Moch and Morse 1977, p. 717). Additionally, these organizations tended to adopt innovations far more readily than those organizations that had less specialization of tasks and more centralization. This finding is supported by several other observers of social change in organizations (Hage and Aiken 1970; Pondy 1970; Zaltman, Duncan and Holbek 1973).

We can think of market research information that is new to a firm's marketing manager as an innovation for that organization. This is in keeping with the Rogers and Shoemaker (1971) conception of innovation as any set of ideas, practices or material artifacts perceived to be new by the relevant unit of adoption. According to Zaltman and Duncan (1977), people change their behavior when they define the situation as being different and requiring new or different behavior. If we then translate the above findings from organization theory into implications for marketing departments of firms, we can hypothesize that departments that are more structured will be less likely to adopt (or use) new research information than departments that are less structured.

However, a problem may exist. Looking once again at new research information as an innovation for the organization, awareness of the information is a function of the extent of work experience of the manager (Zaltman, Duncan and Holbek 1973). Presumably a marketing manager with several years of corporate experience behind him/her will be better able to judge the fit between new research information and its applicability to a specific decision making situation. Hence it is necessary to supplement any inquiry into the impact of structural arrangements on research use with an assessment of managers' work experiences.

These observations bring us to a statement of several formal hypotheses tested in this study. Before these can be stated, a description of the variables and their operationalization is provided.

Variables

Organizational Structure

Although several different concepts have been discussed in the sociology of organizations in measuring organizational structure, this study employs the two dimensions of *formalization* and *centralization*. These dimensions have been studied extensively in the context of organizational adoption of innovations (which is relevant to our case, as discussed earlier), and measures have been developed for these dimensions that have been carefully validated and replicated across a wide variety of different organizations (public and private, large and small, both in the U. S. and in Europe).

Formalization, as defined in the work of Hall, Haas and Johnson (1967), is the degree to which rules define roles, authority relations, communications, norms and sanctions, and procedures. This dimension of organizational structure is an attempt to measure the flexibility that a manager enjoys when handling a particular task (such as the implementation of research recommendations). Centralization, as defined by Aiken and Hage (1968), looks at the delegation of decision making authority throughout an organization and the extent of participation by managers in decision making.

The above concepts have been studied using two methods. The first, espoused in the work of Blau and Schoenherr (1971), Hinings and Lee (1971), Child (1972), and others focuses on "institutional" measures that look at the span of control, worker/supervisor ratios, distribution of employees across functional areas, and other indicants of an organization chart (Payne and Pugh 1976, Pugh et al. 1968). The second method uses questionnaires, with respondents indicating the extent of their agreement or disagreement with a series of statements dealing with issues such as the flexibility allowed in the handling of organizational tasks, the requirement for conformity with rules and guidelines, the amount of decentralization of authority, and so on (Aiken and Hage 1968, Hall 1972). This method has been empirically validated across a series of studies on different firms in both the U. S. and Europe.

However, the two methods do not produce identical results. Using a multitrait-multimethod matrix, Pennings (1973) found a low degree of convergence between institutional and questionnaire measures. He also discovered that although questionnaire measures of formalization and centralization were positively associated, the institutional measures produced negative correlations between the dimensions. These anomalies were criticized by several researchers on the grounds of instrument unreliability (Dewar, Whetten and Boje 1980; Pennings 1973; Seidler 1974). It is conceivable that some instrument bias can occur with self-report measures such as those used in the questionnaire method and also in the use of informants for the institutional method.

An alternative explanation has been suggested by Sathe (1978), who replicated Pennings' (1973) study with some modifications to increase internal validity and instrument reliability and suggested that the two methods were measuring different concepts. Since the institutional measure examines the structure of an organization in terms of the organization chart, the results reflect the structure as it was designed to operate. However, the questionnaire method, since it asks respondent managers to indicate their perceptions of participation in decisions, job flexibility, etc., taps the organizational structure as managers see it operating. "The questionnaire measures tend to reflect the degree of structure experienced by organizational members in work related activities on a day-to-day basis and to the extent that such information is not biased, describe the *emergent* structure" (Sathe 1978, p. 234). This latter method is more pertinent to this study, since we are less interested in how a marketing organization was designed to function than in how an individual manager perceives the organization influencing his or her job. Therefore the questionnaire measures of formalization and centralization were used in this study.

The two structural dimensions are themselves conceptual aggregates of certain independent constructs. Formalization, for instance, is composed of measures tapping the extent to which jobs are codified (Job Codification), the degree to which rules are observed (Rule Observation), and the extent to which the specifics of tasks are stated (Job Specificity). Centralization is composed of the subdimensions of Participation in Decision Mak-

ing and Hierarchy of Authority (the extent to which authority to make decisions affecting the firm is confined to higher levels of the hierarchy). Each of these dimensions is represented by a series of questions measured on a four or five point scale. The questions are displayed in Figure 1.

Utilization of Research Information

Most of the work in defining and measuring what constitutes research use has been in nonmarketing areas, primarily political science and public administration. Robert Rich, in his study of federal policymaking (1977), defines use as specific information coming to the desk of a decision maker, being read, and influencing the discussion of particular policies. In this sense the use of information is analogous to the use of a marketing research report being examined by a manager. Nathan Caplan and his co-workers have also looked largely at this instrumental type of information use in their study of 204 government officials (Caplan, Morrison and Stambaugh 1975). They define use in terms of familiarity of the officials with pertinent research and a consideration of an attempt to apply the research to some relevant policy areas. However, there is still much discussion as to how best to define research information use and the optimal way to measure it (Deshpande 1979, Larsen 1980, Weiss 1980). In this study, use of research information was defined and operationalized in terms of whether a decision could have been made without it or whether the decision, when made without research, would have been very different from the decision for which research information was considered. Two questions were asked to determine research use. The first asked respondents to agree or disagree (on a five point Likert scale) with the statement, "Without this research information, the decisions made would have been very different." And the second, using the same response format, stated, "No decision would have been made without this research information."

Admittedly, several alternative methods of operationalizing research use do exist. Some of the literature cited above indeed defines use in different ways. However, in this study we are most concerned with the *so what?* or *impact* dimension of market research. Has there been any change caused by the presence of new information? Has the research affected managers' decision making

FIGURE 1   Perceptions of Organizational Structure

Formalization Questions	**Centralization Questions**
Response Categories: 1 Definitely true	**Response Categories:** 1 Never
2 More true than false	2 Seldom
3 More false than true	3 Often
4 Definitely false	4 Always
5 Not applicable	

Formalization Questions

Job Codification

(1) First, I felt that I was my own boss in most matters relating to the project.
(2) I could make my own decisions regarding the project without checking with anybody else.
(3) How things were done around here was left pretty much up to me.
(4) I was allowed to do almost as I pleased.
(5) I made up my own rules on this job.

Rule Observation

(6) I was constantly being checked on for rule violations.
(7) I felt as though I was constantly being watched to see that I obeyed all the rules.
(8) There was no specific rules manual relating to this project.
(9) There is a complete written job description for going about this task.

Job Specificity

(10) Whatever situation arose, we had procedures to follow in dealing with it.
(11) Everyone had a specific job to do.
(12) Going through the proper channels in getting this job done was constantly stressed.
(13) The organization kept a written record of everyone's performance.
(14) We had to follow strict operating procedures at all times.
(15) Whenever we had a problem we were supposed to go to the same person for an answer.

Centralization Questions

Participation in Decision Making

(1) How frequently did you usually participate in decisions on the adoption of new products.
(2) How frequently did you usually participate in decisions on the modification of existing products.
(3) How frequently did you usually participate in decisions to delete existent products.

Response Categories: 1 Definitely true
2 More true than false
3 More false than true
4 Definitely false
5 Not applicable

Hierarchy of Authority

(1) There could be little action taken on this project until a superior approved a decision.
(2) If I wished to make my own decisions, I would be quickly discouraged.
(3) Even small matters on this job had to be referred to someone higher up for a final answer.
(4) I had to ask my boss before I did almost anything.
(5) Any decision I made had to have my boss' approval.

in any way? What would have happened to the decisions if the research did not exist? These are the types of issues that this definition of research use attempts to get at. Additionally, the questions on research use are relatively more indirect than operationalizations that ask, "Did you actually use the market research?" As will be seen when the means and standard deviations of variables are described, the tendency toward positive bias is limited by utilizing more inferential methods of measuring use.

Interrelationship Between Concepts

Now that the major concepts have been defined and their operationalizations described, we can proceed to show how they are interrelated. Research in organization behavior indicates that firms that are more decentralized and less formalized are likely to adopt innovations quicker than those that are more structured (Hage and Aiken 1970; Moch and Morse 1977; Zaltman, Duncan and Holbek 1973). Additionally, as mentioned earlier, new research information can be thought of as an innovation that a manager may or may not decide to use (Deshpande and Zaltman 1981). The following propositions, therefore, flow from these considerations:

1. The greater the Job Codification[1] perceived by managers, the lower the utilization of market research information.
2. The greater the Rule Observation perceived by managers, the lower the utilization of market research information.
3. The greater the Job Specificity perceived by managers, the lower the utilization of market research information.
4. The lower the Participation in Decision Making, the lower the utilization of market research.
5. The greater the Hierarchy of Authority, the lower the utilization of market research.

Additionally, looking once again at new research information as an innovation to the organization, awareness of the information is a function of the work experience of the manager (Radnor, Rubinstein and Tansik 1970; Zaltman, Duncan and Holbek 1973). If we consider the years of experience in a firm or an industry as surrogates for work experience, then two further propositions are:

6. The greater the number of years of work experience in the firm, the greater the managers' perceived utilization of research.
7. The greater the number of years of work experience in the industry to which the firm belongs, the greater the managers' perceived utilization of research.

The questionnaire described below asked direct questions concerning number of years of work experience in the firm and the industry. The seven propositions were tested on the sample.

Sample and Research Methodology

Data used here come from a larger study of a sample of 92 managers who were questioned about

[1] Each of the organizational structure measures is briefly described at the end of the earlier section on Variables.

marketing research projects in their companies (Deshpande and Zaltman 1982). The first stage of the study involved personal interviews conducted with 16 individuals in 7 firms (10 managers and 6 research suppliers). All 16 persons were selected on a convenience basis from large firms (all in the Fortune 500 sample) and from leading advertising and research agencies. The questionnaire used for the personal interviews was modified with structured queries for use, after a pilot test, in the mail survey that constituted the second stage of the data collection.

The sampling frame for managers was an *Advertising Age* listing of the 100 largest U. S. advertisers. Five hundred primarily product/brand and marketing managers in marketing divisions and firms (the universe of such managers in the frame) were selected. Firms dealing with industrial products or services were deleted, and out of 249 eligible respondents, 92 (37%) managers responded after one follow-up mailing. This rate of response, though good when compared with those in studies of similar organizations, required a detailed nonresponse analysis. A randomly selected subsample of 50 nonrespondents was contacted directly by telephone to ascertain their reasons for not returning questionnaires. The major reasons for nonresponse concerned the lack of time to fill out the rather lengthy questionnaire. No subject matter related reasons for nonresponse were stated by any of the individuals contacted. Thus the actual replies received can be assumed to represent the valid responses of the total original sample (since the randomly selected nonrespondent's sample is assumed to be representative of all nonrespondents). Eligible respondents did not differ from nonrespondents in terms of organizational demographics or the salience of issues being studied.

Rather than asking managers questions about their general experiences with market research, it was felt necessary to get specific details on one such critical research experience per manager. In this manner it is possible to focus more precisely on the factors contributing to research use for that research project.

Accordingly, after preliminary questions regarding job title and work experience, the questionnaire asked respondents to focus on the *most recently completed* marketing research project with which they had been associated and for which a research report had already been presented.[2] The research incident preferred was to have contributed toward a consumer product (or service) strategy decision, i.e., the addition, modification or deletion of a product from the firm's line of offerings. In addition, the research for the scenario was to have been conducted by a research agency external to the marketing firm.[3] The questions regarding the use of the research were posed toward the end of the questionnaire to limit contamination of earlier responses.

Analysis

In order to improve the face validity of the 23 statements dealing with perceptions of organizational structure, the questions were altered slightly to make them market research project specific. For example, the original statement, "How frequently do you usually participate in decisions on the adoption of new programs?" was modified to read, ". . . in decisions on the adoption of new products?" These modifications are in keeping with recent suggestions to improve instrument validity (Dewar, Whetten and Boje 1980). However, such alterations limit the comparability of these questions to nonmarketing uses of the organizational structure measures. In order to ascertain whether the measures still retained construct validity (i.e., measure what they are supposed to), a factor analysis was conducted. This resulted in five factors explaining 70% of the overall variance. Table 1 shows the variables loading on each of the five factors.

It may be seen that the analysis produces a clean factor structure with items loading on the appropriate factors. With only a few items being deleted because of low or incorrect loading, the measures of the three formalization constructs (Job Codification, Rule Observation, Job Specificity) and the two centralization constructs (Participation in Decision Making, Hierarchy of Authority) show excellent validity. Additionally, internal re-

[2] Clearly the one research project described by each manager may not be entirely representative of that firm's research experience. However, by having each manager describe the most recently completed research project, it is hoped that the sum of all such project experiences across the total sample would be representative for the sample of firms considered in this study.

[3] Since the study described here represents an initial exploration into the area of market research use by private firms, the investigation was not designed to include questions concerning internal marketing research departments or divisions.

liability tests showed strong Cronbach alphas ranging from 0.73 through 0.92.

The next step, following the treatment suggested by developers of the measures, was to form cumulative, equally weighted indices for each of the five measures so as to develop scores for each case. As a further validity check, the sample was split randomly and Cronbach alphas were recalculated for the indices on each subsample. Alphas continued to be excellent with a range of 0.62 to 0.96.

In order to measure Research Utilization, a simple additive index was formed of the responses to the two questions on research use. (The mean response on each of the questions was 2.85 and 3.53 with standard deviations of 1.01 and 1.14, respectively. This indicates first, that managers generally agreed that decisions, if made, would have been very different without the research information, and second, that the tendency for positive bias on these questions was not a major problem. This additive index of Research Utilization was then utilized as the dependent variable in an ordinary least squares regression. The five dimensions of organizational structure (Job Codification, Rule Observation, Job Specificity, Participation and Hierarchy of Authority) and the measures of industry and firm work experience were the predictor variables. Results of the regression analysis are displayed in Table 2. As this Table shows, the overall regression equation explains 67.2% of the total variance, a result that is statistically significant at the 0.001 level. In order to test for internal validity of this result, the sample was split randomly into halves and the regression recomputed. Both standardized betas and the estimate of explained variance remained stable, suggesting the results reported here are not due to chance.

Now going beyond the summary R^2 statistic, a perusal of the contributions of individual independent variables produces interesting findings. First, it appears that the length of work experience of managers (in either the firm or the industry) is not a major determinant of the use of market research information. The standardized beta weights are small and statistically insignificant. This finding substantively rejects Propositions (6) and (7).

Among the formalization variables, Job Codification and Specificity have significant impacts on the utilization of research (betas of −0.38 and −0.38, significant at 0.006 and 0.017 levels, re-

TABLE 1　Factor Analysis of Perceptions of Organizational Structure

	Job Codi- fication	Rule Obser- vation	Job Speci- ficity	Partici- pation in Decision Making	Hierarchy of Au- thority
(1) My own boss	**.67**	.10	.24	.10	.18
(2) Make my own decisions	**.76**	.19	.08	.06	.17
(3) Doing things left up to me	**.91**	.05	.19	.01	.20
(4) Do almost as I pleased	**.89**	.08	.15	−.02	.15
(5) Made my own rules	**.70**	.18	.16	.03	.23
(6) Checked for rule violations	.24	**.82**	.25	−.06	.21
(7) Constantly being watched	.27	**.91**	.24	−.11	.14
(8) No rules manual*	.35	.11	.42	−.03	.07
(9) Complete job description*	.18	.31	.76	−.09	−.02
(10) Procedures for dealing with situations	.22	.10	**.85**	−.01	−.03
(11) Specific job to do	.14	−.09	**.71**	−.25	.07
(12) Going thru proper channels	.05	.08	**.54**	−.03	.30
(13) Written performance record	.15	.17	**.69**	.12	.18
(14) Strict operating procedures	.06	.07	**.80**	.06	.37
(15) Same person for problem referral	.15	.14	**.56**	.08	.33
(16) Adoption of new products	.11	.04	−.06	**.61**	.03
(17) Modification of existent products	.02	−.08	−.01	**.83**	.05
(18) Deletion of existent products	−.04	−.09	−.00	**.81**	.04
(19) Superior approves decision*	.16	.14	.08	.10	.25
(20) Discourage my own decision	.20	.07	.00	.04	**.68**
(21) Refer to superior for small matters	.19	.08	.36	.06	**.63**
(22) Had to ask boss	.22	.07	.25	−.03	**.82**
(23) Boss' approval required	.30	.19	.20	.12	**.76**

Cumulative variance explained by five factors: 70%

* Items deleted due to low item-factor correlations.

TABLE 2　Multiple Regression Analysis of Research Utilization with Perceptions of Organizational Structure and Work Experience

Dependent Variable: Utilization of Research Information

Independent Variables	Standardized Beta	F	Significance
Job Codification	−.38	8.013	.006
Rule Observation	−.07	0.304	.583
Job Specificity	−.38	6.000	.017
Participation in Decision Making	.48	21.650	.001
Hierarchy of Authority	−.38	6.857	.011
Number of years in firm	−.02	0.169	.897
Number of years in industry	.04	0.685	.794

Overall F = 21.45　　significance = .001　　adjusted R^2 = .672
Sample: Managers (N = 92)

spectively). In addition, the signs on these two coefficients are in the hypothesized direction, thus both substantively and statistically confirming Propositions (1) and (3). However, Rule Observation makes a nonsignificant contribution to explaining the variance in the dependent variable. Although the sign on its coefficient is as hypothesized (thus providing some reason for acceptance on substantive grounds), the low beta value of the Rule Observation coefficient leads to a rejection of Proposition (2).

Additionally, among the centralization variables both Participation in Decision Making and Hierarchy of Authority prove to have significant effects on the utilization of market research information. The coefficients of both of these variables are substantial, in the proposed direction, and statistically significant (beta of 0.48, significant at 0.001 and beta of -0.38, significant at 0.011, respectively). This leads us to accept Propositions (4) and (5) as hypothesized.

Discussion

Briefly summarizing the results of the above analysis, it appears that managers who see themselves as operating in firms that are relatively decentralized and have few formalized procedures for carrying out marketing tasks are likely to make extensive use of market research information. This finding is true for this sample regardless of the extent of work experience of the managers. That is, both junior and senior managers in the marketing divisions of firms utilize research information more when they perceive a considerable amount of flexibility in how they should go about their tasks.

It is interesting to note that the extent to which a manager is perceived as observing organizational/departmental rules does not have great importance in determining whether market research information gets used. Managers appear to be less concerned (in a research utilization context) with whether they were constantly being watched to see they obeyed all the rules than with whether or not they were their own boss, whether or not they could make decisions on their own, or whether or not they were required to "go through channels." Looking at the questions constituting job codification and job specificity in Figure 1 it seems that marketing managers tend to utilize research more when they feel that they, rather than

their colleagues or bosses, are in control. These managers felt that they had substantial latitude in defining their roles ("I made up my own rules on this job," "I was allowed to do almost as I pleased"). They felt they were working in organizations or departments where strict operating procedures did not exist or, if they did exist, did not have to be followed with great attention to specific detail.

Although the question was not asked in this study, it is possible that when managers feel they have greater flexibility, they may also believe they have more freedom in doing their jobs. In the marketing research project situation this freedom can translate itself into a manager being more committed to research activity, being more involved with the entire research process, and as a result being more likely to use the findings from the research.

The feeling of freedom while handling marketing tasks is reinforced when managers participate frequently in product (or service) adoption, modification or deletion decisions. Together with the decentralization of decision making authority comes added responsibility, which puts a burden on the managers' shoulders, the burden of eventual accountability for the decisions they make. A decision to launch a million dollar product is not one taken easily. Consequently, it is conceivable that managers will want to get as much corroborative and supportive evidence as possible before making the decision. This evidence is readily available in the form of market research data. Hence, the more decentralized the marketing operation of a company, the more likely it is that its managers will seek out research support.

The logic of the above argument may seem intuitively appealing and also provide a rationale for why managers in decentralized and less formalized marketing firms tend to be greater users of research. But this result is not trivial since it is also possible to argue the opposite. Indeed this is what Wilensky (1967) has done, as mentioned earlier, in posing his "dilemma of centralization." Firms that are highly centralized and also have established more formally structured bureaucratic rules and procedures for handling marketing tasks are also likely to have centralized information systems. One would expect that in such a centralized system, research information goes directly to an individual at a senior level in the management hierarchy and, as such, is highly likely to be used in making strategic or policy decisions. However, when we consider the enormous amount of information that is likely to cross a decision maker's desk in such a centralized management system, we can see immediately the difficulties that will result. As Rich (1975) found in the public policy setting, to avoid information overload problems senior policy makers ask their aides to distill information so that only the most critical and urgent issues reach their offices. In the distillation process a great deal of (perhaps pertinent) information is lost. Additionally, advocacy positions of junior aides are reflected in the reports reaching their superiors. The end product report that arrives on the policy maker's desk is frequently a far cry from the research that was originally executed, and the recommendations bear only a mild resemblance to those originally proposed by researchers.

The above scenario can be contrasted effectively with one where individual line managers are given the flexibility to take products from the R&D laboratory to test market shelves. Along with this flexibility comes, as mentioned earlier, accountability in terms of maintaining profit expectations. The line manager, who also oversees and interacts with the firm's research function, authorizes the collection of data concerning the potential viability of the product in the marketplace. The resulting market research information will then be carefully examined before a decision (largely based upon the information) is pronounced.

Conclusion

Very little empirical attention in marketing has been given to designing organizations or task groups to enhance managerial efficiency. Yet it is clear that even before considering company market transactions, it is important to ensure that the internal marketing operation functions effectively. One area of much significance is that involving the use of marketing research information. Although annual research transactions of larger business firms in the U. S. comprise expenditures of millions of dollars, the management of the research function to increase research productivity has received little scrutiny.

The study reported here indicates that marketing managers making consumer product strategy decisions are more likely to use research information when they see themselves working in a decen-

tralized organization with few formal rules or procedures that must be observed. With increasing sophistication of research operations and growing uncertainties of the economic environment, this study has clear implications for senior marketing management. In order to enhance the efficient use of market research, line marketing managers should be allowed to operate in reasonably flexible task environments. This flexibility would allow managers a generous amount of freedom to participate extensively in product strategy decisions, coupled with accountability for demonstrating desired returns on product investment. The responsibilities with which the managers are entrusted would include overseeing the collection and analysis of marketing research information on the product in their charge. This would permit line marketing managers to be strongly involved in the research process, ensuring that the research information produced would be highly relevant to decisions that need to be made. The final result of the managers' commitment to the marketing research activity would be a more effective utilization of research.

The interesting results this study has provided reflect the perceptions of marketing managers. To complement these perceptions, further investigation in this area might examine marketing researchers working on the same research projects as managers. It would then be enlightening to compare manager and researcher perspectives on factors affecting the utilization of research on the same project. However, the insights provided here are no less important or valid. In fact, from a marketing firm's viewpoint it can be argued that the crucial element in considering the design of its organization is the task reality as seen by its own managers. This reality for marketing research projects has been reflected in the responses of marketing managers in this study. Their insights should help both managers in other companies and scientists and observers of the management of market research activity in an organization.

Future research in this area should replicate this study with an extension to industrial and services marketing firms. Additionally, since the domain of inquiry here concerned product and marketing managers, future work could be directed at a sample of marketing vice presidents or marketing directors. It is not yet known whether the findings from such samples would be similar to those reported here.

References

Aiken, M. and J. Hage (1968), "Organizational Independence and Intra-Organizational Structure," *American Sociological Review*, 33 (December), 912–930.

Bell D. (1976), *The Coming of Post-Industrial Society: A Venture in Social Forecasting*, New York: Basic Books.

Blau, P. M. and R. A. Schoenherr (1971), *The Structure of Organizations*, New York: Basic Books.

Bonoma, T. V., G. Zaltman and W. J. Johnston (1977), *Industrial Buying Behavior*, Cambridge, MA: Marketing Science Institute.

Caplan, N., A. Morrison and R. J. Stambaugh (1975), *The Use of Social Science Knowledge in Policy Decisions at the National Level*, Ann Arbor, MI: Institute for Social Research.

Child, J. (1972), "Organization Structure and Strategies of Control: A Replication of the Aston Study," *Administrative Science Quarterly*, 17 (June), 163–177.

Day, G. S. and Y. Wind (1980), "Strategic Planning and Marketing: Time for a Constructive Partnership," *Journal of Marketing*, 44 (Spring), 7–8.

Deshpande, R. (1979), "The Use, Nonuse, and Abuse of Social Science Knowledge: A Review Essay," *Knowledge: Creation, Diffusion, Utilization*, 1 (September), 164–176.

_____ and G. Zaltman (1981), "The Characteristics of Knowledge: Corporate and Public Policy Insights," in *Government Marketing: Theory and Practice*, M. P. Mokwa and S. E. Permut, eds., New York: Praeger, 270–278.

_____ and _____ (1982), "Factors Affecting the Use of Market Research Information: A Path Analysis," *Journal of Marketing Research*, 19 (February), 14–31.

Dewar, R. D., D. A. Whetten and D. Boje (1980), "An Examination of the Reliability and Validity of the Aiken and Hage Scales of Centralization, Formalization, and Task Routineness," *Administrative Science Quarterly*, 25 (March), 120–128.

Dyer, R. F. and T. A. Shimp (1977), "Enhancing the Role of Marketing Research in Public Policy Decision Making," *Journal of Marketing*, 41 (January), 63–67.

Ernst, L. A. (1976), "703 Reasons Why Creative People Don't Trust Research," *Advertising Age*, 47 (February 10), 35–36.

Greenberg, B. A., J. L. Goldstucker and D. N. Bellenger (1977), "What Techniques Are Used by Marketing Researchers in Business," *Journal of Marketing*, 41 (April), 62–68.

Hage, J. and M. Aiken (1970), *Social Change in Complex Organizations*, New York: Random House.

Hall, R. (1972), *Organizations, Structure and Process*, Englewood Cliffs, NJ: Prentice-Hall.

_____, E. F. Haas and N. F. Johnson (1967), "Organizational Size, Complexity, and Formalization," *American Sociological Review*, 32 (December), 903–911.

Hardin, D. K. (1973), "Marketing Research and Productivity," in *Proceedings of the AMA Educators' Conference*, T. V. Greer, ed., Chicago: American Marketing Association, 169–171.

Hinings, C. R. and G. Lee (1971), "Dimensions of Organization Structure and Their Context: A Replication," *Sociology*, 5 (February), 83–93.

Kover, A. J. (1976), "Careers and Noncommunication: The Case of Academic and Applied Marketing Research," *Journal of Marketing Research*, 13 (November), 339–344.

Krum, J. R. (1978), "B For Marketing Research Departments," *Journal of Marketing*, 42 (October), 8–12.

Kunstler, D. A. (1975), "An Outline of AMA Research Division's Responsibilities," *Marketing News*, 8 (March 14), 12.

Larsen, J. K. (1980), "Knowledge Utilization: What Is It?," *Knowledge: Creation, Diffusion, Utilization*, 1 (March), 421–442.

Luck, D. J. and J. R. Krum (1981), "Conditions Conducive to the Effective Use of Marketing Research in the Corporation," *Report No. 81–100*, Cambridge, MA: Marketing Science Institute (May).

Moch, M. K. and E. V. Morse (1977), "Size, Centralization, and Organizational Adoption of Innovations," *American Sociological Review*, 42 (October), 716–725.

Myers, J. G., S. A. Greyser and W. F. Massy (1979), "The Effectiveness of Marketing's 'R&D' for Marketing Management: An Assessment," *Journal of Marketing*, 43 (January), 17–29.

_____, W. F. Massy and S. A. Greyser (1980), *Marketing Research and Knowledge Development*, Englewood Cliffs, NJ: Prentice-Hall.

Newman, J. A. (1962), "Put Research into Marketing Decisions," *Harvard Business Review*, 40 (March–April), 105–112.

Payne, R. and D. S. Pugh (1976), "Organizational Structure and Climate," in *Handbook of Industrial and Organizational Psychology*, M. D. Dunnette, ed., Chicago: Rand-McNally, 1125–1173.

Pennings, J. (1973), "Measures of Organizational Structure: A Methodological Note," *American Journal of Sociology*, 79 (November), 686–704.

Pondy, L. (1970), "Toward a Theory of Internal Resource Allocation," in *Power in Organizations*, M. Zald, ed., Nashville: Vanderbilt University Press, 270–311.

Pugh, D. S., D. J. Hickson, C. R. Hinings and C. Turner (1968), "Dimensions of Organization Structure," *Administrative Science Quarterly*, 13 (June), 65–105.

Radnor, M., A. Rubenstein and D. Tansik (1970), "Implementation in Operations Research and R&D in Government and Business Organizations," *Operations Research*, 18 (November–December), 967–991.

Rich, R. F. (1975), "An Investigation of Information Gathering and Handling in Seven Federal Bureaucracies: A Case Study of the Continuous National Survey," Ph.D. dissertation, University of Chicago.

_____ (1977), "Uses of Social Science Information by Federal Bureaucrats: Knowledge for Action Versus Knowledge for Understanding," in *Using Social Research in Public Policy Making*, C. H. Weiss, ed., Lexington, MA: D. C. Heath, 199–211.

Rogers, E. M. and F. F. Shoemaker (1971), *Communication of Innovations*, New York: The Free Press.

Sathe, V. (1978), "Institutional Versus Questionnaire Measures of Organizational Structure," *Academy of Management Journal*, 21 (June), 227–238.

Seidler, J. (1974), "On Using Informants: A Technique

for Collecting Quantitative Data and Controlling Measurement Error in Organization Analysis," *American Sociological Review*, 39 (December), 816–831.

Silk, A. J. and M. U. Kalwani (1980), "Measuring Influence in Organizational Purchase Decisions," unpublished working paper No. 1077–79, Sloan School of Management, Massachusetts Institute of Technology.

Spekman, R. E. and L. W. Stern (1979), "Environmental Uncertainty and Buying Group Structure: An Empirical Investigation," *Journal of Marketing*, 43 (Spring), 54–64.

Weiss, C. H., ed. (1977), *Using Social Research in Public Policy Making*, Lexington, MA: D. C. Heath.

_____ (1980), "Knowledge Creep and Decision Accretion," *Knowledge: Creation, Diffusion, Utilization*, 1 (March), 381–404.

_____ and M. J. Bucuvalas (1980), "Truth Tests and Utility Tests," *American Sociological Review*, 45 (April), 302–312.

Wilensky, H. (1967), *Organizational Intelligence: Knowledge and Policy in Government and Industry*, New York: Basic Books, Inc.

Wind, Y. (1980), "From the Editor: Marketing in the Eighties," *Journal of Marketing*, 44 (Winter), 7–9.

Zaltman, G. and R. Duncan (1977), *Strategies for Planned Change*, New York: John Wiley.

_____, _____ and J. Holbek (1973), *Innovations and Organizations*, New York: John Wiley.

A Factorial Analysis of the Authoritarian Personality

PAUL KLINE
COLIN COOPER

The authoritarian personality, like old soldiers, will never die. Since its powerful delineation by Adorno *et al.* (1950) the concept and its related personality test, the F scale (measuring the Fascistic personality), have been subjected to a number of criticisms: acquiescence and social desirability affect the scores on the F scale (Christie & Jahoda, 1954; Brown, 1965); the F scale measures only right-wing authoritarianism (Rokeach, 1960). However, Dixon (1976) convincingly argued that authoritarianism was at the root of military incompetence, despite all these problems.

Naturally advocates of the utility of the authoritarian personality as an explanatory concept have attempted to meet the criticisms: balanced scales have appeared (e.g. Ray, 1974); scales with special item forms designed to reduce response sets (Wilson & Patterson, 1970); and scales with the item content made more general (Rokeach, 1960).

All this has resulted in a plethora of scales all claiming to measure authoritarianism. In addition there are variables which are conceptually similar (rigidity, obsessional personality, anal

"A Factorial Analysis of the Authoritarian Personality," Paul Kline and Colin Cooper, Vol. 75 (1984), pp. 171–176. Reprinted from the *British Journal of Psychology*, published by the British Psychological Society. Paul Kline and Colin Cooper are professors in the Department of Psychology, University of Exeter, Washington Singer Laboratories.

characteristics). Clearly, therefore, an investigation of such scales is necessary, especially since none of them has much evidence of validity beyond item content and item homogeneity.

There is a further difficulty in this area. Factor analysis has been extensively used in the exploratory survey of personality traits. Cattell (e.g. Cattell & Kline, 1977) has claimed to have sampled the whole field. Many other authors make similar claims for their factors (e.g. Comrey, 1970; Eysenck & Eysenck, 1976; Howarth, 1980). Yet in a detailed examination of all these personality factors and their possible meanings which tried to take into account different factor analytic procedures, little evidence for the authoritarian personality could be found. The nearest factor, one of obsessional personality, was not identical in terms of descriptive traits with the authoritarian personality (Kline, 1979). Thus a question is raised: if the authoritarian personality is an important syndrome of personality, why has it not been found in factor structures? This does not mean that putative authoritarian factors cannot be found among items. Such factors are, however, of unproven validity, as we have argued.

Knapp (1976) went some way to answering the questions raised in this introduction by factoring 13 measures of authoritarianism, alienation and related variables. However, this study in fact only used two measures of authoritarian personality, the F scale and a scale which shared items with the F scale, that of Struening & Richardson (1965). Thus it is from our viewpoint not as useful as is desirable. In addition, this study was flawed in principle since no attempt was made to locate any authoritarian factors that might have emerged in a wider personality space. Without this, proper factor identification is simply not possible (see Cattell & Kline, 1977; Kline, 1979). In an attempt to identify the factor structure underlying authoritarian scales and to locate such factors in personality space, an investigation was carried out as described below.

Method

Tests Used

Authoritarianism. It was not possible to use all tests of authoritarianism and related variables simply because many of them utilize common items. However, a selection was made such that: (i) common items were avoided; (ii) balanced scales were utilized; (iii) other similar syndromes were covered; and (iv) other criticisms (e.g. of right-wing content) were taken into account. This resulted in the following scales being used (all of which were satisfactorily reliable):

(1) *Ai3Q (Anal questionnaire).* This test (Kline, 1971) measures anal character or obsessional personality, a syndrome that Dixon (1976) suggested is related to authoritarianism.
(2) *The Wilson-Patterson conservatism scale.* A scale also clearly related to authoritarianism measuring *inter alia* ethnocentrism, anti-hedonism and militarism (Wilson & Patterson, 1970).
(3) *Balanced dogmatism* (Ray, 1974). This is the scale which in terms of item content seems to capture best the spirit of authoritarianism and to take account of the psychometric problems of the F scale.
(4) *Balanced F scale* (Kohn, 1972). This was put in because it incorporated many of the improvements deemed important in our introduction and has reasonable psychometric characteristics, and some evidence for criterion group validity.
(5) *The Machiavellian scales* (Christie & Geis, 1970). These were used since it was possible that they in fact measured variables not that different from authoritarianism.

These scales were used in the expectation that they would allow the factorial nature of authoritarianism to emerge. However, the location of the authoritarian factor or factors in personality space is vital for proper identification. A clear factor, for example, could be simply anxiety under another name. Thus, to locate the factor or factors, two sets of well-established tests were used.

Tests for Locating Factors. Two tests were used:

(1) *The EPQ* (Eysenck & Eysenck, 1975) measuring extraversion, neuroticism and psychoticism together with a measure of social desirability—useful in a study of authoritarianism in the light of our criticisms.
(2) *The Cattell 16PF test* (Cattell et al., 1970) which measures the most clearly defined primary factors and possesses a measure of intelligence (factor B) which is particularly important in the light of Brown's (1965) criticism of the original F scale—that it was negatively related to educational level and thus intelligence.

This selection of tests should therefore result in the proper location of authoritarianism in factor space, provided that simple structure is obtained.

Subjects

Subjects were volunteers from among the students at the University of Exeter. This investigation was part of a larger study of personality and ability tests involving 12 hours of testing per subject in four sessions. All the tests described here were taken during one session. In fact 94 subjects completed the tests (48 males, 46 females; age: $\bar{X} = 20\cdot3$, SD = $1\cdot4$ years). These subjects were spread fairly evenly among the faculties of science, arts, social studies and engineering, and the majority were from socio-economic classes 1 and 2.

Administration of Tests

Tests were administered to 10 subjects at a time in a testing room. Breaks for coffee and biscuits were given. Rapport was good since subjects knew that all results were confidential to the research team and that any subject who wished could have his or her results individually discussed by the senior author.

Statistical Analysis

A procedure generally accepted by factor analysts as yielding replicable simple structure was used (Cattell, 1978; Carroll, 1981): principal factor extraction based upon the Scree test (Cattell, 1966), oblique rotation by direct oblimin (Jennrich & Sampson, 1966) followed by a hyperplane count. We did not use the currently popular confirmatory analysis because this was primarily an exploratory study and we remain unconvinced by the power of the chi-square test (Joreskog, 1969).

Results and Discussion

Table 1 shows the rotated factor structure of the five-factor solution. Note that we split the Wilson-Patterson test into component scales without item overlap. We carried out another analysis using the conservatism scale alone but the loadings were so similar that no extra information could be extracted from this solution.

There are a number of issues in this study that must be discussed before its full implications can be explored.

(A) CONSTITUTION OF THE SAMPLE   A student sample might be expected to be homogeneous for authoritarian traits given their relation to education. However, this study is not concerned with norms or absolute levels of this variable. Rather it is con-

TABLE 1   Rotated Factor Structure and Correlation of Factors: *Direct Oblimin Rotated Factor Structure Matrix*

	Variables	1	2	3	4	5
	Ai3	0·60	−0·11	0·10	−0·04	−0·08
Wilson-Patterson	Military	0·40	0·15	−0·18	0·04	0·08
Wilson-Patterson	Anti-hedonism	0·54	−0·01	0·60	−0·16	−0·15
Wilson-Patterson	Ethnocentrism	0·40	0·11	0·22	−0·14	0·37
Wilson-Patterson	Religion	0·53	−0·06	0·52	0·02	−0·09
Balanced	Dogmatism	0·23	−0·27	0·52	−0·14	0·31
Machiavellian	Tactics	0·16	0·01	−0·75	−0·05	0·00
Machiavellian	Views	0·02	−0·15	−0·49	−0·22	0·15
Machiavellian	Morality	−0·23	−0·08	−0·73	0·02	0·08
	Kohn's F scale	0·84	0·11	−0·01	0·18	−0·20
	EPQ E extraversion	−0·07	0·13	−0·05	0·87	0·12
	EPQ N neuroticism	−0·10	−0·87	0·04	−0·37	−0·13
	EPQ P psychoticism	−0·53	−0·03	−0·34	0·16	0·27
	EPQ L lie scale	0·10	0·06	0·33	−0·28	0·14
	16PF A sociability	0·15	−0·22	0·00	0·27	0·00
	16PF B intelligence	−0·08	0·17	0·07	−0·18	0·24
	16PF C ego-strength	−0·00	0·61	0·18	0·12	0·10
	16PF E dominance	−0·11	0·08	−0·11	0·48	0·70
	16PF F cheerfulness	−0·13	0·03	−0·16	0·73	0·12
	16PF G conscientious	0·67	0·21	0·20	−0·07	−0·03
	16PF H adventurous	−0·05	0·30	0·12	0·69	0·37
	16PF I tough minded	−0·09	−0·43	0·38	0·01	−0·36
	16PF L suspiciousness	−0·14	−0·30	−0·18	0·18	0·47
	16PF M unconventionality	−0·32	0·09	−0·03	0·09	0·09
	16PF N shrewdness	0·23	−0·14	0·08	−0·23	−0·46
	16PF O guilt feeling	−0·08	−0·77	0·07	0·07	−0·23
	16PF Q1 radicalism	−0·31	0·03	−0·36	0·12	0·45
	16PF Q2 independence	−0·00	0·12	−0·10	−0·27	0·00
	16PF Q3 self sentiment	0·55	0·33	0·11	−0·21	−0·15
	16PF Q4 tension	−0·17	−0·76	0·05	0·05	0·07
	% of total variance accounted for	14·2	12·0	11·6	10·0	7·8

Hyperplane count = 41%
Hyperplane width = ±0·1
                   = −0·6

Factor intercorrelations

	1	2	3	4
Factor 1				
Factor 2	−0·06			
Factor 3	0·14	0·09		
Factor 4	−0·05	−0·04	−0·06	
Factor 5	0·00	0·09	−0·14	0·07

cerned with the relationship between authoritarianism and the other variables. Any homogeneity would therefore underestimate the size of correlations. Significant results may be interpreted with more confidence, namely, that in more general samples correlations and factor loadings are likely to be larger. This is particularly true because the variables measured are claimed to be general traits. Despite this, it must be borne in mind that this is a student sample.

(B) THE ADEQUACY OF THE FACTOR ANALYSIS.   We realize that our sample size is small for a study of this type. However, an examination of Table 1 reveals a factor structure so clear and so in accord with expectations for the main personality variables that there can be no fear that our sample size was insufficient. This clarity of structure also persuaded us that the five factors suggested by the Scree test was the optimum solution. The Scree test also revealed a break at eight factors but this solution was not so clear.

(C) IDENTIFICATION OF THE MARKER FACTORS.   Barrett & Kline (1981) on a large Gallup sample showed that extraversion and neuroticism were unequivocal in the EPQ. *Factor 2* is clearly the N factor loading 0·87 on that scale and having high loadings on the relevant Cattell primaries, C, O and Q4. *Factor 4* is clearly the E factor loading 0·87 on that scale and again with high loadings on the relevant primaries in the Cattell test, E, F and H.

Thus the two largest personality factors have emerged as expected. This clarity of structure despite the relatively low hyperplane count is striking (equally in the light of opposition to trait theories of personality). These factors incidentally have very low correlations with each other and are essentially independent. Thus confident that our marker variables are working and that a good simple structure has been obtained we can now meaningfully interpret and locate authoritarianism in the personality sphere.

This factor analysis enables certain questions to be answered:

1. Is there a factor of authoritarianism? The answer must be yes. In our sample factor 1 has loadings on most of the authoritarian scales.
2. Is this factor independent of extraversion and neuroticism? Clearly yes. The authoritarian factor measures variance quite separately from these factors in this sample of students.

The Nature of the Authoritarian Factor

Regarding loadings greater than 0·3 as salient, we shall now describe the authoritarian high scorer as delineated by other loadings on factor 1. The authoritarian is, according to the Cattell scales, conscientious (G+) conventional (M+) conservative (Q1−) and controlled with high willpower (Q3+). Now in the light of the original description of the authoritarian by Adorno and his colleagues (1950), the findings are striking confirmation: the words conventional, adherence to rigid values and, of course, the belief in an iron-will all occur in this description.

The fact that the obsessional trait test Ai3Q, which regularly loads on an obsessional trait factor (Kline, 1979) further supports the notion of authoritarianism as rigid rule following.

Indeed the only loading not in obvious accord with the nature of authoritarianism is the negative loading on P, Eysenck's factor of psychotism. With its tough-minded, inhumane characteristics, a positive loading might have been more sensible. However, the aspect of P, the delight in strange sensations, the open admission of unconventionality is the cause of the negative loading (e.g. item 26, concerned with hurting people you love—the antithesis of the authoritarian's conventional respect for the family and the goodness of mother).

Some other interesting features of authoritarianism are revealed by the tests that do not load on the factor. EPQ-L, the measure of social desirability, does not load on this factor. This indicates that the new authoritarian scales are not as distorted by this response set as was the original F scale (e.g. Brown, 1965). Similarly the fact that B, intelligence, does not load on the factor argues against the claim that being authoritarian is a function of low intelligence (although caution in interpretation is necessary here with our student sample).

Thus the other scale factor loadings clarify well our picture of the authoritarian personality—rigid, conscientious, conventional and obsessional. We shall now turn to two final points: (a) the quality of the authoritarian measures themselves and (b) the proper identification of the authoritarian factor.

The highest loading scale is Kohn's F scale,

the least good being balanced dogmatism. It is certainly not possible to argue that this factor 1 is not the same as Adorno's concept, for Kohn's scale is clearly related to the original. The Wilson-Patterson subscales are all substantially loaded on the factor. Indeed the analysis with the conservatism scale alone produced a loading of 0·786, virtually as high as the F scale.

On these grounds there is little to choose between the Kohn scale and the Wilson-Patterson C scale as measures of authoritarianism. Both are unifactorial. However, the Wilson-Patterson subscales are not: they load also on factor 3 as does balanced dogmatism. Factor 3 is clearly the Machiavellian factor.

Finally the question must be asked concerning the identification of this factor 1. Here, the Ai3Q obsessional scale gives us a useful clue. In a recent study of the Dynamic Personality Inventory (Kline & Storey, 1978), an obsessional trait factor emerged with similar loadings on the Cattell scales. This is a factor which has emerged in various studies with Ai3Q (see Kline, 1979). For these reasons, together with the fact of the G and Q3 loadings, we would interpret this factor 1 as the obsessional trait factor, meaning that, as Dixon (1976) argued, authoritarianism is a close relative of the obsessional personality. That P loads on obsessional personality is perhaps an accident of the variables used in this study: there was no other factor on which it could load and authoritarianism is clearly antithetic to psychoticism.

Conclusions

The conclusions can be stated briefly in view of our discussion:

1. Within a student sample authoritarian scales do load on a factor independent of the main temperamental factors.
2. This factor is probably that of obsessional personality.
3. The factor is not distorted by social desirability or, in our sample, intelligence.
4. Of the scales used, Kohn's F scale and the Wilson-Patterson C scale are the clearest.
5. The factor is independent of the Machiavellian scales.
6. In summary one would argue that these results support the utility of the concept of authoritarianism in the theory and measurement of per-

sonality. Clearly these findings require replication in other less homogeneous samples.

References

Adorno, T. W., Frenkel-Brunswick, E., Levenson, D. J. & Sanford, R. N. (1950). *The Authoritarian Personality*. New York: Harper.

Barrett, P. & Kline, P. (1981). Factors in the EPQ. *Personality and Individual Differences*, **1**, 317–333.

Brown, R. (1965). *Social Psychology*. New York: Free Press.

Carroll, J. B. (1981). Ability factors and processes. Paper given at the NATO Conference on Cross-Cultural Psychology, Kingston, Ontario.

Cattell, R. B. (1966). The Scree test for the number of factors. *Multivariate Behavioral Research*, **1**, 140–161.

Cattell, R. B. (1978). *The Scientific Use of Factor Analysis in Behavioral and Life Sciences*. New York: Plenum.

Cattell, R. B. & Kline, P. (1977). *The Scientific Analysis of Personality and Motivation*. London: Academic Press.

Cattell, R. B., Eber, H. W. & Tatsuoka, M. M. (1970). *Handbook to the 16 Personality Factor Questionnaire*. Windsor, Berks: National Foundation for Educational Research.

Christie, R. & Geiss, F. L. (1970). *Studies in Machiavellianism*. New York: Academic Press.

Christie, R. & Jahoda, M. (eds) (1954). *Studies in the Scope and Method of the Authoritarian Personality'*. Glencoe, IL: Free Press.

Comrey, A. L. (1970). *The Comrey Personality Scales*. San Diego, CA: Educational and Industrial Testing Service.

Dixon, N. F. (1976). *On the Psychology of Military Incompetence*. London: Jonathan Cape.

Eysenck, H. J. & Eysenck, S. B. G. (1975). *The EPQ*. London: London University Press.

Eysenck, H. J. & Eysenck, S. B. G. (1976). *Psychoticism as a Dimension of Personality*. London: Routledge & Kegan Paul.

Howarth, E. (1980). *Howarth Personality Questionnaire*. Alberta: University of Alberta Press.

Jennrich, R. I. & Sampson, P. F. (1966). Rotation for simple loadings. *Psychometrika*, **31**, 313–323.

Joreskog, K. G. (1969). A general approach to confirmatory maximum likelihood factor analysis. *Psychometrika*, **34**, 183–202.

Kline, P. (1971). *Ai3Q Test*. Windsor, Berks: National Foundation for Educational Research.

Kline, P. (1979). *Psychometrics and Psychology*. London/New York: Academic Press.

Kline, P. (1982). Psychometrics: A science with a great future behind it. Paper given at the International Congress of Applied Psychology, Edinburgh.

Kline, P. & Storey, R. (1978). The Dynamic Personality Inventory: What does it measure? *British Journal of Psychology*, **69**, 375–383.

Knapp, R. J. (1976). Authoritarianism, alienation and related variables: A correlational and factor-analytic study. *Psychological Bulletin*, **83**, 194–222.

Kohn, P. M. (1972). The Authoritarian–Rebellion Scale: A balanced F scale with left-wing reversals. *Sociometry*, **35**, 176–189.

Ray, J. J. (1974). Balanced recognition scales. *Australian Journal of Psychology*, **26**, 9–14.

Rokeach, M. (1960). *The Open and Closed Mind.* New York: Basic Books.

Struening, E. L. & Richardson, A. H. (1965). A factor analytic exploration of the alienation, anomia and authoritarianism domain. *American Sociological Review*, **30**, 768–777.

Wilson, G. D. & Patterson, J. R. (1970). *The Conservatism Scale.* Windsor, Berks: National Foundation for Educational Research.

The Application and Misapplication of Factor Analysis in Marketing Research

DAVID W. STEWART

Factor analysis is one of the more widely used procedures in the market researcher's arsenal of analytic tools. Despite its wide-scale usage, factor analysis is not a universally popular technique and has been the subject of no small amount of criticism (Ehrenberg 1968; Ehrenberg and Goodhart 1976). Both the popularity and the criticism of the method can be traced to the availability of computer programs that are easy to use and require little knowledge of the underlying theory or methodology of factor analysis. This rather blind use of factor analysis is a principal cause of the dissatisfaction with the technique.

So widespread are current misconceptions about factor analysis in the marketing community that even its defenders and some prominent reviewers perpetuate misinformation. Many reviewers have discussed the use of factor analysis as a clustering procedure, at best an extreme perversion of the method. The alleged subjectivity of the technique is part of the folklore of the discipline. In a field where precedent often provides justification for action, misinformation about and misapplication of factor analysis tend to be perpetuated. The intent of this article is to review selected issues involved in the use of factor analysis.

"The Application and Misapplication of Factor Analysis in Marketing Research," David W. Stewart, Vol. 18 (February 1981), pp. 51–62. Reprinted from the *Journal of Marketing Research*, published by the American Marketing Association. David W. Stewart is professor of Management, Owen Graduate School of Management, Vanderbilt University.

What Is Factor Analysis?

Factor analysis is a multivariate statistical technique that is concerned with the identification of structure within a set of observed variables. Its appropriate use involves the study of interrelationships among variables in an effort to find a new set of variables, fewer in number than the original variables, which express that which is common among the original variables. Factor analysis establishes dimensions within the data and serves as a data reduction technique. Three general functions may be served by factor analysis.

1. The number of variables for further research can be minimized while the amount of information in the analysis is maximized. The original set of variables can be reduced to a small set which accounts for most of the variance of the initial set.

2. When the amount of data is so large as to be beyond comprehension, factor analysis can be used to search data for qualitative and quantitative distinctions.

3. If a domain is hypothesized to have certain qualitative and quantitative distinctions, factor analysis can test this hypothesis. Thus, if a researcher has an *a priori* hypothesis about the number of dimensions or factors underlying a set of data, this hypothesis can be submitted to a statistical test.

Much of the expressed dissatisfaction with factor analysis can be attributed to its use for purposes other than those stated above.

Comments on the Interpretation of Factors

A basic assumption of the factor analyst is that major differences found in everyday human relationships become part of the language of the culture. At one level factor analysis is concerned with how people use the language, its words, concepts, etc. and the empirical relationships within the language. The underlying assumption is that these empirical relationships within the language will reveal something about human behavior on another level. A factor is a qualitative dimension, a coordinate axis. It defines the way in which entities differ much as the length of an object or the flavor of a product defines a qualitative dimension on which objects may or may not differ. A factor does not indicate how much different various entities are, just as knowing that length is an important physical dimension does not indicate how much longer one object is than another. Quantitative differences may very well be important, but the identification of a particular factor does not provide that information. Factor analysis provides a dimensional structure for data; it indicates the important *qualities* present in the data. This definition of factor analysis is clearly in keeping with both Thurstone's (1942, p. 69–70) and Cattell's (1978, p. 44–6). It also has a bearing on at least two common misconceptions about factors.

Factors are not clusters. The confusion between factors and clusters seems to have arisen because a cluster is "more concrete, immediately evident, and easier to understand than a factor" (Cattell 1978, p. 45). Clusters are defined by relatively contiguous points in space and are not the axes that are factors. The criterion for admission to a cluster is much more arbitrary than is the definition of a factor. There are no well-established rules for the definition of a cluster as there are for a factor. As a result, clusters tend to be weaker theoretical constructs than are factors. Cattell (1978) provides a systematic treatment of the distinction. It is sufficient here to point out that there will almost always be more clusters than factors though a cluster may be close to a particular factor. Because factor analysis provides a set of coordinate axes, numerous clusters can be defined by the position of entities on any one dimension and by various combinations of dimensions. It is also possible to obtain meaningful factors in the absence of clusters. Such a circumstance would occur where the subject coordinates are homogeneously distributed in factor space. The practice of interpreting an otherwise successful factor analysis in terms of clusters of products, or clusters of television shows, misses the point of factor analysis. Such clusters could probably be identified more readily by other means (Ehrenberg and Goodhart 1976).

Few factor analytic studies in the marketing literature have provided such an interpretation. Ehrenberg and Goodhart (1976) give three examples of marketing applications of factor analysis. In two of the examples, one involving purchase frequency of ready-to-eat cereals and the other involving television viewing habits, an attempt is made to interpret the factors as clusters. Ziff (1974) and Pernica (1974) describe studies using factor analyses of psychographic data as a means for identifying market segments, i.e., clusters of persons similar with respect to certain attributes. In each case what the factor analysis has actually done is to identify the important dimensions by which people or products *may* be differentiated. The determination of actionable market segments requires analysis beyond factoring. Horton (1979) provides an example of a more appropriate use of factor analysis. Starting with a large number of personality and demographic variables, he used factor analysis to reduce the data to six basic dimensions. These six dimensions, rather than the original set of variables, were then related to product usage by means of discriminant and canonical variate analyses. Thus, factor analysis provided a means for reducing the number of variables in the study without great loss of information and served to identify the important qualitative distinctions in the data.

Also related to the notion that factors are underlying dimensions involving an examination of the use of language (through there may be applications in which language is less important) is the problem of differences among factors across populations. Differences in factors have been found between samples taken from different cultures, between samples of children and adults, and between samples of men and samples of women. In each case it is reasonable to assume that the individuals who make up the samples use language somewhat differently. Thus, when language differences are suspected to be present, separate factor analyses may be warranted. In most marketing applications they are not. There is no reason to presume that the user of beer (at least when sober) will use the language in a manner different from that of the nonuser. The underlying dimensions should be the same. In fact, when different factors emerge analysis must stop. Apples and oranges cannot be compared beyond mere description. The marketing researcher should hope to find the same factors so that comparisons of magnitude on the same factor can be made through the use of factor scores.

Modes of Factor Analysis

Most market researchers are familiar with two modes of factor analysis, the R technique whereby the relationships among items or variables are examined, and the Q technique whereby the relationships among persons or observations are examined. Several other modes of analysis with which market researchers may be less familiar may be useful to the thoughtful researcher. Table 1 is a description of the several modes of factor analysis. Though not well known to market researchers, the S and T techniques appear particularly suited for the analysis of purchase incidence and message recall data. The P and O techniques might be useful for analyzing the life cycle of a particular product class or promotional campaign. Cattell and Adelson (1951) and Cattell (1953) give examples of the use of P technique for analyzing changes in demographic and economic characteristics of a nation over time.

All of the various modes of factor analysis provide information about the dimensional structure of data. The use of factor analysis as a means for defining types, or clusters, has led to much of the criticism of the method and, in some cases, the outright rejection of factor analysis (Ehrenberg 1968). Q factor analysis is often a principal tool employed in market segmentation studies. Its use has continued despite repeated criticisms and its disappointingly poor record of reliability (Wells and Sheth 1974). The use of Q factor analysis

for the identification of segments or types is one of the paramount abuses of factor analysis.

Applications of Q factor analysis as a clustering tool have involved at least two major misunderstandings. Both might have been avoided by a reading of the original literature of factor analysis rather than the superficial summaries found throughout the literature. The essential concepts of Q factor analysis can be found in Burt's early work (1917, 1937). The issues have recently been restated by Cattell (1978, p. 326–7).

The two main problems in Q technique . . . concern a) the need for introducing sampling methods with variables, and b) the realization that though it correlates people it is not a method for finding types but for finding dimensions. Further, as in all transpose pairs, the dimensions from Q technique are systematically related to, and with proper conditions, identical with, those from the corresponding R technique analysis. . . .

As regards *dimensions*, Q technique tells us nothing we do not know from R technique. . . . The choice of R or its transpose (or any other technique and its transpose) is therefore not a matter of end goal but of convenience and of the ease of meeting statistical requirements. . . .

. . . Rogers' index of similarity of two people (or an ideal and a real person) by Q-sort overlooked that the correlation coefficient is blind to differences of level in their profiles, while other "Q typists" have overlooked that any different choice of variables could lead to a radical difference of correlation between two people and of subsequent sorting.

Thurstone takes a similar position (1942, p. 558).

When an analyst uses Q factor analysis to define types or segments, each dimension defines a segment. Figure 1 provides three examples of the types of distortions this approach may bring about. Figure 1A is a two-dimensional space. The axes represent two factors identified by Q technique. The points represent the vector termini of the subjects in the two-dimensional space. Two very obvious segments are present. These segments differ in terms of the direction and level of association with Factor I. A Q factor analysis used for typing would define two segments, one associated with Factor I and one associated with Factor II. However, Factor II makes no appreciable contribution to the differentiation of the segments. Any attempt to classify subjects on the basis of their relationship to Factor II would distort the actual clusters present in the data. In addition, if only one factor

TABLE 1   Modes of Factor Analysis

Technique	Factors Are Loaded by	Indices of Association Are Computed Across	Data Are Collected on
R	Variables	Persons	one occasion
Q	Persons	Variables	one occasion
S	Persons	Occasions	one variable
T	Occasions	Persons	one variable
P	Variables	Occasions	one person
O	Occasions	Variables	one person

FIGURE 1

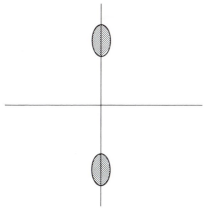

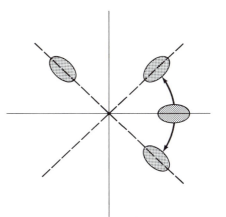

(a) Two dimensions—two clusters

(b) Two dimensions—four clusters (dotted lines represent possible transformation of axes reducing clusters to two

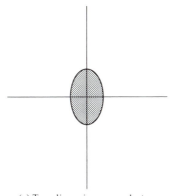

(c) Two dimensions—one cluster

were retained in the Q factor analysis it would suggest only one segment. Indeed, factor analysis could not recover the two segments in the data.

Figure 1B is a set of four clusters in two-dimensional space. Again, the axes are the factors defined by Q technique and the points are the termini of the subject vectors in person space. All four clusters are uniquely defined. A Q factor analysis used for typing would distort the nature of the segments. Because a cluster interpretation of Q factor analysis would require a segment to correspond to each factor, no more than two segments could be identified. Rotation of the axes as is customary in factor analysis would still result in only two clusters. After rotation clusters 1 and 4 would be indistinguishable and cluster 3 would be divided between cluster 2 and cluster 1, 4.

Figure 1C is only one cluster in two dimensions. However, defining clusters as axes would cause the single cluster to be divided in two on the basis of the factor loadings of the subjects. Thus, the analyst might be led to believe a unitary market was composed of two segments.

Some researchers (Levin 1965; Overall and Lett 1972; Tucker 1972) have argued that the factors obtained from Q analysis might be interpretable as ideal or pure types. This approach to the definition of Q factors would be acceptable if most individuals relate primarily to only one ideal type or if one is willing to ignore all subjects who do not load on only one factor. In most marketing applications it is unlikely that many subjects will load on a single factor. In such cases the discussion of ideal types tells the marketer little about the nature of the market segments. Subjects will belong to a different type for each dimension found.

The only appropriate use of Q factor analysis is as the transpose of R factor analysis. When more variables than observations are present, Q factor analysis will result in a more stable factor solution than R factor analysis (Gorsuch 1974). The reason is that the standard error of a correlation is a function of the sample size. Cliff and Hamburger's (1967) results suggest that the standard errors of the elements of a factor loading matrix are even greater than those of the ordinary correlation coefficients. Further, statistical independence of correlation coefficients in an R analysis can be obtained only when the number of subjects is greater than the number of variables. Thus, when faced with a circumstance involving more variables than subjects, an analyst can compute factors by using the Q technique. The solution will be more reliable than that obtained with the R technique. In addition, many current computer software packages will not provide a solution to an R factor analysis if the number of variables is greater than the number of subjects. Both Gorsuch (1974) and Cattell (1978) hold that the use of Q factor analysis is justified only when there are considerably fewer observations than variables. Otherwise, R factor analysis should be used to maximize the stability of the solution.

The relationship between the R mode of factor analysis and the Q mode has been recognized by some marketing researchers (Johnson 1970) and is often capitalized on in the computation of a Q factor analysis. The most commonly used computer programs for Q factor analysis actually begin with the R matrix. The rationale for this approach lies in the computational limitations of computers and the need for a matrix composed of independent correlation coefficients. No computer in use today can decompose a correlation matrix of several thousand individuals; the smaller R matrix is used instead. In addition, the decomposition of a Q matrix based on fewer variables than subjects presents a more fundamental problem of decomposition. In such a case the correlation coefficients in the Q matrix are not independent of one another. To obtain a unique factor solution the R matrix is used. The Eckhart-Young decomposition procedure (1936) can be employed to find Q factors by first decomposing the R matrix. The components of the R matrix can then be used to determine the Q factors. Yet, this is just the case in which the R mode would provide a more stable

result. One reason for the observed unreliability of the results of Q factor analysis is the failure to use it only when the number of variables is greater than the number of observations. Thus, the use of Q factor analysis with large numbers of subjects and small numbers of variables is questionable at best.

In fact there is a well-known, if complex, relationship between the factors obtained from a R analysis and those obtained from a Q analysis (Burt 1937; Cattell 1966a, 1978). If an unrotated factor solution is obtained from the cross-products matrix derived from the original raw score matrix, the R and Q techniques will produce the same factors (Cattell 1966a). If a correlation matrix is used, the first unrotated factor obtained from the R technique is missing in the Q technique and vice versa. The first factor obtained in Q analysis is a "species" factor, that is, a general factor related to the average characteristics of the subjects. If physical measures were involved this first factor would tell only that people have longer arms than fingers. The first general factor obtained from the R technique is a general size factor related to the average differences among subjects. Various transformations have been suggested to eliminate these general factors and to make R and Q analyses of correlation matrices equivalent. Though none of these approaches appears to completely solve the problem, the use of a double-centered matrix whereby all the means and variances of the variables and all the means and variances of the subjects are brought to the same value offers some hope.

Factors obtained from either mode are interpreted the same way, as dimensions. The factors obtained in the R mode are dimensions along which attributes may differ; the factors of Q analysis are dimensions along which persons may differ. Aside from the first factor the unrotated factors of Q and R techniques should be identical. Thus, the choice of which mode to use is a tactical decision dictated by the relationship of attributes to subjects.

Other problems with Q factor analysis make it less desirable than R factor analysis in most cases. For example, in significance testing one would assume a random sampling of variables from a population of variables. Because this assumption would be very difficult to defend, significance tests do not generally apply to Q factor analysis. When the definition of a variable is arbitrary in terms of direction of scoring, the direction selected can affect the resulting correlation coefficient. Cohen (1968, 1969) provides an example in which one of five variables is reversed in scoring and the correlation coefficient between individuals shifts from .67 to −.37. This effect is not found in the R technique because an individual cannot be reversed. Cohen (1968, 1969) suggests a method for addressing this problem. An additional problem associated with the use of Q factor analysis with large numbers of observations is related to the rotation procedure. It is customary to rotate the eigenvectors obtained from the factoring procedure. Given current computer limitations, no more than about 10 Q factors can be rotated when several thousand observations are involved. It is reasonable to assume that more than 10 dimensions are present in many data sets. The arbitrary limitation on rotation can have severe repercussions on the results of a factor analysis. Evidence that rotation of too few factors seriously distorts the final solution is presented hereafter.

A note on the use of factor analysis for defining types would not be complete without at least a mention of the problems associated with using covariance and correlation indices for that purpose. The computation of a covariance eliminates differences associated with the means of variables or observations, whereas the correlation coefficient eliminates differences attributable to both the mean and the dispersion of the variables or observations. The information contained in the correlation coefficient describes only the extent to which variables or observations vary together and the shape of their distributions. Information about mean and variance differences among individuals would be particularly important in a clustering aplication. Though it may be reasonable to assume that the mean and the variance of a *variable* are arbitrary, this assumption is probably not valid with respect to individuals. Computing correlation coefficients among individuals thus removes potentially meaningful information. A more appropriate measure of individual similarity is the Mahalanobis generalized distance, D^2, which incorporates information about mean and variance differences among individuals.

Related to the correlation problem is the common practice of using ipsative data as the starting point for a clustering procedure. Ipsative data contain information about intraindividual differences only. Interindividual differences, which would be of primary interest in market segmentation, are eliminated. Ipsative measures for a given variable

are not comparable from one individual to another. Thus, a score of three on one variable for one person may not have the same meaning as a score of three on the same variable for another person. Ipsative data may arise as a result of data collection procedures, as in Q sorts, or through an arithmetic transformation. Indeed, Q factor analysis implicitly ipsatizes data, thus removing any mean differences among the individuals. The transformation differs from subject to subject so that comparisons among subjects are meaningless. Any data set in which the rows add up to the same value would be said to be ipsative. A comprehensive treatment of the problems posed by ipsative measures is given by Clemans (1956).

The foregoing discussion should not be taken to mean that there is no place in segmentation for factor analysis. Factor analysis, properly employed, may play an important and perhaps essential role in the definition of types. Its role is to identify the basic dimensions of a data set. These dimensions can then be used for further analyses aimed at identifying segments. Gorsuch (1974) and Cattell, Coulter, and Tsujioka (1966) provide excellent overviews of the search for types and the place of factor analysis in defining clusters.

Types of Factor Analysis

Two very general types of factor analysis can be defined by the intended purpose of the analysis. In those cases where the underlying dimensions of a data set are unknown, exploratory factor analysis is appropriate. That type of analysis is most common in marketing research applications. Though exploratory factor analysis is an important tool, all too often it is applied *post hoc* to a set of data that was collected without regard to the assumptions and requirements of factor analysis. Previous factor analytic research is often ignored and factors are given new names by each researcher. Ehrenberg (1968, p. 58) has criticized the use of factor analysis for that reason.

Factor analysis is in fact generally regarded as an exploratory technique, to be used when one happens to know nothing about the subject matter.

Factor analysis need not be merely a data reduction tool or a means for exploratory data analysis. Indeed, it is a particularly appropriate procedure for theory building. Relationships can be hypothesized and tested by use of confirmatory factor analysis, a procedure developed to allow the testing of hypotheses about the structure of a data set. This second type of factor analysis, confirmatory analysis, has been virtually ignored by marketing researchers. The most common of the confirmatory factor analytic procedures is maximum likelihood factor analysis (Jöreskog 1971). This procedure is an attempt to estimate population parameters from sample statistics. Thus, in using the procedure one seeks to provide generalizability from a sample of individuals to a population of individuals (in exploratory procedures one assumes the entire population of interest is represented in the analysis). The procedure is concerned with estimating the population correlation matrix under the assumption that the variables come from a designated number of factors. The procedure has built-in tests of statistical significance for the number of factors.

Anderson, Engledow, and Becker (1979) provide an excellent example of the use of confirmatory factor analysis in a marketing application. These researchers used maximum likelihood factor analysis to test a hypothesis about the underlying dimensions of a set of attitudinal and behavioral data. The hypothesis was generated from an exploratory factor analysis of a data set collected some years earlier.

A second example of the use of factor analysis for hypothesis testing can be found in an article by Jones and Siller (1978). Although these authors did not use the maximum likelihood procedure, they hypothesized that several dimensions related to the content of a daily newspaper could be found in audience exposure data. The authors were able to confirm their prior expectations by using a factor analytic procedure.

Any researcher embarking on the use of factor analysis is confronted with a bewildering array of techniques. Principal components, principal factors, alpha analysis, and maximum likelihood analysis are but a few of the types of factor analysis the researcher may encounter. An argument often advanced by the critics of factor analysis is that the choice of technique is crucial to the final result. Fortunately, the empirical evidence comparing the several types of factor analysis does not support this conclusion (Browne 1968; Gorsuch 1974; Harris and Harris 1971; Tucker, Koopman, and Linn 1969).

An important difference among the several procedures is the nature of the value placed in the diagonal of the covariance or correlation matrix. This issue is actually a problem of communality estimation. Communality estimation is an important theoretical issue and the interested reader can refer to Gorsuch (1974) or Cattell (1978) for a discussion of the problem. Fortunately, the communality issue is of less practical concern for the typical marketing researcher. Studies by Tucker, Koopman, and Linn (1969), Harris and Harris (1971), and Browne (1968) have demonstrated that practically any technique other than the multiple-group analysis (an early development in factor analysis, seldom used today) will lead to the same interpretations. This is nearly always the case when the number of variables is moderately large and the analysis contains virtually no variables expected to have low communalities, e.g., less than .4. The principal components procedure, which involves using ones in the diagonals of the correlation matrix, does tend to produce somewhat inflated loadings in comparison with the other procedures but otherwise yields similar results. When communalities are high there are virtually no differences among the procedures. Thus, the subjective choice of procedure ultimately has little bearing on the results of an analysis.

When Is a Data Set Appropriate for Factor Analysis?

Even when the goal of the researcher is to identify dimensions within a set of data, factor analysis may not be appropriate. The determination of the appropriateness of factor analysis for a particular data set is not always a simple matter. The factors obtained from an analysis are often readily interpretable and intuitively appealing. Armstrong and Soelberg (1968) and Shaycroft (1970) show that an ostensibly acceptable factor structure can be obtained through the application of factor analysis to a correlation matrix based on random normal deviates. Few computer programs provide any test of the appropriateness of a matrix for factoring and even those that do then go on to produce a solution which the researcher may try to interpret.

There are several very useful methods for determining whether a factor analysis should be applied to a set of data. Some are relatively simple; others require some computation. Two of the simplest procedures for determining the appropriateness of a matrix for factoring are the examination of the correlation matrix and the plotting of the latent roots obtained from matrix decomposition.

An examination of communality estimates may also be instructive. Though none of these procedures may be definitive, they may provide some important clues. If the correlation coefficients are small throughout the matrix, factoring may be inappropriate. Factor analysis is concerned with the homogeneity of items. A pattern of low correlations indicates a heterogeneous set of items.

A plot of the latent roots obtained from a factoring procedure should ordinarily contain at least one sharp break. This break may represent the point where residual factors are separated from the "true" factors (details of this procedure are given hereafter). If a plot of the original, unrotated roots results in a continuous, unbroken line, whether straight or curved, factoring may be inappropriate. An examination of the communality estimates should reveal moderate to large communalities. Consistently small values may be an indication that factor analysis is inappropriate.

Another procedure for determining the appropriateness of a correlation matrix for factoring is an inspection of the off-diagonal elements of the anti-image covariance or correlation matrix. The anti-image of a variable is that part which is unique, i.e., cannot be predicted from the other variables. The work of Kaiser (1963) suggests that when the factor-analytic model is appropriate for a set of data, the matrix of the covariances or correlations of the unique parts of the variables should approach a diagonal. Even in the absence of a factor-analytic package that routinely produces the anti-image matrix, such a matrix can be readily obtained by $S_2R^{-1}S^2$, where R^{-1} is the reverse of the correlation matrix and the diagonal matrix S^2 is $(\text{diag } R^{-1})^{-1}$. If the anti-image matrix does have many nonzero off-diagonal entries, the correlation matrix is not appropriate for factoring. Psychometric theory also suggests that the inverse of the correlation matrix, R^{-1}, should be near diagonal of the matrix is appropriate for factoring (Dziuban and Shirkey 1974).

A widely programmed statistical test of appropriateness is Bartlett's test of sphericity (1950, 1951). Tobias and Carlson (1969) recommend that the test be applied prior to factor analysis. It is computed by the formula:

$$-\left[(N-1) - \left(\frac{2P+5}{6}\right)\right] \text{Log}_e|R|$$

where N is the sample size, P is the number of variables, and $|R|$ is the determinant of the correlation matrix. The statistic is approximately distributed as a chi square with $\frac{1}{2}P(P-1)$ degrees of freedom. the hypothesis tested is that the correlation matrix came from a population of variables that are independent. Rejection of the hypothesis is an indication that the data are appropriate for factor analysis. One problem with this procedure has been demonstrated by Knapp and Swoyer (1967): for sample sizes greater than or equal to 200, variables equal to 10, and a .05 level of significance, one is virtually certain to reject the independence hypothesis when the intercorrelations among the variables are as low as .09. Thus, though failure to reject is a clear sign of inappropriateness, rejection of the independence hypothesis may fail to indicate appropriateness.

A final test of the appropriateness of a matrix for factoring is a Kaiser-Meyer-Olkin measure of sampling adequacy, MSA (Kaiser 1970). This measure appears to have considerable utility and has recently been incorporated into some statistical software packages (Dixon 1975). Although it involves additional computation the MSA may be the best of the methods currently available. A measure can be obtained for both the correlation matrix as a whole and for each variable separately. The overall MSA is computed as:

$$\text{MSA} = \frac{\sum\limits_{j \neq k}\sum r_{jk}^2}{\sum\limits_{j \neq k}\sum r_{jk}^2 + \sum\limits_{j \neq k}\sum q_{jk}^2}$$

where q_{jk}^2 is the square of the off-diagonal elements of the anti-image correlation matrix and r_{jk}^2 is the square of the off-diagonal elements of the original correlations. For each variable the formula is:

$$\text{MSA}_j = \frac{\sum\limits_{k \neq j} r_{jk}^2}{\sum\limits_{k \neq j} r_{jk}^2 + \sum\limits_{k \neq j} q_{jk}^2}$$

The MSA provides a measure of the extent to which the variables belong together and are thus appropriate for factor analysis. Kaiser and Rice (1974) give the following calibration of the MSA.

.90+—marvelous
.80+—meritorious
.70+—middling
.60+—mediocre
.50+—miserable
Below .50 —unacceptable

Indiscriminate application of factor analysis to data sets is both unwarranted and misleading. The use of one, or preferably several, of the preceding techniques may prevent the analyst from factor analyzing data that are inappropriate for that purpose. The analyst can then search for alternative methods for answering questions about the data.

The Number of Factors Problem, Revisited

Perhaps no problem has generated more controversy and misunderstanding than the number of factors problem. Ehrenberg and Goodhart (1976) discuss the "relatively arbitrary decision of how many factors are extracted." Careful examination of the literature of factor analysis indicates that the criteria for ceasing to extract factors are both well established and objective. Even when the analyst has no idea about what to expect from a data set, several very useful stopping rules are available.

Bartlett's test (1950, 1951) is one of the most widely used statistical rules for determining the number of factors to extract. It has been incorporated into a number of factor-analytic packages. Gorsuch (1973, 1974) has reviewed this procedure, which tests the hypothesis that the correlation matrix (or subsequent residual correlation matrices) does not depart from an identity matrix except by chance. His conclusion is that this test is useful only for those situations in which the *complete* components model is used. Further, it is applicable only to R analysis and serves only to indicate the *maximum* number of factors that could be extracted with confidence. Gorsuch presents data to support his position and concludes that Bartlett's test should not be applied routinely when one is attempting a factor analysis.

Even more widely used than Bartlett's test is the well-known roots criterion. This procedure, which stops the extraction process when all factors with eigenvalues greater than 1.0 have been removed, is an almost universal default criterion built into most computer programs. The rule is widely misunderstood. First, the rule involves the assumption that population correlations are being considered and that it represents the minimum number of factors in such cases. The introduction of error can substantially change the accuracy of the measure. Second, the roots ≥ 1.0 criterion holds only when unities are in the diagonal of

the correlation matrix being factored (Guttman 1954). When squared multiple correlations are in the diagonal, the criterion should be roots greater than 0.0. The rationale for the rule lies in the psychometric proof that the reliability of the factor scores in association with factors that have roots smaller than 1.0 in the case of unities in the diagonal, or 0.0 when squared multiple correlations are in the diagonal is negative. Because a negative reliability is an absurdity such factors would be meaningless. Though the roots criterion provides an indication of the minimum number of factors, many researchers have used it to determine both the minimum and maximum number of factors.

Horn (1963, 1965) and Hakstian and Muller (1973) criticize this procedure on theoretical grounds, but the most damaging testimony about the roots criterion comes from Monté Carlo work with the technique.

Linn (1968) and Tucker, Koopman, and Linn (1969) show that when a sample correlation matrix is being factored, the roots criterion produces an exorbitant number of factors. Gorsuch (1974) concludes that the roots criterion is at best an approximate procedure. The roots criterion appears to be most accurate when the number of variables is small to moderate and the communalities are high. When large numbers of variables are involved (e.g., >40), the roots criterion seems particularly inaccurate.

Horn (1965) offers a useful if complex and time-consuming approach to the determination of the number factors. Horn suggests the construction of a series of random numbers matrices with characteristics similar to those of the data matrix of interest, e.g., the same dimensions. These random numbers matrices are factored and the roots obtained are plotted. The roots obtained from the data matrix are also plotted and the point where the roots of the data matrix intersect the roots of the random numbers matrices is taken as the point for stopping the factoring. The approach has much to justify it and seems to work well. The time and cost of constructing random numbers matrices have limited its use, however.

A simpler approach is suggested by Cattell (1966b). Cattell's scree test involves the plotting of the roots obtained from decomposition of the correlation or covariance matrix. A large break in the plot of the roots is taken to indicate the point where factoring should stop. The procedure is relatively simple to apply. A straight edge is laid across the bottom portion of the roots to see where they form an approximately straight line. The point where the factors curve above the straight line gives the number of factors, the last factor being the one whose eigenvalue immediately precedes the straight line. Strong support for the efficacy of the scree test is provided by Tucker, Koopman, and Linn (1969), Cattell and Dickman (1962), Cattell and Sullivan (1962), Cattell and Gorsuch (1963), and Cattell and Jaspers (1967). The most definitive study to date of the scree test is that of Cattell and Vogelmann (1977). The authors provide a very complete set of guidelines for the application of the test and report that even relatively naïve individuals are able to obtain reliable and dependable judgments of the number of factors. Because their data involved data sets with known factor structures and included some "tricky" solutions, this finding suggests that the determination of the number of factors is not a subjective decision if one follows the procedure and has a modicum of training. Cattell and Vogelmann also report the scree test to be more accurate than the roots criterion. Woods (1976) recently used the rules for the scree test given by Cattell and Vogelmann (1977) and Cattell (1978) to develop a computer program which automates the test. Such a routine should facilitate the reliability of factor solutions obtained by investigators.

Most authorities in the field recommend a combination of approaches for determining the number of factors to extract (Cattell 1978; Gorsuch 1974; Harman 1976). The use of the roots criterion and the scree test appears to provide an effective means for determining the number of factors.

Several researchers have sought to examine the effects of extracting too few or too many factors on the stability of the factors after rotation (Dingman, Miller, and Eyman 1964; Howard and Gordon 1963; Keil and Wrigley 1960). The findings of these studies suggest that over-factoring by one or two factors has less severe consequences for the final solution than does taking too few factors. Cattell (1952) has long recommended the extraction of an extra factor or two on the grounds that the extra factors become residual factors upon rotation and their presence improves the interpretation of the results. Too many factors will result in factor splitting, however. Having too few factors in the solution appears to seriously distort the rotated solution.

The Rotation Problem

It is perhaps fortuitous that the most common rotation procedure, VARIMAX, has been shown to be among the best orthogonal rotation procedures (Dielman, Cattell, and Wagner 1972; Gorsuch 1974). Equally fortuitous are the findings (Dielman, Cattell, and Wagner 1972; Gorsuch 1970; Horn 1963) that the basic solutions provided by most rotational programs result in the same factors. Thus, the rotation employed should have relatively little impact on the interpretation of results.

Several points should be made about rotation. First, VARIMAX, or any of the procedures based on it e.g., PROMAX, Harris-Kaiser), should not be used when there is a theoretical expectation of a general factor (Gorsuch 1974). Because VARIMAX serves to spread variance evenly among factors, it will distort any general factor in the data. QUARTIMAX (Carroll 1953) is probably the orthogonal rotation procedure of choice when a general factor is expected.

Many marketing researchers do not seem well acquainted with the aim of rotation. Thurstone (1942) sought not only interpretable factors, but also the most parsimonious solution when he proposed the use of simple structure as a criterion for the "goodness" of a factor solution. Interpretation of factors is important certainly, but most marketing researchers seem to stop at the point of meaningful factors; they may not have found the most parsimonious set of factors. Thurstone (1942, p. 335) provides five criteria for evaluating a solution for simple structure.

1. Each variable should have at least one zero loading.
2. Each factor should have a set of linearly independent variables whose factor loadings are zero.
3. For every pair of factors, there should be several variables whose loadings are zero for one factor but not for the other.
4. For every pair of factors, a large proportion of the variables should have zero loadings on both factors whenever more than about four factors are extracted.
5. For every pair of factors, there should be only a small number of variables with non-zero loadings on both.

Cattell (1952) has been a leading proponent of the notion of simple structure. He proposed an additional criterion for simple structure, the hyperplane count. The hyperplane count consists of the number of essentially zero loadings on a factor or set of factors. In practice, the hyperplane count is usually defined as the total number of loading coefficients between ±.10. Such a criterion can be easily programmed for use with a computer and may serve as a means for deciding what rotated solution to use.

The marketing literature, though filled with factor analytic reports, provides few examples of oblique rotations. The orthogonal rotation dominates despite the strong likelihood that correlated factors and hierarchical factor solutions are intuitively attractive and theoretically justified in many marketing applications. The careful researcher should almost invariably perform both an orthogonal and an oblique rotation, particularly in exploratory work. These solutions can be compared to identify the simpler structure and to determine whether the oblique rotation produces a marked increase in the hyperplane count. Oblique solutions have been found particularly useful in the theory building of other disciplines (e.g., psychology, sociology, regional science, biology), and are likely to play a significant role in the development of any theory of consumer behavior.

A Note on the Correlation Coefficient

Most factor analytic studies in marketing start with the correlation matrix. The Pearson product moment correlation coefficient is certainly the most frequently reported measure of association in the marketing literature. It is also poorly understood. Carroll (1961) presents a general discussion of the correlation coefficient and Horn (1973) and Nunnally (1978) also offer useful treatments of the topic. The discussion here is restricted to problems the Pearson r may pose for factor analysis.

Correlation coefficients, and factors derived from such coefficients, can be influenced by departures from normality, departures from linearity, departures from homoskedasticity, restrictions of range, and the form of the distribution regardless of normality. The first three problems are probably not serious unless the departure is great (Nunnally 1978) and even then alternative procedures may be available (e.g., nonlinear factor analysis, data

transformation prior to factor analysis). The latter two problems are much more serious.

The restriction of the range that may occur when sampling procedures are biased with respect to one or more of the variables in the analysis can result in spuriously low correlation coefficients. To the extent that all coefficients are equally affected by the restriction of range, the results of a factor analysis will not be affected appreciably. The problem occurs when some of the variables suffer a restriction of range and others do not. Sampling is an important component of well-planned factor analytic research and serves to prevent the restriction of range of variables in many cases.

A more insidious problem lies in the shape of the distributions of the variables involved in the analysis. If the distributions of variables are shaped differently from one another, the size of the correlation is restricted. Fortunately, these effects are slight in studies of continuous variables (Nunnally 1978). Unfortunately, most of the data available to market researchers are discrete. Carroll (1961) points out that any Pearson product moment correlation coefficient computed between discrete variables will, in part, be a function of the marginal distributions of the variables. The extreme case that illustrates the problem is the use of dichotomous data. Dichotomous data in the form of user-nonuser/agree-disagree information are commonly used in marketing research. The correlation between such variables will range between ±1.0 only when both variables have the same marginal distributions. It should be pointed out that the restriction on the coefficient is due to the *difference* in the distributions. A perfect correlation can be obtained when variables have the same marginal distributions regardless of what those distributions look like. To illustrate the problem consider two items involving the use or nonuse of a particular product. If 50% of the respondents are users of one product and 20% are users of a second product, the correlation coefficient cannot be greater than .50. Also, because it is not possible for the distribution shape of a dichotomous variable to be the same as that of a continuous variable or nondichotomous discrete variable, a perfect correlation cannot exist between such variables. For example, the maximum size of the correlation between a dichotomous variable and a normally distributed variable is about .80, which occurs when the p value of the dichotomous variable is .50. The further p deviates from .50, the greater the restriction on the obtained correlation. Though this problem is most severe with dichotomous variables, it is also serious for other types of discrete variables. The problem generally becomes less severe as the number of event classes for each discrete variable increases.

Because factor analysis involves a correlation matrix composed of many variables, and factors are defined by the relative magnitudes of the various correlation coefficients, it is possible to obtain factors that are based on the marginal distributions of variables rather than an underlying associative relationship. These spurious method factors are of no interest to the market researcher. In some cases, these factors may even yield a plausible marketing interpretation if the analyst does not realize what has transpired.

Avoiding this distribution problem requires a careful examination of the marginal distributions of the variables proposed for factor analysis. Meaningful factor analysis may not be possible in many cases where the variables are a mixture of dichotomous, nondichotomous discrete, and continuous variables. The best recommendation appears to be to avoid variables that are extremely skewed.

One suggestion for addressing the distribution problem in dichotomous data is to use the G-coefficient (Holley 1966; Holley and Guilford 1964). To compute the G-index a hypothetical observation that is the exact opposite of each of the original observations is added to the matrix. This extended matrix is then submitted to factor analysis. Because all variables now have the same marginal distributions, no method factors should emerge. No such procedure has been developed for the case involving a mixture of dichotomous and nondichotomous variables. In such cases the researcher might dichotomize all variables, compute a G-index, and proceed with the factor analysis.

A second approach to the distribution problem has been suggested by Peters and Van Voorhis (1940) and elaborated by Martin (1978). This approach involves the use of correction factors which take into account the differences in scaling. These correction factors, derived by Peters and Van Voorhis, may be useful for removing the effects of the distribution of responses from the correlation coefficients prior to factor analysis. Martin (1978) provides a brief example of the use of the procedure. To date little work has examined the effects the corrected correlation coefficients would have on the subsequent use of multivariate procedures.

Summary and Conclusion

The preceding discussion is an attempt to demonstrate that much of the criticism of factor analysis is based on a misunderstanding and misapplication of the technique. Factor analysis is not a clustering technique. Rather it seeks to establish dimensions within the data. This is true regardless of the particular mode, i.e., R or Q, employed.

Factor analysis need not be the subjective exercise it is often accused of being. By examining the raw data and the correlation matrix, one can determine those instances in which factor analysis is appropriate. Examining the marginal distributions of discrete data and computing the measure of sampling adequacy are the minimum steps a researcher should take before initiating a factor analysis. The type of analysis or rotation does not appear to be critical to the final solution in most situations but the extraction of too few or too many factors may have a dramatic effect on the outcome of the analysis. Use of the roots criterion and the scree test can provide a very reliable and consistent indication of the number of factors to extract, however.

The ultimate goal of any factor analysis should be the identification of not only interpretable factors but also simple structure. Simple structure may often be better after an oblique rotation and researchers would do well to use oblique rotations more often.

Factor analysis is not a procedure for every season. The preceding discussion identifies when it might be most useful.

References

Anderson, R. D., J. L. Engledow, and H. Becker (1979), "Evaluating the Relationships Among Attitude Toward Business, Product Satisfaction, Experience, and Search Effort," *Journal of Marketing Research*, 16 (August), 394–400.

Armstrong, J. S. and P. Soelberg (1968), "On the Interpretation of Factor Analysis," *Psychological Bulletin*, 70 (June), 361–4.

Bartlett, M. S. (1950), "Tests of Significance in Factor Analysis," *British Journal of Statistical Psychology*, 3 (January), 77–85.

_____ (1951), "A Further Note on Tests of Significance in Factor Analysis," *British Journal of Statistical Psychology*, 4 (January), 1–2.

Browne, M. S. (1968), "A Comparison of Factor Analytic Techniques," *Psychometrika*, 33 (September), 267–334.

Burt, C. L. (1917), *Distributions and Relations of Educational Abilities*. London: P. S. King.

_____ (1937), "Correlations Between Persons," *British Journal of Psychology*, 28 (January), 56–96.

Carroll, J. B. (1953), "An Analytic Solution for Approximating Simple Structure in Factor Analysis," *Psychometrika*, 18 (March), 23–38.

_____ (1961), "The Nature of Data, or How to Choose a Correlation Coefficient," *Psychometrika*, 26 (December), 347–72.

Cattell, R. B. (1952), *Factor Analysis*. New York: Harper and Brothers.

_____ (1953), "A Quantitative Analysis of the Changes in Culture Patterns of Great Britain, 1837–1937, by P Technique," *Acta Psychologica*, 9 (January), 99–121.

_____ (1966a), "The Meaning and Strategic Use of Factor Analysis," in *Handbook of Multivariate Experimental Psychology*, R. B. Cattell, ed. Chicago: Rand McNally and Company.

_____ (1966b), "The Scree Test for the Number of Factors," *Multivariate Behavioral Research*, 1 (April), 245–76.

_____ (1978), *The Scientific Use of Factor Analysis in the Behavioral and Life Sciences*. New York: Plenum Press.

_____ and M. Adelson (1951), "The Dimensions of Social Change in the U. S. A. as Determined by P Technique," *Social Forces*, 30 (March), 190–201.

_____, M. A. Coulter, and B. Tsujioka (1966), "The Taxonometric Recognition of Types and Functional Emergents," in *Handbook of Multivariate Experimental Research*, R. B. Cattell, ed. Chicago: Rand McNally and Company.

_____ and K. Dickman (1962), "A Dynamic Model of Physical Influences Demonstrating the Necessity of Oblique Simple Structure," *Psychological Bulletin*, 59 (June), 389–400.

_____ and R. L. Gorsuch (1963), "The Uniqueness and Significance of Simple Structure Demonstrated by Contrasting 'Natural Structure' and 'Random Structure' Data," *Psychometrika*, 28 (March), 55–67.

_____ and J. Jaspers (1967), "A General Plasmode (No. 30-10-5-2) for Factor Analytic Exercises and Research," *Multivariate Behavioral Research Monographs*, 67, 1–212.

_____ and W. Sullivan (1962), "The Scientific Nature of Factors: A Demonstration by Cups of Coffee," *Behavioral Science*, 7 (May), 184–93.

_____ and S. Vogelmann (1977), "A Comprehensive Trial of the Scree and KG Criteria for Determining the Number of Factors," *Multivariate Behavioral Research*, 12 (July), 289–325.

Clemans, W. V. (1956), *An Analytic and Experimental Examination of Some Properties of Ipsative Measures*, unpublished doctoral dissertation, University of Washington.

Cliff, N. and C. D. Hamburger (1967), "The Study of

Sampling Errors in Factor Analysis by Means of Artificial Experiments," *Psychological Bulletin*, 68 (June), 430–45.

Cohen, J. (1968), "rc: A Profile Similarity Coefficient Invariant Over Variable Reflection," *Proceedings of the 76th Annual Convention of the American Psychological Association*, 3, 211.

_____ (1969), "A Profile Similarity Coefficient Invariant Over Variable Reflection," *Psychological Bulletin*, 71 (April), 281–4.

Dielman, T. E., R. B. Cattell, and A. Wagner (1972), "Evidence on the Simple Structure and Factor Invariance Achieved by Five Rotational Methods on Four Types of Data," *Multivariate Behavioral Research*, 7 (April), 223–31.

Dingman, H. F., C. R. Miller, and R. K. Eyman (1964), "A Comparison Between Two Analytic Rotational Solutions where the Number of Factors is Indeterminate," *Behavioral Science*, 9 (January), 76–85.

Dixon, W. J. (1975), *BMDP: Biomedical Computer Programs*. Los Angeles: University of California Press.

Dziuban, C. D. and E. C. Shirkey (1974), "When Is a Correlation Matrix Appropriate for Factor Analysis," *Psychological Bulletin*, 81 (June), 358–361.

Eckhart, C. and G. Young (1936), "The Approximation of One Matrix by Another of Lower Rank," *Psychometrica*, 1 (June), 211–18.

Ehrenberg, A. S. C. (1968). "On Methods: The Factor Analytic Search for Program Types," *Journal of Advertising Research*, 8 (March), 55–63.

_____ and G. J. Goodhart (1976), "Factor Analysis: Limitations and Alternatives," *Marketing Science Institute Working Paper No. 76–116*. Cambridge, Massachusetts: Marketing Science Institute.

Gorsuch, R. L. (1970), "A Comparison of Biquartimin, Maxplane, Promax, and Varimax," *Educational and Psychological Measurement*, 30 (Winter), 861–72.

_____ (1973), "Using Bartlett's Significance Test to Determine the Number of Factors to Extract," *Educational and Psychological Measurement*, 33 (Summer), 361–4.

_____ (1974), *Factor Analysis*. Philadelphia: W. B. Saunders Company.

Guttman, L. (1954), "Some Necessary Conditions for Common Factor Analysis," *Psychometrika*, 19 (June), 149–61.

Hakstian, A. R. and V. J. Muller (1973), "Some Notes on the Number of Factors Problem," *Multivariate Behavioral Research*, 8 (October), 461–75.

Harman, H. H. (1976), *Modern Factor Analysis*. Chicago: University of Chicago Press.

Harris, M. L. and C. W. Harris (1971), "A Factor Analytic Interpretation Strategy," *Educational and Psychological Measurement*, 31 (Fall), 589–606.

Holley, J. W. (1966), "A Reply to Phillip Levy: In Defense of the G Index," *Scandinavian Journal of Psychology*, 7, 244–7.

_____ and J. P. Guilford (1964), "A Note on the G Index of Agreement," *Educational and Psychological Measurement*, 24 (Winter), 749–53.

Horn, J. L. (1963), "Second-Order Factors in Questionnaire Data," *Educational and Psychological Measurement*, 23 (Spring), 117–34.

_____ (1965), "A Rationale and Test for the Number

of Factors in Factor Analysis," *Psychometrika*, 30 (June), 179–85.

———— (1973), *Concepts and Methods of Correlational Analysis*. New York: Holt, Rinehart, and Winston.

Horton, R. L. (1979), "Some Relations Between Personality and Consumer Decision Making," *Journal of Marketing Research*, 16 (May), 233–46.

Howard, K. I. and R. A. Gordon (1963), "Empirical Note on the 'Number of Factors' Problem in Factor Analysis," *Psychological Reports*, 12 (Spring), 247–50.

Johnson, R. M. (1970), "Q Analysis of Large Samples," *Journal of Marketing Research*, 7 (February), 104–5.

Jones, V. J. and F. H. Siller (1978), "Factor Analysis of Media Exposure Data Using Prior Knowledge of the Medium," *Journal of Marketing Research*, 15 (February), 137–44.

Jöreskog, K. G. (1971), "A General Approach to Confirmatory Maximum Likelihood Factor Analysis," *Psychometrika*, 36 (December), 409–26.

Kaiser, H. F. (1963), "Image Analysis," in *Problems in Measuring Change*, C. W. Harris, ed. Madison: University of Wisconsin Press.

———— (1970), "A Second Generation Little Jiffy," *Psychometrika*, 35 (December), 401–15.

———— and J. Rice (1974), "Little Jiffy Mark IV," *Educational and Psychological Measurement*, 34 (Spring), 111–17.

Keil, D. and C. Wrigley (1960), "Effects Upon the Factorial Solution of Rotating Varying Numbers of Factors." *American Psychologist*, 15 (March), 383–9.

Knapp, T. R. and V. H. Swoyer (1967), "Some Empirical Results Concerning the Power of Bartlett's Test of Significance of a Correlation Matrix," *American Educational Research Journal*, 4 (Winter), 13–17.

Levin, J. (1965), "Three-Mode Factor Analysis," *Psychological Bulletin*, 64 (June), 442–52.

Linn, R. L. (1968), "A Monte Carlo Approach to the Number of Factors Problem," *Psychometrika*, 33 (March), 37–71.

Martin, W. S. (1978), "Effects of Scaling on the Correlation Coefficient: Additional Considerations," *Journal of Marketing Research*, 15 (May), 304–8.

Nunnally, J. C. (1978), *Psychometric Theory*. New York: McGraw-Hill Book Company.

Overall, J. E. and C. J. Klett (1972), *Applied Multivariate Analysis*. New York: McGraw-Hill Book Company.

Pernica, J. (1974), "The Second Generation of Market Segmentation Studies: An Audit of Buying Motivations," in *Life Style and Psychographics*, W. D. Wells, ed. Chicago: American Marketing Association.

Peters, C. C. and W. R. Van Voorhis (1940), *Statistical Procedures and Their Mathematical Bases*. New York: McGraw-Hill Book Company.

Shaycroft, M. F. (1970), "The Eigenvalue Myth and the Dimension Reduction Fallacy," paper presented to the American Educational Research Association Annual Meeting, Minneapolis.

Thurstone, L. L. (1942), *Multiple Factor Analysis*. Chicago: University of Chicago Press.

Tobias, S. and J. E. Carlson (1969), "Brief Report: Bartlett's Test of Sphericity and Chance Findings in Factor Analysis," *Multivariate Behavioral Research*, 4 (October), 375–7.

Tucker, L. R. (1972), "Relations Between Multidimensional Scaling and Three-Mode Factor Analysis," *Psychometrika*, 37 (March), 3–27.

————, R. F. Koopman, and R. C. Linn (1969), "Evaluation of Factor-Analytic Research Procedures by Means of Simulated Correlation Matrices," *Psychometrika*, 34 (December), 421–59.

Wells, W. D. and J. N. Sheth (1974), "Factor Analysis," in *Handbook of Marketing Research*, R. Ferber, ed. New York: McGraw-Hill Book Company.

Wrigley, C. (1960), "A Procedure for Objective Factor Analysis," paper presented at the first annual meeting of the Society of Multivariate Experimental Psychology.

Woods, G. A. (1976), "A Computer Program for the Scree Test for Number of Factors," manuscript submitted for publication, University of Hawaii.

Ziff, R. (1974), "The Role of Psychographics in the Development of Advertising Strategy and Copy," in *Life Style and Psychographics*, W. D. Wells, ed. Chicago: American Marketing Association.

CHAPTER SEVEN

Cluster Analysis

Cluster analysis is a technique for grouping individuals or objects into clusters so that objects in the same cluster are more like each other than they are like objects in other clusters. The aim of this chapter is to explain the nature and purpose of cluster analysis, and to guide the analyst in the selection and use of various cluster analysis approaches. Before proceeding further, it will be helpful to review the key terms.

Application of cluster analysis techniques involves three major stages: (1) partitioning, (2) interpretation, and (3) profiling. This chapter explains each of these stages and answers the following questions:

- How is interobject similarity measured?
- What are the various distance measures?
- What clustering algorithm should be used?
- What is the difference between hierarchical and nonhierarchical clustering techniques?
- How many clusters should be formed?
- What are the limitations of cluster analysis?

KEY TERMS

Algorithm   A set of rules or procedures; similar to an equation.

Agglomerative methods   A hierarchical clustering procedure which starts with each object or observation in separate clusters. In subsequent steps, object clusters which are closest together are combined to build a new aggregate cluster(s).

Average linkage   An agglomerative algorithm which uses the average distance from individuals in one cluster to individuals in another cluster as the clustering criterion. This approach tends to combine clusters with small variances.

Centroid method   An agglomerative algorithm in which the distance between two clusters is the distance (typically Euclidean) between their centroids (means). Each time objects are grouped, a new centroid is computed. Thus, cluster centroids migrate, or move, as cluster mergers take place.

Cluster centroid   The average value of the objects contained in the cluster on all of the variables included in the analysis.

Cluster seeds The initial center or starting point of a cluster. This individual value(s) is selected in order to initiate non-hierarchical clustering procedures. Clusters are built around these pre-selected seeds.

Complete linkage An agglomerative algorithm in which the clustering criterion is based on the maximum distance between objects. All objects in a cluster are linked to each other at some maximum distance or minimum similarity.

City-block approach A method of calculating distances based upon the sum of the absolute differences of the coordinates for the objects. This method assumes that the variables are uncorrelated and unit scales are compatible.

Dendrogram A graphical representation (a tree graph) of the results of a clustering procedure in which the vertical axis consists of the objects or individuals and the horizontal axis represents the number of clusters formed at each step of the clustering procedure.

Divisive A clustering procedure which begins with all objects in a single cluster. This is opposite of agglomerative procedures. The procedure starts with a single large cluster that is divided into separate clusters based on objects that are most dissimilar.

Entropy group A group of objects or individuals who are independent of any cluster. That is, they do not fit into any cluster.

Euclidean distance The most commonly used measure of the similarity between two objects. Essentially, it is a measure of the length of a straight line drawn between two objects.

Hierarchical procedures Clustering procedures involving the construction of a hierarch or tree like structure composed of separate clusters.

Inter-object similarity How similar two objects are based on their ratings on variables of interest. Similarity can be measured using "proximity" or "closeness" between each pair of objects. Likewise, "distance" or "difference" can be used.

Mahalanobis distance A standardized form of Euclidean distance. Data are standardized by scaling responses in terms of standard deviations and adjustments are made for inter-correlations between the variables.

Non-hierarchical procedures Instead of using a tree-like construction process as in hierarchical procedures, cluster seeds are used to group objects that are within a pre-specified distance of the seeds.

Normalized distance function A process which converts each raw data score to a standardized variate with zero mean and unit standard deviation. The purpose of this procedure is to remove the bias introduced by differences in scales of several variables.

Optimizing procedure A non-hierarchical clustering procedure which allows for the reassignment of objects to another cluster from the original one on the basis of some overall optimizing criteria.

Profile diagram A graphical representation of data which aids in screening for outliers. Typically, the variables are listed along the horizontal axis and the value scale on the vertical axis. Standardized object responses are then plotted in the graphic plane.

Parallel threshold method   A non-hierarchical clustering procedure which selects several cluster seeds simultaneously in the beginning. Objects within the threshold distances are assigned to the nearest seed. Threshold distances can be adjusted to include fewer or more objects in the clusters.

Sequential threshold method   A non-hierarchical procedure that begins by selecting one cluster seed. All objects within a pre-specified distance are then included in that cluster. Subsequent cluster seeds are selected until all objects are grouped in a cluster.

Single linkage method   A hierarchical clustering procedure based on minimum distance. The procedure finds two objects with the shortest distance and places them in the first cluster. The process continues until all objects are in one cluster.

Vertical icicle diagram   A graphic representation of clusters. The numbers of the objects are shown horizontally across the top and the number of clusters is shown vertically down the left side. This plot is used to aid in determining the appropriate number of clusters in the solution.

Ward's method   A hierarchical clustering procedure. The distance between two clusters is calculated as the sum of squares between the two clusters summed over all variables.

What Is Cluster Analysis?

Cluster analysis is the name of a group of multivariate techniques whose primary prupose is to identify similar entities from the characteristics they possess. It identifies and classifies objects or variables so that each object is very similar to others in its cluster with respect to some predetermined selection criteria. The resulting object clusters should then exhibit high internal (within-cluster) homogeneity and high external (between-cluster) heterogeneity. Thus, if the classification is successful, the objects within clusters will be close together when plotted geometrically, and the objects in different clusters will be far apart.

Cluster analysis has been variously referred to as *Q-analysis, typology, classification analysis*, and *numerical taxonomy*. This variety of names is due in part to the usage of clustering methods in such diverse disciplines as psychology, biology, sociology, and business. Although the names differ across disciplines, they all have a common dimension: classification according to natural relationships. This dimension represents the essence of all clustering approaches. As such, the primary value of cluster analysis lies in the preclassification of data, as suggested by "natural" groupings of the data itself.

Cluster analysis is a useful tool for data analysis in many different situations. For example, a researcher who has collected data via a questionnaire may be faced with a large number of observations that are meaningless unless classified into manageable groups. Cluster analysis can be used to perform this *data reduction* procedure objectively by reducing the information from an entire population or set to information about specific smaller subgroups. In this fashion the researcher has a more concise, understandable description of the observations with minimal loss of information.

Cluster analysis is also useful when a researcher wishes to *develop* hypotheses concerning the nature of the data or to examine previously stated hypotheses. For example, a researcher may believe that attitudes toward the consumption of light versus regular beer could be used to separate beer consumers into logical segments or groups. Cluster analysis can be used to classify beer consumers by their attitudes about light versus regular beer, and the resulting clusters, if any, can be profiled for demographic similarities and/or differences.

How Does Cluster Analysis Work?

The nature of cluster analysis can be illustrated by a graphic presentation of a bivariate example. Suppose a marketing researcher has to determine relevant market segments in a small community. Assume further that a random sample of the population is selected and information tabulated on two criteria:

1. Level of education.
2. Brand loyalty.

All of the respondents are then plotted on the scatter diagram in Figure 7.1.

Examination of Figure 7.1 shows that a definite relationship exists. In fact, the researcher could simply draw a line dividing the two groups and designate each as a cluster. In the terminology of cluster analysis, the researcher has determined *two distinct clusters with an average subject (within-cluster) correlation of perhaps .75.* Furthermore, if the researcher considers a profile representation of each cluster, it may

FIGURE 7.1 Scatter diagram of cluster observations.

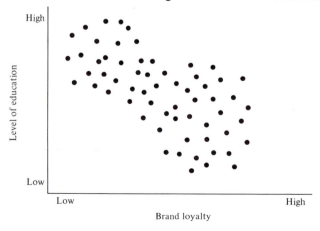

be determined that the two clusters are uncorrelated or are even negatively correlated; that is, they are quite dissimilar. Finally, the handful of respondents who are independent of either cluster are designated to be, in the terminology of cluster analysis, the *entropy group*.

This concept is rather simple in the bivariate case, since the data are two-dimensional. In most marketing research studies, however, more than two variables are measured on each entity, and the situation is much more complex.

To illustrate a more complex situation, suppose the researcher is interested in using several variables. The variables involved may be metric, such as weight, height, income, or age; or they may be categorical (i.e., religion, nationality, race, sex); or some combination of the two. But whatever the situation, the use of cluster analysis becomes more complicated as more variables are added or when mixed data sets with both metric and nonmetric variables are included.

Application of Cluster Analysis

Application of cluster analysis can be divided into three major stages: (1) partitioning, (2) interpretation, and (3) profiling. The *partitioning* stage is the process of determining if and how clusters may be developed. The *interpretation* stage is the process of understanding the characteristics of each cluster and developing a name or label that appropriately defines its nature. The *profiling* stage involves describing the characteristics of each cluster to explain how they may differ on relevant dimensions such as demographics.

Stage One: Partitioning

During the partitioning stage, three major questions need to be considered: (1) How should interobject similarity be measured? (2) What procedure (algorithm) should be used to place similar objects into groups or clusters? (3) How many clusters should be formed? Many different approaches can be used to answer these questions. However, none of them has been evaluated sufficiently to provide a definitive answer to any of these questions, and unfortunately, many of the approaches provide different results for the same data set. Thus, cluster analysis, along with factor analysis, is much more of an art than a science. For this reason, our discussion reviews these issues in a very general way by providing examples of the most commonly used approaches without focusing on their theoretical or practical limitations.

SIMILARITY MEASURES.   Interobject similarity can be measured in a variety of ways. One way is to look at the proximity or closeness between each pair of objects in order to determine their similarity. Another way is to look at the distance or difference between the pairs of objects. Since distance is the complement of similarity, this approach can be used to assess similarity.

The interobject measure of similarity that probably comes to mind first is the correlation coefficient between a pair of objects measured on several variables. In effect, instead of correlating two sets of variables, we invert the objects' X variables matrix so that the columns represent the objects and the rows the variables. Thus, the correlation coefficient between the two columns of numbers is the correlation (or similarity) between the profiles of the two objects. High correlations indicate similarity and low correlations lack of similarity. This procedure is followed in the application of Q-type factor analysis.

Distance measures are the most commonly used measures of similarity between objects. Before discussing them, we must point out that many of the analytical techniques for assessing distance are particularly sensitive to outliers (objects that are very different from all others). Thus, a preliminary screening for outliers is advisable. Probably the easiest way to conduct this screening is to prepare a graphic profile diagram such as that shown in Figure 7.2. The analyst typically standardizes the data first by subtracting the mean and dividing by the standard deviation for each variable (computer programs for this procedure are readily available). Then a profile diagram is developed, listing the variables along the horizontal axis and the standardized value scale along the vertical axis. Each point on the graph represents the value of the corresponding variable, and the points are connected in order to facilitate visual interpretation. Profiles for all objects are then plotted on the graph, and outliers can be identified. Obviously, such a procedure becomes cumbersome with large numbers of objects (observations). Therefore, a step such as eliminating observations more than, for example, 2.5 SD from the mean is necessary. Such a procedure could easily be programmed for preliminary screening purposes.

FIGURE 7.2   Profile diagram.

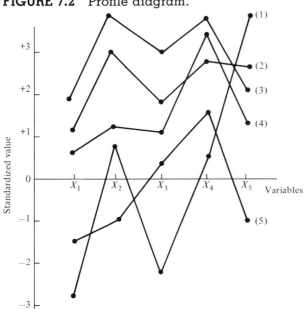

FIGURE 7.3 An example of Euclidean distance between two objects measured on two variables—X and Y.

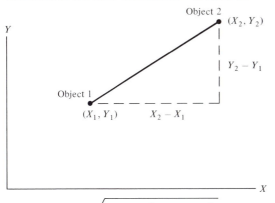

$$\text{Distance} = \sqrt{(X_2 - X_1)^2 + (Y_2 - Y_1)^2}$$

The most commonly used measure of similarity is Euclidean distance. An example of how Euclidean distance is obtained is shown geometrically in Figure 7.3. Suppose that two points in two dimensions have coordinates (X_1, Y_1) and (X_2, Y_2), respectively. The Euclidean distance between the points is the length of the hypotenuse of a right triangle, as calculated by the formula under the figure. This formula is for distance measured on only two variables, but the concept is easily generalized to additional variables. Also, although the coordinates of the two objects are plotted in raw score form in the figure, in practice the analyst typically converts them to their Z-score form prior to calculating distances to eliminate spurious effects due to unequal variances of the variables represented.

Several options other than simple Euclidean distance are available from various computer programs. For example, since calculation of the square root does not change the distance the points are from each other, some programs use the sum of squared differences instead of the Euclidean distance (i.e., the square root is not taken). Another option in some programs involves replacing the squared differences by the sum of the absolute differences of the coordinates. This procedure is referred to as the *absolute* or *city-block* distance function.

The *city-block approach* to calculating distances may be appropriate under certain circumstances [12], but it causes several problems. One is the assumption that the variables are not correlated with each other; if they are correlated, the clusters are not valid. Another problem with the city-block approach arises when the analyst is faced with noncompatibility of units (scales) in which the characteristics are measured. For example, suppose three objects, A, B, and C, are measured on two variables, probability of purchasing brand X (in percentages) and amount of time spent viewing commercials for brand X (in minutes), with the following results:

Object	Purchase Probability (%)	Commercial Viewing Time (Min)
A	60	3.0
B	65	3.5
C	63	4.0

The absolute Euclidean distances are as follows:

AB = 25.25
AC = 10.00
BC = 4.25

If commercial viewing time were measured in seconds, the distances would become:

AB = 61
AC = 153
BC = 40

Object A is now closer to B than to C. Therefore, even when all of the variables are uniquely determined except for scale changes, absolute Euclidean distances do not preserve distance rankings. In order to overcome this problem of noncompatibility, the researcher may wish to select a normalized distance function, such as squared Euclidean distance.

A normalized distance function utilizes a Euclidean distance measure amenable to a normalizing transformation of the raw data. This process converts each raw data score into a standardized variate with a zero mean and a unit standard deviation. This transformation, in turn, eliminates the bias introduced by the differences in the scales of the several attributes or variables used in the analysis.

The most commonly used measure of Euclidean distance that incorporates the standardization procedure is the Mahalanobis distance. The Mahalanobis approach not only performs a standardization process on the data by scaling in terms of the standard deviations but also sums the pooled within-group variance-covariance, which adjusts for intercorrelations among the variables. In short, the Mahalanobis generalized distance procedure computes a distance measure between objects that is comparable to the R^2 in regression. Many packages do not include the Mahalanobis D^2 distance as an option. In such cases, the analyst usually selects the squared Euclidean distance option.

In attempting to select a particular distance measure, the analyst should remember the following caveats. In most situations, different distance measures lead to different cluster solutions. Thus, it is advisable to use several measures and compare the results to theoretical or known patterns. Also, when the variables have different units, one should standardize before running the cluster analysis. Standardization is particularly advisable when the range of one variable is much larger than that of others. Finally, when the variables are intercorrelated (either positively or negatively), the Mahalanobis distance measure is likely to be the most appropriate because it adjusts for intercorrelations and weights all variables equally. Of course, if the analyst wishes to weight the variables unequally, other procedures are available [7,8].

CLUSTERING ALGORITHMS. The second major question to answer in the partitioning phase is, what procedure should be used to place similar objects into groups or clusters? That is, what clustering algorithm or set of rules is most appropriate? This is not a simple question because hundreds of computer programs using different algorithms are available and more are being developed. The essential criterion of all of them, however, is that they attempt to maximize the differences between clusters relative to the variation within the clusters, as shown in Figure 7.4. The ratio of the between-cluster variation to the average within-cluster variation is then comparable to (but not identical) to the F-ratio in analysis of variance.

Most of the commonly used clustering algorithms can be classified in two general categories: (1) hierarchical and (2) nonhierarchical. We will discuss the hierarchical techniques first.

FIGURE 7.4 Cluster diagram showing between- and within-cluster variation.

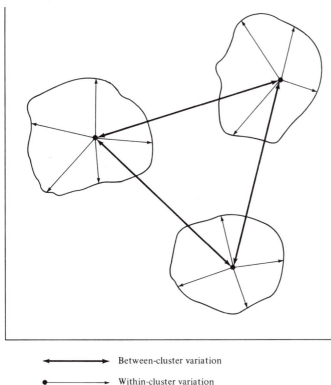

→ Between-cluster variation

●→ Within-cluster variation

HIERARCHICAL CLUSTER PROCEDURES.   *Hierarchical procedures* involve the construction of a hierarch or tree-like structure. There are basically two types of hierarchical clustering procedures—agglomerative and divisive. In the *agglomerative* methods, each object or observation starts out as its own cluster. In subsequent steps, the two closest clusters (or individuals) are combined into a new aggregate cluster, thus reducing the number of clusters by one in each step. In some cases, a third individual joins the first two to form a new cluster. In others, another group of two individuals join together to form a new cluster. Eventually, all individuals are grouped into one large cluster; for this reason, agglomerative procedures are sometimes referred to as *build-up methods*. This process is shown in Figure 7.5; the representation is referred to as a *dendrogram* or *tree graph*.

When the clustering process proceeds in the opposite direction to agglomerative methods, it is referred to as *divisive*. In divisive methods, we begin with one large cluster containing all observations (objects). In succeeding steps, the observations that are most dissimilar are split off and turned into smaller clusters. This process continues until each observation is a cluster in itself. In Figure 7.5, agglomerative methods would move from left to right, while divisive methods would move from right to left. Since most commonly used computer packages use agglomerative methods, we will not discuss divisive methods further.

Five popular agglomerative procedures used to develop clusters are (1) single linkage, (2) complete linkage, (3) average linkage, (4) Ward's method, and (5) centroid. These rules differ in how the distance between clusters is computed.

Single Linkage.   The *single linkage* procedure is based on minimum distance. It finds the two individuals (objects) with the shortest distance and places them in the first cluster. Then the next shortest distance is found, and either a third individual joins the first two to form a

FIGURE 7.5   Dendrogram illustrating hierarchical clustering.

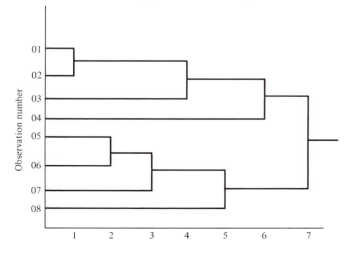

FIGURE 7.6 Points A and B are not very similar.

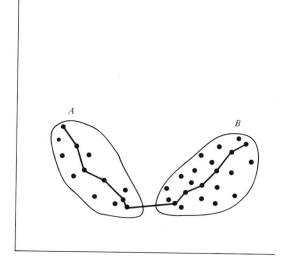

cluster or a new two-individual cluster is formed. The process continues until all individuals are in one cluster. This procedure has also been referred to as the *nearest-neighbor* approach.

The distance between any two clusters is the shortest distance from any point in one cluster to any point in the second cluster. Two clusters are merged at any stage by the single shortest or strongest link between them. Problems occur, however, when clusters are poorly delineated. In such cases, single linkage procedures form long snake-like chains, and eventually all individuals are placed in one chain. Individuals at the end of a chain may be very dissimilar. An example of this arrangement is shown in Figure 7.6.

Complete Linkage. The *complete linkage* procedure is similar to single linkage except that the cluster criterion is based on maximum distance. For this reason, it is sometimes referred to as the *furthest-neighbor* approach. This is a diameter method. The maximum distance between any two individuals in a cluster represents the smallest (minimum-diameter) sphere that can enclose the cluster. This method is called complete linkage because all objects in a cluster are linked to each other at some maximum distance or by minimum similarity. We can say that within-group similarity equals group diameter. This technique eliminates the snaking problem identified with single linkage.

The problem of measuring distance between groups still arises, however. In discriminant analysis we use centroids, but in this case our data may not permit the use of means. Figure 7.7 shows how the shortest and longest distances may not represent true similarity between groups. The use of the shortest distance indicates that the two groups are very similar, whereas the use of the longest distance indicates that they are very dissimilar.

Average Linkage. The *average linkage* method starts out the same

FIGURE 7.7 Comparison of distance measures for single linkage and complete linkage.

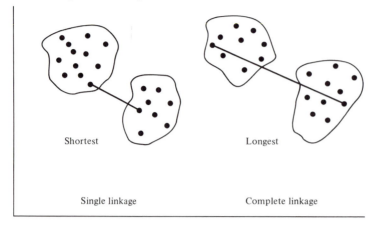

as single linkage and complete linkage, but the cluster criterion is the *average* distance from individuals in one cluster to individuals in another. Such techniques do not use extreme values, as do single linkage or complete linkage, and partitioning is based on all members of the clusters rather than on a single pair of extreme members. Average linkage approaches tend to combine clusters with small variances. They also tend to be biased toward the production of clusters with approximately the same variance.

Ward's Method. In *Ward's method* the distance between two clusters is the sum of squares between the two clusters summed over all variables. At each stage in the clustering procedure, the within-cluster sum of squares is minimized over all partitions (the complete set of disjoint or separate clusters) obtainable by combining two clusters from the previous stage. This procedure tends to combine clusters with a small number of observations. It is also biased toward the production of clusters with approximately the same number of observations.

Centroid Method. In the *centroid method* the distance between two clusters is the distance (typically Euclidean) between their centroids (means). In this method, every time individuals are grouped, a new centroid is computed. Cluster centroids migrate as cluster mergers take place. In other words, there is a change in a cluster centroid every time a new individual or group of individuals is added to an existing cluster. These methods are most popular with biologists but may produce messy and often confusing results. The confusion occurs due to reversals; that is, when the distance between the centroids of one pair may be less than the distance between the centroids of another pair merged at an earlier time (combination). The advantage of this method is that it is less affected by outliers than are other hierarchical methods.

It should be noted that centroid methods require metric data, which may severely limit their application in the social sciences. Other linkage methods do not have this requirement.

NONHIERARCHICAL CLUSTERING PROCEDURES. In contrast to hierarchical methods, nonhierarchical clustering procedures do not involve the tree-like construction process. Instead, the first step is to select a cluster center or *seed*, and all objects (individuals) within a prespecified threshold distance are included in the resulting cluster. Nonhierarchical clustering procedures are frequently referred to as *K-means clustering*.

Nonhierarchical clustering techniques typically use one of the following three approaches [2, p. 428]. The *sequential threshold procedure* starts by selecting one cluster seed, and all objects within a prespecified distance are included. When all objects within that distance are included, a second cluster seed is selected and all objects within the prespecified distance are included. Then a third seed is selected and the process continues as before. When an object is clustered with a seed, it is no longer considered for subsequent seeds. In contrast, the *parallel threshold procedure* selects several cluster seeds simultaneously in the beginning, and objects within the threshold distance are assigned to the nearest seed. As the process evolves, threshold distances can be adjusted to include fewer or more objects in the clusters. Also, in some methods, objects remain unclustered if they are outside the prespecified threshold distance from any cluster seed. The third procedure, referred to as *optimizing*, is similar to the other two except that it allows for reassignment of objects to another cluster from the original on the basis of some overall optimizing criterion.

The SAS FASTCLUS procedure is an example of a nonhierarchical clustering program. It is designed for large data sets. After the user specifies the maximum number of clusters allowed, the FASTCLUS procedure begins by selecting cluster seeds, which are used as initial guesses of the means of the clusters. The first seed is the first observation in the data set with no missing values. The second seed is the next complete observation (no missing data) that is separated from the first seed by a specified minimum distance. The default option is a zero minimum distance. After all seeds are selected, the program assigns each observation to the cluster with the nearest seed. The DRIFT option enables the user to revise (update) the cluster seeds by calculating seed cluster means each time an observation is assigned. If the DRIFT option is not used, new cluster means are not calculated till all observations are assigned. The DRIFT option requires more computer time. (QUICK CLUSTER in the SPSSX package is an approach similar to that of SAS FASTCLUS.)

The major problem with the FASTCLUS and other nonhierarchical clustering procedures is how to select the cluster seed or seeds. For example, with FASTCLUS the initial and probably the final cluster results depend on the order of the observations in the data set, and shuffling the order is likely to affect the results. Specifying the initial cluster seeds as in the parallel threshold procedure can reduce this problem. But the choice process for seeds can still affect the results and thus remains a problem.

SHOULD HIERARCHICAL OR NONHIERARCHICAL METHODS BE USED? A definitive answer to this question cannot be given for two reasons. First, the research problem at hand typically may suggest one method or the other. Second, both methods are evolving rapidly, and what we learn with future applications may suggest one over the other.

In the past, hierarchical clustering techniques were more popular, with Ward's method and average linkage being probably the best available [6]. Hierarchical procedures do have the advantage of being fast and therefore taking less computer time. But they can be misleading because undesirable early combinations may persist throughout the analysis and lead to artificial results. To reduce this possibility, the analyst may wish to cluster-analyze the data several times after deleting problem observations or outliers.

However, nonhierarchical methods appear to be gaining more acceptability. If a practical, objective, and theoretically sound approach can be developed to select the seeds or leaders, these methods should become more widely used. This is particularly true for large data bases, which in the past, by necessity, required the use of hierarchical approaches because of their size.

HOW MANY CLUSTERS SHOULD BE FORMED?   A major issue with all clustering techniques is how to select the number of clusters. There are many criteria and guidelines for approaching the problem. Unfortunately, no standard, objective selection procedure exists. The distances between clusters at successive steps may serve as a useful guideline, and the analyst may choose to stop when this distance exceeds a specified value or when the successive distances between steps make a sudden jump. These distances are sometimes referred to as *error variability measures*. Also, some intuitive conceptual or theoretical relationship may suggest a natural number of clusters. In the final analysis, however, it is probably best to compute solutions for several different numbers of clusters (e.g., two, three, four) and then to decide among the alternative solutions based upon a priori criteria, practical judgment, common sense, or theoretical foundations. Also, one might start this process by saying, "My findings will be more manageable and easier to communicate if I have, for example, three to six clusters," and then solve for this number of clusters and select the best alternative after evaluating all of them.

Stage Two: Interpretation

The interpretation stage involves examining the statements that were used to develop the clusters in order to name or assign a label that accurately describes the nature of the clusters. To clarify this process, let's refer back to the light beer versus regular beer example. Assume that an attitude scale was developed that consisted of statements regarding consumption of beer. Individuals were asked to evaluate these statements on a seven-point scale. Examples of statements are: "Light beer tastes smoother," "Regular beer has a full-bodied taste," "Light beer is less filling," and so forth. Assume further that demographic and beer consumption data were also collected.

When starting the interpretation process, one measure that is used frequently is the cluster's centroid (the average value of the objects contained in the cluster on each of the variables included in the analysis). If the clustering procedure was performed on the raw data, this would be a logical description. If the data were standardized or if the cluster analysis was performed using factor analysis components (component factors), the analyst must go back to the raw scores for the original variables and compute average profiles using these data. The use of modal profiles is also a possibility, as is assessment of the variability within the clusters.

Continuing with our beer example, this stage involves looking at the average score profiles on the attitude statements for the three groups and assigning a descriptive label to each. Average score profiles typically are developed using discriminant analysis. For example, two of the groups (clusters) may have favorable attitudes about light beer and the third cluster negative attitudes. Moreover, of the two favorable clusters, one may exhibit favorable attitudes toward only light beers, while the other may display favorable attitudes toward both light and regular beer. From this analytical procedure, one would evaluate each cluster's attitudes and develop substantive interpretations that would facilitate the labeling of each. For example, one cluster might be labeled the "Health- and Calorie-Conscious" group (likes only light beer), whereas another might be labeled the "Get a Buzz Going" group (likes only regular beer).

Stage Three: Profiling

The profiling stage involves describing the characteristics of each cluster in order to explain how they may differ on relevant dimensions. Just as in stage two, this phase typically involves the use of discriminant analysis or some other appropriate statistic. The procedure begins with the clusters identified (labeled) in stage two. The analyst utilizes data not previously included in the cluster procedure to profile the characteristics of each cluster. These data typically are demographic characteristics, psychographic profiles, consumption patterns, and so forth. Using discriminant analysis, the analyst compares average score profiles for the clusters. The categorical dependent variable is the previously identified clusters, and the independent variables are the demographics, psychographics, and so on. From this analysis, assuming statistical significance, the analyst could conclude, for example, that the "Health- and Calorie-Conscious" cluster from our previous example consists of younger, better-educated, higher-income professionals who are moderate consumers of beer. In short, the profile analysis focuses on describing not what directly determines the clusters but the characteristics of the clusters after they are identified. Moreover, the emphasis is on the characteristics that differ significantly across the clusters, and in fact could be used to predict membership in a particular attitude cluster.

An Illustrative Example

To illustrate the application of cluster analysis techniques, let's turn to the HATCO data base. We begin by cluster-analyzing the first six social-psychological variables. Recalling that the purposes of cluster analysis include data reduction and parsimony, let's review the results for both the SAS and SPSSX cluster procedures. We shall use squared Euclidean distances and Ward's method.

Having decided how to measure interobject similarity (squared Euclidean distances) and which algorithm to use (Ward's), our next question is, how many clusters should we have? Since the data involve social-psychological profiles of salespersons, and since our interest is in identifying types or profiles of salespersons that predict good sales performance, the intuitive feeling is that two, three, or four clusters are most manageable.

Turning first to the results for SAS shown in Table 7.1, we assess the cubic clustering criterion shown on the bottom right-hand side of the table. (Note that with such a small N, this criterion should be interpreted cautiously.) After 2 clusters, all values are negative and decreasing through 10 clusters. This suggests that after two clusters the data are probably unimodal [11].

A *vertical icicle* diagram can also be used to evaluate a cluster solution. This diagram shows the numbers of the objects horizontally across the top and the number of clusters vertically down the left side. The 1 at the top left occurs when all objects are grouped to form one cluster; the 2 shows the two-cluster solution. To see which objects are found in each group of the two-cluster solution, look for the blank spot in the row of X's beside the 2. To do this for the three-cluster solution, look for the blank spots in the row of X's beside the 3 on the left side of the table. The number of objects is obtained by counting the number of objects between the blanks. Looking at the vertical icicle shown in Table 7.2, at two clusters we note that the size of the clusters is approximately equal (23 and 27), whereas when we move on to three clusters the sizes are very unequal (7, 16, and 27). Thus, the data suggest that two clusters is the best solution.

Looking next at the SPSS-X results shown in Table 7.3, let's focus on the coefficient (center column). If this value is plotted against the number of clusters, ranging up to about 10 percent of the number of observations, one notes that after two clusters the tail (line) flattens out and decreases slowly, with no substantial (quick) drops. Again, the two-cluster solution is indicated.

Table 7.4 shows the dendrogram for this solution. This provides a quick visual overview of the clustering process, as well as showing which observations are found in each cluster.

Earlier in the chapter, we stated that different methods provide different results. To demonstrate this fact, compare the information given in Tables 7.5 and 7.6. Table 7.5 shows the cluster membership of each observation for the Ward's method and Table 7.6 the findings for the single linkage method. Recall that this method is supposed to produce solutions (clusters) containing approximately the same number of observations; the results show that this has occurred if one considers the

TABLE 7.1 Ward's Minimum Variance Hierarchical Cluster Analysis

SAS

WARD'S MINIMUM VARIANCE CLUSTER ANALYSIS

SIMPLE STATISTICS

	MEAN	STD DEV	SKEWNESS	KURTOSIS	BIMODALITY
X1	6.8831	2.6795	-0.0956	-0.4180	0.3637
X2	4.6049	2.3532	0.4233	-0.5259	0.4422
X3	15.6631	2.8210	-0.2301	-1.1284	0.5101
X4	10.3749	2.2105	0.2319	0.1329	0.3168
X5	5.7167	1.4246	-0.2636	-0.4946	0.3964
X6	5.2104	1.4871	0.3624	0.3375	0.3203

EIGENVALUES OF THE COVARIANCE MATRIX

	EIGENVALUE	DIFFERENCE	PROPORTION	CUMULATIVE
1	13.7096	6.16922	0.460009	0.46001
2	7.5404	2.87676	0.253008	0.71302
3	4.6636	1.42733	0.155811	0.86883
4	3.2163	2.55394	0.107918	0.97675
5	0.6523	0.63165	0.022224	0.99897
6	0.0307		0.001030	1.00000

ROOT-MEAN-SQUARE TOTAL-SAMPLE STANDARD DEVIATION = 2.22871
ROOT-MEAN-SQUARE DISTANCE BETWEEN OBSERVATIONS = 7.72048

NUMBER OF CLUSTERS	CLUSTERS JOINED		FREQUENCY OF NEW CLUSTER	RMS STD OF NEW CLUSTER	SEMIPARTIAL R-SQUARED	R-SQUARED	APPROXIMATE EXPECTED R-SQUARED	CUBIC CLUSTERING CRITERION
10	CL34	CL15	7	1.361732	0.018410	0.777395	0.806743	-1.8267
9	CL16	CL19	8	1.343619	0.019485	0.757910	0.786694	-1.6956
8	CL14	CL22	10	1.294623	0.026944	0.730966	0.763422	-1.7747
7	CL9	CL21	11	1.486231	0.030787	0.700179	0.735893	-1.8296
6	CL18	CL13	9	1.417715	0.030877	0.669302	0.702503	-1.6135
5	CL6	CL10	16	1.600481	0.046090	0.623212	0.660395	-1.7154
4	CL7	CL8	21	1.611865	0.060762	0.562449	0.605324	-1.6280
3	CL4	CL12	27	1.787203	0.093688	0.468761	0.521993	-1.6252
2	CL5	CL11	23	1.776158	0.095124	0.373637	0.365487	0.2156
1	CL3	CL2	50	2.228711	0.373637	0.000000	0.000000	0.0000

TABLE 7.2 Vertical Icicle for Ward's Method (Distance Measure = Squared Euclidean)

310

TABLE 7.3 Agglomeration Schedule Using Ward Method

STAGE	CLUSTERS COMBINED CLUSTER 1	CLUSTER 2	COEFFICIENT	STAGE CLUSTER 1ST APPEARS CLUSTER 1	CLUSTER 2	NEXT STAGE
1	28	45	.677910	0	0	19
2	1	3	1.605186	0	0	6
3	4	17	2.702909	0	0	23
4	19	21	4.187385	0	0	10
5	2	18	5.817424	0	0	22
6	1	22	7.831538	2	0	15
7	26	33	10.130273	0	0	24
8	32	42	12.715859	0	0	33
9	43	50	15.339000	0	0	18
10	6	19	18.946411	0	4	22
11	31	47	22.602539	0	0	27
12	40	44	26.484909	0	0	29
13	41	48	30.952253	0	0	34
14	13	38	35.683477	0	0	31
15	1	15	40.640533	6	0	37
16	9	11	45.655319	0	0	40
17	34	35	50.706696	0	0	25
18	36	43	55.779938	0	9	36
19	28	39	60.945724	1	0	36
20	10	23	66.264664	0	0	32
21	16	24	71.997482	0	0	37
22	2	6	77.850327	5	10	30
23	4	7	84.130554	3	0	35
24	26	49	90.436722	7	0	28
25	34	46	97.757594	17	0	33
26	8	14	105.262848	0	0	35
27	27	31	112.901443	0	11	34
28	26	30	120.690201	24	0	42
29	12	40	128.558319	0	12	43
30	2	5	138.215693	22	0	37
31	13	37	148.193634	14	0	41
32	10	20	158.193878	20	0	44
33	32	34	168.369415	8	25	38
34	27	41	185.261353	27	13	41
35	4	8	205.234314	23	26	40
36	28	36	226.462433	19	18	42
37	1	16	248.928940	15	21	44
38	29	32	273.484375	0	33	47
39	2	25	298.194824	30	0	48
40	4	9	325.078857	35	16	45
41	13	27	353.533691	31	34	43
42	26	28	392.881104	28	36	46
43	12	13	437.840088	29	41	46
44	1	10	482.930664	37	32	45
45	1	4	550.237793	44	40	48
46	12	26	638.970947	43	42	47
47	12	29	775.786865	46	38	49
48	1	2	914.699951	45	39	49
49	1	12	1460.335205	48	47	0

TABLE 7.4 Dendrogram Using Ward Method

```
                    RESCALED DISTANCE CLUSTER COMBINE

C A S E       0         5        10        15        20        25
LABEL  SEQ    +----------+----------+----------+----------+----------+

       28     -+-+
       45     -+ +----+
       39     ----+     +------+
       43     -+    I        I
       50     -+------+      I
       36     -+           +---------+
       26     -+-+         I         I
       33     -+ +-+       I         I
       49     ----+ +-------+        I
       30     ------+       +------+
       40     -+----+       I      I
       44     -+    +---------+    I      I
       12     ------+      I       I      I
       13     -+----+      +---+   I      I
       33     -+  +------+ I   I   I
       37     ------+    +-+   I
       41     +------+   I         +------------------------+
       48     -+    +------+       I                        I
       31     -+-+  I              I                        I
       47     -+ +----+            I                        I
       27     ---+                 I                        I
       32     -+----+              I                        I
       42     -+  +----+           I                        I
       34     --+-+ I  I           I                        I
       35     -+ +-+  +---------------------+               I
       46     ---+   I                                      I
       29     ----------+                                   I
        2     -+-+                                          I
       18     -+ +-+                                        I
       19     -+ I I                                        I
       21     -+-+ +------+                                 I
        6     -+  I +---------------------------+           I
        5     ------+  I                        I           I
       25     ---------+                        I           I
        9     ---------------+                  I           I
       11     -+         I    +-------------------------+
        4     -+-+  +--------+ I
       17     -+ +---+ I      I
        7     ---+ +---+      I
        8     ---+---+        I
       14     ---+          +---------------+
       10     ---+-+        I
       23     ---+ +--------- --+ I
       20     -----+         I I
        1     -+           +-+
        3     -+           I
       22     -+--------+   I
       15     -+        +-------+
       16     ---+-----+
       24     ---+
```

TABLE 7.5 Cluster Membership of Cases Using Ward Method

		NUMBER OF CLUSTERS		
LABEL	CASE	5	4	3
	1	1	1	1
	2	2	2	②
	3	1	1	1
	4	1	1	1
	5	2	2	②
	6	2	2	②
	7	1	1	1
	8	1	1	1
	9	1	1	1
	10	1	1	1
	11	1	1	1
	12	3	3	3
	13	3	3	3
	14	1	1	1
	15	1	1	1
	16	1	1	1
	17	1	1	1
	18	2	2	②
	19	2	2	②
	20	1	1	1
	21	2	2	②
	22	1	1	1
	23	1	1	1
	24	1	1	1
	25	2	2	②
	26	4	3	3
	27	3	3	3
	28	4	3	3
	29	5	4	3
	30	4	3	3
	31	3	3	3
	32	5	4	3
	33	4	3	3
	34	5	4	3
	35	5	4	3
	36	4	3	3
	37	3	3	3
	38	3	3	3
	39	4	3	3
	40	3	3	3
	41	3	3	3
	42	5	4	3
	43	4	3	3
	44	3	3	3
	45	4	3	3
	46	5	4	3
	47	3	3	3
	48	3	3	3
	49	4	3	3
	50	4	3	3

TABLE 7.6 Cluster Membership of Cases Using Single Linkage

		NUMBER OF CLUSTERS		
LABEL	CASE	5	4	3
	1	1	1	1
	2	1	1	1
	3	1	1	1
	4	1	1	1
	5	1	1	1
	6	1	1	1
	7	1	1	1
	8	1	1	1
	9	1	1	1
	10	1	1	1
	11	1	1	1
	12	1	1	1
	13	1	1	1
	14	1	1	1
	15	1	1	1
	16	1	1	1
	17	1	1	1
	18	1	1	1
	19	1	1	1
	20	②	②	1
	21	1	1	1
	22	1	1	1
	23	1	1	1
	24	1	1	1
	25	③	③	②
	26	1	1	1
	27	1	1	1
	28	1	1	1
	29	④	④	③
	30	1	1	1
	31	1	1	1
	32	1	1	1
	33	1	1	1
	34	1	1	1
	35	1	1	1
	36	1	1	1
	37	⑤	1	1
	38	1	1	1
	39	1	1	1
	40	1	1	1
	41	1	1	1
	42	1	1	1
	43	1	1	1
	44	1	1	1
	45	1	1	1
	46	1	1	1
	47	1	1	1
	48	1	1	1
	49	1	1	1
	50	1	1	1

pattern over the first few clusters. In contrast, the single linkage procedure has produced one large cluster with several single-observation clusters. This is an unacceptable solution from both an intuitive and a practical standpoint. In addition, having used the other method (Ward's), we also know that the single-linkage solution is not valid, since in fact there are latent clusters (at least two) and not an overall unimodal distribution, as suggested.

TABLE 7.7
Group Means and
Significance
Levels for Two-
Group Cluster
Solution

Variable	Means		F-Ratio	Level of Significance
	Cluster 1	Cluster 2		
X_1	4.94	8.82	55.4	.0001
X_2	6.11	3.10	34.1	.0001
X_3	13.52	17.81	68.6	.0001
X_4	10.87	9.88	2.6	.1160
X_5	5.52	5.92	1.0	.3212
X_6	5.35	5.07	0.4	.5183
X_7	8.32	9.87	11.2	.0016
X_8	8.58	10.25	15.9	.0002
X_9	0.20	1.00	96.0	.0001
X_{10}	0.96	0.00	529.0	.0000

Information essential to the interpretation and profiling stages is provided in Table 7.7. On the left-hand side of the table are the columns of means for the two groups identified in the Ward's method of clustering from our previous example. Beside the means are listed the univariate F-ratios and levels of significance comparing the differences between the group means. For the interpretation stage, we would consider only the top six variables because these were the only variables used to determine the two-group cluster solution. Looking at the levels of significance for these six variables, we note that three of the six—X_1 (self-esteem), X_2 (locus of control) and X_3 (alienation) exhibit significantly different patterns. The remaining three variables are not significant. Thus, in interpreting and ultimately labeling the cluster, we focus on the three significant variables and their respective group means. Interpretation of the means of the clusters for these three variables shows that cluster 1 generally exhibits lower self-esteem and an external locus of control and is relatively more alienated; cluster 2 is the opposite. Thus, the group 2 cluster could be labeled as relatively more well adjusted, confident, and in control of their lives.

For the profiling stage, we focus on variables not included in the cluster solution. In the present case, we will consider variables X_7 to X_{10}. Note that the univariate F-ratios show that the group means for all four variables are significantly different. The profiling process here shows that cluster 2 persons are relatively more knowledgeable and more motivated, were trained using a role-playing approach, and tend not to receive the outstanding salesperson award more often than those in cluster one. The clear implication of the findings here, from a HATCO managerial perspective, is that an effort should be made to hire persons who exhibit the characteristics of cluster 2.

Summary

Cluster analysis can be a very useful data reduction technique. But since its application is more an art than a science, it can easily be abused (misapplied) by the analyst. Different interobject measures and different algorithms can and do affect the results. The analyst needs to consider these problems and, if possible, replicate the analysis under varying conditions. If the analyst is cautious, cluster analysis can be very helpful in identifying latent patterns suggesting useful groupings (clusters) of objects.

QUESTIONS

1. What are the three stages in the application of cluster analysis?

2. What is the purpose of cluster analysis, and when should it be used instead of factor analysis?

3. What should the analyst remember when selecting a distance measure to use in cluster analysis?

4. How does the analyst know whether to use hierarchical or nonhierarchical cluster techniques? Under which conditions would each approach be used?

5. How can you decide how many clusters to have in your solution?

6. What is the difference between the interpretation stage and the profiling stage?

7. How do you use a vertical icicle?

REFERENCES

1. Everitt, B., *Cluster Analysis*, 2nd ed. New York: Halsted Press, 1980.
2. Green, P. E., *Analyzing Multivariate Data*. Hinsdale, Ill.: Holt, Rinehart & Winston, 1978.
3. Green, P. E., and J. Douglas Carroll, *Mathematical Tools for Applied Multivariate Analysis*. New York: Academic Press, 1978.
4. McGraw-Hill, *SPSS-X Users's Guide*, 2nd ed. Chicago, 1986.
5. McGraw-Hill, *SPSS-X Advanced Statistics Guide*. Chicago, 1986.
6. Milligan, G., "An Examination of the Effect of Six Types of Error Perturbation on Fifteen Clustering Algorithms," *Psychometrica*, vol. 45, (September, 1980) 325–342.
7. Morrison, D., "Measurement Problems in Cluster Analysis," *Management Science*, vol. 13, no. 12 (August 1967), pp.
8. Overall, J., "Note on Multivariate Methods for Profile Analysis," *Psychological Bulletin*, vol. 61, no. 3 (1964), pp. 195–198.
9. Punj, G. and D. Stewart, "Cluster Analysis in Marketing Research: Review and Suggestions for Application," *Journal of Marketing Research*, vol. 20 (May 1983), pp. 134–148.
10. SAS Institute, *SAS User's Guide: Basics*, Version 5 ed. Cary, N. C., 1985.
11. SAS Institute, *SAS User's Guide: Statistics*, Version 5 ed. Cary, N. C., 1985.
12. Shephard, R. "Metric Structures in Ordinal Data," *Journal of Mathematical Psychology*, vol. 3 (1966), pp. 287–315.

SELECTED READINGS

Cluster Analysis in Marketing Research: Review and Suggestions for Application

GIRISH PUNJ
DAVID W. STEWART

Cluster analysis has become a common tool for the marketing researcher. Both the academic researcher and the marketing applications researcher rely on the technique for developing empirical groupings of persons, products, or occasions which may serve as the basis for further analysis. Despite its frequent use, little is known about the characteristics of available clustering methods or how clustering methods should be employed. One indication of this general lack of understanding of clustering methodology is the failure of numerous authors in the marketing literature to specify what clustering method is being used. Another such indicator is the tendency of some authors to differentiate among methods which actually differ only in name.

The use of cluster analysis has frequently been viewed with skepticism. Green, Frank, and Robinson (1967) and Frank and Green (1968) have discussed problems with determining the appropriate measure of similarity and the appropriate number of clusters. Inglis and Johnson (1970),

"Cluster Analysis in Marketing Research: Review and Suggestions for Application," Girish Punj and David W. Stewart, Vol. 20 (May 1983), pp. 134–148. Reprinted from the *Journal of Marketing Research,* published by the American Marketing Association. Girish Punj is professor of Marketing, School of Business Administration, University of Connecticut. David W. Stewart is professor of Management, Owen Graduate School of Management, Vanderbilt University.

Morrison (1967), Neidell (1970), and Shuchman (1967) have also expressed concern about the use of cluster analysis. More recently, Wells (1975) has expressed reservations about the use of cluster analysis unless very different, homogeneous groups can be identified. Such skepticism is probably justified in the light of the confusing array of names and methods of cluster analysis confronting the marketing researcher. As this confusion is resolved and as additional information about the performance characteristics of various clustering algorithms becomes available, such skepticism may disappear. Recent work on clustering algorithms affords a basis for establishing some general guidelines for the appropriate use of cluster analysis. It is useful to note that many of the problems associated with cluster analysis also plague multivariate statistics in general: choice of an appropriate metric, selection of variables, cross-validation, and external validation.

Two general sets of issues confront the marketing researcher seeking to use cluster analysis. One set of issues involves theoretical properties of particular algorithms. These issues are considered in the literature on cluster analysis (Anderberg 1973; Bailey 1974; Cormack 1971; Hartigan 1975), and are not addressed here. The second set of issues are more practical and pertain to the actual use of clustering procedures for data analysis. These issues are the foci of our article, in which we review applications of clustering methodology to marketing problems, provide a systematic treatment of the clustering options open to the marketing researcher, and use both theoretical and empirical findings to suggest which clustering options may be most useful for a particular research problem.

Cluster analysis has most frequently been employed as a classification tool. It has also been used by some researchers as a means of representing the structure of data via the construction of dendrograms (Bertin 1967; Hartigan 1967) or overlapping clusters (Arabie et al. 1981; Shepard and Arabie 1979). The latter applications are distinct from the use of cluster analysis for classification and represent an alternative to multidimensional scaling and factor analytic approaches to representing similarity data. Whereas classification is concerned with the identification of discrete categories (taxonomies), structural representation is concerned with the development of a faithful representation of relationships. Both uses of cluster analysis are legitimate, but the objectives of these applications are very different. The best clustering algorithm for accomplishing one of these objectives is not necessarily the best for the other objective. We restrict our treatment of cluster analysis to the more common of the two applications, classification.

Cluster analysis is a statistical method for classification. Unlike other statistical methods for classification, such as discriminant analysis and automatic interaction detection, it makes no prior assumptions about important differences within a population. Cluster analysis is a purely empirical method of classification and as such is primarily an inductive technique (Gerard 1957). Though some theorists have not been favorably disposed toward the use of cluster analysis, and criticism of the ad hoc nature of clustering solutions is common, classification is an important and frequently overlooked tool of science. Wolf (1926) has suggested that classification is both the first and last method employed by science. The essence of classification is that certain things are thought of as related in a certain way. Indeed, the final outcome of other methods of study may well be a new classification.

Kemeny (1959) and Kantor (1953), discussing the philosophy of science, point to the fundamental importance of classification. Wolf (1926) holds that verification of laws of science may occur only after classification has been completed. Thus, whether the classification exercise is completed explicitly or implicitly, it must occur. Cluster analysis provides one, empirically based, means for explicitly classifying objects. Such a tool is particularly relevant for the emerging discipline of marketing which is still wrestling with the problems of how best to classify consumers, products, media types, and usage occasions.

Uses of Cluster Analysis in Marketing

The primary use of cluster analysis in marketing has been for market segmentation. Since the appearance of Smith's now-classic article (1956), market segmentation has become an important tool for both academic research and applied marketing. In a review of market segmentation research and methodology, Wind (1978) identifies both the impact of this most fundamental of marketing tools and some rather significant problem areas. Not the least of these problems is the plethora of methods that have been proposed for segmenting markets. This multiplication of techniques has served to confuse many marketers, shift discussions of researchers from more substantive issues to issues of method, and impede the development of meta-research directed at integrating market segmentation research. In concluding his review of the segmentation literature, Wind suggests that one important area of future research should be the "evaluation of the conditions under which various data analytical techniques are most appropriate" (1978, p. 334).

All segmentation research, regardless of the method used, is designed to identify groups of entities (people, markets, organizations) that share certain common characteristics (attitudes, purchase propensities, media habits, etc.). Stripped of the specific data employed and the details of the purposes of a particular study, segmentation research becomes a grouping task. Wind (1978) notes that researchers tend to select grouping methods largely on the basis of familiarity, availability, and cost rather than on the basis of the methods' characteristics and appropriateness. Wind attributes this practice to the lack of research on similarity measures, grouping (clustering) algorithms, and effects of various data transformations.

A second and equally important use of cluster analysis has been in seeking a better understanding of buyer behaviors by identifying homogeneous groups of buyers. Cluster analysis has been less frequently applied to this type of theory-building problem, possibly because of theorists' discomfort with a set of procedures which appear ad hoc. Nevertheless, there is clearly a need for better classification of relevant buyer characteristics. Bettman (1979) has called for the development of taxonomies of both consumer choice task and individual difference characteristics. Cluster analysis is one means for developing such taxonomies. Examples of such use may be found in articles by Claxton, Fry, and Portis (1974), Kiel and Layton (1981), and Furse, Punj, and Stewart (1982).

Cluster analysis has been employed in the development of potential new product opportunities. By clustering brands/products, competitive sets within the larger market structure can be determined. Thus, a firm can examine its current offerings vis-à-vis those of its competitors. The firm can determine the extent to which a current or potential product offering is uniquely positioned or is in a competitive set with other products (Srivastava, Leone, and Shocker 1981; Srivastava, Shocker, and Day 1978). Although cluster analysis has not been used frequently in such applications, largely because of the availability of other techniques such as multidimensional scaling, factor analysis, and discriminant analysis, it is not uncommon to find cluster analysis used as an adjunct to these other techniques. Cluster analysis has also been suggested as an alternative to factor analysis and discriminant analysis. In such applications it is important for the analyst to determine whether discrete categories of products are desirable or whether a representation of market structure is desirable. The latter may be more useful in many market structure applications, in which case cluster analysis would not be used as a classification technique and the analyst would face a different set of issues from those addressed here.

Cluster analysis has also been employed by several researchers in the problem of test market selection (Green, Frank, and Robinson 1967). Such applications are concerned with the identification of relatively homogeneous sets of test markets which may become interchangeable in test market studies. The identification of such homogeneous sets of test markets allows generalization of the results obtained in one test market to other test markets in the same cluster, thereby reducing the number of test markets required.

Finally, cluster analysis has been used as a general data reduction technique to develop aggregates of data which are more general and more easily managed than individual observations. For example, limits on the number of observations that can be used in multidimensional scaling programs often necessitate an initial clustering of observations. Homogeneous clusters then become

the unit of analysis for the multidimensional scaling procedure. Fisher (1969) discussed the use of cluster analysis for data reduction from the perspective of econometrics and argued that cluster analysis is most appropriate whenever the data are too numerous or too detailed to be manageable. Such data simplification and aggregation are carried out for the convenience of the investigator rather than in the interest of theory building.

Table 1 is a brief description of some recent applications of cluster analysis to marketing problems. Although not a complete set of all applications of cluster analysis in marketing, it illustrates several points. First, the array of problems addressed by these studies is striking. Equally striking is the diversity of clustering methods employed. In constructing this table we had difficulty discerning the specific clustering algorithm used by the researchers. Cluster analysis methods were often identified by the name of the program used, e.g., BMDP2M, BCTRY, or Howard and Harris, rather than by the specific clustering algorithm used. Only by consulting a particular program's manual could we identify the method actually employed. For one of the studies cited we could not find any information on the clustering method used.

The lack of specificity about the method of clustering employed in these studies is illustrative of the problems associated with the use of cluster analysis. The lack of detailed reporting suggests either an ignorance of or lack of concern for the important parameters of the clustering method used. Failure to provide specific information about the method also tends to inhibit replication and provides little guidance for other researchers who might seek an appropriate method of cluster analysis. Use of specific program names rather than the more general algorithm name impedes interstudy comparisons.

This situation suggests a need for a sound review of clustering methodology for the marketing researcher. Previous reviews on this subject appeared prior to the publication of much of the research on the performance characteristics of clustering algorithms. Sherman and Sheth (1977) discuss selected similarity measures and clustering algorithms. Though they mention some empirical work on the characteristics of these measures and algorithms, their report is primarily a catalog of techniques and some marketing applications. Relatively little guidance is provided the re-

TABLE 1   Some Recent Applications of Cluster Analysis in Marketing

Application	Purpose of Research	Nature of Data	Clustering Method Used
Anderson, Cox, and Fulcher (1976)	To identify the determinant attributes in bank selection decisions and use them for segmenting commercial bank customers	Determinant attribute scores on several bank selection variables	Iterative partitioning—MIKCA (McRae 1973)
Bass, Pessemier, and Tigert (1969)	To identify market segments with respect to media exposure	Attribute scores on several media exposure variables	Average linkage cluster analysis (Sneath and Sokal (1973)
Calantone and Sawyer (1978)	To examine the stability of market segments in the retail banking market	Attribute scores on several bank selection variables	K-means (Howard and Harris 1966)
Claxton, Fry, and Portis (1974)	To classify furniture and appliance buyers in terms of their information search behavior	Attribute scores on several prepurchase activity measures	Complete linkage cluster analysis (Johnson 1967; Lance and Williams 1967a)
Day and Heeler (1971)	To classify stores into similar strata	Factor scores on several store attributes	(1) Complete linkage analysis (2) Iterative partitioning (Rubin 1965)
Green, Frank, and Robinson (1967)	To identify matched cities for test marketing	Factor scores on several city characteristics	Average linkage cluster analysis (Sneath and Sokal 1973)
Greeno, Sommers, and Kernan (1973)	To identify market segments with respect of personality variables and implicit behavior patterns	Q sorts on 38 product items	Ward's minimum variance
Kernan (1968)	To identify groups of people along several personality and decision behavior characteristics	Scores on several personality and decision traits	Ward's minimum variance method (Ward 1963)
Kernan and Bruce (1972)	To create relatively homogeneous configuration of census traits	Characteristics of census traits	Ward's minimum variance method (Ward 1963)
Kiel and Layton (1981)	To develop consumer taxonomies of search behavior by Australian new car buyers	Factor scores derived from several search variables	Average linkage cluster analysis (Sneath and Sokal 1973)
Landon (1974)	To identify groups of people using purchase intention and self-concept variables	Scores on self-image and purchase intention variables	Iterative partitioning (BCTRY; Tryon and Bailey 1970)
Lessig and Tollefson (1971)	To identify similar groups of consumers along several buyer behavior variables	Scores on several buyer behavior variables	Ward's minimum variance method (Ward 1963)
Montgomery and Silk (1971)	To identify opinion leadership and consumer interest segments	Scores on several interest and opinion leadership variables	Complete linkage cluster analysis (Johnson 1967)

TABLE 1 (Continued)

Application	Purpose of Research	Nature of Data	Clustering Method Used
Moriarty and Venkatesan (1978)	To segment educational institutions in terms of benefits sought when purchasing financial-aid MIS	Importance scores on financial-aid management services	K-means (Howard and Harris 1966)
Morrison and Sherman (1972)	To determine how various individuals interpret sex appeal in advertising	Ratings of advertisements by respondents	Iterative Partitioning (Friedman and Rubin 1967)
Myers and Nicosia (1968)	To develop a consumer typology using attribute data	Scores of supermarket image variables	Iterative partitioning (BCTRY; Tryon and Bailey 1970)
Sethi (1971)	To classify world markets	Macrolevel data on countries	Iterative partitioning (BCTRY; Tryon and Bailey 1970)
Sexton (1974)	To identify homogeneous groups of families using product and brand usage data	Brand and product usage rate data	Type not specified
Schaninger, Lessig, and Panton (1980)	To identify segments of consumers on the basis of product usage variables	Scores on several product usage variables	K-means (Howard and Harris 1966)
Green and Carmone (1968)	To identify similar computers (strata in the computer market)	Performance measures for different computer models	K-means (Howard and Harris 1966)

searcher seeking to discover the characteristics and limitations of various grouping procedures. Indeed, the Sherman and Sheth report may mislead some readers because its categorization of clustering algorithms suggests substantive differences among identical algorithms which differ only in name. Frank and Green (1968) also provide an introduction and review of clustering methodology but make no specific recommendations to guide the user of the methodology. After reviewing the problems and issues facing the user of cluster analytic procedures, we offer clarification of the similarities and differences among various clustering algorithms and some suggestions about their use.

Problems in Using Cluster Analysis

Unlike other data analytic methods, cluster analysis is a set of methodologies that has developed outside a single dominant discipline. Factor analysis and various scaling methods were developed within the discipline of psychology and one would look to that discipline for guidance in the use of these methods. Regression, though used in a variety of disciplines, has tended to be the special province of econometricians, who have developed a large body of literature on the technique. In contrast, no single discipline has developed and retained clustering methodology. Rather, numerous disciplines (econometrics, psychology, biology, and engineering) have independently approached the clustering problem. Often working in parallel, researchers in these disciplines have arrived at similar solutions but have given them different names. For example, Blashfield (1978) reviewed the literature on hierarchical clustering methods and found as many as seven different names for the same technique. This diversity of names for identical techniques has tended to prevent comparisons of algorithms across disciplines. It has also served to confuse the data analyst by implying a much greater number of available clustering methods than actually exists.

Also confronting the potential user of cluster analysis is the problem of cluster definition. There are currently no clear guidelines for determining the boundaries of clusters or deciding when observations should be included in one cluster or another. Cattell (1978) has suggested that clusters are "fuzzy" constructs. The criterion for admission to a cluster is rather arbitrary. There are no well-established rules for the definition of a cluster. The preferred definition of a cluster seems to vary with the discipline and purpose of the researcher.

Clusters have most frequently been defined by relatively contiguous points in space (Stewart 1981). Cormack (1971) suggested that clusters should exhibit two properties, external isolation and internal cohesion. External isolation requires that objects in one cluster be separated from objects in another cluster by fairly empty space. Internal cohesion requires that objects within the same cluster be similar to each other. Everitt (1974) offered a similar concept which he defines as a natural cluster. The requirement of external isolation does not provide for overlapping clusters. Although a few algorithms have been developed for identifying overlapping clusters (Jardine and Sibson 1971; Peay 1975; Shepard 1974), these methods are primarily concerned with the representation of structure rather than classification. Applications of these methods have been few and are not reviewed here.

In the absence of a generally accepted or definitive definition of a cluster, various algorithms have been developed which offer particular operational definitions. Differences among clustering algorithms are frequently related to how the concept of a cluster is operationalized. Thus, to develop a set of recommendations for the application of cluster analysis, we must first develop a recognition of the clustering algorithms available to the marketing researcher and an understanding of the performance of these methods in relation to one another.

Clustering Algorithms

Table 2 provides a description of the more common clustering algorithms in use, the various alternative names by which the algorithms are known, and a brief discussion of how clusters are formed by each of these methods. Table 2 shows clearly that there are relatively few cluster-

ing methods from which to choose, far fewer than one might suspect from a reading of the literature on cluster analysis. Four primary hierarchical methods are available, single linkage, complete linkage, average linkage, and Ward's minimum variance method. Although there are several variations of the average linkage method, only one, simple average linkage, is widely used. In addition, two variants of the average method, the centroid and median methods, have very undesirable properties (Aldenderfer 1977; Sneath and Sokal 1973) which recommend against their use. The weighted average linkage method has been shown to produce results very similar to those produced by the simple average method (Blashfield 1977).

There is more variety among the nonhierarchical methods, though all work on similar principles. These iterative partitioning methods begin by dividing observations into some predetermined number of clusters. Observations are then reassigned to clusters until some decision rule terminates the process. These methods may differ with respect to the starting partition, the type of reassignment process, the decision rule used for terminating clustering, and the frequency with which cluster centroids are updated during the reassignment process. The initial partition may be random or based on some prior information or intuition. One method (MIKCA) uses several different random starting partitions to ensure an efficient solution. Two types of reassignment are generally employed, K-means and hill-climbing. These methods are briefly discussed in Table 2 as are the termination decision rules used with each method. Cluster centroids may be updated after each membership move or only after a complete pass through the entire data set.

Not included in Table 2 are two methods frequently used for cluster analysis: Q factor analysis and automatic interaction detection (AID) (Morgan and Sonquist 1963). Q factor analysis is not included because Stewart (1981) in the marketing literature and Cattell (1978) in the psychology literature have forcefully argued that factor analysis is inappropriate as a method for identifying clusters. Skinner (1979) discusses some relationships between factor analysis and cluster analysis. AID is not included because it operates on a rather different principle than the clustering procedures. AID requires the prior specification of independent and dependent variables and seeks to identify sets of nominal dependent variables which group observations in a manner that minimizes the vari-

TABLE 2 Clustering Methods

Primary Name	Alternative Names	Method of Forming Clusters
Hierarchical methods		
Single linkage cluster analysis	Minimum method (Johnson 1967); linkage analysis (McQuitty 1967); nearest neighbor cluster analysis (Lance and Williams 1967a); connectiveness method (Johnson 1967)	An observation is joined to a cluster if it has a certain level of similarity with at least *one* of the members of that cluster. Connections between clusters are based on links between single entities.
Complete linkage cluster analysis	Maximum method (Johnson 1967); rank order typal analysis (McQuitty 1967); furthest neighbor cluster analysis (Lance and Williams 1967a); diameter method (Johnson 1967)	An observation is joined to a cluster if it has a certain level of similarity and *all* current members of the cluster.
Average linkage cluster analysis	Simple average linkage analysis (Sneath and Sokal 1973); weighted average method (McQuitty 1967); centroid method (Gower 1967); median method (Lance and Williams 1967a)	These are actually four similar methods. In all four methods an observation is joined to a cluster if it has a certain *average* level of similarity with all current members of the clusters. These methods differ in the manner in which the average level of similarity is defined. The weighted average method and median method provide for an *a priori* weighting of the averages based on the number of entities desired in each cluster. The centroid method provides for an initial computation of the centroid of each cluster. Average similarity is based on this centroid. Only the simple average linkage procedure has been widely used.
Minimum variance cluster analysis	Minimum variance method; Ward's method; error sum of squares method (Ward 1963); HGROUP (Veldman 1967)	The minimum variance method is designed to generate clusters in such a way as to minimize the within-cluster variance. Unlike other hierarchical clustering methods, Ward's method optimizes an objective statistic: it seeks to minimize tr **W,** where **W** is the pooled within-clusters sum of squares and cross-products matrix. Ward's method is somewhat similar to the average method in that variance is a function of deviations from the mean. Some authors have included Ward's method as a special case of the average method (Bailey 1974). It is an average linkage method because it does not seek to minimize distance between one member of the cluster and the entity, or all members of the cluster and the entity as in single linkage and complete linkage, respectively, but minimizes the average distance within the cluster.

TABLE 2 (Continued)

Primary Name	Alternative Names	Method of Forming Clusters
Iterative partitioning methods *(nonhierarchical methods)*		These methods begin with the partition of observations into a specified number of clusters. This partition may be on a random or nonrandom basis. Observations are then reassigned to clusters until some stopping criterion is reached. Methods differ in the nature of the reassignment and stopping rules.
K-means		Cases are reassigned by moving them to the cluster whose centroid is closest to that case. Reassignment continues until every case is assigned to the cluster with the nearest centroid. Such a procedure implicitly minimizes the variance within each cluster, tr $\mathbf{W}$.
Hill-climbing methods		Cases are not reassigned to the cluster with the nearest centroid but are moved from one cluster to another if a particular statistical criterion is obtained. Reassignment continues until optimization occurs. The objective function to be optimized may be selected from one of four options, tr $\mathbf{W}$, tr $[(\mathbf{W}^{-1}\mathbf{B}]$, $|\mathbf{W}|$, and the largest eigenvalue of $[(\mathbf{W}^{-1}\mathbf{B}]$, where $\mathbf{W}$ is the pooled within-cluster covariance matrix and $\mathbf{B}$ is the between cluster covariance matrix.
Combined K-means and hill-climbing methods		Uses of combination of K-means and hill-climbing methods.

TABLE 3 Common Clustering Packages/Programs

Name of Package/ Program	Where Available?/Authors	Clustering Methods[a]	Comments
ANDERBERG	In the appendices of book entitled *Cluster Analysis for Applications* by M. R. Anderberg (1973)	S, C, A, W, K, H, KH	1. No missing value treatment 2. Only binary data type with octal coding scheme 3. User manual not available
BCTRY	D. Bailey and R. C. Tryon, Tryon-Bailey Associates, Inc., c/o Mr. Peter Lenz, 2222 S.E. Nehalem St., Portland, OR	K	1. No MANOVA statistics are optimized 2. Initial partition has to be user specified 3. Factor analysis of variables may be performed as well

TABLE 3 (Continued)

Name of Package/ Program	Where Available?/Authors	Clustering Methods[a]	Comments
BMDP	W. J. Dixon (ed.), Health Sciences Computing Facility, School of Medicine UCLA, Los Angeles, CA	S, C, A, K	1. Single and complete linkage available for clustering 2. Binary data not permissible 3. Continuous type similarity measure 4. Method for clustering cases and variables simultaneously is available 5. User cannot supply only similarity matrix for cases
CLUS	H. Friedman and J. Rubin (1967 *JASA* article), IBM SHARE system	K, H	1. Fixed number of clusters 2. Expensive to use
CLUSTAN	D. Wishart, Computer Centre, University College of London, 19 Gordon St., London, WC1H OAH, Great Britain	S, C, A, W, K, H	1. High versatility (38 s/dis measures) 2. Initial partition for iterative partitioning methods has to be user specified 3. Binary and continuous data types 4. Binary data in 3 coding schemes 5. Variable transformations not available 6. Permits overlapping clusters
HARTIGAN	In the appendices of book entitled *Clustering Algorithms* by J. J. Hartigan (1975)	S, A, K	1. Fixed number of clusters for iterative partitioning methods 2. No variable transformations available 3. No user manual available
HGROUP	D. J. Veldman (1967), *FORTRAN Programming for the Behavioral Sciences*	W	1. Part of the University of Texas EDSTAT statistics package
HICLUS	S. C. Johnson (based on 1967 *Psychometrika* article), Bell Telephone Labs, Murray Hill, NJ	S C	1. No user manual available 2. No missing value treatment 3. No standardization of variables 4. No transformation of variables 5. User must supply similarity matrix (hence is versatile in some sense)
HOWD (Howard- Harris)	Britton Harris, F. J. Carmone, Jr., University of Pennsylvania, Philadelphia, PA	K	1. No user manual available 2. No MANOVA statistics 3. Number of clusters fixed
ISODATA	Daviel Wolf, SRI, 333 Ravenswood Avenue, Menlo Park, CA	K	1. No user manual available 2. No MANOVA statistics optimized

TABLE 3   (Continued)

Name of Package/ Program	Where Available?/Authors	Clustering Methods[a]	Comments
MIKCA	D. J. McRae, Coordinator, Testing & Computer Applications, Jackson Public Schools, Jackson, MI	K, H, KH	1. No user manual available 2. 4 MANOVA statistics optimized 3. 3 different distance measures
NT-SYS	F. James Rohlf, John Kishapugh, David Kirk, Dept. of Ecology and Evolution, SUNY at Stony Brook, Stony Brook, NY	S, C, A	1. Permits overlapping clusters 2. Alphanumeric coding scheme for binary data 3. Moderately versatile
OSIRIS	Institute of Survey Research, University of Michigan, Ann Arbor, MI	C	
SAS	James H. Goodnight, SAS Inst. Inc., P. O. Box 10066, Raleigh, NC	C	1. Continuous similarity measure

[a]S = single linkage.
C = complete linkage.
A = average linkage.
W = Ward's minimum variance method.
K = K-means.
H = hill climbing.
KH = joint K-means, hill climbing.

ance of the dependent variable within each group. Cluster analysis procedures require no such *a priori* specification of independent and dependent variables.

These clustering algorithms exist in various forms but most have been programmed. Several software programs are currently available for cluster analysis. They differ in their comprehensiveness and ease of use. Table 3 briefly describes several of the more common clustering software programs, identifies the types of clustering methods available within each program, and cites the original source of the program. Selecting an appropriate cluster analytic method or software package requires some knowledge of the performance characteristics of the various methods.

Empirical Comparisons of Clustering Methods

One method for evaluating clustering methods that has been used with increasing frequency involves comparing the results of different clustering methods applied to the same data sets. If the underlying characteristics of these data sets are known, one can assess the degree to which each clustering method produces results consistent with these known characteristics. For example, if a data set consists of a known mixture of groups, or subpopulations, the efficacy of a cluster solution can be evaluated by its success in discriminating among these subpopulations. This mixture model approach to the evaluation of clustering algorithms has recently been employed by several researchers. Table 4 summarizes the findings of 12 such studies.

The number of clustering algorithms, distance measures, and types of data that might be incorporated in a mixture model study is so large as to preclude any one comprehensive study of the relative efficacy of clustering methods. We can look across the studies in Table 4, however, and begin to draw some conclusions about clustering methods. Three procedures seem to warrant special consideration. Ward's minimum variance method, average linkage, and several variants of the iterative partitioning method appear to outperform all other methods. Ward's method appears to outperform the average linkage method except in the

TABLE 4 Empirical Comparisons of the Performance of Clustering Algorithms

Reference	Methods Examined	Data Sets Employed	Coverage[a]	Criteria	Summary of Results
Cunningham and Ogilvie (1972)	Single, complete, average linkage with Euclidean distances and Ward's minimum variance technique	Normal mixtures	Complete	Measures of "stress" to compare input similarity/dissimilarity matrix with similarity relationship among entities portrayed by the clustering method	Average linkage outperformed other methods
Kuiper and Fisher (1975)	Single, complete, average, centroid, median linkage, all using Euclidean distances and Ward's minimum variance technique	Bivariate normal mixtures	Complete	Rand's statistic (Rand 1971)	Ward's technique consistently outperformed other methods
Blashfield (1976)	Single, complete, average linkage, all using Euclidean distance and Ward's minimum variance technique	Multinormal mixtures	Complete	Kappa (Cohen 1960)	Ward's technique demonstrated highest median accuracy
Mojena (1977)	Simple average, weighted average, median, centroid, complete linkage, all using Euclidean distances and Ward's minimum variance technique	Multivariate gamma distribution mixtures	Complete	Rand's statistic	Ward's method outperformed other methods
Blashfield (1977)	Eight iterative partitioning methods: Anderberg and CLUSTAN K-means methods, each with cluster statistics updated after each reassignment and only after a complete pass through the data; CLUS and MIKCA (both hill-climbing algorithms), each with optimization of tr **W** and **W**	Multinormal mixtures	Complete	Kappa	For 15 of the 20 data sets examined, a hill-climbing technique which optimized **W** performed best, i.e., MIKCA or CLUS. In two other cases a hill-climbing method which optimized tr **W** performed best, CLUS

TABLE 4 (Continued)

Reference	Methods Examined	Data Sets Employed	Coverage[a]	Criteria	Summary of Results
Milligan and Isaac (1978)	Single, complete average linkage, and Ward's minimum variance technique, all using Euclidean distances	Data sets differing in degree of error perturbation	Complete	Rand's statistic and kappa	Average linkage and Ward's technique superior to single and complete linkage
Mezzich (1978)	Single, complete linkage, and K-means, each with city-block and Euclidean distances and correlation coefficient, ISODATA, Friedman and Rubin method, Q factor analysis, multidimensional scaling with city-block and Euclidean metrics and correlation coefficients, NORMAP/NORMIX, average linkage with correlation coefficient	Psychiatric ratings	Complete	Replicability; agreement with "expert" judges; goodness of fit between raw input dissimilarity matrix and matrix of 0's and 1's indicating entities clustered together	K-means procedure with Euclidean distances performed best followed by K-means procedure with the city-block metric; average linkage also performed well as did complete linkage with a correlation coefficient & city-block metric & ISODATA; the type of metric used (r, city-block, or Euclidean distance) had little impact on results
Edelbrock (1979)	Single, complete, average, and centroid, each with correlation coefficients, Euclidean distances, and Ward's minimum variance technique	Multivariate normal mixtures, standardized & unstandardized	70, 80, 90, 95, 100%	Kappa	Ward's method and simple average were most accurate; performance of all algorithms deteriorated as coverage increased but this was less pronounced when the data were standardized or correlation coefficients were used. The latter finding is suggested to result from the decreased extremity of outliers associated with standardization or use of the correlation coefficient.

328

Study	Methods	Data	Levels	Index	Results
Edelbrock and McLaughlin (1980)	Single, complete, average, each with correlation coefficients, Euclidean distances, one-way and two-way intraclass correlations, and Ward's minimum variance technique	Multivariate normal mixtures & multivariate gamma mixtures	40, 50, 60, 70, 80, 90, 95, 100%	Kappa and Rand's statistic	Ward's method and the average method using one-way intraclass correlations were most accurate; performance of all algorithms deteriorated as coverage increased
Blashfield and Morey (1980)	Ward's minimum variance technique, group average linkage, Q factor analysis, Lor's nonhierarchical procedure, all using Pearson product moment correlations as the similarity measure	Multivariate normal mixtures	Varying levels	Kappa	Group average method best at higher levels of coverage; at lower levels of coverage Ward's method and group average performed similarly
Milligan (1980)	Single, complete, group average, weighted average, centroid & median linkage, Ward's minimum variance technique, minimum average sum of squares, minimum total sum of squares, beta-flexible (Lance & Williams 1970a b), average link in the new cluster, MacQueen's method, Jancey's method, K-means with random starting point, K-means with derived starting point, all with Euclidean distances, Cattell's	Multivariate normal mixtures, standardized and varying in the number of underlying clusters and the pattern of distribution of points of the clusters. Data sets ranged from error free to two levels of error perturbation, from two levels of the distance measures, from containing no outliers to two levels of outlier conditions, and from no variables unrelated to the clusters to one or two randomly assigned dimensions unrelated	Complete	Rand's statistic; the point biserial correlation between the raw input dissimilarity matrix and a matrix of 0's and 1's indicating entities clustering together	K-means procedure with a derived point generally performed better than other methods across all conditions 1. Distance measure selection did not appear critical; methods generally robust across distance measures. 2. Presence of random dimensions produced decrements in cluster recovery. 3. Single linkage method strongly affected by error-perturbations; other hierarchical methods moderately so; nonhierarchical methods only slightly affected by perturbations. 4. Complete linkage and Ward's method exhibited noticeable decrements in performance in the outlier conditions; sin-

TABLE 4 (Continued)

Reference	Methods Examined	Data Sets Employed	Coverage[a]	Criteria	Summary of Results
	(1949) r_p, and Pearson r	to the underlying clusters.			gle, group average, & centroid methods only slightly affected by presence of outliers; nonhierarchical methods generally unaffected by presence of outliers. 5. Group average method best among hierarchical methods used to derive starting point for K-means procedure. 6. Nonhierarchical methods using random starting points performed poorly across all conditions
Bayne, Beauchamp, Begovich, and Kane (1980)	Single, complete, centroid, simple average, weighted average, median linkage and Ward's minimum variance technique, and two new hierarchical methods, the variance and rank score methods; four hierarchical methods: Wolfe's NORMIX, K-means, two variants of the Friedman-Rubin procedure (trace $\mathbf{W}$ & $|\mathbf{W}|$). Euclidean distances served as similarity measure.				

| Six parameterizations of two bivariate normal populations | Complete | Rand's statistic | K-means, trace **W**, and \|**W**\| provided the best recovery of cluster structure. NORMIX performed most poorly. Among hierarchical methods, Ward's technique, complete linkage, variance & rank score methods performed best. Variants of average linkage method also performed well but not as well as other methods. Single linkage performed poorly |

* The percentage of observations included in the cluster solution. With complete coverage, clustering continues until all observations have been assigned to a cluster. Ninety percent coverage could imply that the most extreme 10% of the observations were not included in any cluster.

presence of outliers. K-means appears to outperform both Ward's method and the average linkage method if a nonrandom starting point is specified. If a random starting point is used K-means may be markedly inferior to other methods, but results on this issue are not consistent. Nevertheless, the K-means procedure appears to be more robust than any of the hierarchical methods with respect to the presence of outliers, error perturbations of the distance measures, and the choice of a distance metric. It appears to be least affected by the presence of irrelevant attributes or dimensions in the data.

One conclusion in several of the studies is that the choice of a similarity/dissimilarity measure, or distance measure, does not appear to be critical. Despite the considerable attention given such measures (Green and Rao 1969; Morrison 1967; Sherman and Sheth 1977), the selection of a similarity measure appears to be less important for determining the outcome of a clustering solution than the selection of a clustering algorithm. Two cautions should be observed in taking this conclusion at face value, however. First, the number of studies of the relative import of distance measures for determining clustering solutions is small and many types of data have yet to be examined. There may be types of data for which the selection of a distance measure is critical to the clustering solution. Second, clustering algorithms which are sensitive to the presence of outliers (e.g., complete linkage specifically, and more generally all of the hierarchical methods of clustering) seem to produce better solutions when Pearson product moment or intraclass correlation coefficients are used. Such similarity measures tend to reduce the extremity of outliers in relation to Euclidean distance measures. This, in turn, reduces the influence of outliers on the final clustering solution. A similar effect is obtained if data are standardized prior to clustering.

One characteristic of data appears to have a marked decremental effect on the performance of all clustering methods—the presence of one or more spurious attributes or dimensions. A variable that is not related to the final clustering solution, i.e., does not differentiate among clusters in some manner, causes a serious deterioration of the performance of all clustering methods, though this problem is least severe with the K-means procedure and is probably less serious for other iterative partitioning methods as well. This finding indicates the need for careful selection of variables

for use in clustering and the need to avoid "shotgun" approaches where everything known about the observations is used as the basis for clustering. Clearly one cannot know in advance what variables may differentiate among a set of as yet unidentified clusters. Nevertheless, it is not unreasonable for a researcher to have some rational or theoretical basis for selecting the variables used in a cluster analysis.

A final conclusion can be drawn from the empirical findings on the performance of clustering algorithms: as a clustering algorithm includes more and more observations, its performance tends to deteriorate, particularly at high levels of coverage, 90% and above. This effect is probably the result of outliers beginning to come into the solution. Clustering all observations may not be a good practice. Rather the identification and elimination of outliers or the use of a decision rule to stop clustering short of the inclusion of all observations is probably advantageous. Suggestions for identifying outliers are provided hereafter. The K-means procedure has shown less decrement in performance as coverage increases than have the hierarchical methods.

Though a reasonable amount of evidence suggests that iterative partitioning methods are superior to hierarchical methods, particularly if nonrandom starting points are used, it is not yet clear which of the iterative partitioning methods are superior. K-means procedures and tr **W** and |**W**| hill-climbing procedures all appear to perform well. Some evidence (Blashfield 1977) suggests that hill-climbing methods which minimize |**W**| have an advantage over other iterative partitioning methods.

Recommendations for Using Cluster Analysis

It should be clear from the preceding discussion that the research analyst must make several decisions which affect the structure of a cluster solution. These decisions can be grouped in the following broad categories:

1. *Data transformation issues*
   A. What measure of similarity/dissimilarity should be used?
   B. Should the data be standardized? How should nonequivalence of metrics among variables be addressed?

    C. How should interdependencies in the data be addressed?

2. *Solution issues*
    A. How many clusters should be obtained?
    B. What clustering algorithm should be used?
    C. Should all cases be included in a cluster analysis or should some subset be ignored?

3. *Validity issues*
    A. Is the cluster solution different from what might be expected by chance?
    B. Is the cluster solution reliable or stable across samples?
    C. Are the clusters related to variables other than those used to derive them? Are the clusters useful?

4. *Variable selection issues*
    A. What is the best set of variables for generating a cluster analytic solution?

Often these decisions are not independent of one another because the choice of a means for addressing one of these issues may constrain the options available for addressing other issues. For example, choosing to use a Pearson product moment correlation coefficient also determines that the data will be standardized because standardization is implicit in the computation of the correlation coefficient. Thus, it is not possible to offer recommendations for the resolution of any one of these issues without an explicit understanding of the interactions among these decisions.

Data Transformation Issues

Although issues related to the choice of a similarity/dissimilarity measure have received considerable attention (Green and Rao 1969; Morrison 1967) the results of the empirical studies cited above suggest that the choice is not crucial to the final clustering solution. The same appears to be true of the standardization issue. To the extent that a particular measure of similarity or standardization reduces the extremity of outliers, the performance of some algorithms which are sensitive to outliers may be improved. Otherwise the selection of a similarity measure or the standardization of data prior to clustering appears to have minimal effect. We do not suggest that the choice of a similarity measure should be indiscriminant; the measure should be appropriate for the type of data being considered. Rather, the choice of a correlation coefficient, a Euclidean distance, or a city-block metric does not seem to produce much difference in the final outcome of a clustering exercise involving data for which each of the similarity measures is appropriate.

Some measures of similarity/dissimilarity explicitly correct for interdependencies. Other measures do not consider interdependencies. Interdependencies among variables may exist by design or, more often, are the unexpected result of the research design. Careful selection of variables may reduce unwanted interdependencies but the problem is likely to remain even in the best of circumstances. Bailey (1974) provides an illustration of the problem, the effect of which is to weight more heavily certain dimensions along which clustering will be carried out. When this is desirable for some theoretical or practical purpose, correcting for interdependencies is inappropriate. When the researcher desires that all dimensions or attributes be given equal weight in the clustering process, it is necessary to correct for interdependencies. This can be achieved by selecting a similarity measure which corrects for interdependencies, Mahalanobis D^2 or partial correlations. Correction may also be achieved by completing a preliminary principal components analysis with orthogonal rotation. Component scores may then be used as input for the computation of a similarity or distance measure. Skinner (1979) gives an example of this latter approach.

Solution Issues

The selection of the clustering algorithm and solution characteristics appears to be critical to the successful use of cluster analysis. Empirical studies of the performance of clustering algorithms suggest that one of the iterative partitioning methods is preferable to the hierarchical methods. This holds, however, only when a nonrandom starting point can be specified. In addition, iterative partitioning methods require prior specification of the number of clusters desired. Hierarchical methods require no such specification. Thus, the user is confronted with determining both an initial starting point and the number of clusters in order to use the methods that have demonstrated superior performance. Information for determining starting points in the form of *a priori* descriptions of expected clusters may be available. In the absence of such information a means for obtaining starting points and an estimate of the number of clusters is required. A two-stage procedure may be employed to cope with this problem.

In the first step one of the hierarchical methods which has demonstrated superior performance, average linkage or Ward's minimum variance method, may be used to obtain a first approximation of a solution. By examining the results of this preliminary analysis, one can determine both a candidate number of clusters and a starting point for the iterative partitioning analysis. In addition, this preliminary analysis can be used for examining the order of clustering of various observations and the distances between individual observations and clusters. This provides an opportunity for the identification of outliers which may be eliminated from further analysis. The remaining cases may then be submitted to an iterative partitioning analysis for refinement of the clusters. Similar two-stage clustering approaches have been suggested by Hartigan (1975) and Milligan (1980). Figure 1 is a schematic representation of the procedure. Only four cluster analytic software packages provide average linkage or Ward's method and

FIGURE 1 Two-stage clustering

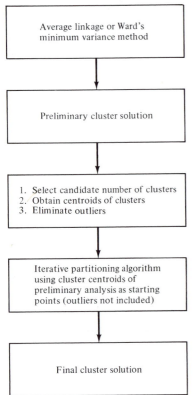

an iterative partitioning algorithm: BMDP, CLUS-TAN, and the Anderberg and Hartigan series.

Validity Issues

Even after careful analysis of a data set and the determination of a final cluster solution, the researcher has no assurance of having arrived at a meaningful and useful set of clusters. A cluster solution will be reached even when there are no natural groupings in the data. This problem is similar to that encountered with a variety of other procedures ranging from factor analysis to regression analysis. Some test or set of tests must be applied to determine whether the solution differs significantly from a random solution. Milligan and Mahajan (1980) and Milligan (1981) reviewed several such methods for testing the quality of a clustering solution and found them wanting on a number of dimensions. A method suggested by Arnold (1979) appears to overcome the problems of other methods. Arnold (1979) proposed using a statistic first suggested by Friedman and Rubin (1967) as a test of the statistical significance of a cluster solution. The statistic is given by

$$C = \text{long } (\text{max}/|\mathbf{T}|/|\mathbf{W}|))$$

where $|\mathbf{T}|$ is the determinant of the total variance-covariance matrix and $|\mathbf{W}|$ is the determinant of the pooled within-groups variance-covariance matrix. A number of iterative partitioning methods seek maximization of the ratio of $|\mathbf{T}|$ to $|\mathbf{W}|$: MIKCA, CLUSTAN, and CLUS. For algorithms which do not optimize $|\mathbf{T}|/|\mathbf{W}|$ the test becomes even more conservative. Arnold generated distributions of the C statistic for 2, 4, and 8 group solutions, 10, 20, 50, 100, 200, 500, and 1000 entities, and 5, 10, 20 attributes and indicated values of C which allow rejection of the null hypotheses that the data arise from unimodal or uniform distributions. He presented data to support the use of the statistics and provided formulas for the derivation of other values.

As with other multivariate statistics, one must demonstrate the reliability and the external validity of a cluster solution as well as its statistical significance. Reliability may be established by cross-validation. External validation requires a demonstration that the clusters are useful in some larger sense. Numerous authors have recommended cross-validating cluster solutions (see, e.g., Sherman and Sheth 1977) and several methods of cross-validation have been proposed. One of the more frequently used methods involves dividing the sample in half and carrying out clustering on each half. Descriptive statistics of the two sets of clusters are compared to determine the degree to which similar clusters have been identified. The problem with such an approach is that no objective measure of reliability is obtained.

Several authors have recommended the use of discriminant analysis for cross-validation (Field and Schoenfeldt 1975; Nerviano and Gross 1973; Rogers and Linden 1973). The approach involves using cluster membership as the group membership variable in a discriminant analysis. After a cluster solution has been developed on one sample, discriminant functions are derived which are applied to a second sample. The degree to which the assignments made with the discriminant functions agree with assignments made by a cluster analysis of the second sample serves as an estimate of the stability of the cluster solution across samples. A coefficient of agreement, such as kappa, may be used to provide an objective measure of such stability. Using discriminant analysis for validating cluster analysis has several drawbacks. Discriminant coefficients may be poor estimates of population values and need to be cross-validated themselves. This procedure is not cost-effective and the sample size available may be insufficient for cross-validating both the cluster analysis and a discriminant analysis.

McIntyre and Blashfield (1980) discussed an alternative approach to cross-validation which is recommended here. The procedure is relatively simple and easy to implement on a computer. Cluster analysis is first carried out on one half of the observations available for analysis. Once a statistically significant clustering solution has been identified, centroids describing the clusters are obtained. Objects in the holdout data set are then assigned to one of the identified clusters on the basis of the smallest Euclidean distance to a cluster centroid vector. The degree of agreement between the nearest-centroid assignments of the holdout sample and the results of a cluster analysis of the holdout sample is an indication of the stability of the solution. A coefficient of agreement, kappa, may be used as an objective measure of stability. If an acceptable level of stability is obtained the data sets may be combined to obtain a final solution.

The demonstration of the statistical significance and stability of a cluster solution is necessary before one can accept and use the classification developed by the methodology. The acceptance of a particular classification system, whether developed through cluster analysis or some other method, requires a further demonstration of utility, however. Classification is only useful if it assists in furthering an understanding of the phenomena of interest. Clusters, or classes, must have demonstrable implications for hypothesis generation, theory building, prediction, or management. The ultimate test of a set of clusters is its usefulness. Thus, the user of cluster analysis should provide a demonstration that clusters are related to variables other than those used to generate the solution. Ideally, only a small number of variables should be required to classify individuals. This classification should then have implications beyond the narrow set of classification variables. The task of classifications is not finished until these broader implications have been demonstrated.

Variable Selection Issues

The findings of empirical studies of cluster methods suggest that attention to initial variable selection is crucial because even one or two irrelevant variables may distort an otherwise useful cluster solution. The basis for classification must be carefully identified to ensure that extraneous characteristics do not distort an otherwise useful cluster analysis. There should be some rationale for the selection of variables for cluster analysis. That rationale may grow out of an explicit theory or be based on a hypothesis. Clearly more attention needs to be paid to this critical issue. As a science develops, researchers must agree on those dimensions which are most relevant to classification for a particular purpose. Much debate in the science of marketing involves the issue of variable selection. Thus, it is not surprising that a variety of different classification systems have been developed for similar phenomena. Indeed, it is probably unrealistic to expect that a single classification system will emerge in any area of marketing in the foreseeable future. Rather, there are likely to be numerous competing systems. The development of diverse systems is healthy and has been observed in other sciences. Experience with rival systems and a comparison of their usefulness ultimately provide a basis for selection of one system over another. Cluster analysis has much to offer as an aid for developing classification systems. To the extent that classification is both the first and last step in scientific investigation, cluster analysis should have increasing application in marketing.

References

Aldenderfer, M. S. (1977), "A Consumer Report on Cluster Analysis Software: (2) Hierarchical Methods," working paper, Department of Anthropology, Pennsylvania State University.

Anderberg, M. R. (1973), *Cluster Analysis for Applications*. New York: Academic Press.

Anderson, W. T., Jr., E. P. Cox, III, and D. G. Fulcher (1976), "Bank Selection Decisions and Market Segmentation," *Journal of Marketing*, 40, 40–45.

Arabie, P., J. D. Carroll, W. DeSarbo, and J. Wind (1981), "Overlapping Clustering: A New Method for Product Positioning," *Journal of Marketing Research*, 18 (August), 310–17.

Arnold, S. J. (1979), "A Test for Clusters," *Journal of Marketing Research*, 16 (November), 545–51.

Bailey, K. D. (1974), "Clusters Analysis," in *Sociological Methodology*, D. Heise, ed. San Francisco: Jossey-Bass.

Ball, G. H. (1970), *Classification Analysis*. Menlo Park, CA: Stanford Research Institute.

Bass, F. M., E. A. Pessemier, and D. J. Tigert (1969), "A Taxonomy of Magazine Readership Applied to Problems in Marketing Strategy and Media Selection," *Journal of Business*, 42, 337–63.

Bayne, C. K., J. J. Beauchamp, C. L. Begovich, and V. E. Kane (1980), "Monte Carlo Comparisons of Selected Clustering Procedures," *Pattern Recognition*, 12, 51–62.

Bertin, J. (1967), *Semiology Graphique*. Paris: Gauthier-Villars.

Bettman, J. R. (1979), *An Information Processing Theory of Consumer Choice*. Reading, MA: Addison-Wesley.

Blashfield, R. K. (1976), "Mixture Model Tests of Cluster Analysis: Accuracy of Four Agglomerative Hierarchical Methods," *Psychological Bulletin*, 83, 377–88.

_____ (1977), "A Consumer Report on Cluster Analysis Software: (3) Iterative Partitioning Methods," working paper, Department of Psychology, Pennsylvania State University.

_____ (1978), "The Literature on Cluster Analysis," *Multivariate Behavioral Research*, 13, 271–95.

_____ and L. C. Morey (1980), "A Comparison of Four Clustering Methods Using MMPI Monte Carlo Data," *Applied Psychology Measurement*, in press.

Calantone, R. J. and A. G. Sawyer (1978), "Stability of Benefit Segments," *Journal of Marketing Research*, 15 (August), 395–404.

Cattell, R. B. (1949), "R_p and Other Coefficients of Pattern Similarity," *Psychometrika*, 14, 279–98.

_____ (1978), *The Scientific Use of Factor Analysis in the Behavioral and Life Sciences*. New York: Plenum Press.

Claxton, J. D., J. N. Fry, and B. Portis (1974), "A Taxonomy of Pre-purchase Information Gathering Patterns," *Journal of Consumer Research*, 1, 35–42.

Cohen, J. (1960), "A Coefficient of Agreement for Nominal Scales," *Educational and Psychological Measurement*, 20, 37–46.

Cormack, R. M. (1971), "A Review of Classification," *Journal of the Royal Statistical Society* (Series A), 134, 321–67.

Cunningham, K. M. and J. C. Ogilvie (1972), "Evaluation of Hierarchical Grouping Techniques: A Preliminary Study," *Computer Journal*, 15, 209–13.

Day, G. S. and R. M. Heeler (1971), "Using Cluster Analysis to Improve Marketing Experiments," *Journal of Marketing Research*, 8 (August), 340–7.

Dixon, W. J. and M. B. Brown (1979), *BMDP: Biomedical Computer Programs, P Series, 1979*. Los Angeles: University of California Press.

Edelbrock, C. (1979), "Comparing the Accuracy of Hierarchical Clustering Algorithms: The Problem of Classifying Everybody," *Multivariate Behavioral Research*, 14, 367–84.

_____ and B. McLaughlin (1980), "Hierarchical Cluster Analysis Using Intraclass Correlations: A Mixture Model Study," *Multivariate Behavioral Research*, 15, 299–318.

Everitt, B. S. (1974), *Cluster Analysis*. London: Halsted Press.

Field, H. S. and L. F. Schoenfeldt (1975), "Ward and Hook Revisited: A Two-Part Procedure for Overcoming a Deficiency in the Grouping of Two Persons," *Educational and Psychological Measurement*, 35, 171–3.

Fisher, W. D. (1969), *Clustering and Aggregation in Economics*. Baltimore, MD: Johns Hopkins Press.

Frank, R. E. and P. E. Green (1968), "Numerical Taxonomy in Marketing Analysis: A Review Article," *Journal of Marketing Research*, 5 (February), 83–98.

Friedman, H. D. and J. Rubin (1967), "On Some Invariant Criteria for Grouping Data," *Journal of the American Statistical Association*, 62, 1159–78.

Furse, D. H., G. Punj, and D. W. Stewart (1982), "Individual Search Strategies in New Automobile Purchase," in *Advances in Consumer Research*, Vol. 9, Andrew Mitchell, ed., 379–84.

Gerard, R. W. (1957), "Units and Concepts of Biology," *Science*, 125, 429–33.

Gower, J. C. (1967), "A Comparison of Some Methods of Cluster Analysis," *Biometrics*, 23, 623–37.

Green, P. E. and F. J. Carmone (1968), "The Performance Structure of the Computer Market: A Multivariate Approach," *Economic and Business Bulletin*, 20, 1–11.

_____, R. E. Frank, and P. J. Robinson (1967), "Cluster Analysis in Test Market Selection," *Management Science*, 13, B-387–400.

_____ and V. R. Rao (1969), "A Note on Proximity Measures and Cluster Analysis," *Journal of Marketing Research*, 6 (August), 359–64.

Greeno, D. W., M. S. Sommers, and J. B. Kernan (1973), "Personality and Implicit Behavior Patterns," *Journal of Marketing Research*, 10 (February), 63–9.

Hartigan, J. A. (1967), "Representation of Similarity Matrices by Trees," *Journal of the American Statistical Association*, 62, 1140–58.

—— (1975). *Clustering Algorithms.* New York: John Wiley & Sons, Inc.

Howard, H. and B. Harris (1966), "A Hierarchical Grouping Routine, IBM 360/65 FORTRAN IV Program," University of Pennsylvania Computer Center.

Inglis, J. and D. Johnson (1970), "Some Observations On, and Developments In, the Analysis of Multivariate Survey Data," *Journal of the Market Research Society,* 12, 75–8.

Jardine, H. and R. Sibson (1971), *Mathematical Taxonomy.* New York: John Wiley & Sons, Inc.

Johnson, S. C. (1967), "Hierarchical Clustering Schemes," *Psychometrika,* 32, 241–54.

Kantor, J. R. (1953), *The Logic of Modern Science.* Bloomington, IN: Principle Press.

Kemeny, J. G. (1959), *A Philosopher Looks at Science.* New York: Van Nostrand.

Kernan, J. B. (1968), "Choice Criteria, Decision Behavior, and Personality," *Journal of Marketing Research,* 5 (May), 155–69.

—— and G. D. Bruce (1972), "The Socioeconomic Structure of an Urban Area," *Journal of Marketing Research,* 9 (February), 15–18.

Kiel, G. C. and R. A. Layton (1981), "Dimensions of Consumer Information Seeking Behavior," *Journal of Marketing Research,* 18 (May), 233–9.

Kuiper, F. K. and L. A. Fisher (1975), "A Monte Carlo Comparison of Six Clustering Procedures," *Biometrics,* 31, 777–83.

Lance, G. N. and W. T. Williams (1967a), "A General Theory of Classificatory Sorting Strategies. I. Hierarchical Systems," *The Computer Journal,* 9, 373–80.

—— (1967b), "A General Theory of Classificatory Sorting Strategies. II. Clustering Systems," *The Computer Journal,* 10, 271–7.

Landon, E. L. (1974), "Self Concept, Ideal Self Concept, and Consumer Purchase Intentions," *Journal of Consumer Research,* 1, 44–51.

Lessig, V. P. and J. D. Tollefson (1971), "Market Segmentation Through Numerical Taxonomy," *Journal of Marketing Research,* 8 (November), 480–7.

McIntyre, R. M. and R. K. Blashfield (1980), "A Nearest-Centroid Technique for Evaluating the Minimum-Variance Clustering Procedure," *Multivariate Behavior Research,* 15, 225–38.

McQuitty, L. L. (1967), "A Mutual Development of Some Typological Theories and Pattern-Analytic Methods," *Educational and Psychological Measurement,* 17, 21–46.

McRae, D. J. (1973), "Clustering Multivariate Observations," unpublished doctoral dissertation, University of North Carolina.

Mezzich, J. E. (1978), "Evaluating Clustering Methods for Psychiatric Diagnosis," *Biological Psychiatry,* 13, 265–81.

Milligan, G. W. (1980), "An Examination of the Effect of Six Types of Error Perturbation on Fifteen Clustering Algorithms," *Psychometrika,* 45, 325–42.

—— (1981), "A Monte Carlo Study of Thirty Internal Criterion Measures for Cluster Analysis," *Psychometrika,* 46, (2), 187–99.

—— and P. D. Isaac (1980), "The Validation of Four Ultrametric Clustering Algorithms," *Pattern Recognition,* 12, 41–50.

—— and V. Mahajan (1980), "A Note on Procedures for Testing the Quality of a Clustering of a Set of Objects," *Decision Sciences,* 11, 669–77.

Mojena, R. (1977), "Hierarchical Grouping Methods and Stopping Rules: An Evaluation," *Computer Journal,* 20, 359–63.

Montgomery, D. B. and A. J. Silk (1971), "Clusters of Consumer Interests and Opinion Leaders' Sphere of Influence," *Journal of Marketing Research,* 8 (August), 317–21.

Morgan, J. N. and J. A. Sonquist (1963), "Problems in the Analysis of Survey Data, and a Proposal," *Journal of the American Statistical Association,* 58, 87–93.

Moriarty, M. and M. Venkatesan (1978), "Concept Evaluation and Market Segmentation," *Journal of Marketing,* 42, 82–6.

Morrison, B. J. and R. C. Sherman (1972), "Who Responds to Sex in Advertising?" *Journal of Advertising Research,* 12, 15–19.

Morrison, D. G. (1967), "Measurement Problems in Cluster Analysis," *Management Science,* 13, B-775–80.

Myers, J. G. and F. M. Nicosia (1968), "On the Study of Consumer Typologies," *Journal of Marketing Research,* 5 (May), 182–83.

Neidell, L. A. (1970), "Procedures and Pitfalls in Cluster Analysis," *Proceedings,* Fall Conference, American Marketing Association, 107.

Nerviano, V. J. and W. F. Gross (1973), "A Multivariate Delineation of Two Alcoholic Profile Types on the 16PF," *Journal of Clinical Psychology,* 29, 370–4.

Peay, E. R. (1975), "Nonmetric Grouping: Clusters and Cliques," *Psychometrika,* 40, 297–313.

Rand, W. M. (1971), "Objective Criteria for the Evaluation of Clustering Methods," *Journal of the American Statistical Association,* 66, 846–50.

Rogers, G. and J. D. Linden (1973), "Use of Multiple Discriminant Function Analysis in the Evaluation of Three Multivariate Grouping Techniques," *Educational and Psychological Measurement,* 33, 787–802.

Rubin, J. (1965), "Optimal Taxonomy Program (7090-IBM-0026)," IBM Corporation.

Schaninger, C. M., V. P. Lessig, and D. B. Panton (1980), "The Complementary Use of Multivariate Procedures to Investigate Nonlinear and Interactive Relationships Between Personality and Product Usage," *Journal of Marketing Research,* 17 (February), 119–24.

Sethi, S. P. (1971), "Comparative Cluster Analysis for World Markets," *Journal of Marketing Research,* 8 (August), 348–54.

Sexton, D. E., Jr. (1974), "A Cluster Analytic Approach to Market Response Functions," *Journal of Marketing Research,* 11 (February), 109–14.

Shepard, R. N. (1974), "Representation of Structure in Similarity Data: Problems and Prospects," *Psychometrika,* 39, 373–421.

—— and P. Arabie (1979), "Additive Clustering: Representation of Similarities as Combinations of Discrete Overlapping Properties," *Psychological Review,* 86, 87–123.

Sherman, L. and J. N. Sheth (1977), "Cluster Analysis and Its Applications in Marketing Research," in *Multivariate Methods for Market and Survey Research,* J. N. Sheth, ed. Chicago: American Marketing Association.

Shuchman, A. (1967), "Letter to the Editor," *Management Science,* 13, B688–96.

Skinner, H. A. (1979), "Dimensions and Clusters: A Hybrid Approach to Classification," *Applied Psychological Measurement*, 3, 327–41.

Smith, W. (1956), "Product Differentation and Market Segmentation as Alternative Marketing Strategies," *Journal of Marketing*, 21, 3–8.

Sneath, P. H. A. (1957), "The Application of Computer to Taxonomy," *Journal of General Microbiology*, 17, 201–26.

———— and R. R. Sokal (1973), *Numerical Taxonomy*. San Francisco: W. H. Freeman.

Stewart, D. W. (1981), "The Application and Misapplication of Factor Analysis in Marketing Research," *Journal of Marketing Research*, 18 (February), 51–62.

Srivastava, R. K., R. P. Leone, and A. D. Shocker (1981), "Market Structure Analysis: Hierarchical Clustering of Products Based on Substitution-in-Use," *Journal of Marketing*, 45 (Summer), 38–48.

————, A. D. Shocker, and G. S. Day (1978), "An Exploratory Study of Usage-Situational Influences on the Composition of Product-Markets," in *Advances in Consumer Research*, H. K. Hunt, ed. Ann Arbor: Association for Consumer Research, 32–38.

Tyron, R. C. and D. E. Bailey (1970), *Cluster Analysis.* New York: McGraw-Hill Book Company.

Veldman, D. J. (1967), *FORTRAN Programming for the Behavioral Sciences*. New York: Holt, Rinehart, and Winston.

Ward, J. (1963), "Hierarchical Grouping to Optimize an Objective Function," *Journal of the American Statistical Association*, 58, 236–44.

Wells, W. D. (1975), "Psychographics: A Critical Review," *Journal of Marketing Research*, 12 (May), 196–213.

Wind, Y. (1978), "Issues and Advances in Segmentation Research," *Journal of Marketing Research*, 15 (August), 317–37.

Wishart, D. (1969), "CLUSTAN IA: A FORTRAN Program for Numerical Classification," Computing Laboratory, St. Andrew's University, Scotland.

Wolf, A. (1926), *Essentials of Scientific Method.* New York: Macmillan Company.

Classification and Validation of Behavioral Subtypes of Learning-Disabled Children

DEBORAH L. SPEECE
JAMES D. McKINNEY
MARK I. APPELBAUM

Hierarchical cluster analysis techniques were used to identify seven distinct behavioral subtypes of young, school-identified, learning-disabled children. The Classroom Behavior Inventory, a teacher rating instrument completed by classroom teachers, was the measure used for clustering. Internal validation techniques indicated that the obtained subtypes were replicable and had profile patterns different from a sample of normally achieving children. The subtypes differed significantly on independent observational measures and on ratings by special education teachers but not on achievement or intelligence. Interpretation of the subtypes emphasized the need for differential identification and intervention strategies that are consistent with a multiple-syndrome view of learning disabilities.

"Classification and Validation of Behavioral Subtypes of Learning-Disabled Children," Deborah L. Speece, James D. McKinney, and Mark I. Appelbaum, Vol. 77 (1) 1985, pp. 67–77. Reprinted from the *Journal of Educational Psychology*, published by the American Psychological Association, Inc. Deborah L. Speece is now at the Department of Special Education, University of Maryland. James D. McKinney is affiliated with the Frank Porter Graham Child Development Center. Mark I. Appelbaum is on the faculty at the University of North Carolina at Chapel Hill.

A persistent problem for research and practice in the field of learning disabilities (LD) has been the great diversity among students who are classified according to the present definition (Keogh, Major-Kingsley, Omori-Gordon, & Reid, 1982; Torgesen & Dice, 1980). The problem of sample heterogeneity has not only frustrated efforts to build a generalizable body of knowledge about LD but has also contributed greatly to the present controversy over misclassification and appropriate education for LD students. Recently, however, several investigators have demonstrated the feasibility of dividing LD samples into more homogeneous subtypes by using empirical classification techniques such as Q-factor analysis and cluster analysis (Fisk & Rourke, 1979; Lyon & Watson, 1981; McKinney, 1984; Satz & Morris, 1981). In general, these techniques group together individuals who show a similar pattern of response or level of response (i.e., elevation) across an array of variables (Everitt, 1980).

Because the field has been influenced greatly by ability deficit theories (Torgesen, 1975), previous research on LD subtypes has focused primarily on neuropsychological and psychoeducational test batteries to classify students. Our interest in studying behavioral subtypes was motivated by a series of studies showing that LD students, as a heterogeneous group, displayed patterns of maladaptive classroom behavior that have been associated with their failure to progress academically (McKinney & Feagans, 1983). Specifically, teachers described LD children as more distractible and dependent than classmates and as less task oriented, independent, and verbally intelligent (Feagans & McKinney, 1981; McKinney, McClure, & Feagans, 1982; McKinney & Speece, 1983). Also, analysis of observational data in the same studies showed that LD children were off-task more often, on-task less in independent work, and interacted more frequently with teachers. Longitudinal research demonstrated that these behavioral differences between LD students and classmates persisted over a 3-year period (McKinney & Feagans, 1982) and predicted academic progress in reading (McKinney & Speece, 1983). Finally, another consequence of their behavior patterns was revealed in the high frequency of managerial interactions between classroom teachers and LD students (Dorval, McKinney, & Feagans, 1982).

In general, the educational importance of these findings about the classroom behavior of LD students is supported by evidence from other investigators (Bryan, 1974; Keogh, Tehir, & Windeguth-Behn, 1974; Routh, 1979) as well as by research linking classroom behavior patterns to achievement in more typical samples of school children (Hoge & Luce, 1979; McKinney, Mason, Perkerson, & Clifford, 1975).

At the same time, given sample heterogeneity, it would not be appropriate to conclude that group data characterize the behavior of individuals or that all LD children show the same pattern of behavior. Accordingly, the aims of the present study were to (a) isolate distinct clusters of LD students based on their pattern of classroom behavior as rated by teachers; (b) determine the extent of overlap among LD students and average achieving classmates in specific patterns of behavior; and (c) validate LD behavioral subtypes on external observational measures and comparable ratings by LD special education teachers.

The second and third objectives, which reflect internal and external validation concerns, are of critical importance in subtyping research. Although empirical classification techniques hold considerable promise for solving many of the theoretical and practical problems associated with heterogeneous samples of LD students, research with these techniques is still at an embryonic stage of development and requires converging evidence

regarding the stability and applied importance of the identified subtypes (McKinney, 1984; Morris, Blashfield, & Satz, 1981).

The primary instrument we selected to classify behavioral subtypes was the Classroom Behavior Inventory (CBI; Schaefer, Edgerton, & Aronson, 1977). This instrument was chosen for both theoretical and empirical reasons in order to facilitate the interpretation of the cluster solution we obtained. Theoretically, the CBI captures Schaefer's (1981) hierarchical model of adaptive behavior that is composed of academic competence, socialization, and temperament. These factors have been supported in previous investigations using the CBI and other measures of classroom behavior (Kohn & Roseman, 1972; Peterson, 1960). Empirically, the CBI has demonstrated high internal consistency (.85–.96) and moderate interrater reliability (.40–.70) as well as cross-sectional and longitudinal validity in predicting achievement, ability, and child adjustment in both normal and learning-disabled samples (McKinney & Feagans, 1980; McKinney & Speece, 1983; Schaefer, 1981). Thus, the CBI provided a stable and interpretable assessment of child behavior from which to subtype the LD children. Observational measures and CBI ratings from a second set of teachers provided the external validation measures and, thus, the means to determine if the resulting LD subtypes were meaningful in terms of actual classroom behavior and a different point of view.

Method

SUBJECTS

The learning-disabled subjects for this study were 63 children in the first and second grades who were newly identified by multidisciplinary placement teams in accordance with state and federal guidelines. In addition to meeting state and federal guidelines for special education placement, children who did not have a verbal or performance score of 85 or above on the Wechsler Intelligence Scale for Children–Revised (WISC–R) were eliminated from the study. Three LD children were dropped from the study because of this criterion. Each LD child was matched by sex and race to a randomly selected normally achieving child (NLD) in the same mainstream classroom ($N = 66$). The WISC–R and the Peabody Individual Achievement Test (PIAT) were administered to verify normal school progress for comparison children. All children were participants in a 3-year

TABLE 1   Subject Characteristics

Characteristic	LD	NLD
n	63	66
Age (in months)	86.3	85.9
Sex (%)		
Male	76	77
Female	24	23
Race (%)		
Black	40	33
White	60	67
Parent occupation level (%)		
Upper	9.1	10.1
Middle	31.0	47.9
Lower Middle	43.7	24.5
Lower	16.3	18.4
Parent education level (%)		
College graduate	18.9	23.3
Some college	21.4	11.5
High school	49.5	58.9
Elementary	10.3	6.6
WISC–R		
Verbal		
M	98.2	105.7
SD	11.6	14.9
Performance		
M	94.9	109.0
SD	12.9	13.0
Full scale		
M	96.1	107.9
SD	11.1	14.0
PIAT[a]		
Math		
M	91.5	103.4
SD	8.8	11.3
Reading recognition		
M	92.6	105.0
SD	7.9	11.1
Reading comprehension		
M	95.2	104.5
SD	7.4	11.5

Note. LD = learning disabled; NLD = non-learning disabled;
WISC–R = Wechsler Intelligence Scale for Children–Revised;
PIAT = Peabody Individual Achievement Test.
[a] Grade standard scores.

longitudinal study; the present study is based on
Year 1 data (McKinney & Feagans, 1982). Table
1 summarizes subject characteristics for both
groups of children.

CLASSROOM BEHAVIOR INVENTORY (CBI)

The CBI is a 60-item teacher rating scale that measures four bipolar dimensions of child behavior:
independence versus dependence, task orientation versus distractibility, extroversion versus

introversion, and considerateness versus hostility
(Schaefer, Edgerton, & Aronson, 1977). Although
three other unipolar scales are included in the
CBI (teacher-perceived verbal intelligence, cre-
ativity/curiosity, and apathy), previous research
has shown that these scales are redundant with
the major bipolar dimensions of the CBI. Therefore, in order to simplify both the analysis and
interpretation, only the bipolar scales were used
in the cluster analysis. Classroom teachers rated
each item on a 5-point continuum from *very much
like* to *not at all like* the target child. In addition,
special education teachers completed the CBI for
the LD children who were placed in their resource
rooms. Cluster analysis was performed on the
classroom teacher's data, whereas the resource
room teacher's data were used in subsequent external validation procedures.

SCHEDULE OF CLASSROOM ACTIVITY NORMS (SCAN)

The SCAN observational system is a time sampling technique in which observers code a child's
classroom behavior into 14 mutually exclusive
categories every 5 s in 10-min blocks. The 14 discrete categories are designed to capture pertinent
classroom behavior in the areas of task-oriented,
social, and affective behaviors (McKinney et al.,
1975; McKinney, Feagans, Ferguson, & Burnett,
1978). Each child was observed a total of 40 min
on 2 consecutive days, which resulted in 480 observations per child. During the observation period, the observers alternated between the LD and
comparison child every 10 min. Thus, LD and
NLD children were observed in similar instruc-
tional activities as well as during the same time
periods.

For the purpose of cluster validation, 11 of
the 14 categories were combined to form four conceptual groupings. On-task behavior was composed of two categories: constructive self-directed
activity (individual work settings) and constructive participation (group setting). Off-task behavior included nonconstructive self-directed
activity, nonparticipation, and distractibility
categories. Interactions were captured by task-
oriented peer conversation, teacher interaction,
and peer social interactions. The final composite
category, problem behavior, included aggression
(verbal/physical abuse, noncompliance), inappropriate gross motor behavior (whole body
movements), and dependency (attention seeking).
A more complete description of the SCAN categories can be found in McKinney et al. (1975, 1982)
and in Feagans and McKinney (1981).

Interobserver reliability was determined by the
percentage of agreement method before the data

were collected. Ninety percent agreement was established with a third observer serving as the standard based on 120 observations; reliabilities for individual categories ranged from 58% to 100%. During data collection, an average interobserver reliability of 86% was obtained for 56 ten-minute observations.

Data Analysis

Cluster analysis represents a family of empirical techniques that has the purpose of identifying homogenous subgroups of subjects within a heterogeneous sample (Everitt, 1980). As the learning-disabled population is often described as heterogeneous in terms of strengths and weaknesses across a wide variety of measurement domains, cluster analysis techniques are well suited for exploring the underlying structure of an LD sample (Satz & Morris, 1981). Hierarchical cluster analysis was used in the present study to classify children based on the classroom teachers' CBI ratings. This method begins with every subject as his/her own cluster and successively merges subjects/clusters until all subjects who show a common response pattern are contained in a single cluster. The formation of clusters is defined by the investigator's choice of a similarity measure and an algorithm that provides the rule for joining observations (Anderberg, 1973). Correlation was chosen for the similarity measure in the present investigation because we were primarily interested in profile shape as opposed to elevation or scatter (Cronbach & Gleser, 1953). A correlational measure groups children who have similar patterns of strengths and weaknesses regardless of level. This decision was based on our experience that teachers identify children for possible special education services by noting, for example, the disparity in competence between academic subject areas or behavioral characteristics. Ward's minimum variance method was the algorithm chosen with the complete linkage algorithm used as a check on the obtained subtypes. Both will be discussed below.

Because cluster analysis is not based on probabilistic statistics, there is no single best solution to a clustering problem. Thus, one does not have specified alpha levels to guide selection of a particular set of clusters from several alternative solutions. Therefore, a three-stage data analysis plan was devised.

First, after examining the cluster solutions resulting from both the complete linkage and Ward's algorithms, candidate solutions from each analysis were plotted by cluster. At this step, individual scores were plotted around cluster means to determine which solutions produced the most cohesive

plots. In addition, the cluster profiles were examined for interpretability from an educational and psychological perspective. These procedures resulted in a six-cluster solution for the complete linkage algorithm and a seven-cluster solution for Ward's method. The difference between these algorithms is that the complete linkage method joins subjects/clusters in terms of the similarity between the most distant elements in each cluster, whereas Ward's method employs a minimum variance function to join observations (Anderberg, 1973; Johnson & Wichern, 1982).

Second, in order to choose one of the two cluster solutions that emerged in the first stage of the analysis, converging evidence was sought on the stability of the cluster solutions. Because random data will produce clusters, this type of internal validation is critical in assessing the adequacy of the cluster solution (Morris, Blashfield, & Satz, 1981). Two methods of validation were used, split sample replication and forecasting. For cluster replication, two thirds of the LD subjects were randomly selected and reclustered. A high degree of membership concordance between the original and replication clusters provides an indication of cluster stability (Morris et al., 1981). The forecasting method involved deriving a discriminant function for each cluster and then classifying the NLD children into one of the LD clusters. If, as assumed, the LD subtypes and NLD group represent different populations in terms of behavioral characteristics, then the majority of NLD children should be classified into clusters that are unremarkable in terms of behavior profiles.

The final step in the analysis was validation of the chosen cluster solution on external criteria. Given that a cluster solution has reasonable internal stability, it is then crucial to demonstrate that the clusters differ on conceptually similar variables that are independent of the measures used for cluster formation. MANOVAS were used with both LD teacher CBI ratings and SCAN data to assess cluster differences. Where appropriate, discriminant function analyses and post hoc comparisons provided follow-up analyses for the MANOVA procedure.

Results

As noted above, two algorithms were used to cluster the data, with Ward's method employed as the primary method. In general, both algorithms resulted in relatively compact and meaningful clusters. In the first part of the Results section, the subtypes obtained from the seven-cluster, Ward's solution are interpreted on the basis of

FIGURE 1 Classroom Behavior Inventory clusters based on classroom teacher ratings of 63 learning-disabled children (IND = independence, EXT = extroversion, TOR = task orientation, CON = considerateness, DEP = dependence, INT = introversion, DIS = distractibility, HOS = hostility).

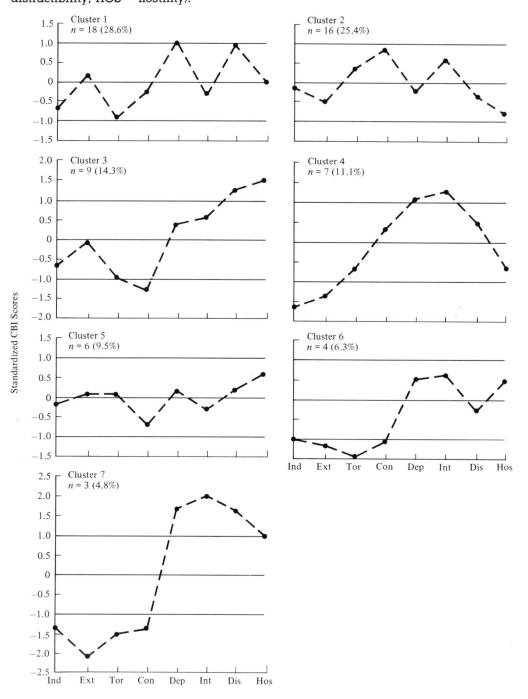

behavioral patterns. The second part focuses on the results of the internal validation techniques that led to the selection of the seven-cluster solution over the six-cluster, complete linkage solution. Finally, the external validation results based on the SCAN observational measures and the LD teachers' CBI ratings are presented.

Cluster Interpretation

In order to facilitate interpretation of the seven subtypes, the cluster profile points for LD students were plotted relative to the CBI mean and standard deviation on each scale for the total sample of LD and NLD children. Although the clusters contained only LD children as members, this method of graphing provided an anchor point from which to view the relative behavioral strengths and weaknesses of the LD clusters. Figure 1 illustrates the cluster means for each of the seven subtypes. The reader should note that the first four variables represent the positive CBI scales so that higher values reflect more favorable teacher ratings. In contrast, because the last four variables are CBI negative scales, higher values represent less favorable ratings.

The first cluster contained 28.6% of the LD sample and was characterized by borderline defi-

ciencies in task-oriented behavior and independence. Children in this cluster were not differentiated from the total sample with respect to ratings of introversion/extroversion or hostility/considerateness as indicated by profile points within ±1 SD. Therefore, this subtype of LD children appeared to be a well-adjusted group of students with mild attention deficits and associated problems with independent work in the classroom.

Approximately one fourth (25.4%) of the sample was classified in the second cluster. Cluster 2 and Cluster 5 (9.5% of sample) appeared to represent two variations of normal classroom behavior as all profile points were within ±1 SD. This was not an unusual result in the subtyping literature, because one would not expect all of the LD children to show atypical performance in a particular domain (e.g., Satz & Morris, 1981). For both of these subtypes, teacher ratings of task-oriented behavior and independence were unremarkable: Cluster 2 was distinguished from Cluster 5 by slightly elevated ratings on considerateness and introversion, whereas Cluster 5 was seen as slightly less considerate and more hostile. As Table 2 shows, Cluster 2 was overly represented by girls, whereas Cluster 5 contained a disproportionate number of boys relative to the sex ratio present in the sample.

Cluster 3 showed a pattern of ratings on task

TABLE 2 Learning-Disabled Subject Characteristics on Demographic and Psychometric Variables by Cluster

Characteristic	Cluster						
	1	2	3	4	5	6	7
n	18	16	9	7	6	4	3
Race (%)							
Black	39	25	22	43	50	75	100
White	61	75	78	57	50	25	0
Sex (%)							
Female	11	50	0	43	17	25	0
Male	89	50	100	57	83	75	100
WISC–R							
Verbal							
M	101.8	91.2	99.0	97.3	103.4	94.5	105.0
SD	10.8	11.6	9.9	14.5	7.2	15.6	7.0
Performance							
M	96.5	87.5	96.6	90.7	104.6	103.0	96.7
SD	14.7	10.1	6.3	17.6	13.1	9.3	6.7
Full scale							
M	99.2	88.4	97.3	93.0	103.0	98.3	101.0
SD	12.7	9.0	8.1	10.7	9.7	11.9	7.0

Note. WISC–R = Wechsler Intelligence Scale for Children–Revised.

orientation and independence that was similar to that seen in Cluster 1. However, in this case mild attention deficits were combined with high ratings on distractibility and hostility as well as with low ratings on considerateness. This subtype was the third largest (14.3%) in the LD sample and could be characterized as a poorly socialized group of students prone to conduct problems and acting out behavior in the classroom. This cluster was exclusively male. In contrast, Cluster 4 could be interpreted as a withdrawn, overly dependent subtype with low ratings on independence and extroversion and high ratings on dependence and introversion. Thus cluster, formed by 11% of the sample, had a disproportionate number of girls.

Cluster 6 was a relatively small (6.3%) subgroup of children who showed a pattern of mixed deficits on the positive CBI scales but little elevation on the negative scales. This pattern was difficult to interpret clinically; however, one might speculate that it represents a mild version of the global behavior disorder seen in Cluster 7. Cluster 7 contained only 3 boys who were rated as seriously impaired on all CBI behaviors. Although the small n could lead to a dismissal of this cluster as representing outliers (Everitt, 1980), the fact that several different clustering procedures yielded this same cluster indicated that this profile, although rare, does seem to describe a subgroup of LD children with serious behavior problems in the classroom.

These seven clusters were significantly differentiated by a MANOVA conducted on the original CBI clustering variables, $F(6,56) = 35.95$, $p < .01$, with all eight univariate tests being significant ($p < .0001$). Because the clusters were formed by minimizing within-cluster variance based on the chosen variables, one expects significant differences at this stage. Thus, this analysis simply provided an indication that further analyses with external variables may be profitable. Concerning psychometric variables, there were no differences between clusters on the PIAT subtests or on the WISC–R subscales. Table 2 presents selected demographic and psychometric descriptive data on each cluster.

Internal Validation

With respect to the split sample validation technique, an average of 73.3% (median = 75.0%) of the children in the replication clusters for the six-cluster solution maintained their original clus-

ter membership. Ten children (23.8%) were placed in four different clusters. The corresponding figures for the seven-cluster solution were 72.9% mean membership similarity with a median value of 100%. Eight children (19.0%) were in four different clusters. In terms of the misclassified children, two clusters from the six-cluster solution had members divided among two replication clusters, whereas misclassified children under Ward's method were only in one other replication cluster. Thus, replication membership for the seven-cluster solution was more consistent with regard to the original solution. These results tended to favor the seven-cluster, Ward's solution although the differences were not great.

In contrast, the forecasting procedure that classified normally achieving children into the LD clusters revealed more striking results. The six-cluster solution classified 57.6% of the NLD children into a "normal" LD cluster, whereas the seven-cluster solution classified 84.8% of the NLD children in this manner. In addition, each cluster solution produced a subtype that showed a pattern of global behavioral impairment. For the six-cluster solution, 22.7% of the NLD children were classified into this maladaptive group, whereas only 1.5% were so categorized in the seven-cluster solution. Although both solutions had cluster patterns depicting similar behavioral strengths and weaknesses as well as similar percentages of LD children falling into essentially equivalent clusters, the forecasting data indicated that the seven-cluster solution using Ward's algorithm was better able to distinguish between normal and maladaptive behavior in regard to the NLD sample. Thus, the seven-cluster solution was chosen for further analyses with external validation procedures.

External Validation

LD TEACHER CBI RATINGS. As one measure of external validation, the LD teachers' CBI ratings provided the dependent measures for a MANOVA between clusters. The omnibus test was statistically significant, $F(6,56) = 7.54$, $p < .05$. Figure 2 graphically presents a comparison of cluster means between the two sets of teacher ratings. To plot the profiles, the CBI was standardized on the LD sample for each group of teachers. As anticipated, profile shapes were consistent across clusters between the two groups of teachers with minor variations in some scale points. Interestingly, the LD teachers did not appear to rate the LD children

FIGURE 2 Comparison of Classroom Behavior Inventory clusters based on classroom teacher ratings with resource room teacher ratings of 63 learning-disabled children (IND = independence, EXT = extroversion, TOR = task orientation, CON = considerateness, DEP = dependence, INT = introversion, DIS = distractibility, HOS = hostility).

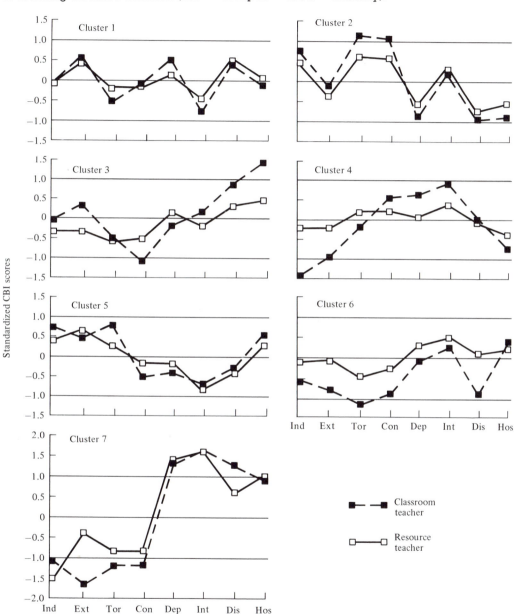

any more or less favorably than did the regular classroom teachers.

SCAN. The second set of MANOVAS used SCAN variables as the dependent measures. The SCAN categories were grouped conceptually to reflect on-task behaviors, off-task behaviors, problem be-

haviors, and teacher and peer interactions. As depicted in Table 3, all four omnibus tests were significant, indicating that the observational measures discriminated among the LD behavioral subtypes.

Follow-up analyses indicated that none of the univariate tests for either on-task behavior or

TABLE 3 Multivariate and Univariate Results for SCAN External Validation of Seven-Cluster Solution

Conceptual group	F	df	p
On-task behavior	2.57	6, 56	.05
Constructive self-directed activity	0.97	6, 56	.45
Constructive participation	0.95	6, 56	.47
Interaction	2.69	6, 56	.05
Task-oriented peer	0.51	6, 56	.80
Teacher	1.00	6, 56	.43
Social peer	2.44	6, 56	.04
Off-task behavior	6.11	6, 56	.05
Distractibility	0.76	6, 56	.60
Nonparticipation	1.07	6, 56	.39
Nonconstructive self-directed activity	5.89	6, 56	.00
Problem behavior	4.60	6, 56	.05
Dependency	0.34	6, 56	.91
Gross motor inappropriate	4.00	6, 56	.00[a]
Aggression	3.08	6, 56	.01[a]

Note. SCAN = Schedule of Classroom Activity Norms.
[a] Indicates significant univariate result at reduced alpha level.

teacher and peer interaction were significant at the appropriate reduced alpha level. These results implied that the specific variables contributed equally to the multivariate results. For off-task behavior, the univariate result for nonconstructive self-directed activity was significant ($p < .01$) but not for distractibility or nonparticipation. Discriminant function analysis supported the importance of this variable in differentiating the subtypes. A single function accounted for a significant proportion of variance ($p = .003$) and was defined primarily by nonconstructive self-directed activity. In terms of problem behavior, the univariate results for gross motor inappropriate behavior and aggression, but not dependency, were significant. One significant function was extracted ($p = .004$) and defined by the two significant univariate variables.

To further analyze these results, post hoc comparisons were conducted to supplement the SCAN multivariate analyses. For SCAN on-task behaviors, the "normal" LD clusters (2 and 5) were contrasted with the other five. The same comparison was made for off-task behavior. In addition, Clusters 4, 6, and 7 were contrasted against the others for off-task behavior because they had the lowest ratings on independence and task orientation. The SCAN grouping of problem behavior captured aggression, dependency, and gross motor inappropriate behavior. Accordingly, Clusters 3, 6, and 7, with the defining features of low task orientation and considerateness and relatively high hostility and distractibility, were compared with the remaining clusters. For the SCAN group consisting of teacher and peer interactions, Clusters 2, 5, and 4 were contrasted against the others. Also, Clusters 7 and 3 were compared with the remaining five clusters.

These contrasts provided more specific information on how the clusters differed on the observational measures. Clusters 2 and 5, which represented variations of normal profiles, were not different from the other clusters in terms of on-task behavior. However, the SCAN off-task behavior comparison representing independent work settings differentiated the normal clusters from the other five ($p < .006$) as well as Clusters 4, 6, and 7 from the other four subtypes ($p < .002$). The general category of problem behavior resulted in a significant difference between Clusters 3, 7, and 6 and the others with respect to gross motor inappropriate behavior ($p < .002$) and aggression ($p < .001$). The two comparisons conducted with teacher and peer interactions were not significant.

Discussion

The behavioral profiles identified in this study present a more detailed portrait of learning disabilities that goes beyond traditional group difference research on classroom functioning. The multivariate strategy employed provided a finer-grained analysis of behavioral patterns present in this LD sample.

First, it is important to note that more than one third of the LD sample did not exhibit a maladaptive pattern of behavior. The interpretation of two clusters as adaptive was supported by the fact that 85% of the randomly selected NLD children were classified into these two subtypes. A finding of relatively normal profiles is consistent with previous research on LD subtypes using different measures (Lyon & Watson, 1981; Satz & Morris, 1981) and serves to illustrate the point that multiple factors may be involved in the identification of LD children and/or account for their failure to profit from mainstream instruction. More specifically, Subtypes 2 and 5 were not differentiated from the other subtypes by either achievement or intelligence. It appeared that classification along the behavioral dimension was relatively independent of academic achievement and aptitude. Thus, it would not be appropriate to speculate that the LD children in the two "normal" clusters were misclassified, because they may show deficiencies in cognitive or linguistic processes that were not measured in this study.

Second, the remainder (65%) of the LD sample clustered into five distinct subtypes that seemed to represent different degrees of attentional problems combined with conduct and personal adjustment problems. Only 15% of the average achievers displayed similar patterns of behavior. Consistent with the literature on LD/NLD group differences (McKinney & Feagans, 1983), all of the atypical behavioral profiles were characterized by relatively low task orientation and independence.

However, previous studies that have compared heterogeneous LD samples with samples of average achievers on CBI teacher ratings have yielded inconsistent findings with respect to the dimensions of introversion/extroversion and hostility/considerateness (Feagans & McKinney, 1981; McKinney & Feagans, 1982; McKinney et al., 1982), even though clinical evidence suggests that personal and social adjustment problems are prevalent among LD children (Lorion, Cowen, & Caldwell, 1974). Data from the present study suggest that hostility was perceived as a problem for 19% of the children in this sample, whereas withdrawn behavior was attributed to 11% of the sample. Only 4.8% (3 children) were seen as seriously behaviorally disordered. Because the composition of these clusters was related to gender, inconsistent findings in previous group comparison studies concerning the prevalence of personal/social problems may be due to the sociodemographic characteristics of the research samples and to the fact that boys are overly represented in the LD school population.

In general, this pattern of findings for the atypical clusters is consistent with previous evidence and theory on the major dimensions that describe deviant classroom behavior. For example, Schaefer (1981) reviewed the literature on behavior problems displayed by exceptional children and proposed a hierarchical model that describes adaptation along three major dimensions—competence (defined by behaviors that reflect task orientation and independence), socialization (defined by considerate vs. hostile behavior), and temperament (defined as extroverted vs. introverted behavior). Similarly, Von Isser, Quay, and Love

(1980) reported three dimensions in a factor analytic study of teacher ratings of deviant behavior that were consistent with Schaefer's model: immaturity defined by short attention span, passivity, and daydreaming; conduct disorders; and anxiety/withdrawal.

Although the subtypes identified in this study must be considered as preliminary due to the need for replication with independent samples, the interpretation of the clusters within this sample was supported by both observational data and independent ratings by special education teachers. With respect to the latter validation data, it is important to note the high degree of similarity in the cluster profiles of special educators and classroom teachers as well as the differences between clusters across the various CBI scales. These data provide support for a multiple-syndrome interpretation of the behavior of LD children.

These different patterns of maladaptive behaviors within the LD group have potential educational importance. Classroom teacher ratings add important information to cognitive assessments of the LD child, yet descriptions of behavioral strengths and weaknesses are typically not included in the identification of LD children for special education services. The behavioral subtypes found in this study were, for the most part, unique to the LD group in that only a small percentage of normally achieving children could be classified into an atypical pattern. Thus, intervention should be targeted at both the cognitive and behavioral differences that exist between LD and normally achieving children. Further, the within-group behavioral differences as defined by the subtypes suggest the need for differential intervention because different forms of deviant behavior may have different etiologies and sequelae. For example, previous research has shown that general intervention strategies were not effective in remediating specific behavioral syndromes (Lorion et al., 1974).

The identification of distinct behavioral clusters of LD children represents an initial step toward clarifying the characteristics of an often ambiguous group of handicapped children. Further research is needed to determine if similar subtypes can be identified in other LD samples and to assess the academic and social consequences associated with cluster membership that was determined early in the child's school career. For example, although achievement did not differentiate the subtypes during the first year of special education services, behavioral differences may serve to impede academic progress in future years for certain behavioral patterns. Research of this type will assist both researchers and practitioners in addressing the difficult questions of identification, intervention, and prevention of learning disabilities.

References

Anderberg, M. R. (1973). *Cluster analysis for applications*. New York: Academic Press.

Bryan, T. S. (1974). An observational analysis of classroom behaviors of children with learning disabilities. *Journal of Learning Disabilities, 7,* 26–34.

Cronbach, L. J., & Gleser, G. C. (1953). Assessing similarity between profiles. *Psychological Bulletin, 50,* 456–473.

Dorval, B., McKinney, J. D., & Feagans, L. (1982). Teacher interaction with learning disabled children and average achievers. *Journal of Pediatric Psychology, 17,* 317–330.

Everitt, B. (1980). *Cluster analysis* (2nd ed.). New York: Halsted Press.

Feagans, L., & McKinney, J. D. (1981). The pattern of exceptionality across domains in learning disabled children. *Journal of Applied Developmental Psychology, 1,* 313–328.

Fisk, J. L., & Rourke, B. P. (1979). Identification of subtypes of learning-disabled children at three age levels: A neuropsychological, multivariate approach. *Journal of Clinical Neuropsychology, 1,* 289–310.

Hoge, R. D., & Luce, S. (1979). Predicting academic achievement from classroom behavior. *Review of Educational Research, 49*(3), 479–496.

Johnson, R. A., & Wichern, D. W. (1982). *Applied multivariate statistics.* Englewood Cliffs, NJ: Prentice-Hall.

Keogh, B., Major-Kingsley, S., Omori-Gordon, & Reid, H. P. (1982). *A system of marker variables for the field of learning disabilities.* Syracuse, NY: Syracuse University Press.

Keogh, B. K., Tehir, C., & Windeguth-Behn, A. (1974). Teachers' perceptions of educationally high-risk children. *Journal of Learning Disabilities, 7,* 367–374.

Kohn, M., & Roseman, B. L. (1972). A social competence scale and symptom checklist for the preschool child: Factor dimensions, their cross-instrument generality and longitudinal persistence. *Developmental Psychology, 6,* 430–444.

Lorion, R. P., Cowen, E. L., & Caldwell, R. A. (1974). Problem types of children referred to a school-based mental health program: Identification and outcome. *Journal of Consulting and Clinical Psychology, 47,* 491–496.

Lyon, R., & Watson, B. (1981). Empirically derived subgroups of learning disabled readers: Diagnostic char-

acteristics. *Journal of Learning Disabilities, 14,* 256–261.

McKinney, J. D. (1984). The search for subtypes of specific learning disability. *Journal of Learning Disabilities, 17,* 43–50.

McKinney, J. D., & Feagans, L. (1980). Learning disabilities in the classroom. Chapel Hill, NC: The Frank Porter Graham Child Development Center.

McKinney, J. D., & Feagans, L. (1982, March). Longitudinal research on learning disabilities. Paper presented at the Association for Children and Adults with Learning Disabilities, Chicago.

McKinney, J. D., & Feagans, L. (1983). Adaptive classroom behavior of learning disabled children. *Journal of Learning Disabilities, 16,* 360–367.

McKinney, J. D., Feagans, L., Ferguson, J., & Burnett, C. K. (1978). The SCAN Manual. Chapel Hill, NC: The Frank Porter Graham Child Development Center.

McKinney, J. D., Mason, J., Perkerson, K., & Clifford, M. (1975). Relationship between classroom behavior and academic achievement. *Journal of Educational Psychology, 67,* 198–203.

McKinney, J. D., McClure, S., & Feagans, L. (1982). Classroom behavior patterns of learning disabled children. *Learning Disability Quarterly, 5,* 45–52.

McKinney, J. D., & Speece, D. L. (1983). Classroom behavior and the academic process of learning disabled students. *Journal of Applied Developmental Psychology, 4,* 149–161.

Morris, R., Blashfield, R., & Satz, P. (1981). Neuropsychology and cluster analysis: Potentials and problems. *Journal of Clinical Neuropsychology, 3,* 79–99.

Peterson, D. R. (1960). The age generality of personality factors derived from ratings. *Educational and Psychology Measurement, 20,* 461–474.

Routh, D. K. (1979). Activity, attention, and aggression in learning disabled children. *Journal of Clinical Child Psychology, 8,* 183–187.

Satz, P., & Morris, R. (1981). Learning disability subtypes: A review. In F. J. Pirozzolo & M. J. Wittrock (Eds.), *Neuropsychological and cognitive processes in reading* (pp. 109–141). New York: Academic Press.

Schaefer, E. S. (1981). Development of adaptive behavior: Conceptual models and family correlates. In M. Begab, H. Garber, & H. C. Haywood (Eds.), *Psychosocial influences in retarded performance: Vol. 1. Issues and theories of child development* (pp. 155–179). Baltimore: University Park Press.

Schaefer, E. S., Edgerton, M., & Aronson, M. (1977). Classroom Behavior Inventory. Chapel Hill, NC: The Frank Porter Graham Child Development Center.

Torgesen, J. (1975). Problems and prospects in the study of learning disabilities. In E. M. Hetherington (Ed.), *Review of child development research* (Vol. 5, pp. 3–31). Chicago: University of Chicago Press.

Torgesen, J., & Dice, C. (1980). Characteristics of research on learning disabilities. *Journal of Learning Disabilities, 13,* 531–535.

Von Isser, A., Quay, H. C., & Love, C. T. (1980). Interrelationships among three measures of deviant behavior. *Exceptional Children, 46,* 272–276.

CHAPTER EIGHT

Multidimensional Scaling

CHAPTER PREVIEW

Multidimensional Scaling is a series of techniques which helps the analyst to identify key dimensions underlying respondents evaluations of objects. For example, in marketing multidimensional scaling is often used to identify key dimensions underlying customer evaluations of products, services or companies. The technique can be used to infer the underlying dimensions from a series of similarity and/or preference judgements provided by respondents about objects. The purposes of multidimensional scaling are to determine: (1) what dimensions respondents use when evaluating objects, (2) how many dimensions they may use in a particular situation, (3) the relative importance of each of the dimensions, and (4) how the objects are related perceptually.

This chapter should enable you to:

- understand the differences between similarities and preference data as they are used in multidimensional scaling.
- distinguish between the various approaches to applying multidimensional scaling.
- appreciate how spatial representation of data can clarify underlying relationships.
- determine the number of dimensions that are represented in the data.
- interpret the spatial maps so the dimensions can be understood.

KEY TERMS

Aggregate analysis An approach to MDS, in which the analyst generates perceptual maps of the respondent's evaluations of stimuli. In an attempt to create fewer maps, the analyst seeks to cluster subjects' responses to find a few "average" or representative subjects. These subjects' responses will then be used to generate perceptual maps.

Appropriateness data A procedure used to obtain respondents' perceptions of which items or attributes are most similar to each other and which are most dissimilar (similarities data). This procedure is similar to confusion data, but derives similarity from a pattern of responses across many stimuli and situational contexts, instead of single frequencies for the pairings of the various stimuli.

Confusion data A procedure used to obtain respondents' perceptions of similarities data. The pairing (or "confusing") of one stimuli with another stimuli is taken to indicate similarity.

Derived measures   A procedure used to obtain respondents' perceptions of similarities data. Derived similarities are typically based on "scores" given to stimuli by respondents. The semantic differential scale is frequently used to elicit such "scores."

Dimensions   Features of a phenomenon (product, service, image, etc.). A particular phenomenon can be thought of as possessing both perceived (subjective) dimensions (i.e., expensive, fragile, etc.) and objective dimensions (i.e., color, price, features).

Disaggregate analysis   An approach to MDS in which the researcher generates perceptual maps on a subject-by-subject basis. The results of this effort may be difficult to interpret. Therefore the analyst may attempt to create fewer maps by some process of aggregation.

Ideal points   The point on a perceptual map which represents the most preferred combination of perceived attributes (according to respondents). This ideal point may reveal the characteristics of an "ideal" product, or "ideal" store. A major assumption is that the position of the ideal point (relative to the other products on the perceptual map) would define relative preference so that products further from the ideal should be preferred less.

Importance/performance grid   A two-dimensional approach for assisting the analyst in labeling dimensions. The respondents' perceptions of the importance (e.g. as measured on an "extremely important" to "not at all important") and performance (e.g., as measured on a "highly likely to perform" to "highly unlikely to perform") for each brand or product/service on various attributes are plotted with "importance" on one axis and "performance" on the opposing axis.

Index of fit   A squared correlation index (R^2) which may be interpreted as indicating the proportion of variance of the disparities (optimally scaled data) that can be accounted for by the MDS procedure. It is a measure of how well the raw data fits the MDS model. This index is used to determine the number of dimensions. Similar to measures of covariance in other multivariate techniques, measures of .60 or greater are considered acceptable.

Initial dimensionality   Before beginning an MDS procedure the researcher must specify how many dimensions or features are represented in the data. The MDS procedure then uses this "initial dimensionality" as a starting point in the selection of the best spatial configuration for the data.

Neighborhood approach   Also called cluster interpretation, the neighborhood approach relies on visual clustering of objects in order to interpret the spatial map produced by the MDS procedure.

Objective dimensions   Physical characteristics of a phenomenon. For example, a product has: size, shape, color, weight, etc.

Perceptual map   The visual representation of a respondent's perceptions of two or more dimensions or features. Usually this map has opposite levels of one dimension on the ends of the X and Y axes, such as "sweet" to "sour" on the ends of the X axis and "high priced" to "low priced" on the ends of the Y axis.

Perceived dimensions Respondents may "subjectively" attach features to a phenomenon. Examples include: "quality," "expensive," "good-looking," etc. These perceived dimensions are unique to the individual and may bear little correspondence to the objective dimensions of the phenomenon (product, service, etc.)

Preference data Preference implies that stimuli are judged by the respondent in terms of dominance relationships; that is, the stimuli are ordered in preference with respect to some property. Direct ranking, paired comparisons, and preference scales are frequently used to determine respondent preferences.

Similarities data When collecting this type of data, the researcher is trying to determine which of the items are most similar to each other and which are the most dissimilar. Implicit in similarities measurement is the ability to compare all pairs of objects. Three procedures used to obtain similarities data are: confusion data, appropriateness data, and derived measures.

Similarity scale An arbitrary scale, say from -5 to $+5$, which allows the representation of an ordered relationship between phenomenon from most similar (closest) to the least similar (farthest apart). This type of scale is only appropriate for representing a single dimension, hence the development of MDS techniques.

Spatial maps MDS techniques enable the researcher to represent respondents' perceptions spatially. That is, to create visual displays that represent the dimensions perceived by the respondents when evaluating stimuli. This spatial map aids in understanding similarities and dissimilarities between objective and perceptual dimensions.

Stress measure A measure of the proportion of the variance of the disparities (optimally scaled data) which is not accounted for by the MDS model. This type of measurement varies according to the type of program and the data being analyzed. The stress measure helps to determine the appropriate number of dimensions to include in the model.

Subjective evaluation A method of determining how many dimensions are represented in the MDS model. The analyst makes a "subjective inspection" of the spatial maps and asks: Does the configuration look reasonable? The objective is to obtain the best fit with the least number of dimensions.

What Is Multidimensional Scaling?

Multidimensional scaling is a large family of procedures for drawing pictures of data so that the researcher can:

- Visualize relationships described by the data more clearly.
- Give clearer explanations of these relationships to others.

The data are typically gathered by having respondents give simple unidimensional responses such as:

- Product A is more similar to B than to C.

  or
- I like product A better than product B.

From these simple responses, a picture may be drawn that reveals a pattern. The following example illustrates this process: A respondent was asked to rank the following 15 pairs of candy bars, where rank 1 is assigned to the pair of candy bars that is *most similar* and rank 15 to the pair that is *least alike*. The respondents had 15 index cards with the labels of the pair of candy bars attached, such as the following:

```
┌─────────────────────────────────────┐
│  ┌──────────────┐                    │
│  │  Mars Bar    │                    │
│  └──────────────┘                    │
│           ┌──────────────┐           │
│           │   Hershey    │           │
│           └──────────────┘           │
└─────────────────────────────────────┘
```

The results for one respondent are these:

Candy Bar	A	B	C	D	E	F
A	—	2	13	4	3	8
B		—	12	6	5	7
C			—	9	10	11
D				—	1	14
E					—	15
F						—

This respondent thought that candy bars D and E were most similar, and that E and F were least similar. If one wanted to illustrate similarity, a first attempt would be to draw a similarity scale and fit all the candy bars to it. For example: placing A, B, and C on an arbitrary scale might yield the following:

		A	**B**							**C**	
−5	−4	−3	−2	−1	0	1	2	3	4	5	6	7	8	9

This shows the distances $\overline{AB}$, $\overline{BC}$, and $\overline{AC}$ in an ordered relationship, as the data suggest. (Note: The line over the pairs of letters—e.g., $\overline{AB}$ —indicates that it refers to the distance between—i.e., similarity—the pair A and B. It also respresents an ordered relationship from most similar $\overline{AB}$—closest—to least similar ($\overline{AC}$)—farthest apart.) For example, $\overline{AB}$ is the closest pair, $\overline{BC}$ is the next closest, and $\overline{AC}$ is the farthest apart. Now try to place D on the chart so that the following ordered relationship is represented—$\overline{AB}$, $\overline{AD}$, $\overline{BD}$, $\overline{CD}$, $\overline{BC}$, $\overline{AC}$.

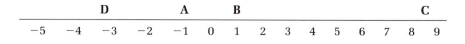

	D		**A**	**B**							**C**
−5	−4	−3	−2	−1	0	1	2	3	4	5	6	7	8	9

This can't be done. In one dimension, if $\overline{AB}$, $\overline{AD}$, $\overline{BD}$, then $\overline{CD}$ cannot be less than $\overline{BC}$.

It is clear that if the person judging the similarity between the candy bars had been thinking of a *simple* rule of similarity, such as amount of chocolate, the pairs could be placed on a single arbitrary scale that should reproduce the perceived single quality (e.g., chocolate) used to judge the pairs.

Since one dimension (scale) does not fit the data well, a two-dimensional solution can be attempted. The procedure is too tedious to attempt by hand, so a computer produced the two dimensional solution shown in Figure 8.1. Examining this solution, we see that in the ordered relationship given by the respondent, $\overline{DE}$, $\overline{AB}$, $\overline{AE}$, $\overline{AD}$, $\overline{BE}$, $\overline{BD}$, $\overline{FG}$, $\overline{AF}$, $\overline{CD}$, $\overline{CE}$, $\overline{CF}$, $\overline{BC}$, $\overline{AC}$, $\overline{DF}$ and $\overline{EF}$ are all preserved. This could lead to several conclusions by the researcher. The first is that the respondent was probably thinking of at least *two features of* candy bars when making this evaluation.

The conjecture that at least two features (dimensions) were considered is based on the inability to represent the respondent's perceptions in one dimension. This visual representation of perception is often called a *perceptual map*. Before going further with the candy bar example, we must discuss the basic concepts and assumptions of multidimensional scaling.

FIGURE 8.1 Illustration of multidimensional "map" of perceptions of six candy concepts.

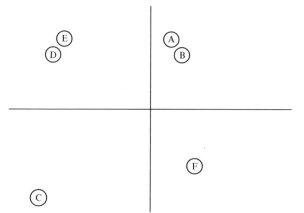

Basic Concepts Any phenomenon (products, services, images, aromas, etc.) can be
and Assumptions thought of as having both *perceived* and *objective* dimensions. For
of example, HATCO's management may see their product (a lawn mower)
Multidimensional as having two color options, a 2-hp motor, and a 24-in. blade. These
Scaling are the objective dimensions. On the other hand, customers may (or
may not) see these attributes. But they may also perceive the HATCO
mower as expensive-looking and/or fragile.These are perceived dimensions. Two products having the same physical characteristics (objective
dimensions) but different brands may be perceived to differ in quality
(a perceived dimension) by many customers. This phenomenon is
known to business people as well as social scientists, and recognition
of it must be a part of business decision making. Two differences between objective and perceptual dimensions are very important:

1. The dimensions perceived by customers may not coincide with (or
   may not even include) the objective dimensions assumed by the
   researcher.
2. The evaluations of the dimensions (even if the perceived dimensions
   are the same as the objective dimensions) may not be independent
   and may not agree. For example, one soft drink may be judged
   sweeter than another because the first has a fruitier aroma, although
   both contain the same amount of sugar.

Multidimensional scaling techniques enable the researcher to represent respondents' perceptions spatially, that is, to create visual displays
that represent the dimensions perceived by the respondents when evaluating stimuli (e.g., brands, objects). This visual representation helps
the researcher to understand more clearly the similarities and dissimilarities between objective and perceptual dimensions. These visual
representations or pictures are often referred to as *spatial maps* (as
illustrated in Figure 7.1). Spatial maps have not been directly proven
to represent perception but have provided insights into this process.
We may also assume (although not prove) that all respondents will
not

1. Perceive a stimulus to have the same dimensionality (although it
   is thought that most people judge in terms of a limited number of
   characteristics or dimensions). For example, some persons might
   evaluate a car in terms of its horsepower, while others do not consider
   this factor at all.
2. Attach the same level of importance to a dimension even if all
   respondents perceive this dimension. For example, two persons may
   perceive a cola drink in terms of its level of carbonation, but one
   considers this dimension unimportant while the other considers it
   very important.
3. Judge a stimulus in terms of either dimensions or levels of importance
   that remain stable over time. In other words, one may not expect
   persons to maintain the same perceptions for long periods of time.

In spite of any weaknesses caused by these assumptions, we will attempt to represent perceptions spatially such that any underlying relationships can be examined. We will limit our discussion to the techniques of multidimensional scaling and touch upon the psychological underpinnings only when this is necessary to qualify a point. You should be cautioned in two areas:

1. Multidimensional scaling computer programs are becoming readily available to persons with access to a computer. The possibilities for abuse are great because of the complex theory underlying the measurement of perception and the assumed relationships among its dimensions, both of which must be understood before using and interpreting multidimensional scaling results.
2. The programs now available are based on many different rationales for scaling. Each program must be carefully scrutinized for the assumptions underlying its successful use.

The variety of computer programs for multidimensional scaling is rapidly expanding. We will use an illustrative program but, by concentrating on the types of input data, the desired types of spatial representations, and interpretational alternatives, we will provide an overview of multidimensional scaling that will allow you to understand readily the differences between these programs.

Input Multidimensional Scaling

When individual perceptions of stimuli are obtained from respondents, a wide variety of techniques or procedures may be used. The resultant data may be generally categorized as preference data, similarities data (proximity data), and ideal points. These data serve as the basis for using multidimensional scaling techniques to derive the perceptual dimensions used by the respondents to judge the similarity of, preference for, and/or ideal stimuli.[1]

Preference Data

Preference implies that stimuli should be judged in terms of dominance relationships; that is, the stimuli are ordered in terms of the preference for some property. For example, brand A is preferred over Brand C. The two most common procedures for obtaining preference data are as follows:

1. Direct ranking: each respondent ranks the objects from most preferred to least preferred; for example:

---

[1] A good deal of the discussion in this chapter is based on the descriptions accompanying the many multidimensional scaling programs and the works of Green with Rao, Carmone, and Robinson [1,2,3]. When a specific reference is made, it is properly footnoted. It would be difficult (impossible?) to write on multidimensional scaling without drawing directly from these sources. If any similarity in presentation is noted, it is acknowledged here.

Rank from most preferred (1) to least preferred (5).
Candy bar A _____
Candy bar B _____
Candy bar C _____
Candy bar D _____
Candy bar E _____

2. Paired comparisons: a respondent is presented with all possible pairs and asked to indicate which pair is most preferred; for example:

The preferred bar in each pair is:

A _____	B _____
A _____	C _____
A _____	D _____
A _____	E _____
B _____	C _____
B _____	D _____
B _____	E _____
C _____	D _____
C _____	E _____
D _____	E _____

Please rate these candy bars on the following scale, where a 10 indicates "strongly preferred" bar and a 1 indicates "preferred very little."

□ □ □ □ □ □ □ □ □ □
1 2 3 4 5 6 7 8 9 10

Ranking is often facilitated by placing the items to be ranked in random order. These approaches to obtaining preferences data are typical, although not exhaustive; other procedures are also used.

Preference data allow the researcher to view the location of objects in a spatial map where distance implies differences in preference. This procedure is useful, since an individual's perception of objects in a preference context may be different from that in a similarity context. That is, a particular dimension may be very useful in describing the differences between two objects but is of no consequence in determining preference. Therefore, two objects could be perceived as different in a similarity-based map but similar in a preference-based spatial map. That is, two objects may be different (e.g., two different brands of candy bars) but may be preferred equally or about the same and may be represented close to each other on a preference map.

Similarities Data[2]

When collecting similarities data, the researcher is trying to determine which items are most similar to each other and which are most dissimi-

---

[2] In this discussion, when similarities are discussed, the reader may transpose the discussion to one of dissimilarities.

lar. Implicit in similarities measurement is the ability to compare all pairs of pairs of objects. If, for example, all pairs of objects of the set A, B, C (i.e., AB, AC, BC) are rank ordered, all pairs of these pairs can be compared. Assume that the pairs were ranked AB = 1, AC = 2, and BC = 3 (where 1 is most similar). Clearly, the pair AB is more similar than the pair AC. The terms *similarities* and *dissimilarities* are sometimes applied to the data collected for multidimensional scaling. This is a semantic issue; the terms *dissimilarities* and *similarities* often are used interchangeably to represent measurement differences.

Several procedures are commonly used to obtain respondents' perceptions of the similarities among stimuli. A few illustrative procedures are presented here. They are all based on the notion that the relative differences between any pair of stimula must be measured such that the researcher can determine if the pair is more or less similar to any other pair. This can be accomplished by either ranking or rating all pairs; for example:

Rank all distinct pairs of stimuli according to their relative similarity. If we have stimuli A, B, C, D, and E, we could rank pairs AB, AC, AD, AE, BC, BD, BE, CD, CE, and DE from most similar to least similar. If, for example, pair AB is given the rank of 1, we would assume that the respondent sees that pair as containing the two stimuli that are most similar, in contrast to all other pairs.

Three procedures used to obtain respondents' perceptions of similarities data are discussed here: confusion data, appropriateness data, and derived measures.

CONFUSION DATA. The pairing (or "confusing") of stimulus I with stimulus J is taken to indicate similarity. A typical procedure for gathering these data is to place the objects whose similarity is to be measured (e.g., 16 candy bars) on small cards, either descriptively or with pictures. The respondent is asked to sort the cards into stacks such that all of the cards in a stack represent candy bars that are similar. Some researchers tell the respondents to sort into a fixed number of stacks; others say, "Sort into as many stacks as you like." In either stiuation, the data results in an aggregate similarities matrix such as the following:

			Product				
	A	B	C				X
A	—	10	3	.	.	.	.	9
B		—	6	.	.	.	.	6
C			—	.	.	.	.	4
.							
.							
.							
X							

These data indicate that products A and B are most similar, since they appeared in the same stack 10 times, while products A and C are least similar, since they appeared together only 3 times.

APPROPRIATENESS DATA. Appropriateness data are similar to confusion data but derive similarity from a pattern of responses instead of single frequencies. For example, when a respondent is asked to indicate similarity between food items, the response may be different depending on the occasion the respondent has in mind. A respondent may consider potato chips more similar to French fries than to cookies when thinking about eating in a sandwich shop. However, when thinking about items to be placed in a picnic basket, cookies and potato chips may be more similar than either is to French fries. To illustrate, the respondents may sort cards representing foods into stacks labeled with specific occasions as follows:

	For Breakfast	For Lunch	For Dinner	For Late Snack
Orange juice	10	7	2	1
Coffee	10	8	2	1
Tea	7	10	10	5
Cola drink	3	10	3	10

In this illustration, orange juice and coffee would be most similar, since their distribution of responses across the four eating occasions is most similar. In addition, the researcher has obtained information on which occasions are most similar by using the same criterion: Similarity of the distribution of beverages for two occasions indicates similarity (within this beverage consumption context).

DERIVED MEASURES. Derived similarities are typically based on scores given to stimuli by respondents. For example, subjects are asked to evaluate K stimuli on two semantic differential scales, such as cherry soda, strawberry soda, and a lemon-lime drink on the following scales:

Sweet   ____ ____ ____ ____ ____ ____ Tart
Light-tasting ____ ____ ____ ____ ____ ____ Heavy

The 2×3 matrix could be evaluated for each respondent (correlation, agreement, etc.) to create similarity measures.

There are two important assumptions here:

1. That the researcher has selected the appropriate dimensions to measure with the semantic differential.
2. That the scales should be weighted (either equally or unequally) to achieve the similarities data for a subject or group of subjects.

In summary, these procedures (preference, similarity, etc.) have the common purpose of obtaining a series of unidimensional responses that represent the respondents' judgments so that they define the underlying multidimensional pattern leading to these judgments. They serve as inputs to the many multidimensional scaling procedures.

Ideal Points

The term *ideal points* has been misunderstood or misleading at times. We can assume that if we locate (on the derived perceptual map) the point that represents the most *preferred combination* of perceived attributes, we have an ideal product. Equally, we assume that the position of this ideal point (relative to the other products on the derived perceptual map) defines relative preference, so that products further from the ideal should be less preferred. Two approaches have generally been used to determine ideal points: explicit and implicit estimation.

Explicit estimation proceeds from the direct responses of subjects. This procedure may involve asking the subject to rate a hypothetical ideal on the same attributes on which the other stimuli are rated. Alternatively, the respondent is asked to include among the stimuli used to gather similarities data a hypothetical ideal stimulus (brand, image, etc.).

When asking respondents to conceptualize an ideal of anything we typically run into problems. Often the respondent will conceptualize the idea at the extremes of the explicit ratings used or as being similar to the most preferred product from among those with which the respondent has had experience. Often these perceptual problems lead the researcher to implicit ideal point estimation.

There are several procedures for implicitly positioning ideal points. The basic assumption underlying most procedures is that derived measures of ideal points' spatial positions are maximally consistent with individual respondents' preferences.

Srinivasan and Shocker [8] assume that the ideal point for all pairs of stimuli is determined so that it violates with least harm the constraint that it be closer to the most preferred in each pair than it is to the least preferred.

Regardless of the analytic logic, implicit estimation positions an ideal point in a defined perceptual space such that the distance from the ideal conveys changes in preference. Consider, for example, Figure 8.2. When preference data were obtained from the person indicated by the asterisk (*) on the six candy bars, the point (*) was positioned such that increasing the distance from it indicated declining preference. One may assume that this person's preference order is C, F, D, E, A, B. To imply that the ideal candy bar is * or that moving beyond * (in the direction shown by the dotted line from the origin) can be extremely misleading. The ideal point simply defines the ordered preference relationship among the set of six candy bars for respondent (*). While ideal points individually may not offer much insight, clusters of them can be very useful in defining segments (many respondents show up

FIGURE 8.2 One respondent's perceptual map of six candy bars; dot indicates a person's ideal point.

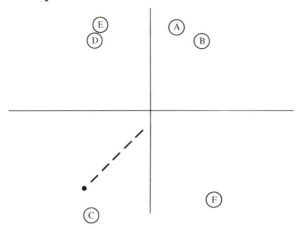

as potential market segments of persons with similar preferences, as indicated in Figure 8.3).

In summary, there are many ways to approach ideal point estimation, and no one best method has been demonstrated. The choice depends on the researcher's skills and on the multidimensional scaling procedure selected for use.

Methods of Multidimensional Scaling

The various approaches to multidimensional scaling can be categorized in regard to:

1. Level of measurement assessed for input and output.
   a. Nonmetric.
   b. Metric.

FIGURE 8.3 Perceptual map of six candy bars for many respondents; dots indicate ideal points for all respondents.

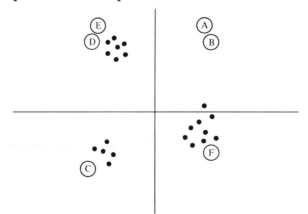

2. Aggregate versus disaggregate (for similarities data).
3. Internal versus external analysis (for preference data).
4. Vector versus point representations.

The following sections discuss and illustrate these points.

Nonmetric Methods

Nonmetric methods assume ordinal input and metric output. That is, the distances output by the multidimensional scaling procedure may be assumed to be approximately intervally scaled. The criterion for determining the location of stimuli (and/or subjects) in t-dimensional space (where t is any dimensionality from 1 to the number of stimuli minus 1) is to find configurations (output perceptual maps) whose rank orders of estimated ratio-scaled distances between all stimuli best reproduce the input rank orders. For each level of dimensionality from t-dimensions to 1 dimensions, a "stress" measure is calculated, and one attempts to find the lowest level of dimensionality (one can usually understand a two-dimensional spatial map more easily than a three-dimensional map) that produces satisfactory "stress."

For example, assume that we have 10 products that have been evaluated by having a person rank all 45 pairs from most similar (1) to least similar (45). We place the 10 points (representing the 10 products) randomly on a sheet of graph paper and measure the distances between every pair of points (45 distances). We then calculate a measure that shows the rank-order agreement between the graph paper distances and the original 45 ranks. (We can use a single rank-order correlation, but most computer programs use a variance-like measure called *stress*.) If the graph paper distances do not agree with the original ranks, move the 10 points and try again. Since this procedure may take several days, we can use a multidimensional scaling computer package and do it in micro-seconds. The major criterion for finding the best representation of the data is preservation of the ordered relationship between the original rank data and the distances between points. The stress measure is simply a measure of how well (or poorly) the ranked distances on the map agree with the ranks given by the respondents. The following illustration provides a good example of a multidimensional scaling application. HATCO is contemplating entering the soft drink market. Uncertain of what type of soft drink to produce for maximum perceived consumer satisfaction, HATCO's management takes a random sample of consumers and elicits their preferences among six soft drinks currently on the market (in identical containers with no brand names) and the consumers' judgments on their similarities. Figure 8.4 contains the responses of one subject.

Figure 8.5 is the perceptual map based on the data in Figure 8.4. The axes are labeled as follows: axis 1—fruit-flavored drinks to cola drinks, axis 2—sweet-tasting to tart-tasting. A, B, and C are cola drinks and D, E, and F are lemon-lime, orange, and strawberry drinks. The various approaches to labeling will be discussed later. In this spatial

FIGURE 8.4 Six soft drinks ranked by similarity.

Soft drink	A	B	C	D	E	F
A	–	3	6	15	1	8
B		–	2	12	4	9
C			–	10	7	13
D				–	14	11
E					–	5
F						–

mapping of similarities, the distances among all points preserve (as well as possible) the rank order of the similarities judged by the respondent. That is, if we had a matrix such as that in Figure 8.4 that revealed the rank similarities between the soft drinks, we could take all of the straight-line distances between all of the points on the map (Figure 8.5) and rank them in the same order as this similarities matrix.

Because we are trying to locate six points in two-dimensional space so that they preserve these 15 relative evaluations, the configuration we derive should give an indication of the dimensions the respondent used in judging the soft drinks. We assume that we have also obtained a spatial representation of the respondents' ideal soft drink—identified by point IP in Figure 8.5. This ideal point is based on positioning

FIGURE 8.5 Two-dimensional map of a single subject's views of the similarities among six soft drinks. A, B, C, D, E, and F indicate the six soft drinks; IP = ideal point for the subject.

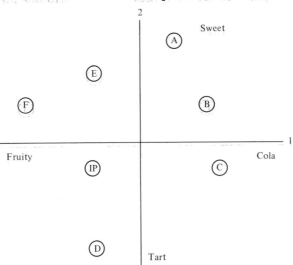

this respondent such that the distances from IP reflect declining preferences. If we assume that these representative perceptions are related to choice, we might conclude that this subject would select from among these drinks in order F, E, D, B, C, A. This represents the ordering of the straight-line (Euclidean) distances from point IP to all of the points representing the soft drinks. We are assuming that the direction of distance from IP is not critical, only the relative distance. In addition, this respondent's ideal soft drink is a slightly tart drink perceived as more fruity than cola drinks. This assumption is often not borne out by further experimentation. For example, Raymond [7] cites an example in which the conclusion was drawn that people would prefer brownie pastries on the basis of degree of moistness and chocolate content. When the food technicians applied this result in the laboratory, they found that their brownies made to the experimental specification became chocolate milk. One cannot always assume that the relationships found are independent, linear, or hold over time, as noted previously. However, multidimensional scaling is a beginning in understanding perceptions and choice that will expand considerably in the next few years.

Metric Methods

Metric methods assume that input as well as output are metric. This assumption allows us to strengthen the relationship between the final output dimensionality and the input data. Rather than assuming that only the ordered relationships are preserved in the input data, we can assume that the output preserves the interval and/or the ratio qualities of the input data. While this assumption is often hard to support with the data available to marketing researchers, the results of nonmetric and metric procedures applied to the same data are often very similar. Typical procedures used in metric multidimensional scaling are regression, factor analysis, and discriminant analyses.

Aggregate versus Disaggregate Analysis

In considering similarities (dissimilarities) data, we are dealing with perceptions of stimuli on a proximity basis of measurement. The output consists of representations of stimulus proximity in t-dimensional space (where t is less than the number of stimuli). The researcher can generate this output on a subject-by-subject basis (producing as many maps as subjects) or can attempt to create fewer maps by some process of aggregation.

The aggregation may take place either before or after scaling the subjects' data. Before scaling, the researcher may cluster-analyze the subjects' responses to find a few average or representative subjects and maps developed for the cluster's "average respondent." The researcher may also develop maps for each individual and cluster the maps based on the coordinates of the stimuli on the maps. The former approach is recommended. In the latter approach, minor rotations of essentially the same map can cause problems in creating reasonable clusters. Some researchers simply map the single average respondent. This technique is not recommended, since it assumes that the average represents all respondents equally well when in fact it may represent none.

Internal versus External Analysis of Preference Data

Internal analysis of preference data refers to the development of a spatial map shared by both stimuli and subject points (or vectors) solely from the preference data. As an example of the flexibility available, Kruskal and Carmone's M-D Scal program [6] allows the user to find configurations of stimuli and ideal points assuming no difference between subjects at all, assuming separate configurations for each subject, or assuming a single configuration with individual ideal points. By gathering preference data, only the researcher can represent both stimuli and respondents on a spatial map.

External analysis of preference data refers to fitting ideal points (based on preference data) to stimulus space developed from similarities data obtained from the same subjects. For example, we might scale similarities data individually, examine the individual maps for commonality of perception, and then scale the preference data for any groups identified in this fashion. This means that the researcher has to gather both preference data and similarities data to achieve external analysis.

Green and Rao [1] believe that external analysis is clearly preferable in most instances. Their conclusion is based on computational difficulties with internal analysis procedures and on the confounding of differences in preference with differences in perception. In addition, the saliences of perceived dimensions may change as one moves from perceptual space (are the stimuli similar or dissimilar?) to evaluative space (which stimuli is preferred?).

Vector versus Point Representations

The discussion of mapping of preference data has emphasized an ideal point that may be used to illustrate the relationship of an individual's preference ordering of a set of stimuli. One might also use a vector to represent the same notion:

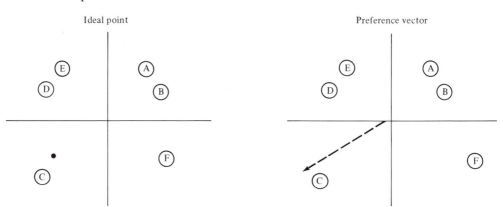

The ideal point presentation shows that C is most preferred and that preference declines as one moves away from that point *in any direction.* The vector shows that C is the most preferred product and that preference declines as one moves *opposite to the direction* in which the vector is pointing. The researcher must use judgment in choosing between these two methods of presentation.

A Generalized Multidimensional Scaling Approach

To illustrate an approach to multidimensional scaling, we present the simplified sequence of steps in multidimensional scaling from a commonly available nonmetric program, M-D-Scal [6].

Step 1. Select an initial configuration of stimuli (Sk) at a desired initial dimensionality (*t*). M-D-Scal either uses a configuration supplied by the researcher based on previous data or generates its own configuration by selecting pseudorandom points from an approximately normal multivariate distribution.

Step 2. Compute the distances between the stimuli points, evaluate the index of fit, and/or evaluate the stress of the configuration. M-D-Scal computes the interpoint distances (dij) in the starting configuration and (for metric data) performs a least squares regression of dij on the original data distances (sij). The regression estimated dij values termed $\hat{d}ij$ are used to calculate the following stress measure:

$$\text{Stress} = \left[\frac{\Sigma(dij - \hat{d}_{ij})^2}{\Sigma(dij - \overline{d})^2} \right]^{1/2}$$

where $\overline{d}$ is the average distance (Σ dij/n) on the map and the $\hat{d}ij$ are chosen to be as close to the dij as possible while remaining in the same order as the original input data.

Step 3. If the stress measure is greater than some small stopping value selected by the researcher, find a new configuration where the stress is further minimized. M-D Scal uses the method of steepest descent to find a new configuration. This essentially involves evaluating the partial derivatives of the stress function to determine the directions in which the best improvement in stress is to be obtained, and moving the points in the configuration in those directions in small increments.

Step 4. The new configurations are evaluated and adjusted until satisfactory stress is achieved.

Step 5. Once satisfactory stress has been achieved, the dimensionality is reduced by one and the process is repeated until the lowest dimensionality with acceptable stress has been reached.

The result of this series of steps is the selection of a spatial configuration, that is, the determination of how many dimensions are represented. Analysts generally utilize three approaches to make this determination: subjective evaluation, index of fit, and stress measures.

The spatial map is a good starting point for the evaluation. The number of maps depends on the number of dimensions specified in the program. A map is produced for each combination of dimensions. The objective of the analyst should be to obtain the best fit with the smallest number of dimensions. The problem here is that measures of fit improve with increased dimensions. A trade-off must be made between the fit of the solution and the number of dimensions. Interpretation of solutions derived in more than three dimensions is extremely difficult and usually is not worth the improvement in fit. The analyst typically makes a subjective inspection of the spatial maps and determines whether the configuration looks reasonable. This must be considered, because at a later stage the dimensions will need to be interpreted and explained.

An index of fit is sometimes used to determine the number of dimensions. The *index of fit* (or R-square) is a squared correlation index that can be interpreted as indicating the proportion of variance of the disparities (optimally scaled data) that can be accounted for by the multidimensional scaling procedure. In other words, it is a measure of how well the raw data fit the multidimensional scaling model. The R-square measure in multidimensional scaling represents essentially the same measure of variance as it does in other multivariate techniques. Therefore, it is possible to use similar measurement criteria; measures of .60 or better are considered acceptable. Of course, the higher the R-square, the better the fit. In particular, the R-square measure of fit is suggested as that measure to rely upon when utilizing the ALSCAL program, which is available on the SAS package.

A third approach is to use a measure of stress. *Stress* measures the proportion of the variance of the disparities (optimally scaled data) that is not accounted for by the MDS model. This measurement varies according to the type of program and the data being analyzed. Kruskal's [6] stress is the most commonly used measure at present for determining the model's goodness of fit. As for the extraction of factors in factor analysis, we can plot the stress value output against the number of dimensions to determine the optimal number of dimensions to utilize in the analysis. For example, in Figure 8.6 the elbow indicates that there is substantial improvement in the goodness of fit when the number of dimensions are increased from one to two. Therefore, the best fit is obtained with a relatively low number (two) of dimensions.

Identifying the Dimensions

As we indicated in Chapter Six, identifying underlying dimensions is often a difficult task. Multidimensional scaling techniques have no built-in procedure for labeling the dimensions. The researcher, having developed the maps in the selected dimensionality, can adopt several procedures:

1. Respondents may be asked to interpret the dimensionality subjectively by inspecting the maps.

FIGURE 8.6

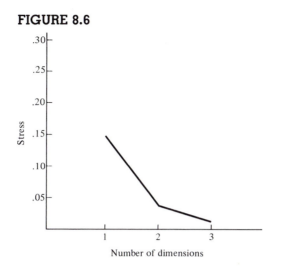

2. The researcher may identify the axes in terms of the objective characteristics of the stimuli.
3. If the similarities (or preference) data were obtained directly, the respondents may be asked (after stating the similarities and/or preferences) to identify the characteristics most important to them in stating these values. The set of characteristics can then be screened for values that match the relationships portrayed in the maps.
4. The subjects may be asked to evaluate the stimuli on the basis of researcher-determined criteria (usually both objective values) and researcher perceived subjective values. These evaluations can be compared to the stimuli distances on a dimension-by-dimension basis for labeling the dimensions.
5. Attribute or ratings data may be collected in the original research to assist in labeling the dimensions. Various methods may help in analyzing the attribute data. Mean ratings may be computed for each product. Then dimension coordinates for the products may be rank-correlated with the means. Importance-performance grids for each product or brand on a two-dimensional approach may also be constructed.

While a dimension can represent a single attribute, usually it does not. The researcher can collect data on several attributes and condense the dimension labels using factor analysis. Many researchers suggest that using attribute data to help label the dimensions is the best alternative. The problem, however, is that the researcher may not include all important attributes in the study.

These procedures, while not representative of all of those suggested, reveal the difficulty of labeling. This task is one that cannot be left until the multidimensional scaling procedures are implemented. The researcher must plan the use of alternative labeling schemes early in the design of the research.

**The Candy Bar
Example
Revisited**

This chapter began with the example of scaling six candy bars based on fitting a two-dimensional representation of the 15 ranks, given all possible pairs of the six candy bars. The resulting perceptual map looks like this:

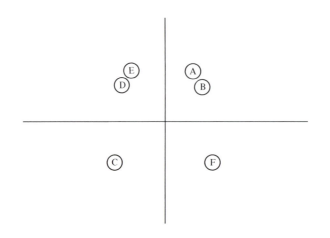

There are two ways of looking at the map:

A *neighborhood* or *cluster interpretation:* products A and B and D and E are seen as similar, while C and F are unique when compared to each other and to A, B, D, and E.

A *dimensional interpretation:* based on knowledge of the characteristics of the candy bars, the horizontal dimension is deduced to be

Not chocolate _____ Chocolate

The vertical dimension is thought to be

All candy   Cookie centers
(Top)   _____   (Bottom)

Both styles of interpretation can provide insight into respondents' perceptions.

A NOTE ON THE NEIGHBORHOOD APPROACH.   The analyst may not be overly concerned with naming the perceived dimensions in a structural sense. Rather, the purpose may be to identify groups of objects and to attempt to determine why they are different. For example, let's consider the candy bar example again (see Figure 8.2). Why are A and B different from the other candy bars? To answer this question, the analyst must first understand that the axes are determined arbitrarily. That is, all of the underlying variables are analyzed through correlation analysis and grouped together along two dimensions, based on positive and negative correlations with each other. The dimensions (axes) may represent positive and negative components of the same dimension (e.g., sweet on one end and sour on the other), or they may represent different dimensions that are negatively correlated (e.g., sweet on one end and lemon-flavored on the other, or even sweet and chocolate-flavored on one end and small and tasty on the other—the latter being an example of multidimensionality). If the analyst has some reason for using the two dimensions, he or she may do so. If not, an alternative is to use the multidimensional scaling procedure as a type of cluster analysis. That is, the analyst may run the MDS procedure and look for visual clusters of objects such as AB and DE. Theory or intuition would then be used to explain the clusters logically without comparing them to the axes. This visual clustering often facilitates interpretation and may make dimensional interpretation of the axes unnecessary.

Summary

Multidimensional scaling is a set of procedures that may be used to display the relationships tapped by data representing similarity (dissimilarity) or preference (dominance). It has been used successfully:

To illustrate market segments based on preference judgments.
To determine which products are more competitive with each other (similar).
To deduce what criteria people use when judging objects (products, companies, ads, etc.).

In summary, multidimensional scaling can reveal relationships that appear to be obscure when one examines only the numbers resulting from a study.

A visual perceptual map does much to emphasize the relationship between the stimuli under study. However, great care must be taken when attempting to use this technique. Misuse is common. The researcher should become very familiar with the technique before using it and should view the output as only the first step in the determination of perceptual information.

QUESTIONS

1. How does MDS differ from cluster analysis? Conjoint?

2. What is the difference between preference data and similarities data, and how does it impact the results of MDS procedures?

3. How are ideal points used in MDS procedures?

4. How do metric and non-metric MDS procedures differ?

5. How can the analyst determine when the "best" MDS solution has been obtained?

6. How does the analyst go about identifying the dimensions in MDS? Compare this procedure with that for factor analysis.

REFERENCES

1. Green, P.E., and Vithala Rao, (1972), *Applied Multidimensional Scaling*, Holt, Rinehart and Winston, Inc.

2. Green, P.E. (1975), "On the Robustness of Multidimensional Scaling Techniques," *Journal of Marketing Research*, Vol. 12 (February), 73–81.

3. Green, P.E. and F. Carmone, "Multidimensional Scaling: An Introduction and Comparison of Nonmetric Unfolding Techniques," *Journal of Marketing Research*, Vol. 7 (August), 330–41.

4. Greenacre, Michael J. (1984), *Theory and Applications of Correspondence Analyses*, Academic Press.

5. Lingoes, James C. (1972), *Geometric Representations of Relational Data*, Mathesis Press, Ann Arbor.

6. Kruskal, Joseph B. and Frank J. Carmone (1967), *How to Use M-D Scal. Version 5-M, and Other Useful Information*, Murray Hill, New Jersey, Bell Laboratories.

7. Raymond, Charles (1974), *The Art of Using Science in Marketing*, Harper & Row: New York, pp. 91–93.

8. Srinivasan, V. and A.D. Shocker (1973), "Linear Programming Techniques for Multidimensional Analysis of Preferences," *Psychometrica*, Vol. 38 (Sept.), pp. 337–369.

SELECTED READINGS

Product Positioning: An Application of Multidimensional Scaling

YORAM WIND
PATRICK J. ROBINSON

Product positioning, a construct frequently referred to by marketing and advertising practitioners, is rarely, if ever mentioned in the professional marketing literature. This has occurred in a situation where there seems to be a marginal acceptance of the relevance (and sometimes even importance of product positioning as a diagnostic device which provides operational guidelines for new product development efforts, redesign of existing products and design of advertising and distribution strategies, while little attention is given to its measurements.

This paper is concerned with just this latter issue—that quality of one research approach—multidimensional scaling and related techniques—to the measurement of product positioning. While many papers [8,9,13] emphasize the methodology of multidimensional scaling, and offer excellent espository discussion of these techniques, this paper offers a discussion of these techniques and emphasizes the position of this methodology to one set of marketing terms—product positioning. This will be done by highlighting a number of studies in which various multidimen-

"Product Positioning: An Application of Multidimensional Scaling," Yoram Wind and Patrick J. Robinson, 1972, pp. 155–175. Reprinted from *Attitude Research in Transition*, published by the American Marketing Association.

Yoram Wind is professor of Marketing, Wharton School, University of Pennsylvania.

Patrick J. Robinson is President, Robinson Associates, Inc.

sional scaling techniques have been used to establish product positioning. This is preceded by a brief discussion on the future and measurement of product positioning. The paper concludes with some comments on the measurement of product positioning and the relevance of positioning studies as guidelines for marketing strategies.

On the Nature and Measurement of Product Positioning

The term product (brand) positioning refers to the place a product occupies in a given market. Conceptually, the origin of the positioning concept can be related to the economist's work on market structure, competitive position of the firm and the concepts of substitution and competition among products. Marketing also has been concerned with such phenomena as product differentiation [14] and market position analysis, an interest which ranges from simple market share statistics to various approaches (such as Markov processes) for forecasting changes in a firm's market position [1].

More recently, increasing attention has been given to product image. This suggests a new perspective on product positioning, one which focuses on *consumers' perceptions* concerning the place a product occupies in a given market. In this context, the word positioning encompasses most of the common meanings of the word position—position as a place (what place does the specific product occupy in its relevant market?), a rank (how does the given product fare against its competitors on various evaluative dimensions?), and a mental attitude (customer attitudes—the cognitive, affective and action tendencies) toward the given product.

Given this view, the product (brand) positioning should be assessed by measuring consumers' or organizational buyers' *perception* and *preference* for the given product in relation to its realistic competitors (both branded and generic). The necessity to determine product positioning, not only on the basis of its perception (perceived similarity to other products) but also on consumers' preferences for it (overall preference as well as preference under various conditions—scenarios) is a key premise of the "ideal" approach to product positioning.[1] This is based on the premise that customer behavior is a function of *both* their perceptions and preferences and the recognition that buyers may differ with respect to both their perception and the preference for a product.

A somewhat different approach has been taken by Stefflre and his associates in their "Market Structure" analysis [2,3,15]. In this new product development procedure, the first few research steps are concerned directly with establishing the market structure, i.e., "determine which items (products and brands) consumers see as constituting a market and the 'position' of each item in the market vis-a-vis the other items."[2] This analysis positions the various brands based on consumers' *perceptions* (similarity) of the various brands and utilizes certain multidimensional scaling procedures. This "positioning" analysis does not utilize preference data in *conjunction with* similarities data but occasionally uses data on patterns of brand-to-brand substitution obtained from large-scale purchase panel data, when these are available.

Whether one uses perceptions, preferences, or both as basis for product positioning, it is essential to start with the appropriate set of competing products. This set, which in many cases includes products outside the immediate product class of the product in question, can be generated by marketing experts based on their experience and analysis of existing information or from unstructured depth interviews with consumers. Identifying a broad set of competing items is quite crucial since it constitutes the stimulus set for the positioning study. In designing the stimulus set it is sometimes desirable to include two types of products and brands—brands in the same product class and products and brands outside the product class

which may be used by consumers as substitutes for the product in question. For example, in a study of soups, one could include a set of different soup brands, forms, types, and flavors as the primary set as well as a set of soup substitutes—sandwich, salad, coffee, various snacks, etc.

Given the stimulus set of brands and products, the next step is to determine consumers' perceived product positioning. This can be done by eliciting (a) consumers' perceptions *via* a variety of available procedures for (overall) similarity measurement, or (b) consumers' preferences—overall and under a variety of usage and purchase conditions —or (c) both consumers' perceptions and preferences.

The data collection procedure employed in each of these cases depends on the number of items in the stimulus set, the total length of the interview, and the researcher's preferences. If one deals with a set of 20 to 40 brands, sorting or rating will be a much more appropriate task than strict ranking which is feasible when a smaller competitive set is included in the study.

Whatever the data collection procedure is, the data should provide in its original or transformed form a $P \times P$ matrix of product similarities (or dissimilarity) at the desired level of analysis—individual, segment, or total market. This matrix serves as the input data for the appropriate set of multidimensional scaling and clustering programs. The basic idea underlying this analysis is that a market (consumers' perceptions of the various brands) can be conceived as a multidimensional space in which individual brands are positioned. A product's positioning is determined from its position on the relevant dimensions of the similarity space, its position on the various product attribute vectors (if a joint space analysis is undertaken), and its position with respect to other brands, as may be obtained *via* cluster analysis.

A prototypical product positioning procedure is summarized in Figure 1. In designing such a study there are a number of research decisions which have to be made concerning the following topics.

Stimulus Definition

The brands and products selected for inclusions in the stimulus set can be defined in a number of ways and may be presented in terms of the

---

[1] In certain empirical studies, because of cost considerations, one may elect to use only preference data as the basis for product positioning.

[2] Barnett [2].

FIGURE 1 Determining product positioning.

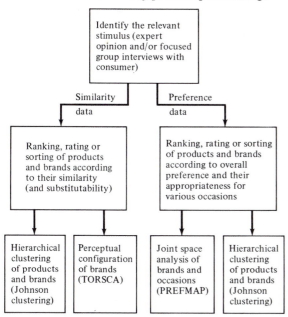

physical objects or services themselves, names of the items, verbalized, profile descriptions of the items, or a mixture of the above. As might be imagined, respondent evaluations may easily differ, depending upon the manner in which the stimuli are described and semantically encoded.

Task Definition

Defining the respondent's task requires an explicit decision. The task could consist of having the respondent react to the stimuli in terms of: overall evaluation of the objects in terms of preference or in terms of their relative similarity; judgments of objects similarity or preference (or other types of orderings) according to a set of prespecified scenarios (problem solving conditions) or according to a set of prespecified attributes (other than overall similarity or preference). Alternatively, one can collect objective data on the characteristics of the stimuli, in which case an objective performance space rather than a subjective brand space will result.

Response Definition

Assuming that the stimulus set and task have been defined, the researcher must still contend with

specifying the nature of the subject's response. Verbal and non-verbal responses may involve ratings judgments, ranking, including strict orderings or weak orderings (those including the possibility of ties) and assignments to prespecified classes.

Even given the restriction of responses to those that represent subjective, verbalized judgments, there are numerous ways for eliciting ratings, rankings, and category assignments.

Some Illustrative Studies

To illustrate the applicability of multidimensional scaling to the study of product positioning we will review briefly the results of a number of commercial studies. All of these studies were originally designed for some other purpose such as evaluating a set of new concepts, evaluating new product designs, or promotional programs.[3] They did include, however, as an integral part of the study an examination of product positioning by each of the relevant market segments.

The studies to be reported cover a wide range of products including calculators, diet products, medical journals, financial services, and retail stores. They also utilize a variety of analytical techniques, and develop maps of product positioning based on perceptions, preferences, or a combination of the two.

Positioning a Calculator

As with any product, calculators can be positioned in an objective space or a subjective space. In a business and scientific calculating machine study, objective data were obtained for 40 calculator models on each of 20 performance characteristics (e.g., control features, maximum number of index registers, minimum and maximum storage size, memory cycle time, output display speed, etc.). From these performance data similarities were computed between each pair of calculator models across the full set of 20 objective variables. The result of this analysis is presented in Figure 2 in a form of a "tree diagram" which resulted from a hierarchical clustering program [11]. The tree

---

[3] Common to all these studies is the concern with *evaluating* alternative stimuli, and not *generating* them. Occasionally, as a by-product of these studies one can get some ideas for new products. Yet, there are more efficient procedures for generating new product ideas which do not require a positioning analysis.

FIGURE 2 Tree diagram of calculator model similarities objective performance evaluation.

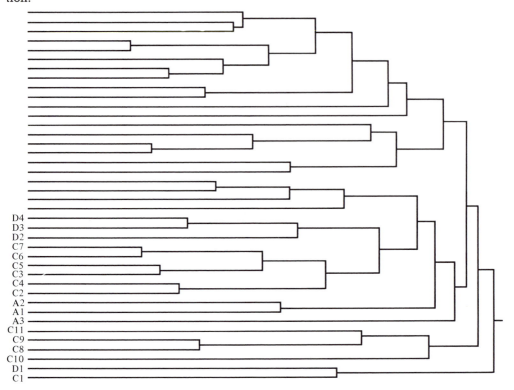

diagram portrays the clusters of models that emerged from this analysis.

This tree diagram may be thought of as being similar to a family tree in the sense that close relations and "branches" of the family that have a closely related pedigree, are shown as being lined in close proximity. That is, the sooner two calculators join, from left to right, the more similar the calculators are in terms of objective performance.

Recognizing that objective positioning does not necessarily correspond to the perceived position of the various calculators, a study was conducted in which various calculator users (representing professional and managerial positions in a variety of using industries) were asked, among other things to group the 40 models according to their impressions of the various calculators regarding their similarities and differences from a performance standpoint. These data were analyzed via various multidimensional scaling [16] and hierarchical clustering procedures. The results of the clustering program are portrayed as a tree diagram

in Figure 3. A comparison of this figure with Figure 2 (the objective space) indicates considerable difference between the subjective configuration and the objective one. The major difference was that some calculator users tended to view all models of a given manufacturer as a group with exception of one model of calculator that was viewed as dissimilar to the others of the same make. Calculators of a given manufacturer or of certain capabilities were grouped together depending on which respondent segment was concerned. When clustered on *objective* performance data the various models did *not* maintain their manufacturer separation nor the same capability configurations. This disparity in the perceived and objective positioning suggests the importance of not restricting positioning analysis to objective performance data base. It may satisfy engineering specifications but buyers and users may see things very differently. Furthermore, the perceived product positioning differed considerably among the various user groups, i.e., Toshiba customers, for example, viewed the calculator market differently from non-

FIGURE 3 Tree diagram of calculator model similarities subjective performance evaluation (segment 1).

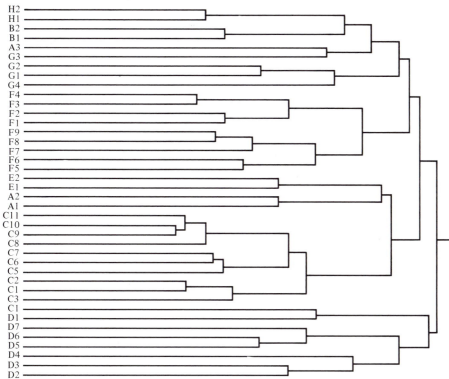

Toshiba customers, suggesting the need for coupling the product positioning analysis with market segmentation.

Among the policy implications stemming from the positioning aspects of this study were distinct sales promotion appeals (emphasizing or deemphasizing the objective product attributes) for certain models aimed at different market segments.

Positioning New Diet Products

The calculator study illustrated a positioning based on contrasting objective and subjective performance evaluations of existing industrial products (calculators). The same approach, viz., gaining better understanding of a product positioning *via* a comparison of its objective and subjective evaluations, was undertaken in a study of diet food items. In this case one of the objectives of the study was to assess the positioning of some new diet products. The respondents were women

who were on a diet and were asked to group food items (some of which were diet products such as Metrecal and some were not, such as pudding, potato chips, milk shakes, etc.) and 13 concepts of new diet products according to their similarity.

In addition they were asked to evaluate the products and concepts according to their overall preference and preference for serving and eating in various eating occasions (scenarios) such as: for lunch, short crash diet to lose weight quickly, to improve appearance, at dinner, when I am by myself, etc. Following this they also rated the products and concepts on a set of 12 attributes including calories, nutrition, taste, and convenience of preparation, vitamins, cholesterol and fillingness. This resulted in three sets of subjective data:

1. Product similarity data
2. Product preference rankings (overall and by scenario)
3. Product rankings on various product attributes

FIGURE 4 Two-dimensional perceptual configuration of twenty-seven food products and thirteen new diet concepts.

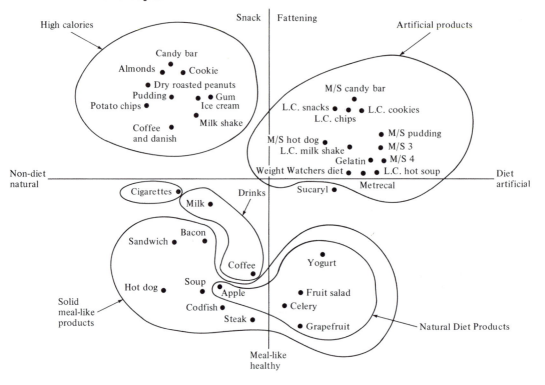

These data were subjected to a variety of multidimensional scaling programs. The 40 × 40 product similarity data across subjects were submitted to the TORSCA multidimensional scaling program and to the Johnson clustering program. The product evaluative data (both preference for overall usage and usage under certain occasions, and attributes) were submitted to Carroll and Chang's joint space (PREFMAP) program [4].

In addition to these data, a group of food technicians evaluated the various products according to their actual "objective" attributes. These data were also submitted to an appropriate set of multidimensional scaling programs enabling a comparison of the subjective-objective maps to be made.

Figure 4 presents the two dimensional configuration of the 40 items as derived from the products similarity data. The product clusters were determined by the Johnson hierarchical clustering program and incorporated into Figure 4. An examination of this figure suggests a number of clusters:

A cluster of high calorie snack/dessert products such as cookies, candies, milk shakes, etc.

A cluster of "non-natural" diet concepts and products including the new diet concepts and Metrecal.

A cluster of natural diet foods such as yogurt, fruit salad, celery sticks, etc.

A cluster of solid, meal-like items.

Examination of any given product or concept reveals its position with respect of other products and concepts as well as its position on the two dimensions. If one is interested in the positioning of the new diet concepts, Figure 4 suggests the following conclusions:

1. The new concepts will compete with other "artificial" diet products such as Metrecal and Weight Watchers Complete Meals.
2. The concepts seem to be positioned opposite their "natural" counterparts. Hence if dimension 2 is reviewed as "fattening-healthy" it may suggest that the "fattening" attribute of the natural products may "rub off" on the diet concepts, leading to overestimation of the fattening attributes of concepts such as a "diet cookie."

FIGURE 5 Joint space configuration: two-dimensional perceptual configuration of forty products and their "perceived" and objective attributes.

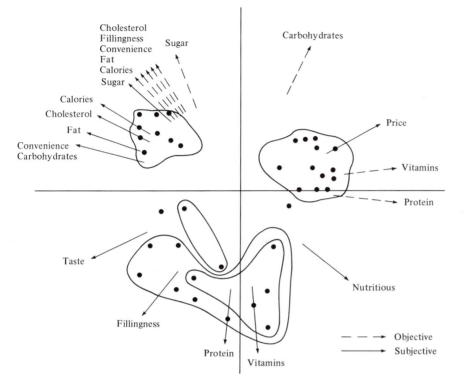

Further insight into the position of each concept and product was gained by an examination of the joint space configuration of the products and concepts and their perceived attributes. Figure 5 presents the results of this analysis. Looking at the solid line vectors (subjective evaluations) and the relation of the various products to them suggest:

"Natural" diet products such as yogurt are perceived as being both nutritious and healthy.

"Natural" meal items such as steak are perceived as filling and rich in proteins.

High calories snacks and desserts, such as potato chips and candies, are perceived as being "fatty" (fat, cholesterol, and carbohydrates), sugary, and high in calories. They are, however, convenient for preparation.

The new diet concepts are perceived as expensive, not tasty, less nutritious than natural diet products and poor on health, protein and fillingness.

A comparison of the "subjective" evaluation configuration with the objective data was done in two stages. First a two and three dimensions "objective" configuration of the 40 products was derived and compared to the subjective configuration. This suggested that dieters do not perceive diet and non-diet products according to their objective attributes.

A more direct comparison was undertaken via a joint space analysis, the results of which are presented in Figure 5. A comparison of the discrepancies between the objective (broken line) and subjective (solid line) vectors suggest that the greatest discrepancy (50% or greater) exists with regard to fillingness, carbohydrates, proteins, and vitamins. Fairly high congruence (less than a 20% discrepancy) is found with respect to calories, sugar, fat, cholesterol, and convenience of preparation. This suggests that dieters are more conscious of and interested—hence knowledgeable with respect to—this latter set of attributes.

Positioning of a Journal

The concept of positioning can apply not only to industrial and consumer products but also to

FIGURE 6 Joint space of journals and (vector directions of) evaluative scenarios—
From PREFMAP Computer Program.

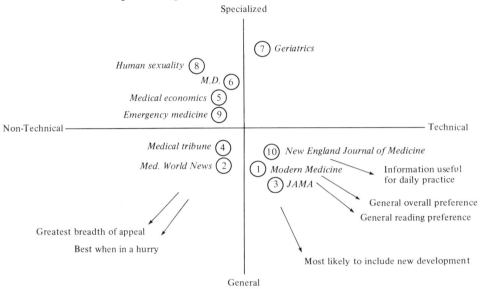

professional journals. A publisher of a medical journal was concerned with the positioning of his journal. A study was undertaken among a sample of physicians who were presented with a set of 10 medical journals and asked to rank them according to 6 scenarios such as "general overall preference," "general reading preference," etc. These data provided the input for a journal dissimilarity matrix, across scenarios.

These data were then submitted to the PREF-MAP joint space multidimensional scaling program which finds vector directions in the perceptual map (obtained in an earlier phase via INDSCAL multidimensional program) [5]. The relationships of the vectors to the journals can be interpreted in terms of preference evaluations. Figure 6 presents the results of this analysis. Interpretation of the dimensionality of this figure shows that the primary attribute by which medical journals are judged are technical versus nontechnical (horizontal axis) and the specialized versus general (vertical axis). We see, for example, that JAMA is perceived as most technical whereas Human Sexuality is seen as the least technical journal.

Observing the six preference vectors illustrates the popularity of JAMA, the New England Journal of Medicine, and Modern Medicine. These journals carrying high general preference are those that are viewed as informative as well. The scenar-

ios, "best when in a hurry" and "greatest breadth of appeal" show directions favorable to the news journals such as Modern World News and Medical Tribune.

The specific position of any of the journals can easily be determined by examining its relations to other journals and its relative position on the two dimensions and the various preference vectors.

In this study one of the objectives was to help the journal in question present itself more effectively to its prospective space-buyers insofar as its relative position and strengths versus other vehicles. There was substantial "surprise value" and yet intuitively satisfying insights for management in the positioning and relative strengths and weaknesses of the various publications. Solicitation policy and various editorial, format, and content changes were clearly identified for improvement and exploitation.

Positioning of Financial Services

The three studies described so far illustrated a number of direct attempts at positioning a product based on consumers' perceptions and preferences as well as pointing out the discrepancy between "objective" and "subjective" product evaluation. Product positioning can also be achieved, how-

FIGURE 7

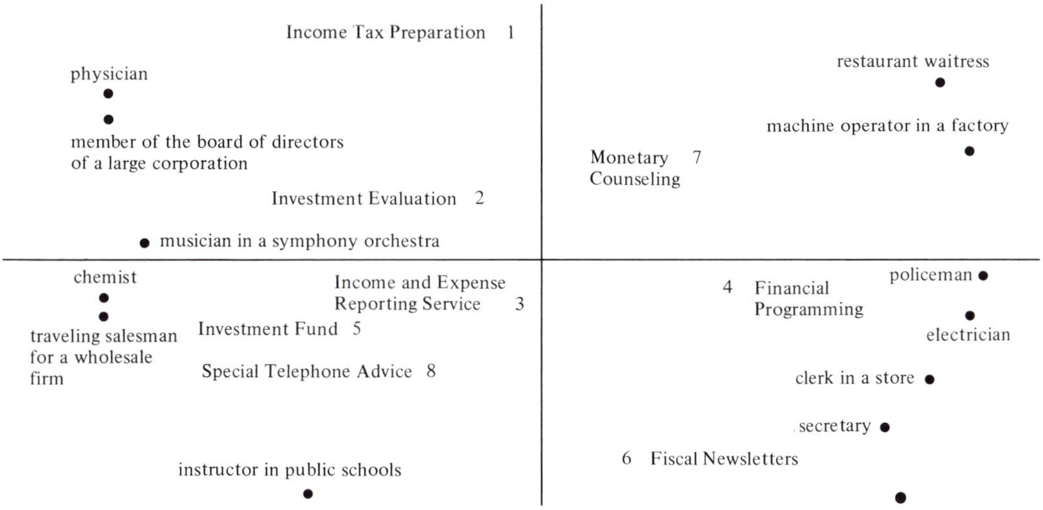

ever, by other more indirect methods. One such procedure is described in the next study. This study was concerned with the evaluation and positioning of a number of new financial services. In this case, in addition to the customary evaluation of new and existing services on a set of evaluative scales, the respondents (male heads of households) were also asked to pick up from a set of 12 occupations—the five occupations whose holders would be most likely to use the given service. Upon selection of the five occupations, respondents were further asked to rank them from most to least likely to use the given service. These data and the similarity configuration of the various financial services (obtained in an earlier phase, via the TORSCA program) were submitted to a joint space program [6]. The results of this analysis are presented in Figure 7.

An examination of this map suggests a clear division of the set of financial services based on their perceived prestige into high prestige services such as "investment fund" or "special telephone advice" as opposed to "financial programming" and "monetary counseling" services which are viewed as low prestige services. In addition a number of services have either a very wide appeal that cuts across all occupations (as might be the case with income expense reporting services).

This indirect service positioning was consistent with the results obtained from the more direct position based on perceived similarity of the concepts and existing services. It provided further guidelines for promotional strategy than the more common approach, by suggesting the most appropriate type of people (occupations) to be portrayed in promoting the service. For example, the use of testimonials of blue-collar workers for one service and testimonials of professionals such as physician and chemist for another.

Positioning of a Retail Store

Our last illustrative study presents a new dimension in positioning studies. Whereas all of the previous studies focus on a product/service/journal positioning at a point in time, the current study is concerned with the changes in a retail store positioning over time. The study was based on housewives' evaluation of various retail stores in a given metropolitan area over a period of two years. The data for the first three months and last three months were grouped separately and analyzed for the overall market and various a priori segments.

Figure 8 presents the results of matching (via the Cliff Match procedure [7] separate TORSCA maps of 9 stores for each of the time periods. The stores in this map have been disguised, but the configuration is the actual configuration that was derived from the actual data. The dimensions of this map could be interpreted as high prestige/relatively expensive vs. low prestige/discount (the horizontal axis) and width of assortment (vertical axis). The change in the perceived position of each store is traced by the dotted line which links

FIGURE 8 Two-dimensional configuration of nine stores.

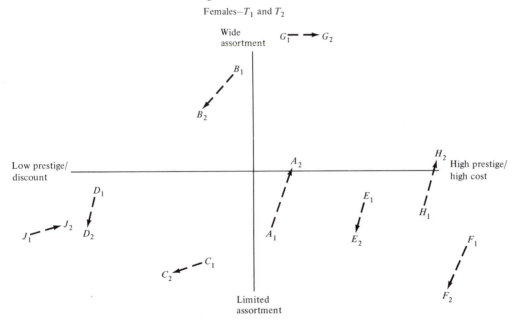

the store position in the two time periods. Examination of the magnitude and direction of changes in the stores' positions can provide management with considerable insight into their market position and changes in it as they occurred over the relevant study period.

As in the other positioning studies, the time series analysis can be extended to cover changes not only in the stores but also in the stores relative to various evaluative scales (such as good service, easy credit, best value, etc.) and/or various products for which the store may be appropriate (e.g., appliances, clothes, etc.). The analytical procedure is quite similar and the only difference being that instead of developing a "super space" (via the Cliff Match program) based on the results of the TORSCA program, the input data to the matching program are the results of the joint space analyses of stores and evaluative scales, or stores and products.

These analyses were of diagnostic significance in revealing relative competing store positions and the extent and direction of movement over a known time interval. As with any time series analysis, certain implications were revealed and of relevance to those concerned with relative standings and trends. More explicit and detailed in-

sights stem from examining store policy image positionings and shifts over time. Clearly, significant differences in one's own position may result not only from one's own but one's competitor's policy shifts and positioning and tracing such moves may suggest a number of strategy implications.

Conclusions

The five studies which were described briefly in this paper illustrate the wide applicability of the concept of positioning to a variety of products (calculators, diet foods, and medical journals), services (financial), and retail stores. They further demonstrate the usefulness of utilizing multidimensional scaling and clustering techniques in these types of studies.

Multidimensional scaling techniques include a set of computer algorithms that permit the researcher to develop perceptual and evaluative "maps" (i.e., geometric configurations) that summarize how people perceive various stimulus objects (products, services, stores, and the like) as being similar or different, and how such stimulus objects are evaluated on a variety of evaluative scales. As such, this set of techniques is especially suited for the portraying of the perceived position of a product.

Moreover, nonmetric multidimensional scaling and clustering techniques utilize data that need only be rank ordered (relative similarity or preference for a set of stimuli) which facilitates the respondent's task and places the burden of analysis on the researcher.

In view of the heterogeneous nature of every market the real value of product positioning is revealed, however, only when the positioning is coupled with an appropriate market segmentation analysis. In the studies reported here and in a variety of other studies which included positioning analyses, a segmentation analysis was always included, enabling one to conduct both an overall and segmented positioning analysis. The scope of this paper did not enable us to elaborate on the differences in product positioning by segment. Yet, in all of the studies reported above the findings were quite conclusive in suggesting that such differences do exist and hence justify the value of conducting separate positioning analysis for each market segment.

Product positioning when coupled with market segmentation can provide useful guidelines for the design and coordination of the firm's marketing strategy.

As with segmentation studies the value of a separate positioning study is quite limited. Applying the concept of positioning requires that each marketing research project which is concerned with evaluating a given marketing strategy (e.g., concept evaluation, advertising evaluation, and the like) should include a section on the perceived positioning of the given product or service. This should provide a useful addition to almost any marketing study.

Finally, it is hoped that employing multidimensional scaling and related techniques in positioning studies will further the ultilization of these and other recently developed techniques in other areas of marketing and image research, hence contributing to improved new directions in the measurement of attitudes and market behavior.

References

Alderson, W. and Green, P. E., *Planning and Problem Solving in Marketing* (Homeward, Ill.: Richard D. Irwin, 1964), especially pp. 170–192.

Barnett, N. L., "Developing Effective Advertising for New Products, *Journal of Advertising Research*, 8 (December 1968), pp. 13–20.

Barnett, N. L., "Beyond Market Segmentation," *Harvard Business Review*, 27 (January–February 1969), pp. 152–166.

Carroll, J. D. and Chang, J. J., "Relating Preference Data to Multidimensional Scaling Solutions via a Generalization of Coombs' Unfolding Model," mimeographed, Bell Telephone Laboratories, Murray Hill, N.J., 1967.

Carroll, J. D. and Chang, J. J., "A New Method for Dealing with Individual Differences in Multidimensional Scaling," mimeographed, Bell Telephone Laboratories, Murray Hill, N.J., 1969.

Chang, J. J. and Carroll, J. D., "How to Use MDPREF, A Computer Program for Multidimensional Analysis of Preference Data," mimeographed, Bell Telephone Laboratories, Murray Hill, N.J., 1969.

Cliff, N., "Orthogonal Rotation to Congruence," *Psychometrika*, 31 (1966), pp. 33–42.

Green, P. E. and Carmone, F. J., *Multidimensional Scaling and Related Techniques in Marketing Analysis* (Boston: Allyn and Bacon, 1970).

Green, P. E. and Rao, V., *Applied Multidimensional Scaling A Comparison of Approaches and Algorithms,* (New York Holt Reinhart Inc., 1972).

Howard, N. and Harris, B., "A Hierarchical Grouping Routine, IBM 360/65 FORTRAN IV Program," University of Pennsylvania Computer Center, Philadelphia, Pa., 1966.

Johnson, S. C., "Hierarchical Clustering Schemes," *Psychometrika*, 32 (1967), pp. 241–54.

Kuehn, A. A. and Day, R. L., "Strategy of Product Quality," *Harvard Business Review*, 40 (November–December 1962), pp. 100–110.

Silk, A. J., "Preference and Perception Measures in New Product Development: An Exposition and Review," *Industrial Management Review*, 11 (Fall 1969).

Smith, W. R., "Product Differentiation and Market Segmentation as Alternative Marketing Strategies," *Journal of Marketing*, 21 (July 1956).

Stefflre, V., "Marketing Structure Studies: New Products for Old Markets and New Markets (Foreign) for Old Products," in Bass, King, and Pessemier (editors) *Applications of the Sciences in Marketing*, (New York: John Wiley and sons, 1968), pp. 251–268.

Young, F. W. and Torgerson, W. S., "TORSCA—A FORTRAN IV Program for Shepart-Kruskal Multidimensional Scaling Analysis," *Behavioral Science*, 12 (1967), p. 498.

Personal Space as a Function of Infant Illness: An Application of Multidimensional Scaling

LARRY H. LUDLOW
SHELDON LEVY

In recent years much research has been devoted to exploring the psychosocial impact of serious childhood illnesses upon the family (Binger, Ablin, Feurstein, Kushner, Zoger, & Mikkelson, 1969; Tew & Lawrence, 1973; Burton, 1975; Satterwhite, 1978; Desmond, 1980; Spinetta & Deasy-Spinetta, 1981; Breslau, Staruch, & Mortimer, 1982; Lewis & Khaw, 1982). The aim of that research has been to identify factors which contribute to healthy familial coping with illness and factors that put family members at risk for psychological difficulties. Researchers have used a variety of instruments to assess family functioning and individual adaptation. These have included questionnaires, clinical interviews, video and audio tapes of various family interactions, naturalistic observations, and several other projective and objective psychometric techniques designed to measure cognitive, personality, and specific psychodynamic functioning.

One of the more intriguing instruments that has come from this research has been the three-dimensional hospital room (Spinetta, Rigler, & Karon, 1974). The instrument is a replica of a hospital room at a scale of ½-inch to 1 foot, with dolls and furniture to scale. Dolls (family members, health care personnel, etc.) are positioned in the room, a child is placed in a hospital bed, and stories about the child are elicited from family members and the child. "Personal space" analyses are conducted upon distances measured from each doll to the child in bed. In some studies the child positions the figures (dolls), other studies required that a "significant other" make the placements.

"Personal Space as a Function of Infant Illness: An Application of Multidimensional Scaling," Larry H. Ludlow and Sheldon Levy, Vol. 9 (3) 1984, pp. 331–347. Reprinted from the *Journal of Pediatric Psychology*, published by the Society of Pediatric Psychology, Plenum Publishing Corporation.
Larry H. Ludlow is on the faculty of Boston College, School of Education.
Sheldon Levy is affiliated with the Cook County Hospital and Michael Reese Hospital and Medical Center, Chicago Illinois.

Theoretically, the placement of figures by a child should be related to the degree of psychological distance the child experiences between himself and significant others. A number of studies have found this to be the case (Spinetta, Rigler, & Karon, 1973; Spinetta et al., 1974; Spinetta & Maloney, 1975, 1978). When leukemic children were asked to place figures where they usually were in the hospital room, the children in subsequent hospital admissions placed figures at a distance significantly farther away than did children at the time of first admission (Spinetta et al., 1974). Also, the placement of figures by someone other than the child should be related to the degree of psychological distance perceived between the person (health care professional or family member) and the child. Desmond (1980), for example, found that spatial placements made by the father of a leukemic girl were related to the father's growing sense of isolation from the child.

Though measuring the distance from various family figures and health personnel to the child has provided useful information to clinicians, the bidirectional nature of the psychological distance does not provide insight into the relations of family members to one another and to health care personnel. This information could be useful for understanding how significant others perceive one another while they interact for a common purpose—the care of the child. A clearer representation of the emotional distance between significant others and the child with a serious medical condition can be obtained when all distances between doll figures are analyzed simultaneously. That is: What is the *pattern* of interrelationships among significant others and the child, rather than how far away are certain members from the child?

The methodological issue, from our perspective, is whether or not the assumption of a one-to-one correspondence between physical distance and psychological distance is necessary. The difficulty with interpreting only the physical distance from a figure to an infant is that subjects are likely to differ in the physical scale of magnitude they employ in such situations. If a criterion was initially specified such that all subjects knew that, say, five "units" of distance away from the infant meant the same thing in psychological terms, then subjects' physical distances would be roughly relative to a common criterion. Without such a criterion there is no guarantee that smaller physical distances given by one subject are indeed psychological distances smaller than those given by another subject. The assumption that everyone employs the same scale of magnitude in their judgments may be avoided by focusing on the pattern of placement—not solely the bidirectional distance.

The purpose for this study was twofold. First, differences in personal space as a function of the infant's condition were investigated. Many clinicians express concern about the physical and psychological separations mothers and infants experience as a result of physical illness in infancy. Thus we compared patterns of figure placements between mothers of infants with either a chronic illness, acute illness, or healthy physical condition. The intent was to determine whether mothers of infants with the same health condition perceived the personal space of their infants in a common framework and whether that framework was unique to the specific health condition. The second purpose was to demonstrate an alternative data-analytic approach that represents personal space in a manner more wholistic than that presented in previous research.

In this study it was assumed that because the healthy infants had not experienced a serious illness, their mother's placement of figures would serve as a baseline against which the placements by mothers of infants with acute or chronic illnesses could be compared. Prior to the analysis it was predicted that each group of mothers would produce a unique framework or configuration of figures. This prediction was based upon the circumplex model of marital and family systems (Olson & McCubbin, 1980). According to this model, healthy families move toward the extremes of adaptability and cohesion when initially confronted with a stressful event but eventually gravitate back toward the moderate ranges as they successfully cope with those events. Thus, mothers of chronically ill infants should differ from mothers of healthy infants but the largest difference should be between the mothers of healthy infants and the mothers of acutely ill infants. At this stage of research, however, it was not possible to predict exactly how the configurations of figures would appear or differ. For example, no predictions were made about whether mothers of acutely ill infants, as opposed to mothers of chronically ill infants, would consider grandparents more important than siblings in contributing to the health of the infant.

Method

Subjects

The subjects were 28 mothers 1-year-old infants. They were drawn from the patient population of the Department of Pediatrics, University of California, Davis Medical Center, and the Department of Pediatrics, Michael Reese Hospital and Medical Center in Chicago. Mothers were chosen for this study because research has pointed out the importance of mothers in the developing behaviors of infants and also that disturbances in the mother–infant relationship might lead to later emotional disturbances in the child (Stern, 1977). The mothers were classified into one of three groups.

Ten mothers had infants who had been diagnosed as having a ventricular septal defect (VSD: a form of heart murmur). The defect must have been diagnosed within the first 9 months of the infant's life and the infant must have been acyanotic. This was the chronic illness group (VSD). An acute illness group consisted of eight mothers of infants who had been diagnosed with an received corrective surgery for an inquinal or umbilical hernia (Hernia). The diagnosis and surgery must have occurred within the first 9 months of the infant's life. A Control group of 10 mothers of infants who had no serious illnesses or had not been hospitalized was also selected. The VSD and hernia defects were chosen because they can occur in complete isolation from other illnesses or defects; their prognosis is good; treatment results in relatively short periods of separation (usually less than a week); and the defects do not necessarily have any long-term deleterious effect upon the child's intellectual, physical, social, or emotional development.

Additional criteria for selection were the following: (a) the infant had no other physical defect; (b) the infants had no other hospitalizations during the first 9 months of life after being taken home from the hospital after birth; and (c) the infants had to be between 11 and 14 months of age at the time of the study. Of the mothers whose infants met these criteria 41% agreed to participate in the study.

Some of the background characteristics of the resulting mother-infant groups are presented in Table I. There were no statistical differences between the three groups on any of the following 10 background variables; age of mother; socioeco-

TABLE 1 Background Characteristics of Mothers and Infants

	VSD (n = 10)	Hernia (n = 8)	Control (n = 10)
Mothers			
Age (years)	28.1	26.8	29.1
	(SD = 4.9)	(SD = 6.6)	(SD = 4.2)
Hollingshead (1965) SES level of head of household			
I & II	2	4	6
III	3	1	1
IV & V	5	3	3
Number of people living in household	4.1	3.5	3.5
	(SD = 1.0)	(SD = .8)	(SD = .7)
Marital status			
Married	7	4	7
Not married	3	4	3
Infants			
Age (months)	12.1	12.7	12.5
	(SD = .7)	(SD = .6)	(SD = 1.1)
Age at diagnosis (months)	1.0	2.1	—
	(SD = 1.6)	(SD = 1.2)	
Sex			
Male	5	7	4
Female	5	1	6
Birth order			
Firstborn	2	4	6
Second born	4	4	2
Third born	4	0	2

nomic status of head of household; number of people living in household; race; religious preference; marital status; illnesses during pregnancy; infant's age at time of diagnosis; infant's age at time of testing; sex of infant; and, birth order of infant.

Procedure

A Spinetta three-dimensional hospital room was constructed (Spinetta et al., 1973). The dimensions were 18 inches by 11 inches with a floor of galvanized steel. Dolls, 4 inches in height, were fitted with magnets on their feet. The dolls represented mother, father, sister, brother, grandparents, nurse, doctor, and clergyman. An infant doll matching the race of the infant was placed in the hospital bed and connected to a miniature intravenous apparatus. The mothers were given the following set of instructions: "I would like you to make up a story about this infant in the hospital. Here are some dolls to represent the people in

the story." The mother, father, sister, brother, grandmother and grandfather dolls are then placed before the mother. The instructions then continue with the following: "Use as many dolls as you would like." Time is allowed for the placement of the dolls. The instructions continue with the following: "I want you to tell me what the situation is in the story, and what the outcome will be, describing the thoughts and feelings of the characters." If not given, there is a probe for the thoughts and feelings of each person represented by a doll placed in the room. When the mother has done this, she is handed the nurse doll and asked to place the doll in the room and describe what she is doing, thinking, and feeling. When the mother has finished the procedure is repeated using the doctor doll and the clergyman doll.

Data

Throughout the protocol a personal space measure was taken of the distance of placement from each

FIGURE 1 Placement grid for hospital room replica.

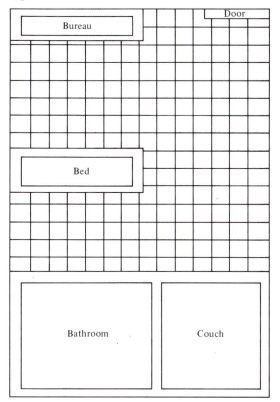

of the dolls to one another and to the infant. Doll placements were recorded on a page (Fig. 1) marked off in grids identical to the grids on the floor of the hospital room replica. Placements were subsequently translated into feet, ½-inch to 1 foot. Distances were measured from the central position of each doll to the head of the infant. In addition, distances were measured between the central positions of all other pairs of dolls, e.g., from mother to father, mother to doctor, etc. All Hernia group mothers were evaluated at the hospital. The Control and VSD mothers were evaluated either in the hospital or their homes.

The data consist of distances measured between pairs of hospital room figures. The 10 figures produced 10 (10 − 1)/2 = 45 distances. The distances were arranged in 10-by-10 symmetric matrices with zero placed on the diagonal. There were 28 subject matrices; one matrix for each mother. Three additional matrices were constructed. These matrices contain the mean distances between figure pairs for each subject group:

one matrix for the Control (n = 10), VSD (n = 10), and Hernia (n = 8) groups.

Most subject matrices contained missing data. Usually this was because there were no siblings in the family. Occasionally the data for a grandparent or a staff person was missing. Of those subjects with missing data, most had two or three, and once, four figures absent. There were no missing data in the group mean matrices, although one "mean" distance was the single distance obtainable from one of eight subjects.

Analyses

The first set of analyses address the extent to which mothers perceive significant others in terms of subsets or clusters of related figures. This issue is relevant because it was predicted that if mothers or groups of mothers differed in their patterns of figure relationships, then those differences would appear as figure clusters positioned differently— not just individual figures. For example, if Mother 1 placed the grandparents in a position relative to the infant different than did Mother 2 (say closer to the infant than were the siblings), then *both* grandparents would be closer than the siblings. Four distinct figure clusters were predicted: (a) infant, mother, father; (b) doctor, nurse, clergyman; (c) brother, sister; and (d) grandmother, grandfather. This a priori expectation is consistent with the family model presented by Minuchin (1974).

A hierarchical clustering technique (Johnson, 1967) was used to reveal whether the predicted figure clusters existed within each of the three group mean distance matrices.[1] This technique operates on the single-linkage principle and gives rise to a property called chaining. The last link in the chain, as a consequence, combines all figures into one large cluster. This particular property is appropriate for these data because it is reasonable to consider the figures as a single general cluster linked by a common medical situation. Thus, it is of interest to understand how clusters of figures initially form and how they merge into larger units.

The second set of analyses investigate mother and group differences. Traditionally, analysis of variance techniques have been applied to test the

[1] The hierarchical cluster analysis was performed using PROC CLUSTER in the SAS statistical package (SAS User's Guide, SAS Institute, 1979).

significance of distance differences. For example, do the mean distances from infant to mother differ significantly between Control, VSD, and Hernia subjects? That approach, however, can become unwieldy when a large number of distances must be analyzed. In addition, as noted earlier, such analyses assume an isomorphic relation between physical and psychological distances.

An alternative approach is available for the analysis of proximity measures. Proximities are either similarity or dissimilarity measures between pairs of objects. Distances are one form of dissimilarity. The approach seeks to reveal geometrically any structure within a set of proximities. Structure, in this sense, refers to whether or not the geometric arrangement of the figures forms a meaningful substantive pattern. The basic output of the analysis is a spatial representation of a configuration of points, each point representing one figure. Such geometric techniques, of which there are many varieties, are subsumed under a general class of analysis referred to as multidimensional scaling (cf. Torgerson, 1958; Shepard, Romney, & Nerlove, 1972; Schiffman, Reynolds, & Young, 1981; Kruskal, 1964).

Weighted multidimensional scaling (WMDS) techniques were applied to the data in this study.[2] WMDS methods are particularly useful for studying individual and group differences. Individual differences were investigated by simultaneously analyzing all 28 subject matrices. Group differences were addressed by the simultaneous analysis of the three mean distance matrices.

The WMDS analysis yields a stimulus (figure) configuration of points and a subject (whether person or group) weight space. The stimulus configuration portrays how figures are perceived to relate to one another in a geometric sense. The subject weight space reveals the relation of the individual to the stimulus space. In an individual analysis each subject is represented by a set of weights that indicate which dimensions in the stimulus space the subject is emphasizing. In a group analysis the weights reveal how groups emphasize particular dimensions.

From the WMDS results it is possible to test for group differences using an analysis of angular

variance technique proposed by Schriffman et al. (1981). The purpose is to test statistically whether subjects in the three groups were differentially weighting the dimensions formed by their composite stimulus space. The analysis tests the direction not the length of the subject weight vectors. Finally, WMDS solutions are obtained for the three group mean distance matrices.[3]

Results

The hierarchical clustering results are shown in Figure 2. From the top downwards the maps are for the Control, VSD, and Hernia groups, respectively. Strings of asterisks indicate which pairs of figures are most closely related from the highest (strongest) level of clustering to the lowest (weakest) cluster. In each of the three maps the four primary clusters occur as predicted. The parents and infant are the A, B, C cluster; health care staff are D, E, H; siblings are F, G; grandparents are I, J.

The groups of mothers differ in how they see the clusters relating to one another. For the Control group the secondary levels of clustering occur as the staff cluster joins the parent cluster, the grandparents join the siblings, and then all clusters merge. In the VSD group the staff merge with the grandparents first and then that group merges with the parents. The final link is with the siblings. In the Hernia group the staff and grandparents again merge first but then merge with the siblings. Here the final link is with the parents.

These patterns of linkage mean that (a) Control group mothers treat the parents and staff as the figures most central to the infant; (b) VSD group mothers distinguish between adults and siblings; and (c) Hernia group mothers focus on the parent-infant relation and consider everyone else to be of secondary importance.

Figure 3 is the three-dimensional (3D) nonmetric WMDS representation for the figures based on the separate perceptions of all 28 mothers (technical details are available from the first author). The figures form identifiable clusters and the clusters are distinguished by their relation to the infant and each other. In this type of representation it is conventional to label the dimensions of the

[2] The 28 subject matrices were analyzed by metric and nonmetric WMDS methods. The matrices were considered conditional and continuous. Metric analyses treated data at the interval level and did not allow negative subject weights to be computed. The initial starting configurations were unspecified and were determined by the ALSCAL computer program (Young & Lewyckyj, 1979).

[3] Program specifications were as above except that initial configurations were (a) determined by the program and (b) read in from the corresponding final solutions for the 28-matrix analyses. Both metric and nonmetric solutions were obtained.

FIGURE 2 Hierarchical cluster maps for the Control, VSD, and Hernia mean distance matrices.

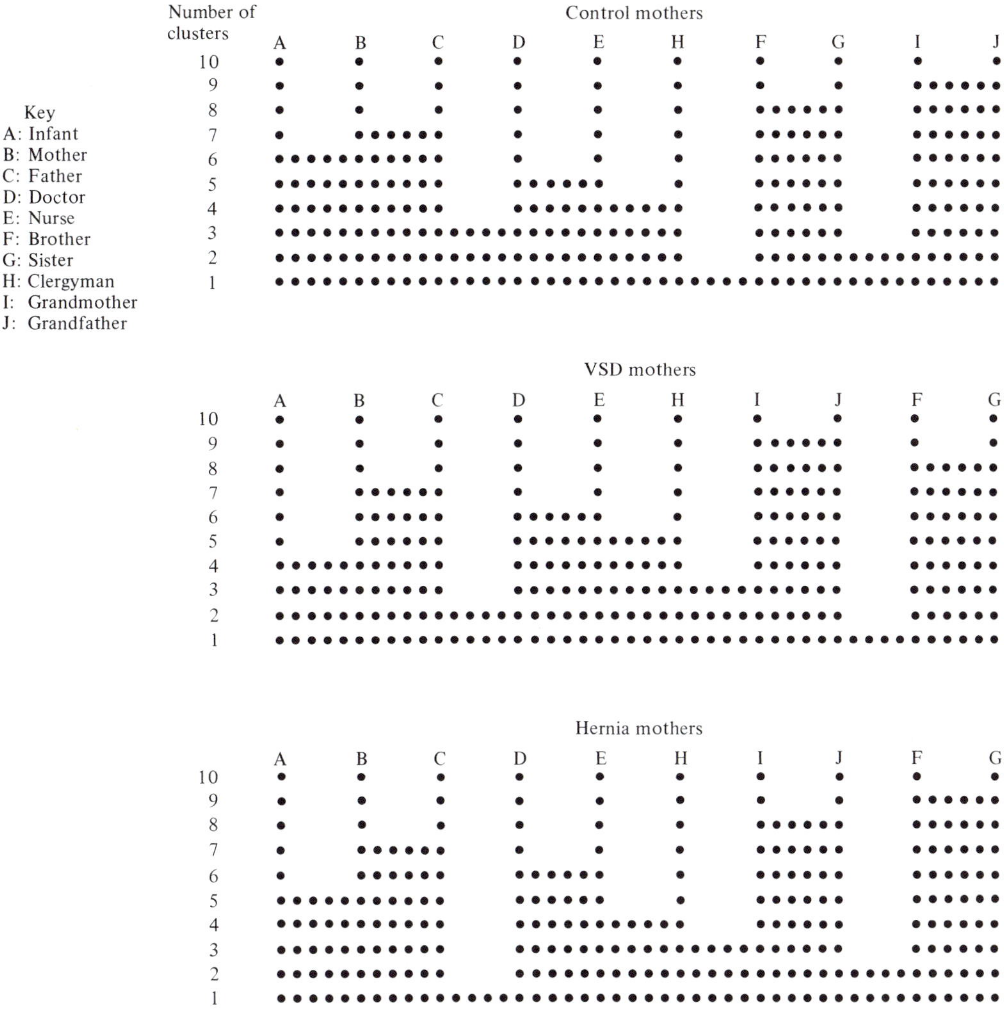

space. The labels we attach to the dimensions are somewhat arbitrary in the sense that a priori knowledge about the dimensions did not exist. They are, however, useful for descriptive purposes and can be validated in follow-up confirmatory research. Dimension I is "immediate caretaking needs of infant." Dimension II is labeled "long-term support of infant." The third-dimension is labeled "overall medical/spiritual support." The relation of the figures to the infant and how that relation changes is the key detail in this and the following spatial representations.

This configuration of figures defines a personal support space around the infant. Dimension I rep-

resents the degree to which figures satisfy the immediate caretaking needs of the infant. If a line is drawn from each figure to the axis representing Dimension I it can be seen that, in general, the parents are nearest the infant, followed by the staff, grandparents, and siblings. Dimension II is interpreted as the degree of support that individual figures provide the infant over time. The staff are nearest the infant, followed by parents, siblings, and grandparents. Dimension III represents the extent of overall medical/spiritual support provided by each figure. The parents are nearest, then come grandparents, siblings, and staff.

The third dimension is particularly interesting

FIGURE 3   Nonmetric 3D, 28 subjects.

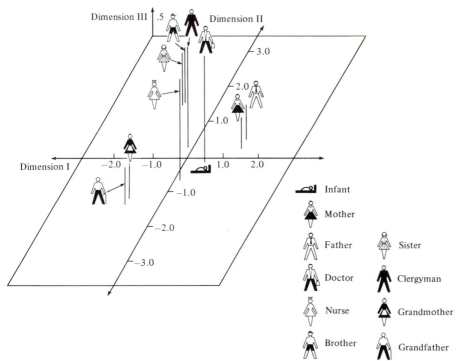

because it indicates that the parents provide the most significant overall medical and spiritual support. The clergyman and health care staff have the largest coordinate loadings but that means they are furthest away, hence, least significant. A two-dimensional representation would not have revealed this relation. We would have seen only that hospitalization does necessitate a close physical proximity between infant and staff. The central medical role played by the staff is seen from their close proximity to the infant on the first two dimensions. The proximity of the parents to the infant on the third dimension, however, suggest it is the parents and not the support staff who are ultimately responsible for the infant's overall care.

An inspection of the subject weight plots (not shown) revealed great variation in the ways mothers used the dimensions. In each of the three bivariate plots for the 3D solution (Dimension 1 weights plotted against Dimension 2 weights, etc.) there were some Control and VSD mothers oriented primarily toward one or the other dimension. It was not the case, however, that most Control mothers weighted heaviest one dimension while most VSD mothers used the other. Mothers were located from

one end of the weight space to the other. No discriminating pattern between those two groups of mothers could be seen.

Most of the Hernia group, however, acted as a homogeneous unit by weighting one dimension more than the other in two of the three plots. The Hernia group mothers weighted Dimension I equally with Dimension II, III more than I, and III more than II. These mothers felt strongly about the overall medical and spiritual attention required by their infant. The entensive heterogeneity within the Control and VSD groups was supported by the analysis of angular variance. The F ratio ($F = .78, df = 4, 50$) is a reflection of the substantial within-group variation even though the Hernia group mothers did react in an identifiable pattern.

Figure 4 presents the WMDS solution for the three group mean distance matrices. Each of the distance matrices now contains the mean distances for the mothers in that group. This configuration is practically identical to the one obtained from the 28 individual mothers. The same dimensions and clusters appear. The same relation between clusters is also seen. The extended family locations form a semicircle. The infant lies in the opening of the semicircle. The staff are in the

FIGURE 4 Nonmetric 3D, three groups.

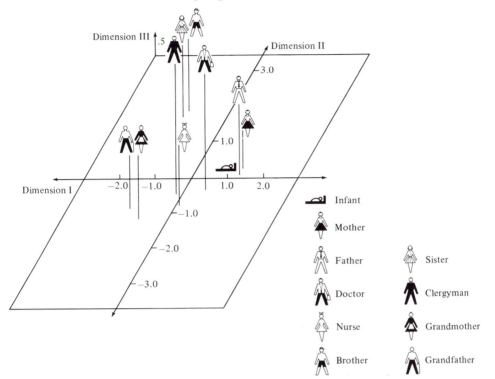

Infant			
Mother			
Father		Sister	
Doctor		Clergyman	
Nurse		Grandmother	
Brother		Grandfather	

middle, yet above everyone else. Of what value is this solution, then, if the 28-subject WMDS solution provided the same information? This solution was expected to provide information about the groups that was not evident in the previous analysis. That is, the lack of homogeneity in VSD and Control mothers to respond as members of common groups is attributed in part to imprecise estimates of subject weights because of missing data. By taking the mean distance we are assuming that a mother with missing data would have given similar figure placements as other mothers in her group, if the missing person had been present in her situation. Thus, if the group stimulus structure from the WMDS mean analysis conformed to the individual stimulus structure, then an analysis of the group weight space might be more meaningful than the subject weight space analysis.

Figure 5 contains the set of group weight plots. The interpretation is based on vector orientation, not length. Group B (VSD) and Group C (Control) regard the three dimensions with about the same degree of orientation. The VSD mothers do, however, weight Dimension II and III higher than the Control mothers. Their real experience with ill-

ness has given them an appreciation for long-term and overall medical/spiritual support; just immediate caretaking action is not sufficient. Group A (Hernia) presents a different pattern. Dimension I is weighted high relative to Dimension II; immediate caretaking action is more important than long-term support. Dimension III is weighted equally with Dimension I; immediate caretaking action and emotional support are both necessary. Dimension III is weighted higher than Dimension II; long-term support is not yet as important as overall medical/spiritual attention. These group results complement and clarify the earlier analysis of the weight space for the 28 mothers. That is, the Control and VSD mothers have similar perceptions but the Hernia mothers are in a situation that requires immediate attention. Their needs lead to a relationship structure that evolves around parental efforts to aid or gain aid for the infant.

Discussion

The purpose of this study was to use multidimensional scaling to compare the impact that chronic

FIGURE 5 Group weights in three dimensions.

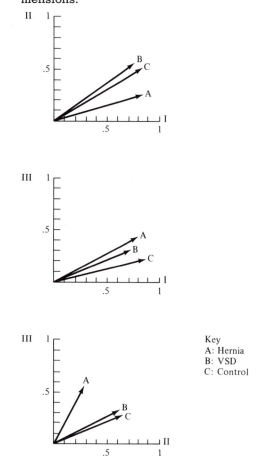

Key
A: Hernia
B: VSD
C: Control

of infant support which the mothers perceive that various figure clusters provide. The degree of support provided by each of the figure clusters depends on the circumstances. The Control mothers imagine how much support the staff and extended family would provide if in fact their baby were hospitalized. That relation is not borne out by the mothers of infants with hernias. They react to the figure groups in a "parents" versus "the rest of them" manner. Their needs are immediate; they are not greatly concerned with long-term support. The mothers of infants with VSD resemble the Control mother's structure but the VSD mothers have a stronger concern for long-term and overall medical/spiritual support.

The labeling of the dimensions, though post hoc, does provide a framework for follow-up confirmatory analyses. One strategy would be to ask each mother to rate on a scale from, say, 1 (little) to 5 (extensive), how much (a) immediate, (b) long-term, and (c) overall medical and spiritual support each significant other contributes to the infant. Each figure would have three mean rating scores. These mean ratings could be correlated with the three-dimensional coordinates. High correlations would provide support for having correctly identified the constructs that led mothers to position figures as they did. Alternative hypothetical constructs, too, could be rated and correlated.

The significance of the present research lies in the demonstration of (a) techniques capable of revealing multiple patterns of interactions and (b) patterns of perceptual differences between mothers of healthy, acute, and chronically ill infants. The observation that mothers of infants with different health conditions have different needs and concerns is important but, perhaps of more importance, one could address how those perceptions change over time. For example, does an acute hernia cause temporary or permanent changes in the relations of others to an infant? Perhaps mothers in general perceive their environment in terms similar to Control mothers, react to any sudden health issue like Hernia mothers and as time progresses they resemble VSD mothers. They might return, finally, full circle to a Control condition. Such an investigation would necessarily be long-term and expensive but could provide valuable information about patterns of illness adaptation.

The following scenario of an illness adaptation process is based on the results of this study: The healthy family structure is perceived as three-dimensional with the relation between parents, si-

or acutely ill infants have upon the personal space perceptions of their mothers. It was predicted that the placement of family and health care figures in a three-dimensional hospital room would yield unique, substantive configurations based upon a family model of adaptation to stress. The differences in those configurations were not predicted in any explicit sense, but it was assumed that differences would correspond roughly to the degree to which mothers perceived significant others contributing support to the infant, as well as the amount of stress experienced by the specific condition of the infant. The differences that were found are consistent with these assumptions.

Mothers treat significant others as four basic identifiable subsets of related members. In addition, the three groups of mothers see the relation of figure clusters to one another in different terms. These relations can be explained by the degree blings, and grandparents represented as semicircular. Everyone contributes a relatively equal degree of support to one another. Health care staff are essential yet exist outside the family structure while they perform important professional functions.

When an acute illness strikes the infant, the parents react and relationships alter. The family structure collapses into a linear relation where the grandparents take on additional responsibility. Immediate professional attention is essential. The siblings are not important support providers.

After the crisis relationships again alter and adapt to the new situation. The care of the infant now passes into a long-term phase. The family begins to relate to one another in the semicircular form with everyone, including siblings, contributing some degree of support. The role of the nurse is enhanced, the doctor and clergyman play a less central role. The postacute illness family returns to, essentially, a healthy stable family unit.

In a general sense, these data suggest how a "normal" family might react to an acute infant illness and the subsequent illness recovery. Although the data in this study cannot fully support such an illness adaptability theory—no family experienced all three phases—such a tentative theory does provide a plausible model for further research. One prospective study would compare larger samples of infants with other illnesses followed longitudinally through periods of illness and recovery. Another avenue of research would be the relationship between how each family member perceives their personal space and the psychological difficulties that each experiences over time in adapting to the illness of the child.

There are limitations to this study. For one, the analysis is based upon a relatively small sample and, hence, may not represent the perceptions of the populations of mothers included in the study. Also, as with other graphical methods, when one looks long enough at a configuration of points there is a potential for "finding" some pattern. To protect against this the researcher must have some a priori idea about the shape the results may take. Otherwise scientific progress is unlikely; everything is fortuitous. But once a baseline configuration can be established and replicated, then other experimental conditions can be compared and contrasted. Surprising departures from the baseline can then be tested in cross-validation studies.

The use of multidimensional scaling, as pre-

sented in this paper, suggests some intriguing possibilities for future research in this and other areas of clinical psychology and psychiatry. The apparent advantage of this technique over previous techniques of analysis is that it gives a pictorial representation of perceived multiple relationships. This type of interactional representation can be useful to clinicians. A picture can depict how family members experience their immediate social environment; it can be used as feedback to families on how they are perceiving themselves and others; it can chart changes in relationships over time; it can offer a means of explaining current relationships with a focus on what might be expected or desired in future relations.

References

Binger, C. M., Abling, A. R., Feurstein, R. C., Kushner, J. H., Zoger, S., & Mikkelson, C. (1969). Childhood leukemia: Emotional impact on patient and family. *New England Journal of Medicine, 280,* 414–418.

Breslau, N., Staruch, M. A., & Mortimer, E. A. (1982). Psychological distress in mothers of disabled children. *American Journal of Diseases of Children, 136,* 682–686.

Burton, L. (1975). *The family life of sick children: Coping with chronic childhood disease.* London: Routledge & Kegan Paul.

Desmond, H. (1980). Two case studies. In J. Kellerman (Ed.), *Psychological aspects of childhood cancer.* New York: McGraw-Hill.

Hollingshead, A. B. (1957). *Two-factor index of social position.* Unpublished manuscript, Yale University.

Johnson, S. C. (1967). Hierarchical clustering schemes. *Psychometrika, 32,* 241–254

Kruskal, J. B. (1964). Nonmetric multidimensional scaling. *Psychometrika, 29,* 1–27; 115–129.

Lewis, B. L., & Khaw, K. (1982). Family functioning as a mediating variable affecting psychosocial adjustment of children with cystic fibrosis. *Journal of Pediatrics, 101,* 636–640.

Minnuchin, S. (1974). *Families and family therapy: A structural approach.* Cambridge: Harvard Press.

Olson, D. H., & McCubbin, H. I. (1980). Circumplex models of marital and family systems: Application to family stress and crises intervention. In H. I. McCubbin (Ed.), *Family stress, coping and social support.* Minneapolis: Burgess.

Satterwhite, B. B. (1978). Impact of chronic illness on child and family: An overview based on five surveys with implications for management. *International Journal of Rehabilitation Research, 1,* 7–17.

Schiffman, S. S., Reynolds, M. L., & Young, F. W. (1981). *Introduction to multidimensional scaling.* New York: Academic Press.

Shepard, R. N., Romney, A. K., & Nerlove, S. (Eds.) (1972). *Multidimensional scaling: Theory and applications in the behavioral sciences* (Vol. 1). Theory. New York: Academic Press.

Spinetta, J. J., & Deasy-Spinetta, P. (Eds.) (1981). *Living with childhood cancer.* Saint Louis: C. V. Mosby.

Spinetta, J. J., & Maloney, M. S. (1975). Death anxiety in the outpatient leukemic child. *Pediatrics, 56,* 1034–1037.

Spinetta, J. J., & Maloney, L. (1978). The child with cancer: Patterns of communication and denial. *Journal of Consulting and Clinical Psychology, 46,* 1540–1541.

Spinetta, J. J., Rigler, D., & Karon, M. (1973). Anxiety in the dying child. *Pediatrics, 52,* 841–845.

Spinetta, J. J., Rigler, D., & Karon, M. (1974). Personal space as a measure of the dying child's sense of isolation. *Journal of Consulting and Clinical Psychology, 42,* 751–756.

Stern, D. N. (1977). *The first relationship: Infant and mother.* Cambridge: Harvard University Press.

Tew, B., & Lawrence, K. M. (1973). Mothers, brothers and sisters of patients with spina bifida. *Developmental Medicine and Child Neurology, 15* (suppl. 29), 69–76.

Torgerson, W. S. (1958). *Theory and methods of scaling.* New York: John Wiley & Sons, Inc.

Young, F. W., & Lewyckyj, R. (1979). *ALSCAL -4 Users Guide* (2nd ed.). Chapel Hill: University of North Carolina, Psychometric Laboratory.

Alternative Perceptual Mapping Techniques: Relative Accuracy and Usefulness

JOHN R. HAUSER
FRANK S. KOPPELMAN

Perceptual mapping has been used extensively in marketing. This powerful technique is used in new product design, advertising, retail location, and many other marketing applications where the manager wants to know (1) the basic cognitive dimensions consumers use to evaluate "products" in the category being investigated and (2) the relative "positions" of present and potential products with respect to those dimensions. For example, Green and Wind (1973) use similarity scaling to

"Alternative Perceptual Mapping Techniques: Relative Accuracy and Usefulness," John R. Hauser and Frank S. Koppelman, Vol. 16 (November 1979), pp. 495–506. Reprinted from the *Journal of Marketing Research,* published by the American Marketing Association.

John R. Hauser is professor of Marketing, Graduate School of Management at Northwestern University.

Frank S. Koppelman is professor of Civil Engineering and Transportation, Technological Institute at Northwestern University.

identify the basic dimensions used in conjoint analysis. Pessemier (1977) applies discriminant analysis to produce the joint-space maps that are used in his DESIGNR model for new product design. Hauser and Urban (1977) use factor analysis to identify consumer perceptions and innovation opportunities in their method for modeling consumer response to innovation. All of these researchers report empirical applications in a number of product and service categories. When used correctly perceptual mapping can identify opportunities, enhance creativity, and direct marketing strategy to the areas of investigation most likely to appeal to consumers.

Perceptual mapping has received much attention in the literature. Though varied in scope and application, this attention has been focused on refinements of the techniques, comparison of alternative ways to use the techniques, or application of the techniques to marketing problems.[1] Few direct comparisons have been made of the three major techniques—similarity scaling, factor analysis, and discriminant analysis. In fact, most of the interest has been in similarity scaling because of the assumption that similarity measures are more accurate measures of perception than direct attribute ratings despite the fact that similarity techniques are more difficult and more expensive to use than factor or discriminant analyses.

In practice, a market researcher has neither the time nor the money to simultaneously apply all three techniques. He/she usually selects one method and uses it to address a particular marketing problem. The market researcher must decide whether the added insight from similarity scaling is worth the added expense in data collection and analysis. Furthermore, if the market researcher selects an attribute-based method such as factor analyses or discriminant analysis he/she wants to know which method is better for perceptual mapping and how such maps compare with those from similarity scaling. To answer these questions, one must compare the alternative mapping techniques.

One way to compare these techniques is theoretically. Each technique has theoretical strengths

[1] See for example Best (1976), Cort and Dominguez (1977), Day, Deutscher, and Ryans (1976), Etgar (1977), Gensch and Golob (1975), Green and Carmone (1969), Green, Carmone, and Wachspress (1977), Green and Rao (1972), Green and Wind (1973), Hauser and Urban (1977), Heeler, Whipple, and Hustad (1977), Jain and Pinson (1976), Johnson (1970, 1971), Pessemier (1977), Singston (1975), Summers and MacKay (1976), Urban (1975), and Whipple (1976).

and weaknesses and the choice of technique depends on how consumers actually react to the alternative measurement tasks. Theoretical hypotheses are established in the next sextion.

Another comparison approach is Monté Carlo simulation. This useful investigative tool has been employed by researchers to explore variations in similarity scaling and other techniques (Carmone, Green and Jain 1978; Cattin and Wittink 1976; Pekelman and Sen 1977). In perceptual mapping Monté Carlo simulation can compare the ability of various techniques to reproduce a hypothesized perceptual map, but it requires that the researcher assume a basic cognitive structure of the individual. Monté Carlo simulation leaves unanswered the empirical question of whether the analytic technique can adequately describe and predict an actual consumer's cognitive structure.

The comparison procedure we use is practical and is based on theoretical arguments and empirical analyses to identify which procedures yield results most useful for marketing research decisions. If the theoretical arguments are supported, researchers can continue to subject a technique to empirical tests in alternative product categories. In this way, one gains insight about the techniques by learning their strengths and weaknesses. If and when the hypotheses are falsified new theories will emerge.

We chose the following guidelines for the comparison.

1. The marketing research environment should be representative of the way the techniques are used empirically.
2. The sample size and data collection should be large enough to avoid exploiting random occurrences and should have no relative bias in favor of the techniques identified as superior.
3. The use of the techniques should parallel as closely as possible the recommended and common usage.
4. The criteria of evaluation should have managerial and research relevance.

To fulfill these criteria, we chose an estimation sample and a saved data sample of 500 consumers each, drawn from residents of Chicago's northern suburbs. The application area is perceptions of the attractiveness of shopping areas in the northern suburbs and the criteria are the ability to predict consumer preference and choice, interpretability of the solutions, and ease of use.

Hypotheses: Theoretical Comparison

We begin with a brief synopsis of the techniques. Because this synopsis is not a complete mathematical description, we have indicated the appropriate references.

Similarity scaling develops measures of consumers' perceptions from consumer judgments with respect to the relative similarity between pairs of products. Though consumers are asked to judge similarity between products, the definition of similarity usually is left unspecified. The statistical techniques select relative values for two, three, or four perceptual dimensions such that distance between products best corresponds to measured similarity. Green and Rao (1972) and Green and Wind (1973) provide mathematical details.

For empirical studies with large sample sizes, common space representations are developed such that each consumer's i perception $\tilde{x}_{ijd}$ of product j along dimension d is a "stretching" of the common representation $\bar{x}_{jd}$. The technique, INDSCAL (Carroll and Chang 1970), simultaneously estimates the common space $\bar{x}_{jd}$ for all i and estimates individual weights v_{id} to "stretch" these dimensions. Effectively, $\tilde{x}_{ijd} = v_{id}\,\bar{x}_{jd}$. For details see Carroll and Chang (1970).

The dimensions are named by judgment or by a regression-like procedure called PROFIT (Carroll and Chang 1964). In PROFIT, consumers are asked to rate each product on specific attributes, e.g., "atmosphere" for shopping centers. The ratings are dependent variables in a regression (possibly monotonic) on the perception measures, $\tilde{x}_{ijd}$, which serve as explanatory variables. The regression weights, called directional cosines, indicate how strongly each perception measure relates to each attribute rating. For details see Carroll and Chang (1964).

Factor analysis begins with the attribute ratings. The assumption is that there are really a few basic perceptual dimensions, x_{ijd}. Many of the attribute ratings are related to each perceptual dimension. Factor analysis examines the correlations among the attributes to identify these basic dimensions. For statistical details see Harmon (1967) and Rummel (1970). Because concern is with the basic structure of perception, the correlations between attribute ratings are computed across products and consumers (sum over i and j). The perceptions of products are measured by "factor scores" which are based on the attribute ratings. The dimensions are named by examining "factor loadings" which are estimates of the correlations between attribute ratings and perception measures. In applications, attribute ratings are first standardized by individual to minimize scale bias.

Discriminant analysis also begins with the attribute ratings, but rather than examining the structure of attribute correlations, discriminant analysis selects the (linear) combinations of attributes that best discriminate between products. See Cooley and Lohnes (1971) and Johnson (1970) for mathematical details. Because concern is with the ability to differentiate products, the dependent measure is "product rated" and the explanatory variables are the attribute ratings. The analysis is run across consumers to find a common structure. The perceptions are measured by "discriminant scores" which are estimates, based on the attribute ratings, of the perceptual dimensions, $\bar{x}_{ijd}$, that best distinguish products. The dimensions are named by examining "discriminant scores" which are the weightings of the attributes that make up a discriminant dimension or by computing correlations that are equivalent to factor loadings. In applications, the discriminant dimensions are constrained to be uncorrelated (also called orthogonal).

Comparison of Similarity Scaling and Attribute-Based Techniques

A major difference between similarity scaling and the attribute-based techniques is the consumer task from which the perceptual measures are derived. Attribute ratings are more direct measures of perceptions than similarity judgments, but may be incomplete if the set of ratings is not carefully developed. Similarity judgments introduce an intermediate construct (similarity) but the judgments are made with respect to the actual product rather than specific attribute scales. *A priori*, if the set of attributes is relatively complete there is no theoretical reason to favor one measure over the other.

Another difference is the treatment of variation among consumers. In the attribute-based techniques a common structure is assumed, but the values of individual measure (x_{ijd} or x_{ijd}') are not restricted. In similarity scaling (INDSCAL) $\bar{x}_{ijd}$ is restricted to be at most a stretching of the common measure, $\bar{x}_{jd}$. This means that although complete

reversals are allowed, no other change in rank order is allowed. For example, INDSCAL will not allow one consumer to evaluate relative sweetness in the order Pepsi, Coke, Royal Crown while another evaluates sweetness as Coke, Royal Crown, Pepsi. This restriction limits the applicability of the similarity scaling.

Finally, similarity scaling is limited by the number of products. At least seven or eight are needed for maps in two or three dimensions (Klahr 1969). There are no such restrictions for factor analysis. The restriction for discriminant analysis is the number of products minus one. This argument favors attribute-based techniques if the number of products in a consumer's evoked set is small; it favors neither technique if the number of products is large. In practice, the evoked set averages about three products (Silk and Urban 1978).

On the basis of these arguments, if the attribute set is reasonably complete, attribute-based techniques should provide better measures of consumer perception than similarity scaling (as implemented by INDSCAL).

Comparison of Factor Analysis and Discriminant Analysis

Factor analysis is based on the correlations across consumers *and* products. Discriminant analysis is limited to dimensions that, on average, distinguish among products. Thus factor analysis should use more attributes than discriminant analysis in the dimensions and therefore produce richer solutions. For example, consider Mercedes Benz and Rolls Royce. Suppose that the true perceptual dimensions are country of origin and reliability and that only reliability affects preference and choice. Suppose that perceptions of country of origin differ among products. Suppose that average perceptions of reliability are the same for both cars but individual perceptions differ among consumers. Discriminant analysis will identify only country of origin. Factor analysis will identify both dimensions.

On the basis of this type of argument, one expects factor analysis to provide a richer perceptual structure than discriminant analysis. It should be able to use more of the attribute ratings and should identify perceptual dimensions that predict preference and choice better than discriminant analysis dimensions.

Empirical Setting: Shopping Locations

The empirical setting is north suburban Chicago and the "product" category is overall image of shopping locations, an increasingly important area of investigation in marketing. From the perspective of retailers, shopping center managers, and community planners, the sensitivity of destination choice behavior to the image or attractiveness of the shopping location provides an important opportunity to develop strategies to attract shoppers.

In terms of an empirical comparison of perceptual techniques, shopping location choice provides a strong test of the model's explanatory and predictive capabilities. Shopping location choice is a complex phenomenon, difficult to model and difficult to understand. If a perceptual technique does well in this category, it is likely to do well in a less complex category. (The complexity of the category should introduce no *relative* bias in the model comparisons.)

We develop models based on seven shopping areas including downtown Chicago and six suburban shopping centers of very different characteristics. The locations represent the types of shopping locations available to the residents of the suburbs north of Chicago including large, medium, and small shopping areas with exclusive, general merchandise, or discount orientations.

The data were collected by a self-administered questionnaire. (For details see Stopher, Watson, and Blin 1974.) The attributes were measured by 16 five-point rating scales chosen in an attempt to get as complete a list as possible without causing consumer wearout. (They were selected and refined on the basis of literature reviews, preliminary surveys, and qualitative research.) The similarity judgments were measured by pairwise comparisons on a seven-point scale. The respondent judged all pairs of the products in his/her evoked set (measured by a knowledge question). Table 1 contains example instructions for the attribute ratings and similarity judgments.

TABLE 1 Example Questions for Attribute Scales and Similarity Judgments

In this question, we would like you to rate each of the shopping centers on these characteristics. We have provided a range from good to poor for each characteristic. We would like you to tell us where *you* feel each shopping center fits on this range.

For example:

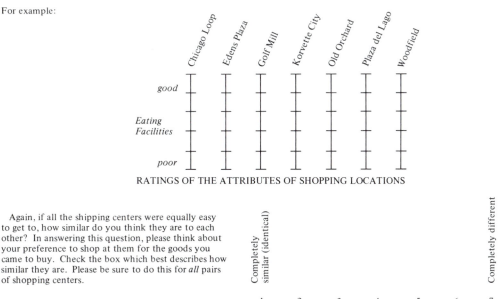

RATINGS OF THE ATTRIBUTES OF SHOPPING LOCATIONS

Again, if all the shipping centers were equally easy to get to, how similar do you think they are to each other? In answering this question, please think about your preference to shop at them for the goods you came to buy. Check the box which best describes how similar they are. Please be sure to do this for *all* pairs of shopping centers.

SIMILARITY JUDGMENTS

The dependent measures for predictive testing were obtained in the same survey. Choice was self-reported frequency of visits. Preference, also called "attractiveness," was rank-order preference in which availability/accessibility was held constant. (Pretests indicated consumers were comfortable with this task.) Availability/accessibility was measured as map distance from the consumer's residence to the shopping location. The model (explained in the next section) was drawn from the retailing and transportation literature and is given as follows.

$$\text{Perceptions} \longrightarrow \text{Preference} \longrightarrow \text{Frequency of Visits}$$

$$\text{Availability/Accessibility}$$

This two-stage model allows managers and researchers to measure the "attractiveness" of a shopping center independently of its location. Thus one can readily investigate strategies, such as improving the atmosphere of a shopping center, that may be more cost effective than relocation strategies. This ability is especially useful when relocation is not an option as is the case in many downtown shopping areas.

In the original data set, 37,500 mailback questionnaires were distributed at four of the shopping locations. Of these, 6,000 were returned as complete and usable questionnaires. Although this low return rate may cause some nonresponse bias, the bias should not affect relative comparisons. Because similarity scaling requires at least seven stimuli, the data were screened to select those consumers who indicated knowledge of and who rated all seven stimuli (1,600 consumers). Any bias introduced by this screening should favor similarity scaling and thus favor falsifying our hypotheses. Finally, 500 of these consumers were randomly selected for an estimation sample and 500 others for saved data testing.

One source of bias is that samples were taken at four locations but judgments were made for seven locations. This sampling technique should produce no bias in the relative predictive ability of the perceptual techniques but could bias the coefficients in the preference and choice models. Fortunately, this bias can be corrected with choice-based sampling (CBS) estimation procedures. Manski and Lerman (1977) give theoretical proofs and Lerman and Manski (1976) provide intuitive arguments. To test CBS empirically for the data, parallel analyses were performed using

only the four sampled shopping locations. Consistent with the theory, all models were statistically equivalent. Koppelman and Hauser (1977) give details of these comparisons.

Empirical Procedure

So that other researchers can replicate the tests, we describe in this section how the measures were developed. In selecting procedures an attempt was made to follow common and recommended usage as closely as possible. When potential variations occurred (rotation vs. nonrotation, direct vs. indirect similarities, three vs. four dimensions) sensitivity tests were performed.

Similarity Scaling

Pairwise similarities were changed to rank order similarities and input to INDSCAL. Three dimensions were selected on the basis of an elbow in stress values. Four dimensions were not possible with the limited number of products.

Factor Analysis

The attributes were standardized by individual. Correlations were computed across consumers and products. Four dimensions were selected by using elbow and interpretability criteria on the common factor solution with varimax rotation. Limited testing was done with three dimensions for comparability to similarity scaling.

Discriminant Analysis

The attributes were standardized by individual. A discriminant analysis was run with the product rated as the dependent variable and the 16 attributes as explanatory variables. The elbow rule and significance statistics suggested at least three dimensions (97.5% of trace), possibly four because the chi square test was still significant. Most analyses were done with four dimensions although limited analyses were done with three dimensions. Varimax rotation improved the interpretations slightly. Dimensions are constrained to be orthogonal.

Preference

The linear compensatory form was chosen because of its widespread use in marketing (Wilkie and Pessemier 1973), because Monté Carlo simulation has shown it reasonably representative of more complex forms such as disjunctive, conjunctive, and additive (Carmone, Green, and Jain 1978), and because many of the nonlinear models such as conjoint analysis (Green and Srinivasan 1978, Green and Wind 1975) require (more intensive) personal interviews rather than the mailback format used by Stopher, Watson, and Blin (1974).

Mathematically, this model is given by:

$$p_{ij} \sim \sum_{l=1}^{m} w_l d_{ijl}$$

where p_{ij} is individual i's rank for product j, $\sim$ indicates monotonic, d_{ijl} is i's perception of j along the lth dimension, and w_l is the importance weight of the lth dimension. (For similarity scaling $d_{ijl} \equiv \bar{x}_{ijd}$, for factor analysis $d_{ijl} \equiv x_{ijd}$, for discriminant analysis $d_{ijl} \equiv x'_{ijd}$.)

Because there is a possibility that any single preference estimation procedure for the w_l's will favor one or another perceptual model, two estimation procedures were simultaneously tested: preference regression and first preference logit. Preference regression is a metric technique which replaces monotonicity ($\sim$) by equality ($=$) and uses ordinary least squares with equation 1. First preference logit is a monotonic technique based on estimating the probability that a consumer will rank a product as first preference. In logit, maximum likelihood techniques are used to determine w_l. For a more complete description, see McFadden (1970). In both models, choice-based sampling variables are used to provide consistent estimates of w_l (Manski and Lerman 1977).

Choice

The multinomial logit model is used to predict choice. (The logit model is based on a probabilistic interpretation of choice. For estimation and prediction equations, see McFadden 1970.) The dependent variable is frequency of visits and the explanatory variables are distance and estimated preference, $\hat{p}_{ij}$, where $\hat{p}_{ij}$ is generated by equation 1 and the estimates, $\hat{w}_l$, of the importance weights. That is, $\hat{p}_{ij} = \Sigma_l \hat{w}_l d_{ijl}$.

Finally, we chose to test the assumption of a two-stage model by using a revealed preference logit model. The dependent variable was frequency and the explanatory variables were distance and the perceptual measures, d_{ijl}. Choice-based sampling variables were included for consistent estimates.

Saved Data Predictions

One begins with the standardized attribute ratings and similarity judgments from the saved data sample. Factor score coefficients from the estimation sample and the attribute ratings from the saved data are used to create factor scores, x_{ijd} (see Rummel 1970). Discriminant score coefficients from the estimation sample and the attribute ratings from the saved data are used to create discriminant scores, x_{ijd}' (see Cooley and Lohnes 1971). INDSCAL is run on the saved data similarity judgments to create similarity scores, $\bar{x}_{ijd}$. If anything, this procedure should favor similarity scaling. Indeed, an alternative procedure of generating the similarity scores from the attribute ratings provided poorer predictive results for similarity scaling and thus is even stronger support for the hypotheses.

For each combination of models (perception, preference, choice) the estimation importance weights, $\hat{w}_l$, from the estimation sample are applied to the d_{ijl}'s to create $\hat{p}_{ij}$. These are rank ordered and compared with the actual p_{ij}. The estimated relative weights of $\hat{p}_{ij}$ and distance then are used (with the relevant logit equation) to estimate the relative frequency of visits. These predictions are compared with reported frequency of visits.

Predictive Tests

Preference prediction was measured by the percentage of consumers who ranked first the shopping center that the model predicted as first. This measure is straightforward and is commonly reported in the literature.

Choice prediction is more difficult to measure. First, the models predict the probability of choice which must be compared with frequency of choice. This problem is resolved by using percentage uncertainty which is an information theoretic statistic that measures the percentage of uncertainty (entropy) explained by the probabilistic model (see Hauser 1978; McFadden 1970; Silk and Urban 1978). The second problem arises because both accessibility and the perceptual dimensions (through preference) are used to predict choice. Only the incremental gains in uncertainty due to the perceptual measures are of interest. Because the information measure is additive, we use the gain in percentage uncertainty achieved by adding the perceptual dimensions to a model based on distance alone.

Other measures including rank order preference recovery, percentage of consumers choosing the maximum probability shopping location, and mean absolute error in market share prediction are not reported here because each of these measures ranked the perceptual model *in the same relative order* as do the reported measures. For these statistics see Koppelman and Hauser (1977).

Empirical Results

Predictive Tests

The result of the predictive tests are reported in Table 2. Preference recovery was not computed

for revealed preference logit which is estimated directly on choice. Table 2 also reports preference and choice prediction for models using distance and the 16 standardized attribute scales as explanatory variables. These statistics are computed to examine whether the perceptual dimensions provide more or less information than a full set of attribute ratings.

To better understand Table 2, it is useful to consider some base level values. The maximum value for preference recovery is 100%, but most empirical models do not obtain values even near 100%. Rather these measures should be compared with that attainable by purely random assignment, i.e., all locations equally likely to be chosen, and that attainable by assigning consumers to shopping centers in proportion to market share. These values are 14.3% preference recovery for the equally likely model and 26.7% preference recovery for the market share model. Rigorous statistical tests are not applicable in comparing one perceptual technique with another because the preference models are based on different explanatory variables. But, intuitively, differences in preference recovery can be compared via the maximum standard deviation, 2.2, which would result under

TABLE 2   Predictive Tests[a]

	Goodness of Fit Tests		Saved Data Tests	
	Preference Recovery	% Uncertainty Explained (Choice)	Preference Recovery	% Uncertainty Explained (Choice)
Similarity scaling				
Preference regression	36.6	0.9	23.1	−32.4
1st preference logit	34.8	1.1	22.7	−41.3
Revealed preference logit	—	1.5	—	−69.5
Factor analysis				
Preference regression	50.6	3.8	47.3	4.3
1st preference logit	55.0	4.7	50.8	5.3
Revealed preference logit	—	5.9	—	6.0
Discriminant analysis				
Preference regression	35.5	1.7	38.1	1.9
1st preference logit	35.3	1.9	40.3	2.0
Revealed preference logit	—	2.9	—	3.0
Attribute scales				
Preference regression	39.6	3.2	41.4	−3.9
1st preference logit	55.6	5.2	51.6	6.4
Revealed preference logit	—	6.6	—	6.9

[a] The preference test is ability to recover first preference, the choice test is percentage of uncertainty explained relative to a model with distance alone. To get total uncertainty, add 32.6 to these measures. All uncertainty measures are significant at the .01 level.

the assumption that for a given model each observation (consumer) has an equal probability of being correctly classified.

The significance of the choice models can be tested because the uncertainty explained is proportional to a chi square statistic (Hauser 1978). All models are significant at the .01 level.[2] The chi square's cutoff, which is less than 0.4, can also be taken as an intuitive measure in comparing the alternative perceptual techniques.

First examine the predictive tests. Factor analysis is superior to similarity scaling and discriminant analysis for all preference models and both prediction measures. This evidence supports the hypotheses. These results are particularly significant because any incompleteness in the attribute scales would favor similarity scaling over discriminant analysis and factor analysis. Furthermore, the sample was screened to favor similarity scaling.

The fact that factor analysis does well in comparison with models based on attribute scales indicates that very little information is lost by using the reduced factors rather than the full set of attributes. The poor showing of preference regression on the attribute scales is the result of multicollinearity. The poor showing of discriminant analysis in relation to the attribute scales supports the hypothesis that concentrating on variations between products neglects dimensions that are important in preference and choice.

Next, examine the saved data test. Factor analysis is still superior to both discriminant analysis and similarity scaling. Both of the attribute-based measures hold up well. The small improvement in the statistics may be attributed to random variation between data sets. Similarity scaling predicts poorly on saved data, doing worse than the naïve model which assigns consumers proportional to market share (26.7% preference recovery). The model does so poorly in relation to distance alone that it adds large amounts of uncertainty (negative uncertainty explained) in choice prediction.

Thus the saved data tests support the hypotheses.

Table 2 is one empirical comparison. Replications, sensitivity, and threats to validity are exam-

[2] After frequency adjustment, $2N$ * relative percentage uncertainty is a chi square statistic if the null model, i.e., distance alone, is a special case of the tested model, i.e., perceptions and distance. The degrees of freedom equals the difference in the number of variables. For the three similarity dimensions the cutoff is .031. For the other models which contain four dimensions, the cutoff is .037.

ined in the next section, but first a more qualitative analysis is used to determine whether these comparisons are consistent with managerial interpretations of the techniques.

Interpretability

First examine how the perceptual dimensions from each technique relate to the attribute scales. These relationships are reported in Table 3. The underlined numbers (directional cosines, factor loadings, or discriminant coefficients) indicate a strong relationship between the underlined attribute (row) and the dimension (column). The dimension names are composites of the underlined attributes.

Despite marked superficial similarities, the different models demonstrate striking differences in interpretation. Factor analysis uses all the attribute scales whereas discriminant analysis uses only 10 of the 16 scales, and only five of those 10 have discriminant scores greater than 0.5. This outcome is consistent with the theoretical hypothesis that concentrating on variance between products neglects information. Note also that the third discriminant dimension, "value vs. satisfaction," is a mixed dimension positively related to value but negatively related to satisfaction. Compare this to the factor analysis solution which has no positive/negative mixed dimension. This difference probably arises because value and satisfaction may be negatively correlated among existing shopping locations (discriminant analysis) but not in the way consumers rate these locations (factor analysis).

These differences are important to the manager who wants as much control over the market as possible. For example, many of the variables left out of the discriminant solution (e.g., "layout of the store," "return and service," "sales assistants") could be affected by relatively low-cost changes in shopping center operation. Because discriminant analysis uses fewer variables the manager can analyze fewer strategies. Also, because of the mixed dimension it is difficult to evaluate strategies that increase both satisfaction and value.

Comparing factor analysis with similarity scaling, one again sees a mixed dimension, "quality vs. value," but the similarity scaling solution uses all 16 attributes. This mixed dimension probably results from the INDSCAL assumption which con-

TABLE 3 Structural Comparison of Perceptual Models

	Factor Analysis—Factor Loadings				Discriminant Analysis—Discriminant Coefficients			
Fundamental Attributes	Variety	Quality and Satisfaction	Value	Parking	Fundamental Attributes	Variety	Quality and Satisfaction	Value vs. Satisfaction	Parking
1. Layout of store	.267	.583	.156	.200	1. Layout of store	.067	−.110	−.085	−.023
2. Return and service	.095	.528	.343	.255	2. Return and service	−.094	.254	.287	.134
3. Prestige of store	.338	.822	−.001	−.058	3. Prestige of store	.156	.366	−.318	−.137
4. Variety of merchandise	.665	.327	.309	−.185	4. Variety of merchandise	.335	.153	.295	−.042
5. Quality of merchandise	.307	.810	.037	−.074	5. Quality of merchandise	−.196	1.114	.020	−.216
6. Availability of credit	.159	.337	.487	.049	6. Availability of credit	−.008	.170	.352	.025
7. Reasonable price	.067	−.063	.599	.113	7. Reasonable price	−.108	−.008	.586	−.009
8. "Specials"	.223	.074	.739	.008	8. "Specials"	−.101	−.171	.225	−.115
9. Free parking	−.15	.068	.043	.811	9. Free parking	.065	−.066	−.007	1.714
10. Center layout	.03	.308	.074	.560	10. Center layout	−.065	−.233	−.334	.082
11. Store atmosphere	.080	.658	.034	.400	11. Store atmosphere	−.158	.087	−.055	−.053
12. Parking available	.145	.105	.108	.841	12. Parking available	−.288	−.020	.018	.171
13. Center atmosphere	.244	.694	.040	.404	13. Center atmopshere	.109	−.123	−.510	.284
14. Sales assistants	.173	.561	.147	.319	14. Sales assistants	−.138	.015	−.168	−.084
15. Store availability	.619	.320	.204	.034	15. Store availability	.089	.035	.084	−.065
16. Variety of stores	.829	.288	.160	−.173	16. Variety of stores	1.291	−.123	−.022	.145

Similarity Scaling-Directional Cosines

Fundamental Attributes	Variety and Value	Quality vs. Value	Parking and Satisfaction
1. Layout of store	.217	.497	.840
2. Return and service	.318	.122	.940
3. Prestige of store	.297	.804	.515
4. Variety of merchandise	.929	.360	.084
5. Quality of merchandise	.295	.811	−.505
6. Availability of credit	.880	−.085	−.468
7. Reasonable price	.485	−.853	.192
8. "Specials"	.786	−.594	.173
9. Free parking	−.294	−.550	.782
10. Center layout	−.447	.036	.894
11. Store atmosphere	−.199	.452	.869
12. Parking available	−.463	−.478	.747
13. Center atmosphere	−.099	.480	.872
14. Sales assistants	−.052	.411	.910
15. Store availability	.872	.429	−.236
16. Variety of stores	.921	.385	−.054

FIGURE 1   Comparison on perceptual maps.

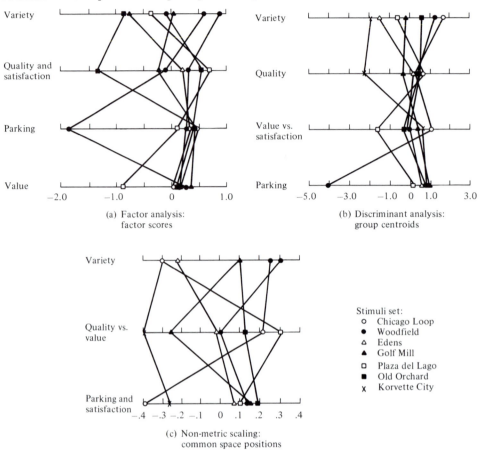

(a) Factor analysis:
factor scores

(b) Discriminant analysis:
group centroids

(c) Non-metric scaling:
common space positions

Stimuli set:
- o   Chicago Loop
- ●   Woodfield
- △   Edens
- ▲   Golf Mill
- □   Plaza del Lago
- ■   Old Orchard
- x   Korvette City

centrates on differences between products at the expense of variations in consumer perceptions. But in this case the "quality vs. value" is easier to understand and operate on than "satisfaction vs. value."

Thus, of the three models, factor analysis has the most managerially useful structure. Because all models have strong face validity and all use the basic five constructs—variety, quality, satisfaction, value, and parking—we believe that the predictive and saved data superiority of factor analysis is due to better structure in identifying dimensions.

The second test of interpretability is the visual maps produced by each method (see Figure 1). These maps help managers identify how each shopping location is "positioned" in the marketplace. Thus a manager can know the relative strengths and weaknesses of each shopping location and can identify opportunities in the market.

Careful inspection of Figure 1 shows consistency among the models when the measured constructs are the same. For example, note the low scores for Korvette City on quality (satisfaction) and for Chicago Loop on parking or the high scores for Woodfield on variety and for Plaza del Lago on quality (satisfaction). These and many other "positionings" have strong face validity and are consistent with prior beliefs about the images of the seven shopping locations. But there is an important difference. Factor analysis is better for strategy development because it separates the dimensions in such a way that ambiguous interpretations ("quality vs. value" and "satisfaction vs. value") are avoided.

Ease of Use

A final consideration is the investment in analysis time required by the techniques. Managers and

TABLE 4 Summary of the Results of Empirical Comparison[a]

Criteria			Results		
Theory	Factor analysis	>	Discriminant analysis	≥	Similarity scaling
Goodness of fit	Factor analysis	>	Discriminant analysis	~	Similarity scaling
Prediction					
(saved data)	Factor analysis	>	Discriminant analysis	>	Similarity scaling
Interpretation	Factor analysis	≥	Discriminant analysis	~	Similarity scaling
Ease of use	Factor analysis	~	Discriminant analysis	>	Similarity scaling

[a] > indicates superior, ≥ indicates probably superior, ~ indicates no major difference.

researchers may be willing to accept more approximate models if the approximations allow great cost savings. Our experience indicates that the attribute-based techniques are relatively easy to use whereas similarity scaling is somewhat more difficult with resepct to both data collection and data analysis. Both factor and discriminant analyses use only the attribute ratings whereas similarity scaling requires, in addition, judged similarities which can be a difficult consumer task. Once the data have been collected, factor and discriminant analyses are readily available on standard statistical packages (e.g., BMDP, SPSS), cost[3] about $10–20 to run, and require little professional time. In sharp contrast, the special programs for similarity scaling require many exploratory runs, special FORTRAN programs for data transfer, and a series of statistical manipulations and data handling to develop a common space, estimate individual weights, and compute directional cosines. A single set of runs costs about $40 in computer time, but because various starting configurations and dimensions must be checked, the effective cost is about $150. Though the added direct computer costs are acceptable, the programming and analysis require significant professional time and could be very costly to organizations not familiar with the programs.

Summary

In this section representative attribute-based and similarity scaling techniques are compared to determine which perceptual mapping techniques is most appropriate for marketing research applications. The empirical results, which support the hypotheses, suggest that the most popular tech-

nique, similarity scaling, may not be the best; rather, factor analysis may be better on predictability, interpretability, and ease of use (see Table 4). These results are subject to confirmation or qualification in other empirical applications but, at the very least, this empirical test raised issues worth further investigation by other marketing researchers.

Threats to Validity, Sensitivity, and Replications

Empirical comparisons are difficult. Only by using the state of the art in each technique can one be fair to that technique. To attempt to extrapolate these comparisons, one must examine causes that threaten to make the results specific to the empirical sample.

Threats to Validity

Nonresponsive bias, choice-based sampling, representativeness of the set of attributes, and screening on the seven stimuli in the evoked set have been discussed. They should introduce no relative bias favoring the hypotheses.

One threat to the hypotheses is potential halo effects in the attribute scales, i.e., the attribute scales may contain an affective as well as a cognitive component (Beckwith and Lehmann 1975). This threat is minimized by standardized of the attribute scales prior to analysis and by using attribute scales designed to measure only the cognitive component. Another threat is that the stimuli set contained at most seven products. Although this constraint is typical of real-world applications, it does suggest future comparisons with larger stimuli sets.

---

[3] Costs vary by computer. These costs represent $510 per cpu hour on a CDC 6400. See Dixon (1975) for BMDP and Nie et al. (1975) for SPSS.

Sensitivity

The comparisons among techniques are found to remain consistent when limited changes are made in each technique. Rotation of discriminant solutions improved interpretability but not predictive ability (the rotated solution is reported). Use of indirect similarity measures did not improve either interpretability or predictive ability. The factor and discriminant analyses used four dimensions whereas similarity scaling was limited to three dimensions. Though this is a minor change in the degrees of freedom in the preference and choice models (490 vs. 489), it could have an effect. Limited testing showed that dropping the least significant dimension in the four-dimensional solutions caused very little shrinkage in prediction. (For example, with preference regression and discriminant analysis the shrinkage was about one-tenth of one percent in preference recovery.) Finally, use of alternative factor analysis solutions such as principal components or three-way analyses might improve prediction, but such improvement would only add support to the hypotheses.

Replications

Since the original study, the hypotheses have been tested on two other data sets. Using a sample of 120 graduate students, Simmie (1978 found preference recoveries of 67.4% for factor analysis, 51.9% for discriminant analysis, and 14.0% for similarity scaling. Simmie also found greater consistency across groups of students with factor analysis than with similarity scaling. Her product category was management schools. Using a random sample of Evanston residents. Englund, Hundt, and Lee (1978) found preference recoveries of 67.1% with factor analysis and 48.6% with discriminant analysis. Their product category was transportation mode choice (bus, walk, or car). The implications of both studies are consistent with our results.

Implications and Future Research

Perceptual mapping is an important marketing research tool used in new product planning, advertising development, product positioning, and many other areas of marketing. Strategies based on perceptual maps have led to increased profits, better market control, and more stable growth. Furthermore, much research is based on implications of market structure as identified by perceptual maps. Because of this interest and use, it is crucial that the best mapping technique available be employed in these applications.

We provide and support hypotheses that factor analysis is superior to both similarity scaling and discriminant analysis for developing measures of consumer perceptions. In particular, factor analysis is likely to be superior in categories where:

1. The number of products in the average consumer's evoked set is relatively small (seven stimuli or less).
2. There is variation in the way consumers perceive products in the category.
3. Qualitative research has identified a set of attributes likely to represent the product category.

The presence or absence of these characteristics does not ensure the superiority of one technique, but without evidence to the contrary they can serve as guidelines.

The results of this one theoretical and empirical comparison are hoped to raise the issue of and the need for continued research to identify whether factor analysis is always superior or, if not, under what conditions the alternative mapping techniques should be used.

Confirmation of these comparisons awaits replication in other categories. In addition, further exploration may be appropriate for other preference models. For example, nonlinear models such as conjoint analysis should theoretically order the perceptual models in the same way, but empirical tests are warranted. We have used disaggregate preference models, i.e., models based on individual consumers (p_{ij} rather than p_j, d_{ijk} rather than d_{jk} where i indexes consumers). In most applications these disaggregate models have proven superior to aggregate models which blur individual differences. Because aggregate models still are used in some marketing applications, these might be tested. For example, PREFMAP, which is a form of preference regression, can be used at both the aggregate (Pessemier 1977) and disaggregate level (Beckwith and Lehmann 1975). We hypothesize that because the superiority of factor analysis over discriminant analysis is based on individual differences, predictive comparisons might shift and discriminant analysis might do well as factor analysis when analyses are limited to the aggregate level.

Other research might be of a more proactive nature, searching for improvements for the weaknesses of each technique. Theoretical developments might make it practical to relax the INDSCAL assumption. Further research could expand the similarity scaling solutions to more stimuli via use of concept statements. Segmentation on perceptions could be used prior to similarity scaling or discriminant analysis to ensure that there is little variation among consumers. These and other methodological developments are suggested by the examination and comparison of the alternative perceptual mapping techniques.

This area of comparative model development is important to marketing and deserves attention from marketing researchers.

References

Beckwith, N. E. and D. R. Lehmann (1975), "The Importance of Halo Effects in Multi-Attribute Attitude Models," *Journal of Marketing Research*, 12 (August), 265–75.

Best, R. J. (1976), "The Predictive Aspects of a Joint Space Theory of Stochastic Choice," *Journal of Marketing Research*, 13 (May), 198–204.

Carmone, F. J., P. E. Green, and A. K. Jain (1978), "The Robustness of Conjoint Analysis: Some Monté Carlo Results," *Journal of Marketing Research*, 15 (May), 300–3.

Carroll, J. D. and J. J. Chang (1964), "A General Index of Nonlinear Correlation and Its Application to the Interpretation of Multidimensional Scaling Solutions," *American Psychologist*, 19.

_____ and _____ (1970), "Analysis of Individual Differences in Multidimensional Scaling via an N-way Generalization of the Eckart-Young Decomposition," *Psychometrika*, 35, 283–319.

Cattin, P. and D. R. Wittink (1976), "A Monté Carlo Study of Metric and Nonmetric Estimation Methods for Multiattribute Models," Research Paper No. 341, Graduate School of Business, Stanford University (November).

Cooley, W. W. and P. R. Lohnes (1971), *Multivariate Data Analysis*. New York: John Wiley & Sons, Inc.

Cort, S. G. and L. V. Dominguez (1977), "Cross Shopping and Retail Growth," *Journal of Marketing Research*, 14 (May), 187–92.

Day, G. S., T. Deutscher, and A. B. Ryans (1976), "Data Quality, Level of Aggregation, and Nonmetric Multidimensional Scaling Solutions," *Journal of Marketing Research*, 13 (February), 92–7.

Dixon, W. J. (1975), *BMDP: Biomedical Computer Programs*. Los Angeles: University of California Press.

Englund, D., F. Hundt, and Y, Lee (1978), "An Empirical Comparison of Factor Analysis and Discriminant Analysis for Non-Work Trips in Evanston," Technical Report, Transportation Center, Northwestern University (August).

Etgar, M. (1977), "Channel Environment and Channel Leadership," *Journal of Marketing Research*, 14 (February), 69–76.

Gensch, P. H. and T. F. Golog (1975), "Testing the Consistency of Attribute Meaning in Empirical Concept Testing," *Journal of Marketing Research*, 12 (August), 348–54.

Green, P. E. (1975), "On the Robustness of Multidimensional Scaling Techniques," *Journal of Marketing Research*, 12 (February), 73–81.

_____ and F. J. Carmone (1969), "Multidimensional Scaling: An Introduction and Comparison of Nonmetric Unfolding Techniques," *Journal of Marketing Research*, 7 (August), 330–41.

_____, _____, and D. D. Wachspress (1977), "On the Analysis of Qualitative Data in Marketing Research," *Journal of Marketing Research*, 14 (February), 52–9.

_____ and V. Rao (1972), *Applied Multidimensional Scaling*. New York: Holt, Rinehart and Winston, Inc.

_____ and V. Srinivasan (1978), "Conjoint Analysis in Consumer Behavior; Status and Outlook," *Journal of Consumer Research*, 5 (September), 103–23.

_____ and Yoram Wind (1973), *Multiattribute Decisions in Marketing*. Hinsdale, Illinois: The Dryden Press.

_____ and _____ (1975), "New Way to Measure Consumer's Judgments," *Harvard Business Review* (July–August).

Harmon, H. H. (1967), *Modern Factor Analysis*. Chicago: University of Chicago.

Hauser, J. R. (1978), "Testing the Accuracy, Usefulness, and Significance of Probabilistic Choice Models: An Information Theoretic Approach," *Operations Research*, 26 (May–June), 406–421.

_____ and G. L. Urban (1977), "A Normative Methodology for Modeling Consumer Response to Innovation," *Operations Research*, 25 (July–August), 579–619.

Heeler, R. M., T. W. Whipple, and T. P. Hustad (1977), "Maximum Likelihood Factor Analysis of Attitude Data," *Journal of Marketing Research*, 14 (February), 42–51.

Jain, A. K. and C. Pinson (1976), "The Effect of Order of Presentation of Similarity Judgments on Multidimensional Scaling Results: An Empirical Comparison," *Journal of Marketing Research*, 13 (November), 435–9.

Johnson, R. M. (1971), "Market Segmentation: A Strategic Management Tool," *Journal of Marketing Research*, 8 (February), 13–18.

_____ (1970), "Multiple Discriminant Analysis Applications to Marketing Research," Market Facts, Inc. (January).

Jolson, Marvin A. and Walter F. Spath (1973), "Understanding and Fulfilling Shoppers' Requirements," *Journal of Retailing*, 49 (Summer).

Klahr, David (1969), "A Monté Carlo Investigation of the Statistical Significance of Kruskal's Non-metric Scaling Procedure," *Psychometrika*, 34 (September), 319–30.

Koppelman, Frank S. and John Hauser (1977), "Consumer Travel Choice Behavior: An Empirical Analysis of Destination Choice for Non-Grocery Shopping

Trips," Technical Report, Transportation Center, Northwestern University (March).

Lerman, S. R. and C. F. Manski (1976), "Alternative Sampling Procedure for Disaggregate Choice Model Estimation," *Transportation Research Record*, 592, Transportation Research Board.

MacKay, David B. and Richard W. Olshavsky (1975), "Cognitive Maps of Retail Locations; An Investigation of Some Basic Issues," *Journal of Consumer Research*, 2 (December).

Manski, C. F. and S. R. Lerman (1977), "The Estimation of Choice Probabilities from Choice Based Samples," *Econometrica*, 45 (November).

McFadden, Daniel (1970), "Conditional Logit Analysis of Qualitative Choice Behavior," in *Frontiers in Econometrics*. P. Zaremblea, ed. New York: Academic Press, 105–42.

Nie, N. H., G. H. Hull, J. G. Jenkins, K. Steinbrenner, and D. H. Bent (1975), *SPSS: Statistical Package for the Social Sciences*, 2nd edition. New York: McGraw-Hill Book Company.

Pekelman, D. and S. Sen (1977), "Improving Prediction in Conjoint Measurement," Working Paper, Graduate School of Management, University of Rochester (January).

Pessemier, E. A. (1977), *Product Management: Strategy and Organization*. New York: Wiley/Hamilton.

Rummel, R. J. (1970), *Applied Factor Analysis*. Evanston, Illinois: Northwestern University Press.

Silk, A. J. and G. L. Urban (1978). "Pretest Market Evaluation of New Packaged Goods: A Model and Measurement Methodology," *Journal of Marketing Research*, 15 (May), 171–91.

Simmie, Patricia (1978), "Alternative Perceptual Models: Reproducibility, Validity, and Data Integrity," *Proceedings of the American Marketing Association Educators Conference*, Chicago (August).

Singston, Ricardo (1975), "Multidimensional Scaling Analysis of Store Image and Shopping Behavior," *Journal of Retailing*, 51, 2.

Stopher, P. R., P. L. Watson, and J. J. Blin (1974), "A Method for Assessing Pricing and Structural Changes on Transport Mode Use," Interim Report to the Office of University Research, U.S. Department of Transportation, Transportation Center, Northwestern University (September).

Summers, J. O. and D. B. MacKay, (1976), "On the Validity and Reliability of Direct Similarity Judgments," *Journal of Marketing Research*, 13 (August), 289–95.

Urban, G. L. (1975), "PERCEPTOR: A Model for Product Positioning," *Management Science*, 8 (April), 858–71.

Whipple. T. W. (1976), "Variation Among Multidimensional Scaling Solutions: An Examination of the Effect of Data Collection Differences," *Journal of Marketing Research*, 13 (February), 98–103.

Wilkie, W. L. and E. A. Pessemier (1973), "Issues in Marketing's Use of Multi-Attribute Attitude Models," *Journal of Marketing Research*, 10 (November), 428–41.

CHAPTER NINE

Conjoint Analysis

CHAPTER PREVIEW

Conjoint analysis is rooted in traditional experimentation. For example, a chemist in a bar soap manufacturing plant may want to know the effect of the temperature and pressure in the soap-making vats on the density of the resulting bar of soap. The chemist could conduct a laboratory experiment to measure these relationships. In situations involving human behavior, we often need to measure the effect on the choice of variables that we control (e.g., should the bar soap be slightly or highly fragrant? should it be promoted as a cosmetic or a cleaner/deodorizer?). Conjoint analysis developed from a need to analyze the effect of predictor variables that are often qualitatively specified or weakly measured. This approach to experimental design and analysis has two objectives:

1. To determine the contributions of the predictor variables and their respective levels to the desirability of the combinations of variables. For example, how much does fragrance contribute to the willingness to buy the soap? Which fragrance level is best? How much change in willingness to buy the soap can be accounted for by differences between levels of fragrance?
2. To examine the validity of the model selected for use by the respondent in making judgments. Do the respondent's choices indicate a simple linear relationship between the predictor variables and the choices? Does a model without interaction terms fit the choice process?

This chapter should enable you to

- Explain the many managerial uses of conjoint analysis.
- Formulate an experimental plan for simple conjoint analysis.
- Examine the validity of a main effects model in comparison to one involving interaction terms in a small experiment.
- Understand the difference between the application of a conjoint approach based on rank choice and approaches using other measures of the outcome variable.

KEY TERMS   **Composition rule**   The rule used in combining attributes to produce a judgement of relative value or utility. For illustration, a person sees four objects and is asked to evaluate them. The person is assumed to evaluate the attributes of the objects and create some overall relative value for each of them. This may be as simple as creating a mental weight for each perceived attribute and adding the weights to produce the overall weight, or it may be a more complex procedure.

Interaction effects   In perceiving value, a person may attribute unique value to combinations of features. For example, the person may attribute a certain perceptual value to the color red when judging mouthwash. A red mouthwash connotes health for this person. When considering flavor, the person places a high value on a mint flavor because of its refreshing quality. However, a red mouthwash with a mint flavor may receive a very poor evaluation because the combination red color/mint flavor is confusing and does not allow the respondent a clear perception of the two features that would normally have high appeal. Mint is expected to be a shade of green. Red color is associated with cherry or strawberry, not mint.

Main effects   The effect of each treatment variable on the dependent variables.

Treatment   The variable that the researcher manipulates to measure the effect (if any) on another variable(s). By definition, the treatment must have levels, such as the following:

1. Color: red versus blue
2. Sweetner: 1, 2, or 3 grams of sugar
3. Size: 10, 12, or 14 oz.
4. Fabric softener additive: present or absent

In conjoint analysis the predictors can be metric, nonmetric, or both.

What Is Conjoint Analysis?   A definition of *conjoint analysis* must proceed from its underlying assumption that a composition rule may be established to predict a response variable from two or more predictor variables. In behavioral research, predictor variables are often not measured in an absolute fashion, nor is the response. Conjoint measurement attempts to find scales that relate the predictor variables to the response variable using the selected composition rule such that this rule may be examined for credibility.

To illustrate, suppose that HATCO wanted to produce a new detergent. In developing the package, the designers suggested a box with a clear plastic window through which the customer could see the colored crystal additives in the detergent. In addition, the packaging consultants wanted to switch from the traditional blue and white boxes that HATCO had used for years to a new red and white box that would complement the color crystals. HATCO's management decided to run a simple experiment involving all four packaging combinations to assess the impact of packaging alternatives on consumer choice. The alternatives were as follows:

No window/blue and white package
No window/red and white package
Window/blue and white package
Window/red and white package

A sample of two hundred qualified respondents were shown the four packages (randomized order for each respondent) and given $1.79 to purchase one box of detergent. Each box actually cost the respondent $2.19, so a small element of commitment was introduced. Afterward, the respondents were asked to select their second and third choices in order, so that all four packages were ranked by each respondent.

Management posited the simple additive composition rule that preference is determined by the preference order: $b1$ (red) + $b2$ (blue) + $c1$ (window) + $c2$ (no window), where the estimates $b1$, $b2$, $c1$ and $c2$ assessed the contribution to the choice of the box color and design (window or no window). The two values for b (red part worth and blue part worth) and the two values for d (window and no window) summed to values that preserved the original choice distribution. When the values were derived, HATCO tested the adequacy of the simple choice model for each respondent. The question to be answered was, what was the marginal contribution of the designers' suggestions to the choice of HATCO's product? The model does not predict preference in an absolute sense, but rather the preference distribution when a choice is to be made from a discrete set. This example will not be developed further until the logic of conjoint analysis is elaborated in the following sections, which divide the discussion into three major areas of consideration:

1. The purpose of conjoint analysis.
2. Experimental designs that are appropriate to use with conjoint analysis.
3. Methods of analysis using conjoint analysis.

The Purposes of Conjoint Analysis

Rather than discuss the purposes of conjoint analysis from the viewpoint of statistical estimation, we will discuss it in term of marketing management's use of it. It may be assumed that any set of objects (e.g. brands, companies) or concepts (e.g. positioning, benefits) is evaluated as a bundle of attributes. For example, a new personal computer may be thought of as composed of levels of the following attributes (among others):

Brand/model
Bytes of random access memory (RAM)
Color of display
Operating system

Types of data storage devices
Capacity of data storage devices
Available software
Price

The list is much longer than this, of course, but it indicates that the purchaser of a personal computer very likely evaluates some or all of these features and forms some "composition rule" for putting together the values attributed to each of them in order to choose from among the many personal computers that are likely to be considered.

It should also be obvious that the level of measurement for many of these features offers different considerations in analysis. Bytes of RAM are typically measured in metric units such as 4K, 16K, 64K, 128K, or 256K, while availability of software is a highly subjective measurement made that includes elements of both the number and suitability of programs. This feature of evaluation is likely to be so idiosyncratic that the value to the purchaser is highly dependent on the purchaser's perception of the vendor's claim. The vendor does not have a unit of measurement for the software feature, but rather a series of claims to which the respondent must react.

Marketing researchers must often determine the contribution of each of the features to the consumer's choice and use these results to

1. Define the product with the optimum set of features.
2. Show the relative contributions of each feature to the choice process (in a sense, the importance of the feature).
3. Estimate market shares among products with differing sets of features (other things held constant).
4. Isolate groups of potential customers who place differing values on the features in order to define high or low potential segments.
5. Show the marketing opportunies by exploring the potential of product feature combinations that are not currently available.

In order to meet this change the researcher typically assumes the following:

1. There is a common composition rule for all respondents in the experiment. This assumption can be examined as the analysis proceeds and rejected if it is not supported by the data.
2. The variables and their levels are easily communicated. The optimum situation is one in which the respondent can see, touch, or use prototypes that represent the various combinations of variables being tested. As the research moves away from this to visual or (a last choice) verbal communication of the combinations, the respondent's ability to perceive the choices becomes questionable and warrant's great care.
3. There is stability of evaluation across all variables and all levels of variables. The respondent's ability to assess the value of the levels of a feature should be reasonably stable. A set of features should

not contain features or levels that a respondent is not familiar with or not capable of judging adequately because they are outside of a normally accepted range. The assessment a respondent gives is expected to be stable over time and relevant to the real-world choice situation.

In summary, the purpose of conjoint analysis is to determine the contributions of variables (and each level of the variables) to the choice order over combinations of variables that represent realistic choice sets (products, concepts, service, companies, etc.). Again, the prediction is not of an absolute preference but of a relative preference within a defined set. If a more absolute measure is desired, the researcher/manager is counseled to use a scale of preference measure rather than ranks as the basis of choice. If scales are used, the following discussion of experimentation is applicable and the choice of analytical techniques is even wider.

An Experiment Using Conjoint Analysis

Up to this point, we have not discussed how to decide which combinations of features are presented to the respondent to be evaluated. In the simplest situation, the respondent may be given all possible combinations, but in more complex ones, a subset of all possible combinations is presented. The key issues that must be addressed in designing the experiment include the number of variables, the number of levels of each variable, the complexity of the variables, the ability to communicate the variables to the respondent, the compositional rule posited for the respondent's choice process, and the style of presentation to the respondent. These issues are illustrated in the following cases.

Case 1: Three Simple Variables, Each at Two Levels

HATCO is trying to decide what combination of attributes leads to choice in the canned dog food market. The variables of interest are as follows:

Brand name: Arf versus Mr. Dog
Ingredients: all meat versus meat and fiber supplements
Can size: 6 oz versus 12 oz

These three variables are easily communicated and familiar to all current dog owners. Since there are three variables at two levels, each respondent could be presented with all eight combinations of the variable levels and asked to evaluate them. At this point, the researcher must decide what compositional rule is to be examined.
If the simplest rule

$$\text{Preference} = b\ (\text{brand}) + c\ (\text{ingredients}) + d\ (\text{can size})$$

is posed, the researcher is assuming that there are no interactions among the variables that affect choice and that a simple additive rule is used by the respondent to determine the relative value of the eight combinations of features that make up a hypothetical product. If the researcher has a firm conviction that this model holds, a shortcut design may be used that is referred to as a *pairwise trade-off* design. For this situation, the design would appear as three grids for which the respondent does three sets of rankings.

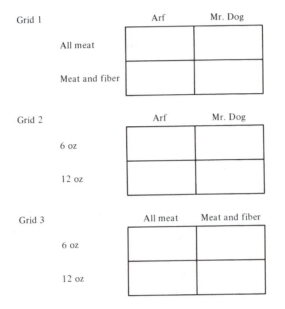

The key assumption underlying these sets of rankings is that the respondents place a value on each of the attribute levels and simply add the values to arrive at a relative preference. If this assumption is true, the presence of all variables is not necessary for the evaluation process, since no predictor variable has any impact on any other predictor variable. As a consequence of this choice of presentation, it is impossible to determine if other compositional rules might hold. Only the simple rule posed here can be examined. If the researcher has any doubts, it is better to present the respondent with the completely crossed design (all eight combinations) so that other composition rules can be examined. For example, if it is thought that can size and ingredients have an interactive effect and that the true model is

$$Preference = b (brand) + c (ingredients) + d (can size) + e (ingredient*size)$$

the simple pairwise trade-off will not allow the estimation of this model of the composition rule. In this simple case, the conservative researcher will probably use all eight combinations in the experiment.

Case 2: Four Variables, Each at Four Levels

If the researcher is interested in assessing the impact of four variables with four levels of each variable, there are 256 combinations that could be created to represent attribute bundles to be ranked or evaluated. This is obviously too many for one respondent to evaluate and give consistent, meaningful answers. The researcher has two choices: use a main effects only model or use a model that estimates some of the interactions.

Main Effects Only Models

As suggested previously, a pairwise trade-off could be used. With four variables, one would use six choice matrices. Each respondent would make 16 choices in each matrix, for a total of 96 decisions. Since the variables are considered two at a time, this should be a relatively simple task. As noted before, this assumes that all respondents adhere strictly to a simple additive protocol.

Another approach to a main effects only model is a fractional design. Because this subject is beyond the scope of this book, we will introduce only the basic idea. With four variables, each at four levels, we need 256 combinations to assess the impact of each of the 4 main effects and the 11 possible interaction effects. However, if we know that none of the interactions exist, we can select 16 of the 256 combinations as measures in the four main effects. The other 240 combinations are necessary only if interactions exist. How the 16 are selected is well documented in the references.

Estimating Selected Interactions

As you might guess from the preceding discussion, a design larger than 16 combinations yet smaller than the total 256 might yield a measurement of some of the interactions. One must still have prior knowledge that some of the interactions do not exist. Any number of combinations less than the total 256 will not allow separation of many of the interaction effects from each other. For example, in a design with 128 combinations, the interaction of variables 1 and 2 cannot be distinguished from the interaction of variables 3 and 4. However, if one knows that variables 3 and 4 do not interact, one can measure the interaction of variables 1 and 2.

Articles and texts on designing experiments are provided in the References.

Procedures Using Conjoint Analysis

There are many procedures for fitting experimental data to a compositional rule for choice. In this discussion, we examine the simplest additive main effects compositional rule and then briefly outline more complex assumptions about how people make choices.

To illustrate the simple additive main effects model, assume that the dog food experiment outlined previously was conducted with 100 respondents who owned dogs and usually fed them canned dog food. (For this illustration we will not be concerned with more detailed definitions of a relevant population such as purchasers of one brand or the other.) Each respondent was given eight dummy cans of dog food in a simulated display and asked to rank their choices in order of preference for purchase. The cans are described in the following table, along with the ranks given by two respondents.

	Description		First Respondent	Second Respondent
			Ranks for Two Respondents	
−6 oz	All meat	Arf	1	1
−6 oz	All meat	Mr. Dog	2	2
−6 oz	Meat and fiber	Arf	5	6
−6 oz	Meat and fiber	Mr. Dog	6	5
−12 oz	All meat	Arf	3	4
−12 oz	All meat	Mr. Dog	4	3
−12 oz	Meat and fiber	Arf	7	7
−12 oz	Meat and fiber	Mr. Dog	8	8

The simple compositional rule of additive main effects

$$\text{Preference} = b \text{ (brand)} + c \text{ (contents)} + d \text{ (size)}$$

will be estimated. If the model holds strictly, a simple difference from the mean (ANOVA) should apply. Since the average rank (of the eight possibilities) is 4.5, we can examine the differences for each main effect from this average. The average ranks and the deviations are as follows:

	Average	Deviation
6 oz	3.5	−1
12 oz	5.5	+1
Meat	2.5	−2
Meat and fiber	6.5	+2
Arf	4.0	−.5
Mr. Dog	5.0	+.5

Since a small number implies a lower rank or a higher choice, we can follow a convention often used in conjoint estimation of reversing all the signs and converting the deviations to coefficients so that when they are squared, they sum to the total number of factor levels (in this case six). The sum of squared deviations is 10.5, so we multiply each of the deviations squared by 6/10.5 and take the square root. This process yields the following coefficients:

6 oz	12 oz	Meat	Meat and fiber	Arf	Mr. Dog
.756	−.756	1.51	−1.51	.38	−.38

These are the same coefficients that would be found using the popular MONANOVA conjoint coefficients analysis program. This will be discussed in more detail later, but we will estimate the model for the second respondent. The second respondent's average, deviations, and coefficients are as follows:

	Average	Deviation	Coefficients
6 oz	3.5	−1	.77
12 oz	5.5	+	−.77
Meat	2.5	−2	1.55
Meat and fiber	6.5	+2	−1.55
Arf	4.5	0	0
Mr. Dog	5.5	0	0

To examine the compositional model, we predict the preference order by summing the coefficients for the appropriate combinations of variable levels and rank ordering the resulting scores. For example, the score for the combination Arf, 12 oz, meat is .00 − .77 + 1.55 = .78 for respondent 2. This score is ranked against all other scores, as shown. Using the two sets of coefficients to predict the original ranks, the following results are found:

Original Ranks	Respondent 1 Prediction	Respondent 2 Prediction
1	1	1.5
2	2	1.5
3	3	3.5
4	4	3.5
5	5	5.5
6	6	5.5
7	7	7.5
8	8	7.5

Since weights of zero were attached to the brand name for the second respondent, the compositional rule is incapable of predicting a difference between brands within the contents and can size combinations. For example, both combinations of 6 oz with meat were considered equal, since the only difference, between the two was the brand name, which carried zero weight in determining the choice. It appears that brand name may be considered a random choice given size and contents. There is no interaction between brand and size or between brand and contents, so in both cases the simple compositional model seems appropriate.

We can now posit a situation where the respondent makes choices in which interactions appear to influence the choices.

	Ranks Given All Combinations		
	6 oz		**12 oz**
Arf	Meat	1	2
	Meat and fiber	3	4
Mr. Dog	Meat	7	8
	Meat and fiber	5	6

These ranks were formed by assuming that this respondent preferred Arf and normally preferred meat over meat and fiber. However, a bad experience with Mr. Dog made the respondent select meat and fiber over meat only if it was Mr. Dog. The respondent was extremely brand loyal and thought size less important than contents and brand.

The coefficients for the simple model are as follows:

		Coefficients
Size	6 oz	+.42
	12 oz	−.42
Contents	Meat	0
	Meat and fiber	0
Brand	Arf	1.68
	Mr. Dog	−1.68

Using these coefficients to create scores for the combination yields

Actual rank	1	2	3	4	5	6	7	8
Predicted	1.5	3.5	1.5	3.5	5.5	7.5	5.5	7.5

The more serious lack of correspondence between the predicted and actual ranks than previously seen is often an indication of the effect of interactions when none are assumed to exist. To fit a model with interaction terms, one may use any number of procedures including dummy variable regression, multivariate classification analysis, or special programs such as POLYCON. The researcher must specify the correct model and verify its appropriateness.

Checking for Interactions

To examine for first-order interactions is a reasonably simple task. For the previous interactive data, we can form the three two-way matrices of ranks and rank sums.

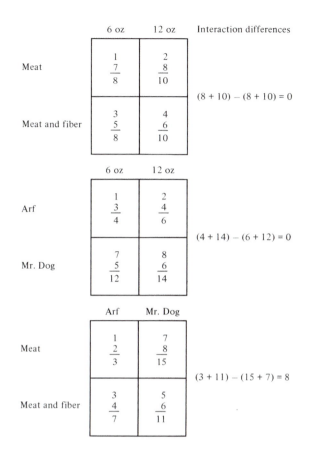

The cross-cell interaction differences should sum to zero when the data are free of interactions.

MONANOVA

A reference to the MONANOVA procedure was made earlier in this chapter. The name refers to MONOTONIC ANOVA (or rank-order ANOVA). This procedure is often cited in the early literature on conjoint analysis. MONANOVA starts by calculating a solution exactly as we have done. It then examines the fit of the solution to the original data. If the fit is poor, as in the preceding interactive example, an iterative procedure is used to alter the coefficients to try to improve the fit. Applying MONANOVA to the preceding example we find:

		Coefficients
Size	6 oz	.006
	12 oz	−.006
Contents	Meat	0
	Meat and fiber	0
Brand	Arf	1.732
	Mr. Dog	−1.732

Applying the MONANOVA model:

Actual rank	1	2	3	4	5	6	7	8
Predicted	1.5	3.5	1.5	3.5	5.5	7.5	5.5	7.5

This is the same result we found using the simple model. Our experience suggests that if the simple model does not give a good fit, the problem is not solved by applying a procedure like MONANOVA. Rather, the problem is in the lack of correspondence between the respondent's behavior and the model posed by the researcher. The researcher may state that an additive model and respondent does not behave that way. The researcher may also say that the respondent is capable of consistently evaluating all the levels of a variable when, in fact, some of the levels are not treated consistently by the respondent. For example, above some threshold the respondent behaves as expected, but below the threshold the respondent behaves quite differently, although we are referring to levels of one variable (price is a good example: the respondent may have a "backward-bending" demand curve below some threshold of price, while above the threshold there is a normal downward-sloping demand curve).

Aggregate versus Disaggregate Analysis

The customary approach to conjoint analysis is disaggregate. That is, each respondent is modeled separately and the fit of the model is examined for each respondent. This approach allows the researcher to appraise the behavior of each respondent relative to the assumptions of the model. It also allows for exclusion from further analysis of respondents who give data that show such poor structure as to suggest that they did not perform the task expected of them.

Aggregate analysis fits one model to the aggregate of the responses. As you may expect, this process generally yields poor results when one is trying to estimate what any one respondent would do and poor results when trying to interpret the values that each of variables have for any one respondent. Unless the researcher is definitely dealing with a population that is very homogeneous in behavior with respect to the experimental variables, aggregate analysis has little to recommend it. The preferred method is individual analysis.

Management's Use of Conjoint Analysis

Typically, the individual models (one for each respondent) are used to:

1. Group respondents with similar values in order to examine possible segments. In the dog food example, we might find a group for which the brand is the most important feature, while another group might value the contents most highly. We are interested in knowing about the presence of such groups and their relative magnitude.

2. Estimate share of various products within segments. For example, among the brand-loyal group of dog owners, what share would each size–contents combination capture? This simulation is done by calculating the rank orders for each combination for each person and assuming whatever they rank first is their most likely first choice. The next step is to count the number of people who would buy each combination first from among all combinations.
3. Do a marginal profitability analysis. If you know the cost of each feature when it is incorporated into a product and the number of people who would buy each combination (step 2), you can calculate the marginal return to management of each combination relative to the others. This might point to a combination with a smaller share as the most profitable because of an increased profit margin due to the low cost of producing the particular combination.

Summary

This chapter has discussed the logic of conjoint analysis without getting too involved in computation details. The primary focus was on the purposes of the results, the importance of the assumed composition model, the necessity of a rigorous experimental design to test the model properly, and the availability of a wide range of computation procedures for analyzing the model. Conjoint analysis places more emphasis on the ability of the researcher/manager to theorize about the behavior of choice than it does on analytical technique. The appropriateness of the exerpimental design is more critical than the choice of analysis technique. The data have been analyzed by means of MONANOVA, dummy variable regression, or various other techniques.

QUESTIONS

1. Ask three of your classmates to evaluate choice combinations based on these variables and on levels relative to the choice of a textbook for a class and specify the compositional rule you think they will use.

Depth: Goes into great depth on each subject
Introduces each subject in a general overview
Illustrations: Each chapter includes humorous pictures
Illustrative topics are presented
Each chapter contains graphics to illustrate the numeric issues
References: Each chapter includes specific references for the topics covered
General references are included at the end of the textbook

How difficult was it for respondents to handle the wordy and slightly abstract concepts they were asked to evaluate? Try both the pairwise and completely crossed designs.

2. Using either the differences model or a conjoint program available at your institution, analyze the data from the preceding experiment. Examine for interactions.

3. Make up a problem with at least four variables and two levels of each variable that is appropriate to a marketing decision. Set up the compositional rule you will use, the experimental design, and the analysis method. Use at least five respondents to support your logic.

SELECTED READINGS

A Hybrid Utility Estimation Model for Conjoint Analysis

PAUL E. GREEN
STEPHEN M. GOLDBERG
MILA MONTEMAYOR

As industrial applications of conjoint analysis continue to grow in both number and level of sophistication, industry researchers are facing a dilemma. On the one hand, the estimation of individual utility functions is essential to the simulation of new products/services—a step that almost invariably accompanies applications of conjoint methodology. On the other hand, the relatively large data requirements that are needed to estimate individual utilities, are becoming increasingly burdensome on respondents.

The problem becomes even more acute in the case of stimulus designs that are capable of estimating selective utility *interactions* as well as all average (or main) effects. Put simply, the trick is to develop a data collection procedure that can maintain individual differences in utility estimation (for input to choice simulators) without placing undue strain on the data supplying capabilities of the respondent. To cope with this problem the authors propose a hybrid data collection and analysis procedure that combines the simplicity and

"A Hybrid Utility Estimation Model for Conjoint Analysis," Paul E. Green, Stephen M. Goldberg, and Mila Montemayor, Vol. 45 (Winter 1981), pp. 33–41. Reprinted from the *Journal of Marketing*, published by the American Marketing Association.

Paul E. Green is S. S. Kresge Professor of Marketing at the University of Pennsylvania.
Stephen M. Goldberg is President of Global Marketing Solutions, Inc.
Mila Montemayor is Manager of Marketing Research for the Ortho Pharmaceutical Corporation, Raritan, NJ.

speed of self-explicated, compositional utility models with the realism and greater generality of regression-like, decompositional models (Wilkie and Pessemier 1973).

We first describe the model. This is followed by an empirical application of the model to a study of physicians' preferences for various attributes describing a new antibiotic (disguised product class) drug. The paper concludes with a brief discussion of possible extensions of the approach.

The Hybrid Utility Estimation Model

Before discussing the actual model, some preliminary notation is considered. First, we represent the hth ($h = 1,H$) stimulus profile (e.g., product description) by the vector:

$$\underline{x}^{(h)} = (x_{i_1}^{(h)}, x_{i_2}^{(h)}, \ldots, x_{i_j}^{(h)}, \ldots, x_{i_J}^{(h)}) \quad (1)$$

where $x_{ij}^{(h)}$ denotes level i ($i = 1,I_j$) of attribute j ($j = 1,J$). For example, in the context of automobiles, we could have $J = 4$ separate attributes—miles per gallon, acceleration time, engine type, and manufacturer—and the vector:

$$\underline{x}^{(h)} = (18 \text{ mpg, } 12 \text{ seconds to reach } 55 \text{ mph,}$$
$$\text{diesel engine, manufactured}$$
$$\text{by General Motors}) \quad (2)$$

Note that the attribute levels can be either discrete values of a continuous attribute or a particular category of an unordered polytomy.

The Self-Explicated Model

Stage one of the proposed model makes use of an old procedure for measuring preference functions, namely the self-explicated utility model (Huber, Sahney and Ford 1969; Huber 1974). Briefly described, self-explicated models entail the following steps:

A respondent is shown J sets of attribute levels, one set of attributes at a time. For example, one attribute could be car make, with the levels General Motors, Ford, Chrysler, Volkswagen, Datsun, and Toyota. Generally, some four to eight levels of each attribute are used.

For each set of attribute levels the respondent is shown a 10-point scale. After choosing the most preferred level and assigning it a weight of 10, the respondent is asked to rate the remaining attribute levels in terms of their desirability (relative to the most preferred) on the 1 to 10 desirability scale. Ties are permitted.

After rating all J sets of attribute levels, the respondent is asked to rate the relative importance of the J attributes themselves on, e.g., a 5-point scale, ranging from unimportant (rated 1) to extremely important (rated 5). Alternatively, the respondent may be asked to distribute 100 points across the attributes in such a way as to reflect their overall importances to him/her.

The respondent's self-explicated utility for the hth stimulus profile is given by a simple additive model:

$$U_h = \sum_{j-1}^{J} w_j u_{ij}^{(h)} \qquad (3)$$

where U_h is the total utility of alternative h, w_j is the self-explicated importance weight of attribute j, and $u_{ij}^{(h)}$ denotes the fact that alternative h has a desirability score of u on level i of attribute j.

Essentially, then, the w_j's are simply stretching or shrinking constants that place all of the separate desirability ratings on a common scale so that addition of the separate utilities is meaningful.

Notice that the preceding model is main effects, additive; that is, no interaction terms appear. The model is also compositional in the sense that each of the components, w_j and u_{ij}, is estimated explicitly by the respondent and U_h is derived rather than given directly by the respondent. Also note that the task is quite simple and direct since the respondent is only required to evaluate one set of attribute levels at a time and then to evaluate

the relative importances of the attributes themselves.

The Conjoint Model

In contrast, the conjoint model is decompositional. In this case the respondent is shown a set of complete stimulus profiles: $\bar{x}^{(1)}, \bar{x}^{(2)}, \ldots, \bar{x}^{(h)}, \ldots$ and asked to rate each *overall* profile on some evaluative scale (e.g., on a 1 to 10 scale of buying intentions). The evaluative response to the h-th alternative is assumed to be given by:

$$V_{i_1 i_2 \ldots i_J} \cong \sum_{j=1}^{J} v_{ij} + \sum_{j<j'} t_{iji_j} \qquad (4)$$

where $V_{i_1 i_2 \ldots i_J}$ denotes the respondent's overall evaluation of a stimulus profile with level i of attributes 1, 2, . . . , J and $\cong$ denotes least squares approximation. Note that this model includes all main effects and (selected) two-way interaction terms —the v's and t's, respectively, in equation (4).

However, the main distinction between the compositional model represented by equation (3) and the decompositional model of equation (4) is in parameter estimation. In equation (3) w_j and u_{ij} are given directly by the respondent and one *computes* U_h as a simple weighted sum.

In equation (4) the evaluative responses and the stimulus profiles are known (the latter expressed in terms of dummy variables for parameter estimation), and one *solves for* the v's and t's via some type of regression-like procedure—metric or nonmetric as the case may be.[1]

The Hybrid Model

The proposed hybrid model collects both kinds of data—self-explicated part worths and regression-derived part worths. The following steps are entailed:

Stage one of the data collection is identical to that described under the self-explicated model.

---

[1] In the hybrid model to be described, we assume, for convenience, that parameters are estimated via ordinary least squares (i.e., by means of metric regression). Even if the original data are only rank ordered, considerable evidence has been adduced to support OLS regression as a reasonably good approximation to nonmetric estimation methods (Green and Srinivasan 1978).

Stage two of the data collection involves presenting each respondent with a *limited* set (usually eight or nine) complete (all-attribute) stimulus profiles. These stimulus profiles are, in turn, drawn from a much larger master design (usually ranging between 64 and 256 profiles) that permits orthogonal estimation of all main effects *and* selected two-way interactions. Moreover, profiles can be "balanced" *within* respondent by means of various blocking designs.[2] The respondent then evaluates each complete stimulus profile on some type of likelihood of purchase or intentions-to-buy scale. Call each of these responses Y_h.

The matrix $\bar{S}$ of utility functions, of order N by $\Sigma_{j=1}^{J}$ for the N respondents, as obtained from the self-explicated task of stage one, is row centered.[3] That is, each respondent's specific set of $w_j u_{i_j}$'s in equation (3)—there are $\Sigma_{j=1}^{J} I_j$ of these for each respondent—are expressed as deviations from his/her mean. Hence the sum of each row is zero. (Row centering mitigates response bias problems.) The respondents are then clustered on the basis of similarities in their *self-explicated* utility functions. Assume that K clusters are found.

The hybrid model's parameters are then separately estimated for each cluster by means of OLS regression. The hybrid model is defined as follows:

$$Y_h = Y_{i_1 i_2 \dots i_j}$$
$$\cong a + b\, U_{i_1 i_2 \dots i_j} + \sum_{j=1}^{J} v_{ij} + \sum_{j<j'} t_{ijij'} \qquad (5)$$

where each $U_{i_1 i_2 \dots i_j}$ is separately computed (and then centered) for each respondent via equation (3); a is an intercept term, b is a regression parameter representing the contribution of the self-explicated utility to Y, and the v's and t's are also regression parameters, estimated at the cluster level.

As can be seen, the model of equation (5) weights the self-explicated utility by b (this parameter being assumed common over all members of the kth cluster) and finds *cluster-based v's and t's* as regression parameters associated with the dummy-variable representation of each $x^{(h)}$. The underlying rationale is that respondents who give similar responses on the first-stage, self-explicated task are also likely to give similar kinds of responses on the second-stage, overall profile evaluation stage. (This assumption is empirically testable *if* sufficient conjoint data can be collected at the individual level to estimate all main effects and at least a small set of two-way interactions.)

Note that the last two terms of equation (5) carry the effects of full profile presentation *beyond* that predicted by a linear function of the self-explicated utilities. Hence, it is a simple matter to run a models comparison test for each cluster to see which of the v's and t's (if any) are needed to account for *additional* variance in the Y's. In this way the researcher can develop the most parsimonious model consistent with its ability to account for variation in Y.

Note further that the v's and t's are utility-function parameters whose arguments are the dummy-variable representations of the original attributes, as illustrated in equation (1). On the other hand the argument of the slope parameter b is the *derived* utility (U) obtained from the self-explicated task. While this value is, in turn, a function of the original attributes, it need not bear any simple relationship to v_{ij}; indeed, the correlation of U with the dummy-variable representations of the original attributes could be (but need not be) quite low.

A Pilot Application of the Hybrid Model

Data were available from a sample of 404 French physicians, stratified by specialty, on their evaluations of a new antibiotic (disguised product class) drug. After being shown a brief pharmacological description of the new drug, each respondent evaluated its possible attribute levels and attribute importances in the manner described in the preceding section. The attributes and their levels are shown in Table 1. Following this, each respondent was shown eight profile descriptions (drawn from a master set of 64 profiles, with levels "balanced" within respondent) and asked to evaluate each profile on a 1–10 likelihood of trying scale. The

---

[2] By "balanced" is meant that within the set of stimulus profiles received by a particular respondent, each pair of attribute levels appears with a frequency given by the product of the relative marginal frequencies of each level and the total number of profiles in the set. For example, if level i of attribute 1 appears in six out of a total of nine profiles and level i^* of attribute 2 appears in three out of the nine, then this pair of levels should appear in precisely $2/3 \cdot —II = \cdot 9 = 2$ out of the nine profiles.

[3] In some applications the researcher may wish to standardize the self-explicated utilities to zero mean and unit standard deviation.

TABLE 1 Attributes and Levels Used in a Pilot Application of the Hybrid Model

A. Sore mouth or tongue
  1. Equal to other antibiotics
  2. No sore mouth or tongue

B. GI (gastrointestinal) upsets
  1. GI upset in 30% of patients
  2. GI upset in 10% of patients
  3. No GI upsets

C. Absorption by the body
  1. Cannot be taken with meals
  2. Can be taken with meals

D. Skin rash during therapy
  1. Transient rash in 25% of patients
  2. Transient rash in 15% of patients
  3. No skin rash

E. Cold extremities
  1. Some effect in 30% of patients
  2. Absence of cold extremities

F. Loss of hair at initiation
  1. Transient loss in 30% of patients
  2. Transient loss in 10% of patients
  3. No hair loss

G. Dosage regimen
  1. Three times a day
  2. Two times a day
  3. Once a day

H. Price per daily dosage
  1. $0.50
  2. $0.75
  3. $1.00
  4. $1.25

(personal) interview concluded with the collection of background data on current antibiotic drug prescribing, nature of the physician's practice, attitudes toward the use of antibiotics, psychographics, and demographics. We confine our discussion here to the utility estimation model. The nature of the attribute levels in Table 1 is such that most physicians were expected to prefer the levels:

No sore mouth or tongue
No GI upsets
Can be taken with meals
No skin rash
Absence of cold extremities
No hair loss
Once a day dosage
$0.50 per day cost.

Self-Explicated Data Analysis

The first step in parameterizing the hybrid model was to analyze the stage-one (self-explicated) desirability and importance ratings. As noted from Table 1, the total number of attribute levels is: $\Sigma_{j=1}^{J} I_j = 22$. Hence, the resulting matrix of importances-times-desirability ratings is of order 404×22. This matrix S was row-centered and the data clustered via the Howard and Harris (1966) algorithm.

The 4-group solution provided a good separation among the clusters (based on the algorithm's decrease in pooled within-groups' variance). The cluster sizes were:

Cluster	Number of Respondents
1	82
2	111
3	117
4	94

Examination of the stage-one cluster centroids indicated complete inter-cluster agreement on the average *ordering* of utilities within attribute. However, some differences were noted across clusters in terms of relative utility ranges. We found that the cluster 1 centroid exhibited relatively high utility for the absence of cold extremities while cluster 2 showed high utility for the drug's ability to be taken with meals. Cluster 3 appeared to be an economically motivated group who placed high utility on low cost per daily dosage. Cluster 4 placed high utility on no sore mouth or tongue and no GI upsets.

In summary, all four cluster averages exhibited the same ordering of levels within attribute; the main source of cluster differences was reflated to the relative utility ranges (that primarily reflected differences in importance weights). At this point a 22-component self-explicated utility function was available for each respondent in each of the four clusters.

Step-Two Estimation

For each of the four clusters separately, a regression model derived from equation (5) was set up;

the dependent variable was Y_h, the respondent's evaluation of each of the eight complete profiles he/she received in stage two. For each respondent in each cluster his/her self-explicated utility function (mean-centered) was used to find eight U's corresponding to the specific eight profiles that he/she received.

Next, a design matrix was prepared that expressed all of the main effects and three selected two-way interactions: $A \times C$, $A \times E$, and $C \times E$ (which previous research suggested as possibly important) as appropriate combinations of dummy variables. Four separate models were fitted for each cluster. Referring to equation (5), if we ignore the intercept term a, the first term on the right of the $\cong$ sign denotes the self-explicated utility (U), the second term denotes the main-effect utilities (the v's), and the third term denotes the selected interaction parameters (the t's).

The four models fitted to the data of each cluster entailed consideration of:

Only the self-explicated utilities (firm term);
The self-explicated utilities plus the main effects (first and second terms);
The main effects and interactions only (second and third terms);
All effects (first, second, and third terms)

The (adjusted) R^2's associated with each model were as follows:

explicated model) are needed. As indicated above, in this example all three sets of parameters are needed.

It should be emphasized that the preceding comparisons are specific to the *particular* data set of this study. As stated earlier, one of the attractive features of the model is its flexibility for determining the *simplest* model that is consistent with the data.[4]

Table 2 shows a summary of the stage-two, derived utilities, which are simply regression coefficients (i.e., the v's and t's). We note that in all four clusters the ranking of utilities within attribute is the same.

For ease of comparisons across clusters, all of the utilities in Table 2 have been scaled so that the utility ranges (for main effects) across the eight attributes sum to 10.0. Also, the scaling of regression coefficients (the part worths) has been set so that the lowest part worth within any attribute is assigned the value of zero. As expected, the highest part worths are associated with no sore mouth or tongue, no GI upsets, etc., surmised earlier. The interaction effects of Table 2 are all positive; for example, in cluster 1 a drug that does not cause sore mouth or tongue *and* can be taken with meals exhibits a part worth of 0.71 beyond the sum of their separate main effects.

While the ranking of part worths within attribute is the same across clusters, the scale separa-

Cluster	(1) First Term Only	(2) First and Second Terms	(3) Second and Third Terms	(4) All Three Terms
1	0.369	0.482	0.282	0.528
2	0.393	0.567	0.317	0.621
3	0.358	0.539	0.363	0.603
4	0.387	0.591	0.345	0.652

Model comparison tests (Green, with Carroll, 1978) were run (for each cluster separately) on model (1) versus model (4), model (2) versus model (4), and model (3) versus model (4). All comparisons were significant at the 0.05 level, or better. Note that the first comparison tests whether the stage-two decompositional part of the hybrid model is needed. The second comparison tests whether the selected interactions are needed. finally, the third comparison tests whether individual utilities (provided by the stage-one, self-

tions differ. For example, we note the following cluster characteristics:

[4] An additional models comparison was also run. First, for each *separate* cluster the self-explicated model:

$$Y^*_{i_1 i_2 \ldots ij} \cong a^* + b^* U_{i_1 i_2 \ldots ij}$$

was fitted and residuals $(Y^*_{i1i2 \ldots 1j} - \hat{Y}^*_{i1i2 \ldots ij})$ computed. Following this, a models comparison test was run in order to see if the v^*'s and t^*'s differed across the four clusters, when the *residuals* were regressed on the dummy-variable representations of the profiles. The results were significant at the 0.01 level, suggesting that cluster individuality should be maintained.

TABLE 2 Part Worths, By Cluster, From Equation (5) (See Table 1 for Attribute-Level Descriptions

Attribute Level	Part Worths			
	Cluster 1	Cluster 2	Cluster 3	Cluster 4
A-1	0	0	0	0
2	0.86	0.14	0.11	1.08
B-1	0	0	0	0
2	0.28	0.25	0.11	0.31
3	2.02	2.60	1.90	2.99
C-1	0	0	0	0
2	0.30	1.03	0.73	0.49
D-1	0	0	0	0
2	1.01	0.85	1.10	0.87
3	1.16	1.17	1.64	1.08
E-1	0	0	0	0
2	2.91	2.55	2.10	2.52
F-1	0	0	0	0
2	0.35	0.25	0.19	0.21
3	0.53	0.53	0.51	0.66
G-1	0	0	0	0
2	0.23	0.28	0.38	0.66
3	1.44	0.39	1.05	0.85
H-1	0.78	1.59	1.96	0.33
2	0.30	0.53	0.48	0.21
3	0.25	0.11	0.24	0.12
4	0	0	0	0
Interactions				
A_2 and C_2	0.71	0.35	0.32	0.38
A_2 and E_2	0.10	0.08	0.08	0.27
C_2 and E_2	0.20	0.30	0.16	0.09

Cluster 1 exhibits relatively high utility for no cold extremities, once-a-day dosage, and the interaction of no sore mouth or tongue with the drug's ability to be taken with meals.

Cluster 2 places relatively high utility on the drug's ability to be taken with meals and the interaction of the ability to be taken with meals with the absence of cold extremities.

Cluster 3 shows relatively high utility for the absence of skin rash and low price.

Cluster 4 places relatively high utility on no sore mouth or tongue, no GI upsets, and the interaction of no sore mouth or tongue with the absence of cold extremities.

While these inter-cluster differences in stage two are not identical to those reported earlier for stage one, several similarities can be noted, indicating some consistency between the self-explicated responses and the conjoint analysis responses.

At this point, an estimating equation (equation 5) had been developed for each separate cluster. The terms of the equation contained both self-explicated utilities and cluster-estimated part worths. While not shown here, the estimating equations can be used in choice simulators in the same manner as individual utility functions (based entirely on conjoint procedures) are conventionally used. Since each respondent's self-explicated desirabilities and importance weights are employed, the Y-values of equation 5 will differ across individuals in each cluster, thus maintaining idiosyncratic variation.

Validation Problems

A full test of the proposed model would entail the collection of individual conjoint data based on much more comprehensive experimental designs than are commonly used today. For example, to estimate all main effects and an appreciable number of two-way interactions in Table 1, could easily require 64 or more observations *at the individual level*. Clearly, this is not practical—at least in commercial surveys. A more tractable approach is to collect self-explicated and limited conjoint data (as was done here) for large enough samples to provide a holdout sample for cross-validation. That is, parameters computed in the calibration sample would be applied to the holdout sample in order to predict, for example, first choices or the preference ratings in the conjoint phrase (entailing eight or so stimulus profile evaluations).

Some limited testing of this type has already been carried out on a different data bank (Goldberg 1980). In this case the stimuli (in the conjoint phase) were 7-element profiles of telecommunication devices; all seven underlying attributes contained three levels each. Following the self-explicated tasks, each respondent evaluated nine of the master set of 81 profiles (levels were balanced within subject) on a likelihood-of-purchase scale.

A calibration sample of 324 respondents was randomly selected from the full data set; a holdout sample of 100 respondents was used for cross validation. The cross validation measures consisted of: (a) the percentage of correct first-choice predictions across the 100 respondents in the holdout sample; (b) the average Kendall tau correlation; and (c) the average product moment correlation between the respondent's actual evaluations of the nine conjoint profiles versus the predicted evaluations derived from the model of interest.

Three models were compared: (a) the self-explicated model; (b) the (group-level) conjoint model; and (c) the hybrid model of equation 5. The adjusted R^2's (for the calibration sample) and the three cross validation measures, noted above, were as follows:

However, we do note that the hybrid model performs considerably better than the self-explicated model on all three cross validation measures. In contrast, its practical superiority over the group-level conjoint model is modest, to say the least. What has happened in this particular data bank is that the systematic part of the data shows high homogeneity across individuals; hence, the idiosyncratic variation captured by the self-explicated part of the hybrid model is not contributing much to either accounted-for variance in the calibration sample or to cross-sample validity.[5]

This example illustrates the general difficulty of "validating" the hybrid model. The situation is simply that its performance, compared to either the self-explicated or group-level (or subgroup-level) conjoint models is obviously dependent on the specific characteristics of the data bank.

Of course, a more definitive test would be to compare the cross validation performance of the hybrid model with the conventional conjoint model, with all parameters of this latter model estimated at the *individual* level. This test could be carried out if the number of attributes and levels were kept small and one were to estimate only main effects, so that the number of stimuli could be kept relatively small, e.g., 16–20 profiles. In this case each model would be used to predict a set of individually evaluated holdout profiles.

Extensions of the Basic Model

Now that the basic hybrid model has been described, it may be useful to describe ways in which the approach can be extended. Two specific areas of research interest are (a) the possibility of employing different variables in the stage-two (decompositional) versus the stage-one (composi-

Model	Adjusted R^2 (Calibration Model)	Percent of Correct First-Choice Predictions	Average Tau Correlation	Average Product Moment Correlation
Self-Explicated (Eq. 3)	0.074	22	0.20	0.27
Conjoint (Eq. 4)	0.170	30	0.32	0.43
Hybrid (Eq. 5)	0.183	32	0.33	0.43

The first observation to be drawn from these results is that the data are considerably noisier than those collected in the antibiotic study; adjusted R^2's are about one-half or less of those found in the antibiotic study.

[5] The high homogeneity noted across respondents in the conjoint task was also found in the self-explicated task. No well-delineated subsets were found in the cluster analysis of self-explicated utilities; hence, the hybrid model was fitted to the total calibration sample.

tional) parameter estimation phases, and (b) extensions of the model's structure, including the incorporation of threshold-like responses. Each of these features is considered briefly, as follows.

Different Variables in Stage One Versus Stage Two

In the pilot study reported here, the attributes and their specific levels evaluated in the self-explicated stage were the same as those evaluated in the conjoint stage. However, this one-to-one correspondence is not necessary if certain safeguards are maintained. For example, suppose the researcher has a set of continuous attributes—gasoline mileage, acceleration time, dealer price, etc.—for which the utility functions at the conjoint stage may be adequately described by general quadratic functions. Assuming that a suitable quadratic interpolation procedure (e.g., Stirling's method) is available, the researcher could employ (say) seven or eight levels of each attribute in the self-explicated stage and a different number of levels (e.g., five) in the conjoint stage. Central composite designs (Myers 1971) are useful for developing general quadratic functions fitted to 5-level attributes.

The model would be estimated by using quadratic interpolation to provide commensurate predictor-variable values for the self-explicated utilities, prior to fitting the quadratic model to the group-level data obtained from the conjoint stage.[6]

In other situations the researcher may wish to omit certain attributes, such as price, in the self-explication stage but introduce them in the conjoint stage. If so, the self-explicated utility relates to only a portion of the profile, and one assumes that the parameters estimated for omitted attributes are representative of individuals' utilities. This device should be used cautiously since the self-explicated utility estimates are vulnerable to specification bias; the extent of the bias depends on the number of variables omitted in the self-explication stage and their correlations with variables appearing in the conjoint stage.

Finally, in some cases the researcher may wish to fix the levels of particular attributes in the conjoint evaluation stage (across subsets of respondents), so as to reduce information overload on any single respondent. (Which attributes to fix for any given respondent could be determined by an appropriate experimental design.) The self-explicated utilities can then carry the individual variation needed to utilize the model in various kinds of new product simulations. It should be mentioned, however, that all of the preceding extensions of the approach require additional research on their practical feasibility, as well as their reliability and validity.

Extensions of the Hybrid Model

As recalled, the present formulation of equation (5) provides a single self-explicated term U which, in turn, is given by equation (3). In some applications the researcher may wish to expand the role of the self-explicated utilities by the following extension of equation (5):

$$Y_h = Y_{i_1 i_2 \ldots i_j}$$
$$\cong = a + \sum_{j=1}^{j} b_j(w_j u_{i_j}) + \sum_{j=1}^{j} v_{i_j}$$
$$+ \sum_{j<j'} t_{i_{j_{i_{j'}}}}.$$

$$(6)$$

In this version of the model a separate b is estimated (in the regression) for *each* of the J attributes. It should be mentioned, however, that this equation introduces more parameters to estimate and will probably lead to higher correlation between the $w_j u_{i_j}$'s and the dummy variables representing the profiles in stage two (for which the v's and t's are regression coefficients). However, if equation (6) is modified so that the v_{i_j} terms are dropped, the model seems more attractive. In this case we assume that all main effects are *already* included in the self-explicated portion; only the two-way interactions are estimated by the decomposition part of the procedure, resulting in reduced multicollinearity between self-explicated utilities and dummy variable representations.

The present model of equation (5) can also be extended to deal with a type of conjunctive decision model.[7] In this case the first stage (self-explicated) task proceeds as before, with the addition

---

[6] Of course, the researcher could employ the same five levels in both stages. However, interpolation would *still* be required when a model is applied to new product designs whose attribute levels are of limited to those originally used in the respondent evaluation task.

[7] By "conjunctive" is meant that a profile, to be acceptable, must exhibit levels that all equal or exceed some set of cutoff values. Failure for any level to do so is sufficient to reject the profile. Actually, the model described here is a compensatory *approximation* to a conjunctive model.

that for any specific attribute level the respondent can indicate whether it is *totally unacceptable*, irrespective of what levels the profile may exhibit on other attributes. Only those levels rated as not unacceptable are then rated on the desirability scale.

In the second stage of data collection the respondent is shown a small set of full profiles (as before). In this case, however, the respondent is first asked to assign the profile to a reject versus consider-for-further-evaluation group. In the third stage, only those profiles in the latter group are rated on a purchase intentions scale.

This type of model is considerably more complicated than equation (5), and several approaches to parameter estimation are being studied. These include both multistage estimation procedures and iterative methods. The latter start out with initial utility estimates for unacceptable attribute levels and subsequently modify these in a manner similar to algorithms used for estimating missing data values in alternating least squares procedures.

In sum, the hybrid model and its various extensions would appear to offer considerable flexibility in combining the speed and convenience of self-explicated utility measurement with the greater power and generality of decompositional methods. Perhaps even more important is the ability of the model to maintain individual differences in utility estimation—a useful advantage in the usual case in which a consumer choice simulator is employed. However, future research (some of which is already in progress) is needed to ascertain the model's internal validity and cost/benefit characteristics relative to current conjoint data collection methods.

References

Goldberg, Stephen M. (1980), "An Empirical Comparison of Hybrid and Non-Hybrid Utility Estimation Models," working paper, Wharton School, University of Pennsylvania.

Green, Paul E., with contributions by J. Douglas Carroll (1978), *Analyzing Multivariate Data*, Hinsdale, Il.: Dryden Press, 71–73.

_____ and V. Srinivasan (1978), "Conjoint Analysis in Consumer Behavior: Status and Outlook," *Journal of Consumer Research*, 5 (September), 103–23.

Howard, Nigel and Britton Harris (1966), "A Hierarchical Grouping Routine, IBM 360/65 FORTRAN IV Program," University of Pennsylvania Computing Center.

Huber, G. P. (1974), "Multiattribute Utility Models: A Review of Field and Field-like Studies," *Management Science*, 20 (June), 1393–1402.

_____, V. K. Sahney, and D. L. Ford (1969), "A Study of Subjective Evaluation Models," *Behavioral Science*, 14 (November), 483–89.

Myers, Raymond H. (1971), *Response Surface Methodology*, Boston: Allyn and Bacon, Inc.

Wilkie, William L. and Edgar A. Pessemier (1973), "Issues in Marketing's Use of Multiattribute Attitude Models," *Journal of Marketing Research*, 10 (November), 428–41.

Reliability and Validity of Conjoint Analysis and Self-Explicated Weights: A Comparison

THOMAS W. LEIGH
DAVID B. MACKAY
JOHN O. SUMMERS

Over the past decade, conjoint analysis has become a popular method for analyzing the preferences of individual consumers and consumer groups. Cattin and Wittink (1982) estimate that more than 1000 applications have been reported. The popularity of conjoint analysis appears to derive, at least in part, from its presumed superiority in reliability and validity over simpler, less expensive techniques such as collecting self-explicated attribute weights.[1] Conjoint analysis techniques also are considered to be more "flexible" in that they allow for such options as the modeling of interactions among the attributes (Akaah and Korgaonkar 1983). However, when considered in empirical studies, these interaction terms frequently have been found to be of marginal value (Akaah

"Reliability and Validity of Conjoint Analysis and Self-Explicated Weights: A Comparison," Thomas W. Leigh, David B. MacKay, and John O. Summers, Vol. 21 (November 1984), pp. 456–462. Reprinted from the *Journal of Marketing Research*, published by the American Marketing Association.

Thomas W. Leigh is professor of Marketing at the Pennsylvania State University.

David B. MacKay is professor of Marketing and Geography at Indiana University.

John O. Summers is professor of Marketing at Indiana University.

[1] Conjoint analysis data recently have been combined with self-explicated weights in hybrid models (cf. Green, Goldberg, and Montemayor 1981; Cattin, Gelfand, and Danes 1983).

and Korgaonkar 1983; Neslin 1981). Furthermore, the claims made for the greater realism of the respondent's task in conjoint analysis (Green, Goldberg, and Montemayor 1981), like modeling flexibility, should be manifested in the increased reliability and validity of these techniques. Ultimately, a major goal of conjoint analysis of consumer preferences is to enable the marketer to predict behavior in the marketplace more accurately.

Green, Carroll, and Goldberg (1981) have identified the assessment of the reliability and validity of conjoint techniques as "perhaps the most important area of all for future research." Though several recent studies have focused on measuring the reliability and/or validity of one or more conjoint measurement approaches (Davidson 1973; Malhotra 1982; McCullough and Best 1979; Parker and Srinivasan 1976; Segal 1982; Wittink and Montgomery 1979) and the issue of the correspondence between conjoint part-worths and self-explicated weights has been addressed (Cattin and Weinberger 1979; Scott and Wright 1976), we were unable to discover a single study comparing the test-retest reliability and/or predictive validity (in the traditional sense of the term)[2] of conjoint analysis results with that of self-explicated attribute weights. The primary purpose of this article is to report the results of such a comparison based on data collected in a recent empirical study.

---

[2] Some conjoint analysis researchers have referred to various "goodness-of-fit" indices (e.g., the correlation between the values predicted by the model and the actual data from which the model's parameters were estimated) and/or cross-validation indices (e.g., the correlation between the model's predictions and actual preference scores for a holdout sample) as measures of predictive validity (Akaah and Korgaonkar 1983). However, we use the term "predictive validity" only as it has been used traditionally in the literature, to refer to the correlation between predictions based on the scores on some "test" (e.g., estimated attribute weights) and some "outside criterion variable of interest" (e.g., choice behavior in the marketplace). Parker and Srinivasan (1976) examined the relationship between actual choice behavior and conjoint analysis results, but in their study, unlike Wittink and Montgomery's (1979), the choices occurred before rather than after the preference data were collected. See Campbell (1960) for a classic discussion of the traditional meaning of "predictive validity."

Further complicating the literature is the fact that other researchers have used the terms "external validity" and "internal validity" in reference to the results of conjoint analysis even though these terms traditionally have been defined in reference to experiments rather than to the measurement of "traits" of individuals (e.g., the subjective utilities individuals attach to particular attribute levels). See Campbell and Stanley (1966) or Cook and Campbell (1979) for a discussion of the traditional meanings of these terms. What has occasionally been referred to as the "external validity" of conjoint analysis results might more appropriately be called "predictive validity" (Green and Srinivasan 1978).

Method

Each subject in a convenience sample of 122 undergraduate business students participated in two experimental sessions, approximately two weeks apart, concerning their preferences for hand-held calculators. During the first session the subjects produced "test" data for the conjoint analysis task to which they were randomly assigned and then provided self-explicated attribute weights. After corresponding "retest" measures were collected in the second session,[3] the subjects participated in a raffle for a calculator of their choice from a predetermined set of ten.

Hand-held calculators were chosen as the product of interest because of their obvious relevance to the subject population. Five dichotomous attributes, defined in terms of the presence or absence of a particular product feature, were selected for inclusion in the study on the basis of their perceived importance as evaluated by a separate pretest convenience sample of 50 drawn from the subject population: (1) algebraic parentheses to assist calculation, (2) rechargability, (3) financial functions, (4) statistical functions, and (5) warranty.[4] This allowed for a maximum of $2^5 = 32$ experimental stimuli for the full-profile "data collection method."

Conjoint Measurement Tasks

All of the conjoint measurement tasks included in the study involved full-profile written descriptions of the experimental stimuli. A part-worth preference model was assumed and its parameters estimated via ANOVA techniques or MONANOVA,[5] depending on whether or not the response scale could be assumed to be metric. The alternative conjoint tasks differed in (1) stimulus set construction (i.e., level of fractionation of the factorial design; full, half, and quarter)[6] and (2) measurement scale for the judgments (i.e., rank order,

---

[3] The subjects were not initially aware that a retest was to be conducted.

[4] These attributes were defined in a precise manner within the experimental materials to ensure the subjects' comprehension.

[5] Though other nonmetric estimation techniques such as LINMAP are available, MONANOVA was selected for use in the study because it is the most widely reported algorithm in the marketing literature.

[6] In the full factorial design all 2^5 possible stimulus descriptions were presented to the subjects. See Green, Carroll, and Carmone (1978) for a detailed discussion of the role of fractional factorials when using the full-profile approach to stimulus set construction.

TABLE 1   Summary of Results

| Experimental Condition | n | Conjoint Time | Conjoint Input | Average Test-Retest Adj. R^2 | | | |
| | | | | Est. Stim. Utilities | | Est. Attr. Weights | |
				Conjoint	Self-Expl.	Conjoint	Self-Expl.
Rank order							
Full	10	11.4	.708	.764	.848	.653	.419
Half	11	10.4	.689	.709	.825	.426	.376
Quarter	11	8.2	.651	.767	.829	.379	.421
(Mean)		(9.9)	(.682)	(.748)	(.833)	(.483)	(.402)
Paired comparisons							
Half	11	15.5	.780	.790	.894	.488	.599
Quarter	11	7.4	.717	.788	.853	.492	.440
(Mean)		(11.5)	(.749)	(.789)	(.874)	(.490)	(.519)
Graded paired comp.— dollar metric							
Half-PBIB	9	15.2	.643	.884	.813	.559	.463
Quarter	9	11.3	.574	.827	.810	.384	.287
(Mean)		(13.3)	(.609)	(.856)	(.812)	(.472)	(.380)
Graded paired comp.— rating scale							
Half-PBIB	11	10.5	.456	.773	.834	.534	.470
Quarter	8	10.6	.566	.726	.869	.548	.325
(Mean)		(10.5)	(.502)	(.753)	(.848)	(.540)	(.414)
Rating scale							
Full	10	9.7	.668	.879	.841	.394	.362
Half	10	8.5	.610	.837	.887	.484	.524
Quarter	9	6.4	.707	.758	.928	.360	.678
(Mean)		(8.3)	(.660)	(.827)	(.884)	(.415)	(.515)
Av. across all methods		(10.4)	(.649)	(.792)	(.853)	(.475)	(.452)

paired comparisons, graded paired comparisons-dollar metric, graded paired comparisons-rating scale, and single stimulus-rating scale). Twelve of the $3 \times 5 = 15$ possible different conjoint tasks were included in the study. Combinations of the full factorial with paired comparisons, graded paired comparisons-dollar metric,[7] and graded paired comparisons—rating scale were omitted because of the excessive number of judgments they require ($C (32, 2) = 496$).

Furthermore, a partially balanced incomplete block cyclical design was used with the graded paired comparisons approaches whenever they were combined with the one-half fractional factorial stimulus set in order to reduce the number of stimulus pairs to be evaluated from $C (16, 2) = 120$ to 48. This reduction appeared necessary because of the increased time and effort these judgments require over traditional paired comparisons.

[7] See Pessemier et al. (1971) and Dykstra (1958) for detailed discussions of graded paired comparisons.

The rank-order instructions were patterned after those presented by Green and Wind (1973, p. 261–2). Subjects completing the paired comparisons task were given randomly ordered series of paired profiles and asked to check the preferred one in each pair. Those respondents in the graded paired comparisons conditions completed the additional task of indicating the magnitude or intensity of their preference. In the dollar metric version, respondents were asked to specify how much more (in dollars) they would be willing to pay for the preferred calculator. Those in the graded paired comparisons—rating scale condition circled a number on a scale from 1 (slightly prefer) to 11 (greatly prefer). Finally, those assigned to the single stimulus—rating scale condition individually rated each calculator profile on an 11-point scale (1 = least preferred to 11 = most preferred). To reduce anchoring problems, subjects were instructed to read all profiles before rating any of them and to assign "1" to the least preferred and "1" to the most preferred profile.

Mean Sq. Difference Est. Attr. Weights		Predictive validity					
		Percent Correct		Actual Less Expected		Mean Sq. Difference	
Conjoint	Self-Expl.	Conjoint	Self-Expl.	Conjoint	Self-Expl.	Conjoint	Self-Expl.
.436	.371	12.0	31.3	.050	.263	.951	.804
.522	.370	45.5	45.5	.355	.409	.399	.739
.465	.376	20.4	29.0	.178	.314	1.153	.477
(.476)	(.372)	(26.8)	(36.7)	(.200)	(.339)	(.809)	(.689)
.442	.288	36.4	26.5	.264	.291	.625	.880
.412	.340	35.0	45.0	.260	.380	1.208	.974
(.428)	(.314)	(35.7)	(35.3)	(.262)	(.333)	(.902)	(.924)
.317	.391	22.2	22.2	.122	.211	1.401	1.660
.385	.424	55.6	33.3	.456	.211	.513	1.387
(.351)	(.407)	(38.9)	(27.8)	(.289)	(.211)	(.957)	(1.524)
.442	.374	37.5	44.8	.275	.438	.643	.258
.515	.340	28.6	55.6	.186	.533	1.041	.547
(.473)	(.361)	(33.3)	(49.4)	(.233)	(.479)	(.829)	(.382)
.321	.366	25.0	38.3	.170	.370	.604	.330
.381	.281	55.6	50.0	.456	.388	.969	1.704
.456	.216	44.4	19.4	.344	.187	1.209	.892
(.384)	(.291)	(41.1)	(35.5)	(.318)	(.315)	(.916)	(.925)
(.425)	(.344)	(34.9)	(36.3)	(.260)	(.328)	(.880)	(.897)

The self-explicated attribute weights were obtained by requesting the subjects to specify how much more (in dollars) they would be willing to pay for a calculator with each feature than for one without it (ceteris paribus).[8]

The raffle was conducted immediately after the retest. The subjects were instructed to select one of 10 calculator descriptions, each with exactly three of the five desirable features, and informed that should they win the raffle they would receive a calculator of the description they selected. The purpose of restricting the choice set was to present the subjects with alternatives spanning a relatively narrow range of desirability. This condition in turn provided a more sensitive test of predictive validity.

[8] This self-explicated weights approach can be adapted to cover attributes with three or more levels. When there is no clear a priori order, subjects first would be instructed to provide their rank-order preferences for the attribute levels. The respondents then would be requested to evaluate the various levels in relation to the least valued level.

Analysis

Conjoint Task Time

Table 1 is a summary of the study results. First, note that the self-reported time the subjects spent on their assigned task, including reading instructions, is generally greater for the less fractionated designs, but it does not rise proportionally with the increase in the number of required judgments. For example, subjects in the rank order—full factorial condition spent an average of 11.4 minutes ranking 32 stimuli whereas those in the quarter factorial took an average of 8.2 minutes with only eight stimuli. Similarly, those in the paired comparisons—half factorial condition allocated an average of 15.5 minutes to making C (16, 2) = 120 paired comparisons (7.7/min.) in contrast to an average of 7.4 minutes for C (8, 2) = 28 judgments for the quarter factorial (3.8/min.). It seems clear that when faced with a large number of judgments, subjects will give considerably less time to each individual judgment.

Estimation of Attribute Weights

Monotone analysis of variance (Kruskal 1965) was used to estimate the part-worths for the nonmetric methods (i.e., rank order and paired comparisons). The paired comparison data first were converted by TRICON (Carmone, Green, and Robertson 1968) into the predominant rank orders required by the MONANOVA algorithm. ANOVA techniques were used to derive the part-worths for the metric methods (i.e., graded paired comparisons—dollar metric, graded paired comparisons—rating scale, and single stimulus—rating scale).[9] In the case of graded paired comparisons, the observations were modeled in terms of the difference in the first-order parameters of the paired profiles. See Dykstra (1958) for a discussion of the analysis of graded paired comparisons data. To make the unit of measurement for the set of five attribute weights comparable across methods (i.e., MONANOVA part-worths, ANOVA part-worths, and self-explicated weights) all weights were scaled so that their sum of squares for each respondent was equal to the number of attributes.

Test-Retest Reliability

Most past studies assessing the reliability of conjoint measurement techniques have utilized some type of correlational approach (Acito 1977; Green, Carmone, and Wind 1972; Jain et al. 1979; McCullough and Best 1979; Parker and Srinivasan 1976). Within-subject test-retest correlations can be calculated at several different "levels" and three of these levels were included in the analysis: (1) input data, (2) estimated stimulus utilities, and (3) estimated part-worths. When a subject attaches approximately equal values to all the attributes, it is possible for the test and retest weights to match closely yet produce a low test-retest correlation. For this reason, the mean squared differences between the test and retest attribute weights were included as a fourth measure of test-retest reliability.

The average within-subjects test-retest r^2s for

[9] With the metric data, tests were conducted at the individual-subject level on the statistical significance of the two-way interactions. Because the number of significant interactins was not substantially different from what one would expect from chance alone, these terms were omitted from all the respondents' models to maintain consistency.

the conjoint *input data* appear to favor paired comparisons, which have an overall average adjusted r^2 of .749, though an initial one-way ANOVA F-test failed to reject the null hypothesis of equal mean test-retest r^2s for the 12 conjoint measurement tasks ($p > .05$). However, the results were significant ($p = .04$) when Fisher's Z-transformation of r was used. This was the only test for which the results differed for the transformed correlations.

One-way ANOVA F-tests also were conducted to investigate potential differences across the 12 conjoint tasks for the other three reliability measures: test-retest r^2 for the estimated stimulus utilities, test-retest r^2 for the estimated part-worths, and mean squared difference between the test and retest part-worths. All three of these reliability measures demonstrate fairly high within-cell variance and none of these F-tests are statistically significant at the .05 level. Hence, one cannot reject the null hypothesis of equally reliable output for these 12 conjoint measurement approaches.

The lack of statistically significant differences suggests the appropriateness of combining the data for the 12 conjoint tasks when comparing the reliability of conjoint analysis output with that of self-explicated weights. This provides for a more powerful test of the differential reliability of these two basic measurement approaches than if the conjoint tasks were considered separately.

The self-explicated weights produce a higher average test-retest r^2 for the *estimated stimulus utilities* than do the combined conjoint procedures (.853 vs. .792) and they exceed the average reliability of each of the conjoint measurement scales with the single exception of the graded paired comparisons—dollar metric. Furthermore, a two-tailed t-test of this difference is statistically significant at the .05 level.

The average test-retest r^2 for the *estimated part-worths* for the self-explicated weights is not statistically different ($p > .05$) from that for the combined conjoint tasks. However, the results for the *mean squared difference* between the test and retest *part-worths* suggest higher reliability (i.e., a lower mean squared difference) for the self-explicated weights (.344 vs. .425). This difference is statistically significant at the .01 level.

Predictive Validity

According to Campbell (1960) the predictive validity of a "test" refers to the correlation between current "scores" on that test and future values of some relevant criterion variable. In the study, the subject's choice of a calculator in the raffle represented the criterion variable for the purpose of assessing predictive validity. Unlike stated preferences, the raffle choice had potential implications of a significant nature for the respondents. The winning subject received a calculator which cost approximately $45 and incorporated the features contained in her raffle choice.

Three measures of predictive validity were considered: (1) the percentage of subjects whose raffle choices were predicted correctly from their respective estimated attribute weights, (2) the actual number of correct predictions minus the expected number of correct predictions under chance divided by the number of predictions,[10] and (3) the mean squared difference between the estimated utilities for the subjects' predicted and actual raffle choices. In this context, the predicted choice was the calculator with the highest estimated utility. All predictions were based on the subjects' test (as opposed to retest) results. The first measure, the percentage of correct predictions, was adjusted for ties (i.e., if two calculator profiles were tied for the highest estimated utility and one of them was the subject's raffle choice, it was counted as one half of a correct prediction). The second measure is automatically adjusted for ties because the expected number of correct "chance" predictions depends on the number of profiles tied for first.

There are no statistically significant differences ($p > .05$) among the 12 conjoint tasks for any of the three predictive validity measures. Furthermore, because of high within-cell variance and small cell sizes, it is impossible to discern any systematic patterns for these measures across either levels of fractionation or measurement scales.

Again, the lack of statistically significant differences among the various conjoint approaches suggests that their results can be aggregated for comparison with those of the self-explicated weights to provide for a more powerful statistical test. These two basic methods for assessing attribute weights are very close for all three predictive validity measures. The self-explicated weights edge out the conjoint part-worths in the percentage of correctly predicted raffle choices (36.3% vs. 34.8%) and the actual minus expected proportion of correct choices (.328 vs. .260), whereas the combined conjoint analysis techniques result in a very

[10] This "scoring rule" was suggested in a different but equivalent form by an anonymous *JMR* reviewer.

slightly smaller mean squared utility difference between the predicted and actual choices (.880 vs. .897). None of the differences are statistically significant ($p > .05$).

Discussion

The results fail to provide support for the presumed greater reliability and validity of conjoint analysis in comparison with self-explicated attribute weights. Instead, in the two cases in which the results are statistically significant (i.e., the test-retest r^2 for the estimated stimulus utilities and the mean squared differences between the estimated attribute weights), they favor the self-explicated weights. With the single exception of Fischer's Z-transformed input data correlations, no statistically significant differences in any of the test-retest reliability or predictive validity measures are found among the 12 conjoint approaches included in the study. This outcome is likely to be due in part to the combination of large within-group variances and limited cell sizes.

Perhaps a different selection of conjoint analysis techniques, including other nonmetric algorithms such as LINMAP, would have produced more favorable results for this basic approach. Notably missing is any representation of Johnson's (1974) tradeoff procedure. However, most of the conjoint analysis procedures utilized in this investigation are commonly applied in industry (Cattin and Wittink 1982), and Segal's (1982) study suggests that tradeoff analysis does not produce more reliable results than the full-profile approach.

Of course, the reliability and validity of self-explicated weights and the various conjoint analysis techniques may vary over different subject populations and stimulus types (e.g., the number and complexity of attributes). Furthermore, these measures may be affected by the method of stimulus presentation (e.g., verbal, pictorial, or three-dimensional representation).

Rather than asking whether conjoint measurement invariably produces superior results, it may be more appropriate to inquire as to the conditions under which self-explicated weights will suffice. For example, one might speculate that the self-explicated weights approach might be sufficient, perhaps even better than conjoint analysis, when (as in this study) the attributes are dichotomous (e.g., features which are either present or absent), for attributes whose values are uncorrelated, and for products and attributes with which the respondents are very familiar. In other situations, it may be necessary to rely on conjoint measurement techniques. Establishing the reliability and validity of a methodological approach is necessarily an on-going process requiring a substantial number of empirical studies.

The fact that the self-explicated weights task always followed the conjoint task is a potentially important methodological limitation of the study. One could argue that providing the conjoint data first had a detrimental effect on the subjects' performances on the self-explicated weights task by causing them to become bored and/or fatigued. Conversely, it is possible that the subjects "learned" (or gained insights) about their preferences by completing the conjoint tasks and that this learning enhanced their abilities to provide reliable and valid self-explicated attribute weights. Because one would not expect the various conjoint tasks to have the same "learning" effect on the self-explicated weights, one can develop some indication of whether this effect is important by examining the self-explicated weights results for those respondents assigned to different conjoint measurement tasks. In particular, to the degree to which it occurs, learning relevant for the self-explicated task should be greatest for those assigned to the graded paired comparisons—dollar metric task because this conjoint task is most similar to providing self-explicated weights (i.e., it most directly encourages the respondent to evaluate the dollar value of the individual attributes). It follows that this group should have the most favorable self-explicated weights results when learning effects are dominant. That this is not the case is apparent from Table 1. In fact, the opposite occurs; the self-explicated weights results for this group are the worst on all six criteria considered. However, the issue of order effects cannot be resolved adequately without randomizing the order of presentation of the tasks.

Conclusion

The managerial value of conjoint analysis appears to be based largely on its ability to predict events in the marketplace accurately. However, though a large number of conjoint analysis studies have been conducted over the past decade, the literature includes very little material *directly* relevant to

the question of whether and under what conditions conjoint analysis performs better than self-explicated attribute weights on this criterion. In this context, goodness-of-fit and cross-validation measures are inadequate. Given a sufficient number of parameters to estimate, one usually can obtain a reasonably good fit of most models to preference data. Furthermore, using conjoint analysis of stated preferences for one set of stimuli to predict stated preferences for a second (holdout) set of stimuli has a built-in advantage (i.e., shared method variance) over the use of self-explicated weights. Basically, this cross-validation procedure serves primarily to indicate the internal consistency of the conjoint analysis data provided by the respondent during a single experimental session.

Though the greater number of judgments typically inherent in conjoint analysis approaches as well as the apparently realistic character of the judgments (i.e., evaluating multidimensional stimuli) seem to suggest higher test-retest reliability and predictive validity than are obtainable for self-explicated weights, our results fail to support this contention. Hence, the assumption that the additional time, effort, and money spent conducting a conjoint study will *invariably* be rewarded by better results than can be obtained from self-explicated weights appears unjustified. Future research directed toward identifying those conditions under which conjoint measurement may produce more accurate results is warranted.

Given the minimal additional cost associated with collecting self-explicated weights in a conjoint analysis study, it appears prudent to make this a standard industry practice. Careful evaluations over time of the comparative quality of the *marketplace* predictions produced from these two basic approaches can provide the marketer with the basic information from which to determine accurately which measurement techniques work best under various conditions. This knowledge, in turn, will enable the marketer to avoid using an expensive, complex, and time-consuming methodological technique when a simple, inexpensive, easy-to-understand approach might provide as good, or perhaps better, data. Conversely, it can identify those conditions under which adopting the more sophisticated conjoint analysis procedures is prudent. Marketing practitioners need to become more aware of the cost-effectiveness of incorporating inexpensive methodological studies into their research programs.

References

Acito, Franklin (1977), "An Investigation of Some Data Collection Issues in Conjoint Measurement," *Proceedings*, American Marketing Association, 82–5.

Akaah, Ishmael P. and Pradeep K. Korgaonkar (1983), "An Empirical Comparison of the Predictive Validity of Self-Explicated, Huber-Hybrid, Traditional Conjoint, and Hybrid Conjoint Models," *Journal of Marketing Research*, 20 (May), 187–97.

Campbell, Donald T. (1960), "Recommendations for APA Test Standards Regarding Construct, Trait, or Discriminant Validity," *American Psychologist*, 15 (August), 546–53.

_____ and Julian Stanley (1966), *Experimental and Quasi-Experimental Designs for Research*. Chicago: Rand McNally, Inc.

Carmone, F. J., Paul E. Green, and D. J. Robertson (1968), "TRICON—An IBM 360-165 Fortran IV Program for the Triangularization of Conjoint Data," *Journal of Marketing Research*, 5 (May), 219–20.

Cattin, Philippe, Alan E. Gelfand, and Jeffrey Danes (1983), "A Simple Bayesian Procedure for Estimation in a Conjoint Model," *Journal of Marketing Research*, 20 (February), 29–35.

_____ and Marc Weinberger (1980), "Some Validity and Reliability Issues in the Measurement of Attribute Utilities," in *Advances in Consumer Research*, Vol. 7, Jerry Olson, ed. Ann Arbor, MI: Association for Consumer Research, 780–3.

_____ and Dick R. Wittink (1982), "Commercial Use of Conjoint Analysis: A Survey," *Journal of Marketing*, 46 (Summer), 44–53.

Cook, Thomas D. and Donald T. Campbell (1979), *Quasi-Experimentation: Design and Analysis Issues for Field Settings*. Chicago: Rand McNally College Publishing Company.

Davidson, J. D. (1973), "Forecasting Traffic on STOL," *Operations Research Quarterly*, 24 (December), 561–9.

Dykstra, Otto, Jr. (1958), "Factorial Experimentation in Scheffé's Analysis of Variance for Paired Comparisons," *American Statistical Association Journal*, 53, 529–42.

Green, Paul E., Frank J. Carmone, and Yoram Wind (1972), "Subjective Evaluation Models and Conjoint Measurement," *Behavioral Science*, 17 (May), 288–99.

_____, J. Douglas Carroll, and Frank J. Carmone (1978), "Some New Types of Fractional Factorial Designs for Marketing Experiments," in *Research in Marketing*, Vol. I, J. N. Sheth, ed. Greenwich, CT: JAI Press, 99–122.

_____, _____, and Stephen M. Goldberg (1981), "A General Approach to Product Design Optimization Via Conjoint Analysis," *Journal of Marketing*, 45 (Summer), 17–37.

_____, Stephen M. Goldberg, and Mila Montemayor (1981), "A Hybrid Utility Estimation Model for Conjoint Analysis," *Journal of Marketing*, 45 (Winter), 33–41.

_____ and V. Srinivasan (1978), "Conjoint Analysis in Consumer Research: Issues and Outlook," *Journal of Consumer Research*, 5 (September), 103–23.

_____ and Yoram Wind (1973), *Multiattribute Decisions in Marketing: A Measurement Approach*. Hinsdale, IL: The Dryden Press.

Jain, Arun K., F. Acito, N. Malhotra, and V. Mahajan (1979), "A Comparison of the Internal Validity of Alternative Parameter Estimation Methods in Decomposition Multiattribute Preference Models," *Journal of Marketing Research*, 16 (August), 313–22.

Johnson, R. M. (1974), "Trade-off Analysis of Consumer Values," *Journal of Marketing Research*, 11 (May), 121–7.

Kruskal, Joseph E. (1965), "Analysis of Factorial Experiments by Estimating Monotone Transformations of the Data," *Journal of the Royal Statistical Society*, Series B, 27, 251–63.

Malhotra, Naresh K. (1982), "Structural Reliability and Stability of Nonmetric Conjoint Analysis," *Journal of Marketing Research*, 19 (May), 199–207.

McCullough, James and Roger Best (1979), "Conjoint Measurement: Temporal Stability and Structural Reliability," *Journal of Marketing Research*, 16 (February), 26–31.

Neslin, Scott A. (1981), "Linking Product Features to Perceptions: Self-Stated Versus Statistically Revealed Importance Weights," *Journal of Marketing Research*, 18 (February), 80–6.

Parker, Barnett R. and V. Srinivasan (1976), "A Consumer Preference Approach to the Planning of Rural Primary Health-Care Facilities," *Operations Research*, 24, 991–1025.

Pessemier, Edgar A., Philip Burger, Richard Teach, and Douglas Tigert (1971), "Using Laboratory Brand Preference Scales to Predict Consumer Brand Purchases," *Management Science*, 17 (February), 371–85.

Scott, Jerome E. and Peter Wright (1976), "Modeling an Organizational Buyer's Product Evaluation Strategy: Validity and Procedural Considerations," *Journal of Marketing Research*, 13 (August), 211–24.

Segal, Madhav N. (1982), "Reliability of Conjoint Analysis: Contrasting Data Collection Procedures," *Journal of Marketing Research*, 19 (February), 139–43.

Wittink, Dick R. and David T. Montgomery (1979), "Predictive Validity of Trade-off Analysis for Alternative Segmentation Schemes," *Proceedings*, American Marketing Association, 68–73.

Levels of Aggregation in Conjoint Analysis: An Empirical Comparison

WILLIAM L. MOORE

Modeling consumer preferences for multiattributed alternatives has been an important consumer research topic for more than a decade. Conjoint analysis, one of the most useful techniques in this area, has been growing in popularity the last several years.

In a recent review of conjoint analysis, Green and Srinivasan (1978) discuss and cite empirical results on several issues such as form of preference model, data collection methods, stimulus set construction, stimulus presentation, measurement scales, and estimation methods. Another important issue they touch upon, but do not fully develop, is methods and levels of aggregation. Specifically, they mention segmentation by clustering and componential segmentation but they do not attempt to compare them or relate them to other levels of aggregation.

The purpose of this article is to develop the aggregation issue more fully and to present an empirical example of the predictive power of conjoint analysis at different levels of aggregation.

Background

Conjoint measurement was developed as a model of individual choice behavior which consisted of a series of tests used to determine whether a person's rank order preferences could be described by a formal composition rule applied to a set of independent variables (Luce and Tukey 1964; Krantz and Tversky 1971).

In contrast, applied work has involved the assumption that some composition rule holds—possibly with some error. The focus of this research has been on estimating the relationship between independent and dependent variables. Green and Srinivasan (1978) call this group of estimation

"Levels of Aggregation in Conjoint Analysis: An Empirical Comparison," William L. Moore, Vol. 17 (November 1980), pp. 516–523. Reprinted from the *Journal of Marketing Research*, published by the American Marketing Association.

William L. Moore is professor of Business, Graduate School of Business, Columbia University.

procedures "conjoint analysis" to distinguish it from its axiomatic roots.

Though several estimation procedures can be used (Jain et al. 1979), they all attempt to find a set of part-worth utilities that relate the attribute levels of an object to overall preference or utility for that object—i.e.,

$$(1) \qquad Y_i = \sum_{l=1}^{N} \hat{\beta}_l X_{il} + e.$$

Y_i is the preference or utility for the ith object. The objects are described in terms of $K_j + 1$ ($j = 1, \ldots, J$) levels of J attributes. The ith object can be represented by a vector of dummy variables $(X_{i1}, \ldots, X_{iN})$ where $N = \Sigma_{j=1}^{J} K_j$. $\hat{\beta}_l$ is the part-worth utility estimate of the lth attribute level.

Individual Level Models

The early applications of conjoint analysis (Green and Rao 1969, 1971; Green and Wind 1972, 1973, 1975; Johnson 1974) indicated that a separate utility function was estimated for each individual. Several researchers have used these estimated individual part-worth utilities to predict preferences for a set of validation objects (i.e., objects not used in the estimation procedures). They have found the predictive ability of the individual level models to be very good. For example, Johnson (1974) found a median rank correlation of .80 between predicted and stated preferences for a set of validation objects. Huber (1975) estimated utilities for different levels of sugar and strength of tea, used them to predict preference for another group of teas, and found a root mean square correlation of approximately .8 between predicted and actual preferences using a variety of estimation techniques. Wittink and Montgomery (1979) were able to predict 63% of students' actual job choices on the basis of utilities estimated from hypothetical job offers and the students' perceptions of actual job offers.

Though individual models have demonstrated good predictive power, the output of the estimation procedure—a separate set of utility weights for each individual—makes managerial analysis and understanding difficult when the number of respondents is large. Both Johnson (1974) and Green and Wind (1975) have suggested that these individual utilities can be put into a choice simulator. The choice simulator uses the estimated part-worth utilities to predict the utility for each object in a set of real and/or hypothetical objects. If one assumes that each individual chooses the object with the highest utility, these simulators can be used to predict market shares for each of a number of new or improved products. Furthermore, a search routine can be employed to find the one product configuration that would have the highest predicted market share for a given set of competitors. Though these simulators have certain desirable properties, they do have the aura of a "black box." Also, though the estimated market share for a number of different concepts would give some understanding about what the market "wants," it is not as easy to use as a single set of utility weights would be. Finally, individual level analysis requires that enough information be collected from each person to estimate separate utility functions. Thus, several good reasons can be given for not analyzing the data at the individual level; however, they need to be balanced against the very good predictive power of the individual models.

Aggregate Models

At the other extreme of the aggregation continuum is the case in which the preference ratings are pooled across all respondents and one overall utility function is estimated. These pooled utility estimates are the same as the average of the individual utility estimates (if each of the respondents has evaluated the same set of objects; McCann 1974). The results of this type of analysis are easy to explain (e.g., on average, travel time is much more important than trip cost in determining a transportation mode).

A potential problem with pooled analysis is termed the "majority fallacy" by Kuehn and Day (1962). It occurs when the item chosen by the "average" customer is not the one chosen most often. The majority fallacy is caused by heterogeneity of preferences; for example, if half of the people like large cars and the other half like small cars, the "average" person may like medium-sized cars best, even though no real person wants one. This problem has been demonstrated in a conjoint analysis setting by Huber and Moore (1979). As expected, heterogeneity of preference reduces the predictive power of the aggregate model; for example, Wittink and Montgomery (1979) found that a pooled model could predict only 46% of the students' job choices (compared with 63% with individual models). In spite of this limitation, this level of aggregation has been used in a number of commercial studies. Similarly, most applications of probabilistic choice models such as logit (McFadden 1976) have pooled the responses of a number of respondents rather than obtaining repeated observations from one person (see Batsell 1979 for an individual level analysis using a logit model). Furthermore, these probability models require some heterogeneity across respondents because if one object were preferred by all respondents over one other object, the underlying probabilistic choice model would not hold. Thus, though these models have the desirable property of producing estimated choice probabilities, they are also subject to the majority fallacy.

Segmented Models

Optimally, then, one would like a model that combines the most desirable properties of the two extreme levels of aggregation and avoids the problems of each. Specifically, one would like a model having predictive power close to that found in individual level models and a small number of utility functions that are easily explained to managers. Furthermore, models employing an intermediate level of aggregation have the benefit that managers may feel more comfortable working with segments than with either of the extreme levels of aggregation. For example, in estimating the impact on preferences of an ad campaign designed to convince people that Toyota makes high quality automobiles, a manager may be more comfortable estimating the effect of the ad campaign on a segment-by-segment basis than on an individual-by-individual basis or at the aggregate level. It therefore seems appropriate to consider various methods of segmenting individuals into homogeneous groups.

In a recent review of market segmentation, Wind (1978) refers to two traditional methods of segmentation, *a priori* and clustering, and two newer methods, flexible and componential. The clustering and componential segmentation methods, which are also mentioned by Green and Srinivasan (1978), are discussed further.

CLUSTERING SEGMENTATION. Benefit clusters or segments (Haley 1968) are formed by grouping the respondents into segments that are homogeneous with respect to the benefits they want from the product or service class. Green and Srinivasan (1978) suggest that they have achieved better discrimination by clustering on the vector of part-worth utilities ($\hat{\beta}$) rather than on the most preferred level of each attribute.

A criticism of this general approach is that it is tautological, meaning that whenever one attempts to cluster data, clusters are formed. Furthermore, this clustering occurs even when there is no real structure to the data. However, researchers who formed benefit segments by clustering on preference found that the resulting segments differ significantly and in expected ways in terms of AIO's (Ginter and Pessemier 1978). Also segment preferences have been found to be good predictors of the segment's actual purchases (Moore, Pessemier, and Little 1979).

A potential advantage of this method of segmentation over componential segmentation is that people in different segments need not differ in terms of background variables. That is, if there is one group of people who like small cars and another group who like big cars, componential (and a priori) segmentation will uncover the two groups only if they differ in terms of background variables. Though it is useful to be able to differentiate segments by background variables, it can also be valuable just to know that there are two distinct groups (in terms of preference for size of car) even if it is not possible to determine which person is in which group.

COMPONENTIAL SEGMENTATION. An alternative approach is componential segmentation (Green 1977; Green, Carrol, and Carmone 1976, 1977; Green and DeSarbo 1977, 1979). This technique focuses on the effect of interactions between the product profile, X, and the person profile, Z (a vector of dummy variables that describes the person in terms of his/her background variables) on preference for the product. Through this mechanism one is able to predict how a person with a certain set of background characteristics will react to a particular product. Thus, a person's reaction to a product is broken into the sum of two pieces (1) the average part-worth utilities due to the attribute levels of that product, pooled across all respondents, and (2) the interactions between the person's background variables and the attribute levels. The part-worth utilities and interactions are estimated by the following equation.

$$(2) \quad Y_i = \sum_{j=1}^{N} \hat{\beta}_j X_{ij} + \sum_{k=1}^{m} \sum_{j=1}^{N} \hat{\gamma}_{jk} X_{ij} Z_k + e$$

where Y, $\hat{\beta}_j$, and X_{ij} are as before, $Z = (Z_1, \ldots, Z_m)$ is a vector of dummy variables that is used to describe the person's background variables, and $\hat{\gamma}_{jk}$ is the interaction between the attribute level of the ith product represented by X_{ij} and the background variable level represented by Z_k. Rather than applying the dummy variable coding typically used in econometric texts (e.g., Johnston 1972; Theil 1971), Green et al. used effects-type coding of the variables (Cohen and Cohen 1975). Under this scheme, which makes the regression directly analogous to ANOVA, two-level variables are coded as 1 or -1 instead of 1 or 0 and three-level variables are coded as (1,0), (0,1), or $(-1,-1)$ instead of (1,0), (0,1), or (0,0). Obviously the method of coding has no effect on the amount of explained variance; however, the meaning of the regression coefficients associated with the main effects and interaction terms is slightly altered by the meaning of coding. Kerlinger and Pedhazur (1973) discuss different coding schemes. Significant interactions between background variables (e.g., sex) and product attributes (gas mileage) indicate that people with different background variables have different utilities for levels of an attribute (men are more interested in good gas mileage than women). The increase in explanatory power achieved through the addition of the interaction variables gives some indication about the segmentability of the market.

A set of data was analyzed by the four methods described.

Data

Eighty-seven graduate students took part in the conjoint analysis study. They were asked to forecast what their personal situation would be after graduation in terms of four background variables: sex (there were no expected changes), marital status, place of residence, and number of days each week they expected to drive a car. Then they were asked to assume these forecasts were correct and to rate their preferences on a 1 to 10 scale for each of 24 hypothetical automobiles (18 calibration and 6 validation) to be purchased upon graduation. The attributes used to describe the cars and the respondent background variables are given in Figure 1.

FIGURE 1

Background Variables and Levels

Sex	Marital status
Male	Single
Female	Married
	Other[a]
Residence	Driving days per week
Center city	Every day
Suburbs	Three days
Rural[b]	Two days or less

Product Attributes and Levels

Gas mileage	Price
15 MPG	$3000
25 MPG	$4500
35 MPG	$6000
Place of origin	Top speed
American	90 MPH
Japan	120 MPH
Europe	
	Number of seats
	4
	6

[a] The "single" and "other" categories were combined in subsequent analysis because of the small number of people in "other."
[b] The "suburban" and "rural" categories were combined in subsequent analysis because of the small number of people in "rural."

Two days after filling out the questionnaire, the participants were asked about their preferred level on each of the attributes and about the relative importance of each of the attributes. Individual level analysis indicated that the conjoint analysis model did an excellent job of representing the most preferred level on all attributes and the relative importance of the three most important attributes.

Results

Calibration Objects

The 18 calibration objects were analyzed by using the four models previously discussed: aggregate or pooled model, individual level model, clustered segmentation model, and componential segmentation model. In each case a dummy variable regression was used to estimate the various part-worth utilities. Each subject's ratings were nor-

malized (Bass and Wilkie 1973) to eliminate main effects due to subjects.

POOLED REGRESSION. The model was estimated by running one regression across all the respondents. This is the most efficient way of estimating the average part-worth utilities. The estimates are given in the first column of Table 1.

The average respondent wanted a 4-seat $4500 European automobile that got 35 MPG and had a top speed of 120 MPH. Gas mileage was by far the most important attribute.[1] Place of origin, cost, number of seats, and top speed (listed in declining order of importance) were much more equal in terms of relative importance.

INDIVIDUAL REGRESSIONS. A different regression was run for each respondent to estimate individual level part-worth utilities. The average of the individual part-worth utilities taken across all respondents is equal to the pooled estimates given in column 1 of Table 1. However, the heterogeneity of the group can be seen by examining column 2.

The second column gives some indication why gas mileage was by far the most important attribute in the aggregate regression. Not only did the majority (83%) of the people view it as one of the two most important attributes, but they also were in substantial agreement about the most preferred level of it. In contrast, though many respondents thought price or place of origin was one of the two most important attributes (63% and 35%, respectively), there was considerably more disagreement on the most preferred level of these two attributes. More people wanted a $3000 car than any other price level, yet the pooled model indicates that the "average" person wanted a $4500 automobile. There was considerable agreement on the most preferred levels of the last two attributes, top speed and number of seats, but neither attribute was considered very important by a large number of respondents. This analysis implies that two conditions must be met if an attribute is to be seen as important at the aggregate level: (1) a large number of people must judge it to be important on an individual level and (2) there must

[1] The importance of each attribute is determined by calculating the difference in utilities between the level with the highest utility and the level with the lowest. In the case of gas mileage, this is 2.55 − 0.00 = 2.55. Note that importance is dependent on the range of levels of each of the attributes. In this example, the range of gas mileage is very large in relation to that of price when compared with the figures for cars currently on the market.

TABLE 1 Estimation Results from Four Levels of Aggregation

Attributes	Pooled Regression[a]	Individual Regression[b]	Clustered Segmentation Clusters			Componential Segmentation[c]
			First[a]	Second[a]	Third[a]	
Gas milage						
15 MPG	0.00	3%	0.00	0.00	0.00	0.00
25 MPG	1.83	16	1.05	3.47	.78	1.92
35 MPG	2.55	81	1.60	4.78	1.91	2.62
Price						
$3000	0.00	48	0.00	0.00	0.00	0.00
$4500	.18	30	−.50	.15	1.20	.17
$6000	−.50	22	−1.49	−1.02	1.61	−.61
Place of origin						
America	0.00	31	0.00	0.00	0.00	0.00
Japan	−.20	14	−.59	.23	−.24	−.20
Europe	.40	55	−.20	.69	.96	.42
Top speed						
90 MPH	0.00	29	0.00	0.00	0.00	0.00
120 MPH	.43	71	.38	.39	.56	.43
Number of seats						
4	0.00	63	0.00	0.00	0.00	0.00
6	−.51	37	−.06	−.38	−1.33	−.51

Interactions in Componential Segmentation				
Price/sex		$3000	$45000	$6000
	Male	−.34	.00	.34
	Female	.34	.00	−.34
Residence/gas mileage		15 MPG	25 MPG	35 MPG
	Center city	.09	.00	−.09
	Suburban/rural	−.09	.00	.09
Residence/price[d]		$3000	$4500	$6000
	Center city	−.15	−.13	.28
	Suburban/rural	.15	.13	−.28
Driving days/place of origin		America	Japan	Europe
	Daily	−.15	.15	.00
	Three	.00	.00	.00
	Two or Less	.15	−.15	.00
Driving days/gas mileage		15 MPG	25 MPG	35 MPG
	Daily	.00	.00	.00
	Three	−.15	.15	.00
	Two or Less	.15	−.15	.00

[a] Part-worth utilities for each level of each attribute. Most preferred level is underlined.
[b] Percentage of respondents preferring each level of each attribute.
[c] Part-worth utilities—main effects coefficients have been transformed (as the zero point is arbitrary) to permit easy comparison with other columns.
[d] This interaction term is the sum of two significant interaction terms.

be substantial agreement on its most preferred level. The analysis again suggests that the pooled results may not accurately capture the part-worth utilities for sizable segments of the population.

CLUSTERED SEGMENTS. Benefit segments were formed by clustering respondents into groups that were homogeneous with respect to the individual part-worth utilities, β's, estimated in the preceding subsection. The clustering was done by using the Howard-Harris algorithm (1966; descriptions are also given by Green and Rao 1972 and Green and Wind 1973), a hierarchical teardown routine that maximizes the ratio of the among-to-within-cluster variance at each step. A three-cluster solution was chosen as most appropriate on the basis of (1) change in the among-to-within-cluster variance and (2) relative size of the clusters. The part-worth utilities for the clusters are listed in columns 3, 4, and 5 of Table 1.

The picture of the groups' heterogeneity is similar to that found in column 2 of Table 1. There is substantial agreement across segments on the preferred levels of gas mileage, top speed, and number of seats, but there are differences with respect to price and place of origin. Furthermore, even though most respondents were interested in gas mileage and price, there was a fair amount of disagreement about the relative importances of the attributes. Thus, even without any information on the background variables of people in each of the clusters, this middle level of aggregation gives some insight about what different people want and does so in a format that is easy to communicate.

A discriminant analysis was run to determine whether there were any differences among the three groups in terms of background variables. These differences were significant at the .05 level and the relationships between background variables and utility functions made sense. Group 1 was found to contain a higher than average proportion of women and tended to drive two days a week or less—this segment wanted a low priced, high gas mileage car. Group 2 also contained more women than average, but they tended to drive every day—this group was much more interested in good gas mileage than any other attribute and was willing to buy a slightly higher priced car. Group 3 had a much higher portion of men than average and the group tended to live in the center city—this group was most interested in small, expensive European cars.

COMPONENTIAL SEGMENTATION. Componential segmentation was performed in a manner similar to that suggested by Green and DeSarbo (1977, 1979). First, a pooled regression was run, with preference as the dependent variable and end effects coding of attributes as the independent variables, to estimate the aggregate part-worth utilities. Then a separate regression was run for each of the background variables with the residuals from the pooled regression as the dependent variable and the interactions between the object profiles and the particular background variable as the independent variables. These regressions were significant for three of the four background variables (.05

level). Marital status was the one variable whose overall interaction regression was not significant. In the three significant regressions, eight variables were significant individually. Finally, a stepwise regression was run with preference as the dependent variable and the product profiles and the eight significant interactions from the previous interaction regressions as the independent variables. All the product variables were entered in the regression in step one, and the interaction variables were allowed to enter as long as they were significant. In this case, six of the eight interactions significant in the earlier interaction regressions were also significant in the overall regression. The part-worth utilities are given in column 6 of Table 1 and the interactions are shown in the bottom part of the table.

As might be expected, the results are similar to those found when discriminant analysis was used to differentiate the people in the three benefit segments. Men wanted more expensive cars than women. People planning to live in suburban/rural areas placed a greater weight on higher gas mileage than those planning to live in cities. People planning to drive three days a week saw a bigger difference between cars getting 15 and 25 MPG than people planning to drive no more than two days a week.

Predictive Validity

Next, the four models were compared on their ability to predict each person's preferences for six validation objects. The average correlations between each person's predicted and stated preferences are given in Table 2.

As expected, the individual model had the highest predictive power and the aggregate model had the lowest predictive power. The two segment

TABLE 2 Comparison of Predictive Powers of Four Models

Average correlations between predicted and stated preferences		
Pooled regression		.471
Componential segmentation		.509
Cluster 1	.468	
Cluster 2	.697	
Cluster 3	.707	
Clustered segmentation		.613
Individual level analyses		.822

models were in the middle and the clustered segmentation model gave considerably better predictions than the componential segmentation model.

Discussion

In this particular instance, the clustered segmentation model has a higher predictive power than the componential segmentation model. Also, surprisingly, the componential segmentation model provides only a marginal improvement over the aggregate model. This relatively small gain appears to be due to the fact that although there are several significant interactions between product profiles and background variables, there are no crossover interactions, i.e., none of the interactions change the *order* of the part-worth utilities for the levels of any attributes. In contrast, the clustered segments show differences in the preferred levels of both price and place of origin.

Stated differently, the part-worth utilities are not *strongly* related to background variables. Both the individual level and clustered segmentation analyses indicate that there are substantial differences in part-worth utilities across people. However, the relationship between these differences and background variables, though significant, is not large in relation to the main effects. Perhaps the differences in utility for attributes like place of origin and price are more closely related to attitudinal variables than to the background variables measured in the study. For example, the men who were planning to live in the center city and buy small, expensive European cars might buy the same car even if they decided not to live in the city. Possibly the small gain in predictive power of the componential segmentation model in relation to the aggregate model is similar to the low R^2's that have been found when demographic variables have been related to various measures of purchasing behavior (Frank and Massy 1968). However, as Bass, Tigert, and Lonsdale (1968) point out, the low R^2's are due to within-group heterogeneity. The usefulness of a segmentation scheme should be based on the between-group difference.

This discussion is *not* to suggest that clustered segmentation will always have a higher predictive power than componential segmentation. Wittink and Montgomery (1979) compared *a priori* and clustered segmentation approaches and found that the *a priori* method gave better predictions of ac-

tual job choice. The *a priori* method consisted of hypothesizing how a person's background variables would affect his/her preferred level of a particular attribute. When they found significant differences in the hypothesized direction, they allowed people with different backgrounds to have different utility functions for that variable. They found that marital status was related to preference for amount of travel, the part of the country where a person had lived the longest was related to geographic preferences, and area of concentration at Stanford was related to preference of functional area of a job. This is very similar to what componential segmentation accomplishes through its use of interaction terms. The primary difference is that componential segmentation is empirically based and *a priori* is theoretically based.

In comparing these two studies for possible clues about when one method of segmentation may work better than the other, one should keep in mind that the samples were probably comparable in terms of homogeneity. Wittink and Montgomery found that people with different concentrations did have different preference orderings for functional areas of specialization. Similarly, people from different regions of the country had different preference orderings for regions of the country in which they wanted to work. Thus, these data would have generated a number of crossover interactions with componential segmentation. Furthermore, to model all combinations of the regions of the country (five regions including international students) and concentration (six areas of concentration) one would have needed 30 clusters to get the level of sensitivity to background variables that could have been accomplished through componential segmentation. Thus, from these two studies, one might hypothesize that componential segmentation will provide a higher predictive power when the following two conditions are met: (1) background variables are strongly related to utilities, e.g., area of concentration in business school to functional activity in job sought and (2) there is a large number of attributes or a large number of levels on a smaller number of attributes.

However, rather than argue which of these methods of segmentation is best, the purpose of the article is to demonstrate the amount of useful information, provided in an easy-to-use form, that can be obtained by using this level of aggregation. Both of the methods provide a large amount of information in a form that is much more usable than a list of each person's utility weights.

Another, potentially even greater, advantage of the segmented analyses over individual level analyses in data collection. To analyze data at the individual level, one must require each respondent to rate enough concepts to estimate individual utilities. This is an arduous task with only five attributes and two or three levels of each attribute, and it becomes much harder and hence increases data collection costs when the number of attributes or levels per attribute is raised. One way to decrease this burden, as suggested by Green at recent ACR and AMA conferences, is to require the respondents to rate only a subset of the concepts and estimate more aggregate level utilities. However, as shown in this article, it is desirable to allow for the modeling of individual differences. Green's development of componential segmentation offers a way of decreasing the respondent burden and, at the same time, allowing for individual differences in utilities. This could not be done with clustered segmentation as used in this study, because the individual level utility estimates were used to form the clusters. However, it is possible to cluster respondents on the basis of preferences for products currently in the product class, stated optimal levels of attributes, or on stated attribute importances. Then segment utilities could be estimated without requiring each individual to rate all of the concepts.

References

Bass, Frank M., Douglas Tigert, and Ronald T. Lonsdale (1968), "Market Segmentation: Group Versus Individual Behavior," *Journal of Marketing Research*, 5 (August), 264–70.

_____ and William L. Wilkie (1973), "A Comparative Analysis of Attitudinal Predictions of Brand Preference," *Journal of Marketing Research*, 10 (August), 262–71.

Batsell, Richard R. (1979), "A Methodology for the Derivation of Consumer Resource Allocation Models at the Individual Level," Marketing Department Working Paper Series #79-002, Wharton School, University of Pennsylvania.

Cohen, Jacob and Patricia Cohen (1975), *Applied Multiple Regression/Correlation Analysis for the Behavioral Sciences*. Hillsdale, New Jersey: Lawrence Erlbaum Associates.

Frank, Ronald E. and William F. Massy (1968), "Market Segmentation Research: Findings and Implications," in *Application of the Sciences in Marketing Management*, Frank M. Bass, Charles W. King, and Edgar A. Pessemier, eds. New York: John Wiley & Sons, Inc.

Ginter, James L. and Edgar A. Pessemier (1978), "Analysis of Brand Preference Segments," *Journal of Business Research*, 6, 111–31.

Goldberger, Arthur S. (1964), *Econometric Theory*. New York: John Wiley & Sons, Inc.

Green, Paul E. (1977), "A New Approach to Market Segmentation," *Business Horizon*, 20, 61–73.

_____, Douglas J. Carroll, and Frank J. Carmone (1976), "Superordinate Factorial Design in the Analysis of Consumer Judgments," *Journal of Business Research*, 4, 281–95.

_____, _____, and _____, (1977), "Design Considerations in Attitude Measurement," *Moving Ahead with Attitude Research*, Y. Wind and M. G. Greenberg, eds. Chicago: American Marketing Association, 9–18.

_____ and Wayne S. DeSarbo (1977), "Demographic Stereotypes and Brand Preferences: An Application of Componential Segmentation," working paper, Wharton School, University of Pennsylvania.

_____ and _____ (1979), "Componential Segmentation in the Analysis of Consumer Tradeoffs," *Journal of Marketing*, 43, 83–91.

_____ and Vithala R. Rao (1969), "Nonmetric Approaches to Multivariate Analysis in Marketing," working paper, Wharton School, University of Pennsylvania.

_____ and _____ (1971), "Conjoint Measurement for Quantifying Judgmental Data," *Journal of Marketing Research*, 8 (August), 355–63.

_____ and _____ (1972), *Applied Multidimensional Scaling*. New York: Holt, Rinehart and Winston, Inc.

_____ and V. Srinivasan (1978), "Conjoint Analysis in Consumer Research: Issues and Outlook," *Journal of Consumer Research*, 5, 103–23.

_____ and Yoram Wind (1972), "Recent Approaches to the Modeling of Individuals' Subjective Evaluations," presented at 1972 Attitude Research Conference, Madrid, Spain.

_____ and _____ (1973), *Multiattribute Decisions in Marketing: A Measurement Approach*. Hinsdale, Illinois: The Dryden Press.

_____ and _____ (1975), "New Way to Measure Consumers' Judgments," *Harvard Business Review*, 53, 107–17.

Haley, Russell I. (1968), "Benefit Segmentation: A Decision Oriented Research Tool," *Journal of Marketing*, 32, 30–35.

Howard, N. and B. Harris (1966), "A Hierarchical Grouping Routine, IBM 360/65 Fortran IV Program," University of Pennsylvania working paper.

Huber, Joel C. and William L. Moore (1979), "A Comparison of Alternative Ways to Aggregate Individual Conjoint Analyses," *1979 Educators' Conference Proceedings*, Neil E. Beckwith et al., eds. Chicago: American Marketing Association, 64–8.

Jain, Arun K., Franklin Acito, Naresh K. Malhotra, and Vijay Mahajan (1979), "A Comparison of the Internal Validity of Alternative Parameter Estimation Methods in Decompositional Multiattribute Preference Models," *Journal of Marketing Research*, 16 (August), 313–22.

Johnson, Richard M. (1974), "Tradeoff Analysis of Consumer Values," *Journal of Marketing Research*, 11 (May), 121–7.

Johnston, J. (1972), *Econometric Methods*, 2nd edition. New York: McGraw-Hill Book Company.

Kerlinger, Fred N. and Elzar J. Pedhazur (1973), *Multiple Regression in Behavioral Research*. New York: Holt, Rinehart and Winston.

Krantz, David H. and Amos Tversky (1971), "Conjoint Measurement Analysis of Composition Rules in Psychology," *Psychological Review*, 778, 151–69.

Kuehn, A. A. and R. L. Day (1962), "Strategy of Product Quality," *Harvard Business Review*, 40, 100–10.

Luce, R. Duncan and John W. Tukey (1964), "Simultaneous Conjoint Measurement: A New Type of Fundamental Measurement," *Journal of Mathematical Psychology*, 1, 1–27.

McCann, John M. (1974), "A Comparison of Pooling Time-Series and Cross-Section Data," presented at AMA Educators' Conference, Portland, Oregon.

McFadden, Daniel (1976), "Quantal Choice Analysis: A Survey," *Annals of Economic and Social Measurement*, 5, 363–90.

Moore, William L., Edgar A. Pessemier, and Taylor E. Little (1979), "Predicting Brand Choice Behavior: Marketing Application of the Schönemann and Wang Unfolding Model," *Journal of Marketing Research*, 16, (May), 203–11.

Rao, Vithala R. and Frederick W. Winter (1978), "An Application of the Multivariate Probit Model to Market Segmentation and Product Design," *Journal of Marketing Research*, 15 (August), 361–8.

Theil, Henri (1971), *Principles of Econometrics*. New York: John Wiley & Sons, Inc.

Wind, Yoram (1978), "Issues and Advances in Segmentation Theory," *Journal of Marketing Research*, 15, (August), 317–37.

Wittink, Dick R. and David B. Montgomery (1979), "Predicting Validity of Trade-Off Analysis for Alternative Segmentation Schemes," in *1979 Educators' Conference Proceedings*, Neil E. Beckwith et al., eds. Chicago: American Marketing Association, 69–73.

INDEX

A

A priori criterion, 246–47
Agglomerative method, 293, 302
Aggregate analysis, 349, 361, 363–64, 418, 437
AID (automatic interaction detection), 322
Algorithm, 293, 301, 322–26
All possible subset regression, 40, 42
Analysis of variance (A-NOVA), 145, 147–49, 154–57, 161–62, 179–85
  assumptions of, 163
Analysis sample, 73, 82–83
Anti-image matrix, 285
Appropriateness data, 349, 357–58
Artificial dichotomies, 81
Assignment matrix, 74
Attribute vectors, 108, 111
Average linkage, 293, 303–4, 322–23

B

Backward elimination, 40
Balanced design, 422
Bartlett's V, 286
Bartlett's test of sphericity, 285
Bayesian regression, 43

Bayesian statistical models, 87
Beta coefficient, 20, 40
Between-group variance, 154
Binary variable, 73
Bivariate analysis, 4
Bivariate partial correlation, 1

C

Canonical correlation, 95–96, 187, 191–92, 196, 201
Canonical correlation analysis (CCA), 5, 187–231
Canonical cross-loadings; See Loadings
Canonical loadings; See Loadings
Canonical roots, 187, 192, 195
Canonical structure matrix; See Loadings
Canonical variates, 188, 190–91, 203
Canonical weights, 198–99, 201
Categorical variable, 73
Cell means, 150, 153, 156, 160
Centroid, 74, 76, 96, 109
Centroid method (CA), 293, 304
Chance criterion, maximum, 89, 100, 105

Chance criterion, proportional, 89–90, 99–100, 105–08
Chance models, 73, 84, 89
Chi-square, 85, 105
City block approach, 294, 299, 333
Classification accuracy, 84, 88, 90
Classification analysis; See Cluster analysis
Classification matrix, 73–74, 84–85, 87–88, 96–99, 105–08, 416
Cluster analysis, 6, 239–40, 293–348, 363, 369, 423, 437, 440–41
Cluster centroid, 293, 307
Clusters, 279
Cluster seeds, 294, 305
Coding of variables, 438
Coefficient of determination, 17, 36–37
Collinearity, 18
Column sum of squared factor loadings, 253
Common factor analysis, 233, 237, 240–41, 247, 255–57
Common variance, 240, 257
Communalities, 233, 254, 256
Communality, 234, 283–85
Communality index, 251

Comparison tests; *See* Multiple comparison tests
Complete linkage, 294, 303, 322–23, 341
Component analysis, 234, 237, 240–41, 247, 252–255, 256
Componential segmentation, 438, 441, 443
Composition rule, 408
Confidence intervals, 38–39
Confirmatory factor analysis, 283–84
Confusion data, 349, 357–58
Confusion matrix, 74
Conjoint analysis, 6, 204, 407–44
  reliability of, 428–35
  validity of, 428–35, 441–42
Conjoint measurement, 408
Conjunctive, 427
Constant variance of the error term; *See* Error
Correlated variables, 27, 30–31
Correlation, 31
Correlation coefficient, 18, 288, 297–98, 333
Correlation matrix, 234, 239, 252, 284–85, 288
Covariance analysis, 145, 162–63, 170–72
Covariate analysis; *See* Covariance analysis
Covariates; *See* Covariance analysis
Criterion variable; *See* Dependent variable
Critical value, 85, 148, 154
Cross loadings, 198, 200–01
Cross-validation, 82, 335, 426, 434
Cubic clustering criterion, 308
Cutting score, 74, 85, 98
Cutting score, optimum, 85–87

D
Degrees of freedom, 18
Dendrogram, 294, 302, 308, 312, 318, 373–75
Dependence technique, 1, 7
Dependent variable, 1, 21, 187
Derived measures, 350, 357–58
Dichotomous data, 288
Dimensions, 350
Dimensional approach, 368
Direct ranking, 355
Disaggregate analysis; *See* Multiple discriminant analysis
Disaggregate analysis (conjoint), 418
Discriminant analysis; *See* Multiple discriminant analysis
Discriminant coefficient, 91, 96
Discriminant function, 74
Discriminant loadings, 74, 91, 96, 100, 117
Discriminant score, 74, 394
Discriminant variates, 74
Discriminant weight, 91, 100
Dissimilarities data, 356–57
Divisive method, 294, 302
Dummy variable, 1, 7, 18, 50

E
Eckhart-Young decomposition procedure, 282
Eigenvalue, 192, 234, 253
Equimax, 245
Entropy Group, 294–297
Error, 18, 32
  constant variance of the error term, 34
  independence of terms, 34
  lack of fit, 33–34
  normality of term distribution, 34
  pure, 33–34

Error—*Continued*
  standard error of coefficient, 46
  standard error of the estimate, 45–46
Error variability measures, 306
Error variance, *See* Residual variance
Euclidean distance, 94, 294, 299–301, 308, 333
Experimental design, 147
Experimental variable, 146
Explicit estimation, 359
Exploratory factor analysis, 283–84
Extreme value (outlier), 298

F
F-distribution, 158
F-ratio, 37–38
F-test, 33, 37–38, 85
F-value, 154
Factor, 146, 161, 234, 279
Factor analysis (FA), 6, 221–231, 233–291, 392–406
  R type, 235, 237, 239, 280–83
  Q type, 235, 237, 239–40, 298, 280–83, 322
  S type, 280
  T type, 280
  P type, 280
  O type, 280
Factor interpretation, 250–52
Factor loadings, 234, 249
Factor loading pattern, 254
Factor matrix, 234, 252
Factor rotation, 234, 241–46
Factor scores, 234, 238, 259–60, 393
Factorial design, 56, 146, 161
Factors, naming of, 257–58
Features based method (CCA), 204–05
First discriminant function (MANOVA), 158

Fisher's method, 94
Fixed effects, 27–28
Fixed predictor variable, 27
Follow-up tests; *See* Multiple comparison tests
Fractional design, 413
Functional relationships, 25
Furthest neighbor; *See* Complete linkage

G
G-coefficient, 289
Global test; *See* Multiple comparison tests
Greatest characteristic root (grc), 146, 154, 158

H
Heterogeneity, 295
Hierarchical clustering techniques, 294, 301–06, 333–34, 338
Hill-climbing methods, 324
Hit ratio, 73, 74, 85, 88, 105
Hold-out sample, 74, 82–83, 134, 426
Homogeneity, 295
Homoscedasticity, 18–19, 26, 27
Hotelling's T, 146, 148–51, 157
Hotelling's trace, 158

I
Ideal point, 350, 355, 359, 362–63
Implicit estimation, 359
Importance performance grid, 350
Independence of error terms; *See* Error
Independence of predictors, 29–30
Independent variable, 2, 21, 188
Index of fit, 350, 365–66

Initial dimensionality, 350
Interacting variables, 30
Interaction, 29–31
Interaction effect, 146, 408, 413, 415–17, 420–21, 438
Interactions-based method (CCA), 205–206
Intercept, 19
Interdependence technique, 2, 7, 235
Inter-object similarity, 294, 297, 308
Interpretation (CA), 91, 100, 108, 297, 306–07
Interval scales, 10
Ipsative data, 283
Iterative partitioning method, 324

J
Jackknife analysis, 135–143

K
K-means clustering, 324, 332
Kaiser-Meyer-Olkin measure of sampling adequacy, 285
Kappa, 335

L
Latent root criterion, 246–47, 286
Latent roots, 234, 284–85
Level of significance, 194
Linear combination, 74, 187, 203
Linear composites, 74, 103–04, 187
Linearity, 19, 32
Loadings, 187, 191, 196, 199, 201, 209

M
Mahalanobis' D, 283–85, 294, 300–01

Mahalanobis procedure, 94–101, 300–01
Main effects, 146, 408, 413, 420–21
MANCOVA; *See* Multivariate analysis of covariance
MANOVA; *See* Multivariate analysis of variance
Maximum likelihood criterion; *See* Wilks' lambda
Measurement scales, 9
Metric analysis (MDS), 363–65
Metric data, 2, 9, 360
Metric variable, 74
Minimum variance cluster analysis, 323
Mixed data sets, 114
MONANOVA, 415, 417, 429, 432
Multicollinearity, 42
Multidimensional scaling (MDS), 6, 349–406
   index of fit, 365
   metric, 363–65
   nonmetric, 361–63
   stress, 351, 361, 365–66
Multiple comparison tests, 151, 155, 159, 161
Multiple discriminant analysis (MDA), 5, 73–144, 392–406, 441
Multiple discriminant function, 74
Multiple R, 45
Multiple regression (MR), 4, 17
Multiple regression analysis, 20–71
Multiple t-tests; *See* t-test
Multivariate analysis, 2, 3, 4
Multivariate analysis of covariance (MANCOVA), 5, 162–63, 170–172

Multivariate analysis of variance (MANOVA), 5, 145, 147, 149–53, 157–62, 179–85
  assumptions of, 163
Multivariate Normal Distribution, 146

N
Nearest neighbor; See Single linkage method
Neighborhood approach, 350, 368
Nominal, 9, 73
Non-hierarchical clustering techniques, 294, 301, 305–06
Nonmetric analysis, 361–63
Nonmetric data, 2, 9, 73, 360–63, 322
Normality of error term distribution; See Error
Normalized distance function, 294
Null hypothesis, 146
Numerical taxonomy; See Cluster analysis

O
Objects-based method (CCA), 204–05
Objective dimensions, 350, 354, 374
Oblique factor, 238
Oblique factor rotation, 241, 244–45, 259, 274, 288
Oblique factor solutions, 234, 259
Optimizing procedure, 294, 305
Ordinal scales, 9
Orthogonal, 188, 234
Orthogonal array, 56
Orthogonal factors, 238

Orthogonal factor rotations, 241, 244–45, 259, 288
Orthogonal factor solutions, 234, 259, 288
Orthonormal, 120
Outlier, 298
Overall tests; See Multiple comparison tests

P
Pairwise comparisons, 356, 395, 430
Parallel threshold method, 295, 305
Parameter, 19
Part-worth, 424, 429, 436, 439
Partial correlation, 46–47
Partial correlation coefficient, 19, 41
Partial F values, 19, 41, 46, 92, 117–19
Partitioning stage (CA), 297
Partitioning the error, 33
Perceptual map, 351, 353, 392, 402, 404
Perceived dimensions, 351, 354, 374
Pillai's criterion, 158
Point representations, 361, 364–65
Polar extremes, 81
Post hoc tests; See Multiple comparison tests
Potency index, 119
Prediction matrix, 74
Predictor variable; See Independent variable
Preference data, 351, 355–56, 373
  external and internal analysis of, 361, 364
Principle component analysis, 237, 284
Profile analysis, 92, 115, 126, 297, 307
Profile diagram, 294, 298

Q
Q analysis; See Cluster analysis
"Q" factor analysis; See Factor analysis
Qualitative variable, 73
Quartimax, 245–46, 287

R
"R" factor analysis; See Factor analysis
R^2 statistic, 85
Random effects, 27–28
Random predictor variable, 27
Random variable, 25
Randomized design, 153
Rao's approximation, 194
Ratio scale, 10–11
Reduced correlation matrix, 255
Redundancy index, 188, 195, 209–210
Redundancy measure, 191–92, 195–97
Regression coefficient, 19, 24
Residual; See Error
Residual variance, 154, 240
Ridge regression, 43
Rotation, 104, 241–46
Row sum of squared factor loadings, 254

S
Sample division, 82
Scheffe's test, 146, 155, 159
Scree test, 274, 276, 286–87
Scree test criterion, 247
Segmented models, 437–38
Sequential threshold method, 295, 305
Significance levels, 148
Significance testing, 158
Simple regression, 21, 23
Similarities data, 351, 355, 356–57, 363, 373

Similarity scale, 351, 392–406

Simulations, 134, 393

Simultaneous discriminant function, 83–84, 92, 100

Single linkage method, 295, 302, 308, 314, 322–23

Spatial maps, 351, 354

Specific variance, 240

Squared error criterion, 25

Standard error, 148

Standard error of coefficient; See Error

Standard error of the estimate; See Error

Standardization, 20, 40

Statistical estimation, 56

Statistical relationship, 25–26

Statistical significance, 36, 84

Step-down analysis, 159

Stepwise method, 43–52, 83–84, 92, 100

Stepwise forward estimation, 40–41

Stretching, 110–12

Structural correlations, 91, 191, 199

Structure loadings matrix, 102

Subjective evaluation, 351, 365

Sum of squared errors, 22

T

t-test, 36, 46, 88, 148

Taxonomic variable, 73

Territorial map, 102, 112–13, 122

Tests of coefficients, 36

Tolerance, 75

Trace, 234

Trace, percentage of, 253–54, 256–57

Trade-off analysis, 412, 433

Treatment, 2, 146

Treatment levels, 408

Tree-graph; See Dendrogram

True dichotomies, 81

Tukey's HSD (honestly significant difference), 184

Two-dimensional factor diagram, 242

Two-group randomized design, 147

Type I error, 146, 148

Type II error, 146

Typology; See Cluster analysis

U

U-method, 135–43

U statistic; See Wilks' lambda

Unique variance, 249

Univariate analysis, 3

Univariate analysis of variance; See Analysis of variance

Unrotated factors, 241–46

Unrotated factor solutions, 241–46

V

Validation, 84–90, 96, 134, 341–44, 426

Validation sample, 74

Variable selection, 81–82

Variance–covariance matrix, 163

Variance criterion, percentage of, 247

Variates, 190–92

Varimax, 101, 104, 120, 245–46, 254–55, 256, 287

Vector, 110, 146, 364–65

Vector representations, 361

Vertical icicle diagram, 295, 308, 310

W

Ward's method, 295, 304, 308–15, 322, 341

Weights, 198

Wilks' lambda, 102, 146, 154, 158

Within-group variance, 154

Z

Zero slope, 20